Foundations for microwave engineering

McGraw-Hill Physical and Quantum Electronics Series

Hubert Heffner and A. E. Siegman, Consulting Editors

Beam: Electronics of Solids

Collin: Foundations for Microwave Engineering

Elliott: Electromagnetics

Johnson: Field and Wave Electrodynamics

Louisell: Radiation and Noise in Quantum Electronics

Moll: Physics of Semiconductors

Papas: Theory of Electromagnetic Wave Propagation

Siegman: Microwave Solid-state Masers

Smith, Janak, and Adler: Electronic Conduction in Solids

Smith and Sorokin: The Laser

White: Basic Quantum Mechanics

Foundations for microwave engineering

Robert E. Collin *Professor of Electrical Engineering*
Case Institute of Technology

McGraw-Hill Book Company *New York   St. Louis
San Francisco   Toronto   London   Sydney*

Foundations for microwave engineering

*Copyright © 1966 by McGraw-Hill, Inc. All Rights Reserved.
Printed in the United States of America. This book, or parts
thereof, may not be reproduced in any form without permission
of the publishers.*
Library of Congress Catalog Card Number 65-21572
07-011801-9

10 11 12 KP KP 7 9 8 7 6

To my mother and father

Preface

This book is written to fill the need for a comprehensive, up-to-date text covering the fundamentals of microwave engineering. It is designed for use in senior-level and also for beginning graduate courses. Teaching an adequate one-semester course in microwave engineering requires that students be familiar with Maxwell's equations and the elements of wave propagation. At Case Institute of Technology we now require two semesters of work in electromagnetic fields as a prerequisite to the microwave engineering course. Thus Chapters 2 and 3, covering electromagnetic theory and the field analysis of waveguides, are not entirely unfamiliar to the students. These two chapters can be covered quite rapidly and thus leave considerable time to treat, with reasonable depth, topics that can be more intimately associated with the engineering aspects of the microwave field.

Not all topics covered in the text would be included in a senior-level course. A number of more advanced topics are included in separate sections, marked with a star, in order to make the book more complete and also suitable for a beginning graduate course. For a senior-level course these sections can be omitted without disrupting the continuity of treatment. Even with the elimination of these special sections, the book contains more material than can be covered in a one-semester course. However, the instructor will have no difficulty in eliminating topics, as he chooses. I have felt it to be more desirable to aim at a reasonable degree of completeness rather than to tailor-make the book for a specific one-semester course.

Most students at the senior level have been introduced to the elements of matrix algebra and the Fourier transform. These tools are therefore freely used, although in only an elementary way, so that students without prior knowledge of them will not be particularly hampered.

My greatest difficulty in writing this book has been the selection of material. My aim has been to ensure the inclusion of all major topics that form the foundations for microwave engineering. Chapter 2 provides a detailed review of electromagnetic theory. A notable feature is the treatment of polarization effects in materials, which is covered in greater detail than usual, and includes a discussion of energy storage in dispersive media. Chapter 3 covers the field analysis of transmission lines and waveguides. Chapter 4 develops the basic microwave equivalent-circuit theory, and Chapter 5 presents a treatment of transmission-line circuit analysis, use of the Smith chart, and impedance matching with stubs, quarter-wave transformers, and tapered lines. Chapter 6 is devoted to a discussion of passive microwave devices, including a treatment of microwave propagation in ferrites and a discussion of a variety of ferrite devices. Chapter 7 treats the subject of microwave cavities. Periodic structures and filters are covered in considerable detail in Chapter 8.

Chapters 9 to 11 are devoted to microwave electronics and deal with tubes, masers, and parametric amplifiers, respectively. The klystron amplifier, reflex oscillator, and the traveling-wave tube are analyzed from the one unified approach based on space-charge wave theory. These three chapters, together with material from Chapters 7 and 8 dealing with cavities and periodic structures, could be used for a one-semester course on microwave electronics.

Many topics, including microwave measurements, antennas, and propagation, had to be omitted in order to keep the book within manageable size. However, it is believed that the student who masters the material presented will be in a good position to proceed with more detailed study of special topics.

Acknowledgment is made of the aid of Professor R. Bolz, Engineering Division Head, Case Institute of Technology, in providing facilities for preparation of the manuscript. The manuscript was typed by Mrs. Flo Antol, to whom I express my sincere thanks and appreciation. Her efforts significantly reduced the burden of preparing the final draft of the manuscript.

Robert E. Collin

Contents

Preface ix

1 Introduction 1

1.1 Microwave frequencies and uses 1
1.2 Microwave circuit elements and analysis 4

2 Electromagnetic theory 11

2.1 Maxwell's equations 11
2.2 Constitutive relations 17
2.3 Static fields 22
2.4 Wave equation 25
2.5 Energy and power 28
2.6 Boundary conditions 34
2.7 Plane waves 38
2.8 Reflection from a dielectric interface 43
2.9 Reflection from a conducting plane 47
2.10 Potential theory 51
★2.11 Derivation of solution for vector potential 54
2.12 Lorentz reciprocity theorem 56
  Problems 59
  References 63

3 Transmission lines and waveguides 64

3.1 Classification of wave solutions 65
3.2 Transmission lines (field analysis) 73
3.3 Transmission lines (distributed-circuit analysis) 80
★3.4 Transmission-line parameters 84
3.5 Terminated transmission line 89
3.6 Rectangular waveguide 95

3.7 Circular waveguides 107
★3.8 Surface waveguides 113
3.9 Power orthogonality 121
★3.10 Attenuation for degenerate modes 124
3.11 Wave velocities 132
Problems 138
References 143

4 Circuit theory for waveguiding systems 144

4.1 Equivalent voltages and currents 145
4.2 Impedance description of waveguide elements and circuits 148
★4.3 Foster's reactance theorem 153
★4.4 Even and odd properties of Z_{in} 155
4.5 N-port circuits 157
4.6 Two-port junctions 160
4.7 Scattering-matrix formulation 170
4.8 Scattering matrix for a two-port junction 176
4.9 Transmission-matrix representation 179
★4.10 Excitation of waveguides 183
★4.11 Waveguide coupling by apertures 190
Problems 197
References 202

5 Impedance transformation and matching 203

5.1 Smith chart 203
5.2 Impedance matching with reactive elements 207
5.3 Double-stub matching network 212
5.4 Triple-stub tuner 215
5.5 Waveguide reactive elements 217
5.6 Quarter-wave transformers 221
5.7 Theory of small reflections 224
5.8 Approximate theory for multisection quarter-wave transformers 226
5.9 Binomial transformer 227
5.10 Chebyshev transformer 229
★5.11 Chebyshev transformer (exact results) 233
5.12 Tapered transmission lines 237
★5.13 Synthesis of transmission-line tapers 240
★5.14 Chebyshev taper 248
★5.15 Exact equation for the reflection coefficient 251
Problems 254
References 258

6 Passive microwave devices 259

6.1 Terminations 259
6.2 Attenuators 262

6.3   Phase changers   265
6.4   Directional couplers   270
6.5   Hybrid junctions   282
6.6   Microwave propagation in ferrites   286
6.7   Faraday rotation   296
6.8   Microwave devices employing Faraday rotation   300
6.9   Circulators   304
6.10 Other ferrite devices   309
      Problems   309
      References   312

7 Electromagnetic resonators   313

7.1   Resonant circuits   313
7.2   Transmission-line resonant circuits   317
7.3   Microwave cavities   321
7.4   Equivalent circuits for cavities   329
7.5   Fabry-Perot resonators   337
★7.6 Field expansion in a general cavity   344
★7.7 Oscillations in a source-free cavity   351
★7.8 Excitation of cavities   356
      Problems   359
      References   362

8 Periodic structures and filters   363

8.1   Capacitively loaded transmission-line–circuit analysis   364
8.2   Wave analysis of periodic structures   369
8.3   Periodic structures composed of unsymmetrical two-port networks   372
8.4   Terminated periodic structures   373
8.5   Matching of periodic structures   375
8.6 k_0-β diagram   377
8.7   Group velocity and energy flow   379
8.8   Floquet's theorem and spatial harmonics   381
8.9   Periodic structures for traveling-wave tubes   383
8.10 Sheath helix   392
8.11 Some general properties of a helix   396
8.12 Introduction to microwave filters   398
8.13 Image-parameter method of filter design   399
8.14 Filter design by insertion-loss method   403
8.15 Specification of power loss ratio   405
8.16 Some low-pass-filter designs   407
8.17 Frequency transformations   410
8.18 Impedance inverters   415
8.19 A transmission-line filter   422
8.20 Quarter-wave-coupled cavity filters   424
8.21 Direct-coupled cavity filters   428
      Problems   430
      References   433

xiv   Contents

9 Microwave tubes 434

9.1   Introduction 434
9.2   Electron beams with d-c conditions 435
9.3   Space-charge waves on beams with confined flow 439
9.4   Space-charge waves on unfocused beams 446
9.5   A-c power relations 452
9.6   Velocity modulation 455
9.7   Two-cavity klystron 462
9.8   Reflex klystron 470
9.9   Magnetron 473
9.10 O-type traveling-wave tube 476
9.11 M-type traveling-wave tube 482
9.12 Other types of microwave tubes 484
9.13 Noise in microwave tubes 485
      Problems 489
      References 491

10 Microwave masers 493

10.1 Some quantum-mechanical fundamentals 494
10.2 Absorption and emission of radiation 501
10.3 Description of a maser amplifier 507
10.4 Energy levels in ruby 510
10.5 Analysis of maser action 514
10.6 Macroscopic magnetic susceptibility 520
10.7 Equivalent circuit of a maser amplifier 524
10.8 Gain of a maser amplifier 529
10.9 Maser noise 532
10.10 Traveling-wave maser 534
10.11 Lasers 537
      Problems 539
      References 540

11 Parametric amplifiers 541

11.1 p-n junction diodes 542
11.2 Manley-Rowe relations 546
11.3 Linearized equations for parametric amplifiers 549
11.4 Parametric up-converter 551
11.5 Negative-resistance parametric amplifier 555
11.6 Noise properties of parametric amplifiers 562
      Problems 569
      References 570

Appendix

I   Useful relations from vector analysis 571
II   Bessel functions 575
III Physical constants and other data 579

Index 581

1
Introduction

The purpose of this introductory chapter is to provide a short, and admittedly incomplete, survey of what the microwave engineering field encompasses. The first section presents a brief discussion of a good many of the varied and sometimes unique applications of microwaves. This is followed by a second section in which an attempt is made to show in what ways microwave engineering differs from the engineering of communication systems at lower frequencies. In addition, a number of microwave devices are introduced to provide examples of the types of devices and circuit elements that are examined in greater detail later on in the text.

1.1 Microwave frequencies and uses

Microwave engineering is the engineering of information-handling systems in the frequency range from about 10^9 to 10^{12} cps, corresponding to wavelengths from 30 cm down to 0.3 mm. The characteristic feature of this branch of engineering is the short wavelengths involved, these being of the same order of magnitude as the circuit devices employed. That is why the descriptive term microwaves is used. The short wavelengths involved in turn mean that the propagation time for electrical effects from one point in a circuit to another point is comparable with the period of the oscillating currents and charges in the system. As a result, conventional low-frequency circuit analysis based on Kirchhoff's laws and voltage-current concepts no longer suffices for an adequate description of the electrical phenomena taking place. It is necessary instead to carry out the analysis in terms of a description of the electric and magnetic fields associated with the device. In essence, it might be said, microwave engineering is applied electromagnetic fields engineering. For this reason the successful engineer in this area must have a good working knowledge of electromagnetic field theory.

The great interest in microwave frequencies arises for a variety of reasons. Basic among these is the ever-increasing need for more radio-frequency-spectrum space and the rather unique uses that microwave frequencies can be put to. When it is noted that the frequency range 10^9 to 10^{12} cps contains a thousand sections like the frequency spectrum from 0 to 10^9 cps, the value of developing the microwave band as a means of increasing the available usable frequency spectrum may be readily appreciated.

At one time (during World War II and shortly afterward), microwave engineering was almost synonymous with radar (*RA*dio *D*etection *A*nd *R*anging) engineering because of the great stimulus given to the development of microwave systems by the need for high-resolution radar capable of detecting and locating enemy planes and ships. Even today radar, in its many varied forms, such as missile-tracking radar, fire-control radar, weather-detecting radar, missile-guidance radar, airport traffic-control radar, etc., represents a major use of microwave frequencies. This use arises predominantly from the need to have antennas that will radiate essentially all the transmitter power into a narrow pencillike beam similar to that produced by an optical searchlight. The ability of an antenna to concentrate radiation into a narrow beam is limited by diffraction effects, which in turn are governed by the relative size of the radiating aperture in terms of wavelengths. For example, a parabolic reflector-type antenna produces a pencil beam of radiated energy having an angular beam width of $140°/(D/\lambda_0)$, where D is the diameter of the parabola and λ_0 is the wavelength. A 90-cm (about 3-ft) parabola can thus produce a 4.7° beam at a frequency of 10^{10} cps, i.e., at a wavelength of 3 cm. A beam of this type can give reasonably accurate position data for a target being observed by the radar. To achieve comparable performance at a frequency of 100 Mc would require a 300-ft parabola, a size much too large to be carried aboard an airplane.

In more recent years microwave frequencies have also come into widespread use in communication links, generally referred to as microwave links. Since the propagation of microwaves is effectively along line-of-sight paths, these links employ high towers with reflector or lens-type antennas as repeater stations spaced along the communication path. Such links are a familiar sight to the motorist traveling across the country because of their frequent use by highway authorities, utility companies, and television networks. A further interesting means of communication by microwaves is the use of satellites as microwave relay stations. The first of these, the Telstar, launched in July, 1962, provided the first transmission of live television programs from the United States to Europe. The diameter of the Telstar satellite is only $34\frac{1}{2}$ in., and hence an efficient antenna system was possible only in the microwave frequency band.

The economic transmission of a large number of television programs across the country depends on the ability to modulate all these programs onto a single carrier and to transmit them over one communication link.

Since each black-and-white television program requires a bandwidth of about 6 Mc, a total bandwidth of 600 Mc is required for a hundred such programs. For the ready processing and handling of a modulated carrier, modulation sidebands can be only a few percent of the carrier frequency. It is thus seen that the carrier frequency must be in the microwave range for efficient transmission of many television programs over one link. Without the development of microwave systems, our communications facilities would have been severely overloaded and totally inadequate for present operations.

Even though such uses of microwaves are of great importance, the applications of microwaves and microwave technology extend much further, into a variety of areas of basic and applied research, and including a number of diverse practical devices, such as microwave ovens that can cook a 10-lb roast in just a few minutes. Some of these specific applications are briefly discussed below.

Waveguides periodically loaded with shunt susceptance elements support slow waves having velocities much less than the velocity of light, and are used in linear accelerators. These produce high-energy beams of charged particles for use in atomic and nuclear research. The slow-traveling electromagnetic waves interact very efficiently with charged-particle beams having the same velocity, and thereby give up energy to the beam. Another possibility is for the energy in an electron beam to be given up to the electromagnetic wave, with resultant amplification. This latter device is the traveling-wave tube, and is examined in detail in a later chapter.

Sensitive microwave receivers are used in radio astronomy to detect and study the electromagnetic radiation from the sun and a number of radio stars that emit radiation in this band. Such receivers are also used to detect the noise radiated from plasmas (an approximately neutral collection of electrons and ions, e.g., a gas discharge). The information obtained enables scientists to analyze and predict the various mechanisms responsible for plasma radiation.

Molecular, atomic, and nuclear systems exhibit various resonance phenomena under the action of periodic forces arising from an applied electromagnetic field. Many of these resonances occur in the microwave range; hence microwaves have provided a very powerful experimental probe for the study of basic properties of materials. Out of this research on materials have come many useful devices, such as some of the nonreciprocal devices employing ferrites, several solid-state microwave amplifiers and oscillators, e.g., masers, and even the coherent-light generator and amplifier (laser).

The development of the laser, a generator of essentially monochromatic (single-frequency) coherent-light waves, has stimulated a great interest in the possibilities of developing communication systems at optical wavelengths. This frequency band is sometimes referred to as the *ultramicro-*

wave band. With some modification, a good deal of the present microwave technology can be exploited in the development of optical systems. For this reason, familiarity with conventional microwave theory and devices provides a good background for work in the new frontiers of the electromagnetic spectrum.

It is not possible here to give a complete account of all the applications of microwaves that are being made. The brief look at some of these, as given above, should convince the reader that this portion of the radio spectrum offers many unusual and unique features. Although the microwave engineering field may now be considered a mature and well-developed one, the opportunities for further development of devices, techniques, and applications to communications, industry, and basic research are still excellent.

1.2 Microwave circuit elements and analysis

At frequencies where the wavelength is several orders of magnitude larger than the greatest dimensions of the circuit or system being examined, conventional circuit elements such as capacitors, inductors, resistors, electron tubes, and transistors are the basic building blocks for the information transmitting, receiving, and processing circuits used. The description or analysis of such circuits may be adequately carried out in terms of loop currents and node voltages without consideration of propagation effects. The time delay between cause and effect at different points in these circuits is so small compared with the period of the applied signal as to be negligible. It might be noted here that an electromagnetic wave propagates a distance of one wavelength in a time interval equal to one period of a sinusoidally time-varying applied signal. As a consequence, when the distances involved are short compared with a wavelength λ_0 (λ_0 = velocity of light/frequency), the time delay is not significant. As the frequency is raised to a point where the wavelength is no longer large compared with the circuit dimensions, propagation effects can no longer be ignored. A further effect is the great relative increase in the impedance of connecting leads, terminals, etc., and the effect of distributed (stray) capacitance and inductance. In addition, currents circulating in unshielded circuits comparable in size with a wavelength are very effective in radiating electromagnetic waves. The net effect of all this is to make most conventional low-frequency circuit elements and circuits hopelessly inadequate at microwave frequencies. In fact, they do not work at all.

If a rather general viewpoint is adopted, one may classify resistors, inductors, and capacitors as elements that dissipate electric energy, store magnetic energy, and store electric energy, respectively. The fact that such elements have the form encountered in practice, e.g., a coil of wire for an inductor, is incidental to the function they perform. The construction used in practical elements may be considered just a convenient way to build these devices so that they will exhibit the desired electrical properties. As is well known, many of these circuit elements do not behave in the desired manner at high frequencies. For example, a coil of wire may be an excellent inductor at 1 Mc, but at 50 Mc it may be an equally good capacitor because of the predominating effect of interturn capacitance. Even though practical low-frequency resistors, inductors, and capacitors do not function in the desired manner at microwave frequencies, this does not mean that such energy-dissipating and storage elements cannot be constructed at microwave frequencies. On the contrary, there are many equivalent inductive and capacitive devices for use at microwave frequencies. Their geometrical form is quite different, but they can be and are used for much the same purposes, such as impedance matching, resonant circuits, etc. Perhaps the most significant electrical difference is the generally much more involved frequency dependence of these equivalent inductors and capacitors at microwave frequencies.

Low-frequency electron tubes are also limited to a maximum useful frequency range bordering on the lower edge of the microwave band. The limitation arises mainly from the finite transit time of the electron beam from the cathode to the control grid. When this transit time becomes comparable with the period of the signal being amplified, the tube ceases to perform in the desired manner. Decreasing the electrode spacing permits these tubes to be used up to frequencies of a few thousand megacycles, but the power output is limited and the noise characteristics are poor. The development of new types of tubes for generation of microwave frequencies was essential to the exploitation of this frequency band. Fortunately, several new principles of operation, such as velocity modulation of the electron beam and beam interaction with slow electromagnetic waves, were discovered that enabled the necessary generation of microwaves to be carried out. These fundamental principles with applications are discussed in a later chapter.

One of the essential requirements in a microwave circuit is the ability to transfer signal power from one point to another without radiation loss. This requires the transport of electromagnetic energy in the form of a propagating wave. A variety of such structures have been developed that can guide electromagnetic waves from one point to another without radiation loss. The simplest guiding structure, from an analysis point of view, is the transmission line. Several of these, such as the open two-conductor line, coaxial line, and shielded strip line, illustrated in Fig. 1.1, are in common use at the lower microwave frequencies.

At the higher microwave frequencies, notably at wavelengths below 10 cm, hollow-pipe waveguides, as illustrated in Fig. 1.2, are preferred to transmission lines because of better electrical and mechanical properties. The waveguide with rectangular cross section is by far the most common type. The circular guide is not nearly as widely used, although present

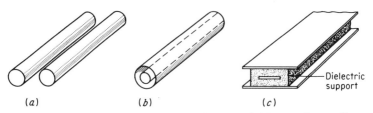

Fig. 1.1 Some common transmission lines. (*a*) Two-conductor line; (*b*) coaxial line; (*c*) shielded strip line.

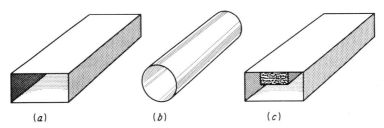

Fig. 1.2 Some common hollow-pipe waveguides. (*a*) Rectangular guide; (*b*) circular guide; (*c*) ridge guide.

trends indicate that it will come into widespread use for long-distance communication lines because of the very low attenuation that can be achieved. The ridge-loaded rectangular guide illustrated in Fig. 1.2c is sometimes used in place of the standard rectangular guide because of better impedance properties and a greater bandwidth of operation. In addition to these standard-type guides, a variety of other cross sections, for example, elliptical, may also be used.

Another class of waveguides, of more recent origin, is surface waveguides. An example of this type is a conducting wire coated with a thin layer of dielectric. The wire diameter is small compared with the wavelength. Along a structure of this type it is possible to guide an electromagnetic wave. The wave is bound to the surface of the guide, exhibiting an amplitude decay that is exponential in the radial direction away from the surface, and hence is called a surface wave. Applications are mainly in the millimeter-wavelength range since the field does extend a distance of a wavelength or so beyond the wire, and this makes the effective guide diameter somewhat large in the centimeter-wavelength range. A disadvantage of surface waveguides and open-conductor transmission lines is that radiation loss occurs whenever other obstacles are brought into the vicinity of the guide.

A unique property of the transmission line is that a satisfactory analysis of its properties may be carried out by treating it as a network with distributed parameters and solving for the voltage and current waves that may propagate along the line. Other waveguides, although they have several properties similar to transmission lines, must be treated as electromagnetic boundary-value problems, and a solution for the electromagnetic fields must be determined. Fortunately, this is readily accomplished for the common waveguides used in practice. For waveguides it is not possible to define unique voltage and current that have the same significance as for a transmission line. This is one of the reasons why the field point of view is emphasized at microwave frequencies.

Associated with waveguides are a number of interesting problems related to methods of exciting fields in guides and methods of coupling energy out. Three basic coupling methods are used: (1) probe coupling, (2) loop coupling, and (3) aperture coupling between adjacent guides. They are illustrated in Fig. 1-3, and some of them are analyzed later. These coupling devices are actually small antennas that radiate into the waveguide.

Inductive and capacitive elements take a variety of forms at microwave frequencies. Perhaps the simplest are short-circuited sections of transmission line and waveguide. These exhibit a range of susceptance values from minus to plus infinity, depending on the length of the line, and hence may act as either inductive or capacitive elements. They may be connected as either series or shunt elements, as illustrated in Fig. 1-4. They are commonly referred to as stubs and are widely used as impedance-

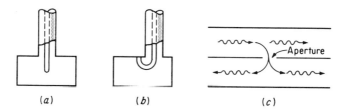

Fig. 1.3 Basic methods of coupling energy into and out of waveguides. (a) Probe coupling; (b) loop coupling; (c) aperture coupling.

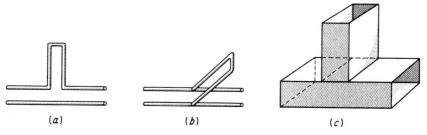

Fig. 1.4 Stub-type reactive elements. (a) Series element; (b) shunt element; (c) waveguide stub.

matching elements. In a rectangular guide thin conducting windows, or diaphragms, as illustrated in Fig. 1-5, also act as shunt susceptive elements. Their inductive or capacitive nature depends on whether there is more magnetic energy or electric energy stored in local fringing fields.

Resonant circuits are used both at low frequencies and at microwave frequencies to control the frequency of an oscillator and for frequency filtering. At low frequencies this function is performed by an inductor and capacitor in a series or parallel combination. Resonance occurs when there are equal average amounts of electric and magnetic energy stored. This energy oscillates back and forth between the magnetic field around the inductor and the electric field between the capacitor plates. At microwave frequencies the LC circuit is replaced by a closed conducting enclosure, or cavity. The electric and magnetic energy is stored in the field within the cavity. At an infinite number of specific frequencies, the resonant frequencies, there are equal average amounts of electric and magnetic energy stored in the cavity volume. In the vicinity of any one resonant frequency the input impedance to the cavity has the same properties as for a conventional LC resonant circuit. One significant feature worth noting is the very much larger Q values that may be obtained, these being often in excess of 10^4, as compared with those obtainable from low-frequency LC circuits. Figure 1-6 illustrates a cylindrical cavity that is aperture-coupled to a rectangular guide.

When a number of microwave devices are connected by means of sections of transmission lines or waveguides, we obtain a microwave circuit. The analysis of the behavior of such circuits is carried out either in terms of equivalent transmission-line voltage and current waves or in terms of the amplitudes of the propagating waves. The first approach leads to an equivalent-impedance description, and the second emphasizes the wave nature of the fields and results in a *scattering-matrix* formulation. Both approaches are used in this book. Since transmission-line circuit analysis forms the basis, either directly or by analogy, for the analysis of all microwave circuits, a considerable amount of attention is devoted to a fairly complete treatment of this subject early in the text. This material, together with the field analysis of the waves that may propagate along waveguides and that may exist in cavities, represents a major portion of the theory the microwave engineer must be familiar with.

The microwave systems engineer must have some understanding also of the principles of operation of various microwave tubes, such as klys-

(a)

(b)

Fig. 1.5 Shunt susceptive elements in a waveguide. (a) Inductive window; (b) capacitive window.

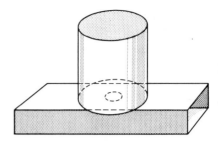

Fig. 1.6 Cylindrical cavity aperture-coupled to a rectangular guide.

trons, magnetrons, and traveling-wave tubes, and of the newer solid-state devices, such as masers and parametric amplifiers. This is required in order to make intelligent selection and proper use of these devices. In the text sufficient work is done to provide for this minimum level of knowledge of the principles involved. A treatment that is fully adequate for the tube designer is very much outside the scope of this book.

The microwave communication-systems engineer also requires a knowledge of antennas and microwave propagation. These two topics are somewhat specialized and too extensive to be included in the text. Rather than give a brief and incomplete treatment of these subjects, we suggest that the reader consult some of the available texts devoted entirely to these topics.†

In the light of the foregoing discussion, it should now be apparent that the study of microwave engineering should include, among other things, at least the following:

1. Electromagnetic theory
2. Wave solutions for transmission lines and waveguides
3. Transmission-line and waveguide circuit analysis
4. Resonators and slow-wave structures
5. Microwave oscillators and amplifiers
6. Antennas
7. Microwave propagation
8. Systems considerations

Apart from the last three, these are the major topics covered in the text. It is not possible to discuss in any great detail more than a few of the

† See, for example:

D. E. Kerr (ed.), "Propagation of Short Radio Waves," McGraw-Hill Book Company, New York, 1951.

S. Silver (ed.), "Microwave Antenna Theory and Design," McGraw-Hill Book Company, New York, 1949.

J. D. Kraus, "Antennas," McGraw-Hill Book Company, New York, 1950.

E. C. Jordan, "Electromagnetic Waves and Radiating Systems," Prentice-Hall, Inc., Englewood Cliffs, N.J., 1950.

S. A. Schelkunoff and H. J. Friis, "Antennas: Theory and Practice," John Wiley & Sons, Inc., New York, 1952.

many microwave devices available and in current use. Therefore only a selected number of them are analyzed, to provide illustrative examples for the basic theory being developed. The available technical literature may be, and should be, consulted for information on devices not included here. Appropriate references are given throughout the text.

The number of topics treated in this text represents a good deal more than can be covered in a one-semester course. However, rather than limit the depth of treatment, it was decided to separate some of the more specialized analytical treatments of particular topics from the less analytical discussion. These specialized sections are marked with a star, and can be eliminated in a first reading without significantly interrupting the continuity of the text.† The student or engineer interested in the design of microwave devices, or in a fuller understanding of various aspects of microwave theory, is advised to read these special sections.

As in any engineering field, measurements are of great importance in providing the link between theory and practice at microwave frequencies. Space does not permit inclusion of the subject of microwave measurements in this text. A number of excellent texts devoted entirely to microwave measurements are available, and the reader is referred to them.‡

† Problems based on material in these sections are also marked by a star.
‡ See, for example:
  C. G. Montgomery, "Technique of Microwave Measurements," McGraw-Hill Book Company, New York, 1947.
  E. L. Ginzton, "Microwave Measurements," McGraw-Hill Book Company, New York, 1957.
  M. Sucher and J. Fox, "Handbook of Microwave Measurements," 3d ed., John Wiley & Sons, Inc., New York, 1963.
  Engineering Staff of Microwave Division, Hewlett-Packard Company, "Microwave Theory and Measurements," Prentice-Hall, Inc., Englewood Cliffs, N.J., 1962.

2
Electromagnetic theory

2.1 Maxwell's equations

Electric and magnetic fields that vary with time are governed by physical laws described by a set of equations known collectively as Maxwell's equations. For the most part these equations were arrived at from experiments carried out by several investigators. It is not our purpose here to justify the basis for these equations, but rather to gain some understanding of their physical significance and to learn how to obtain solutions of these equations in practical situations of interest in the microwave engineering field. The electric field $\boldsymbol{\mathcal{E}}$ and magnetic field $\boldsymbol{\mathcal{B}}$ are vector fields and in general have amplitudes and directions that vary with the three spatial coordinates x, y, z and the time coordinate t.† In mks units, which are used throughout, the electric field is measured in volts per meter and the magnetic field in webers per square meter. Since these fields are vector fields, the equations governing their behavior are most conveniently written in vector form.‡

The electric field $\boldsymbol{\mathcal{E}}$ and magnetic field $\boldsymbol{\mathcal{B}}$ are regarded as fundamental in that they give the force on a charge q moving with velocity $\mathbf{v}$; that is,

$$\mathbf{F} = q(\boldsymbol{\mathcal{E}} + \mathbf{v} \times \boldsymbol{\mathcal{B}}) \tag{2.1}$$

where $\mathbf{F}$ is the force in newtons, q is the charge measured in coulombs, and $\mathbf{v}$ is the velocity in meters per second. This force law is called the Lorentz force equation. In addition to the $\boldsymbol{\mathcal{E}}$ and $\boldsymbol{\mathcal{B}}$ fields, it is convenient to introduce two auxiliary field vectors, namely, the electric displacement $\boldsymbol{\mathcal{D}}$ and the magnetic intensity $\boldsymbol{\mathcal{H}}$. These are related to $\boldsymbol{\mathcal{E}}$ and $\boldsymbol{\mathcal{B}}$ through

---

† Boldface script type is used to represent vector fields having arbitrary time dependence. Boldface roman type is used later for the phasor representation of fields having sinusoidal time dependence.

‡ It is assumed that the reader is familiar with vector analysis. However, for convenient reference, a number of vector formulas and relations are summarized in Appendix I.

the electric and magnetic polarization of material media, a topic covered in the next section. In this section we consider fields in vacuum, or *free space*, only. In this case the following simple relationships hold:

$$\mathcal{H} = \frac{1}{\mu_0} \mathcal{B} \tag{2.2a}$$

$$\mathcal{D} = \epsilon_0 \mathcal{E} \tag{2.2b}$$

where $\mu_0 = 4\pi \times 10^{-7}$ henry/m and is called the permeability of vacuum, and $\epsilon_0 = 10^{-9}/36\pi = 8.854 \times 10^{-12}$ farad/m and is known as the permittivity of vacuum.

One of the basic laws of electromagnetic phenomena is Faraday's law, which states that a time-varying magnetic field generates an electric field. With reference to Fig. 2.1, let C denote an arbitrary closed curve that forms the boundary of a nonmoving surface S. The time rate of change of total magnetic flux through the surface S is $\partial \left(\int_S \mathcal{B} \cdot d\mathbf{S} \right) / \partial t$. According to Faraday's law, this time rate of change of total magnetic flux is equal to the negative value of the total voltage measured around C. The latter quantity is given by $-\oint_C \mathcal{E} \cdot d\mathbf{l}$. Hence the mathematical statement of Faraday's law is

$$\oint_C \mathcal{E} \cdot d\mathbf{l} = -\frac{\partial}{\partial t} \int_S \mathcal{B} \cdot d\mathbf{S} \tag{2.3}$$

The line integral of $\mathcal{E}$ around C is a measure of the circulation, or "curling up," of the electric field in space. The time-varying magnetic field may be properly regarded as a vortex source that produces an electric field having nonzero curl, or circulation. Although (2.3) is in a form that is readily interpreted physically, it is not in a form suitable for the analysis of a physical problem. What is required is a differential equation that is equivalent to (2.3). This equation may be obtained by using Stokes' theorem from vector analysis, which states that the line integral of a vector around a closed contour C is equal to the integral of the normal component of the curl of this vector over any surface having C as its boundary. The curl of a vector is written $\nabla \times \mathcal{E}$ (Appendix I), and hence

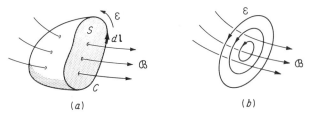

Fig. 2.1 Illustration of Faraday's law.

(2.1) becomes

$$\oint_C \pmb{\mathcal{E}} \cdot d\mathbf{l} = \int_S \nabla \times \pmb{\mathcal{E}} \cdot d\mathbf{S} = -\frac{\partial}{\partial t}\int_S \pmb{\mathcal{B}} \cdot d\mathbf{S}$$

Since S is completely arbitrary, the latter two integrals are equal only if

$$\nabla \times \pmb{\mathcal{E}} = -\frac{\partial \pmb{\mathcal{B}}}{\partial t} \tag{2.4}$$

which is the desired differential equation describing Faraday's law. The curl is a measure of the circulation of a vector field at a point.

Helmholtz's theorem from vector analysis states that a vector field is completely defined only when the curl, or circulation, of the field, and also its divergence, are given at every point in space. Now the divergence (or convergence) of field lines arises only if a proper source (or sink) is available. The electric field, in addition to having a curl produced by the vortex source $-\partial \pmb{\mathcal{B}}/\partial t$, has a divergence produced by electric charge. Gauss' law states that the total flux of $\pmb{\mathcal{D}} = \epsilon_0 \pmb{\mathcal{E}}$ from a volume V is equal to the net charge contained within V. If ρ represents the charge density in coulombs per cubic meter, Gauss' law may be written as

$$\oint_S \epsilon_0 \pmb{\mathcal{E}} \cdot d\mathbf{S} = \int_V \rho \, dV \tag{2.5}$$

This equation may be converted to a differential equation by using the divergence theorem to give

$$\oint_S \epsilon_0 \pmb{\mathcal{E}} \cdot d\mathbf{S} = \int_V \nabla \cdot \epsilon_0 \pmb{\mathcal{E}} \, dV = \int_V \rho \, dV$$

Since V is arbitrary, it follows that

$$\nabla \cdot \epsilon_0 \pmb{\mathcal{E}} = \nabla \cdot \pmb{\mathcal{D}} = \rho \tag{2.6}$$

where $\nabla \cdot \pmb{\mathcal{D}}$ is the divergence of $\pmb{\mathcal{D}}$, that is, a measure of the total outward flux of $\pmb{\mathcal{D}}$ from a volume element, divided by the volume of the element, as this volume shrinks to zero. Since both the curl and divergence of the electric field are now specified, this field is completely determined in terms of the two sources, $\partial \pmb{\mathcal{B}}/\partial t$ and ρ.

To complete the formulation of electromagnetic phenomena we must now relate the curl and divergence of the magnetic field to their sources. The vortex source that creates the circulation, or curl, of the magnetic field $\pmb{\mathcal{H}}$ is the current. By current is meant the total current density, the conduction current density $\pmb{\mathcal{J}}$ measured in amperes per square meter, the displacement current density $\partial \pmb{\mathcal{D}}/\partial t$, and the convection current $\rho \mathbf{v}$ consisting of charge in motion if present. Convection current is not included in this chapter. However, in the chapter dealing with microwave tubes, convection current plays a central role and is discussed in detail there. The displacement current density $\partial \pmb{\mathcal{D}}/\partial t$ was first introduced by Maxwell, and leads to the possibility of wave motion, as will be seen. Mathematically, the circulation of $\mathcal{H}$ around a closed contour C bounding a surface S as in Fig. 2.1 is given by

$$\oint_C \mathcal{H} \cdot d\mathbf{l} = \int_S \frac{\partial \mathcal{D}}{\partial t} \cdot d\mathbf{S} + \int_S \mathcal{J} \cdot d\mathbf{S} \tag{2.7}$$

Application of Stokes' law to the left-hand side yields

$$\int_S \nabla \times \mathcal{H} \cdot d\mathbf{S} = \int_S \frac{\partial \mathcal{D}}{\partial t} \cdot d\mathbf{S} + \int_S \mathcal{J} \cdot d\mathbf{S}$$

from which it may be concluded that

$$\nabla \times \mathcal{H} = \frac{\partial \mathcal{D}}{\partial t} + \mathcal{J} \tag{2.8}$$

Since magnetic charge, as the dual of electric charge, does not exist in nature, it may be concluded that the divergence of $\mathcal{B}$ is always zero; i.e., the flux lines of $\mathcal{B}$ are always closed since there are no charges for them to terminate on. Thus the net flux of $\mathcal{B}$ through any closed surface S is always zero; i.e., just as much flux enters through the surface as leaves it. Corresponding to (2.5) and (2.6) we thus have

$$\oint_S \mathcal{B} \cdot d\mathbf{S} = 0 \tag{2.9}$$

$$\nabla \cdot \mathcal{B} = 0 \tag{2.10}$$

Conduction current, of density $\mathcal{J}$, is the net flow of electric charge. Since charge is conserved, the total rate of flow of charge out of a volume V is equal to the time rate of decrease of total charge within V, as expressed by the equation

$$\oint_S \mathcal{J} \cdot d\mathbf{S} = -\frac{\partial}{\partial t} \int_V \rho \, dV \tag{2.11}$$

This is the continuity equation, and it may be converted to a differential equation by using the divergence theorem in the same manner as was done to derive (2.6) from (2.5). It is readily found that

$$\nabla \cdot \mathcal{J} + \frac{\partial \rho}{\partial t} = 0 \tag{2.12}$$

This equation may also be derived from (2.8) and (2.6). Since the divergence of the curl of any vector is identically zero, the divergence of (2.8) yields

$$0 = \frac{\partial \nabla \cdot \mathcal{D}}{\partial t} + \nabla \cdot \mathcal{J}$$

Using (2.6) converts this immediately into the continuity equation (2.12). If the displacement current density $\partial \mathcal{D}/\partial t$ had not been included as part of the total current density on the right-hand side of (2.8), that equation would have led to the conclusion that $\nabla \cdot \mathcal{J} = 0$, a result inconsistent with the continuity equation unless the charge density was independent of time.

In summary, the four equations, known as Maxwell's equations, that describe electromagnetic phenomena in vacuum are

$$\nabla \times \mathcal{E} = -\frac{\partial \mathcal{B}}{\partial t} \tag{2.13a}$$

$$\nabla \times \mathcal{H} = \frac{\partial \mathcal{D}}{\partial t} + \mathcal{J} \tag{2.13b}$$

$$\nabla \cdot \mathcal{D} = \rho \tag{2.13c}$$

$$\nabla \cdot \mathcal{B} = 0 \tag{2.13d}$$

where in (2.13b) the convection current $\rho \mathbf{v}$ has not been included. The continuity equation may be derived from (2.13b) and (2.13c), and hence contains no additional information. Although $-\partial \mathcal{B}/\partial t$ may be regarded as a source for $\mathcal{E}$, and $\partial \mathcal{D}/\partial t$ as a source for $\mathcal{H}$, the ultimate sources of an electromagnetic field are the current $\mathcal{J}$ and charge ρ. For time-varying fields, that charge density ρ which varies with time is not independent of $\mathcal{J}$ since it is related to the latter by the continuity equation. As a consequence, it is possible to derive the time-varying electromagnetic field from a knowledge of the current density $\mathcal{J}$ alone.

It is not difficult to show in a qualitative way that (2.13a) and (2.13b) lead to wave propagation, i.e., to the propagation of an electromagnetic disturbance through space. Consider a loop of wire in which a current varying with time flows as in Fig. 2.2. The conduction current causes a circulation, or curling, of the magnetic field around the current loop as in Fig. 2.2a (for clarity only a few flux lines are shown). The changing magnetic field in turn creates a circulating, or curling, electric field, with field lines that encircle the magnetic field lines as in Fig. 2.2b. This changing electric field creates further curling magnetic field lines as in

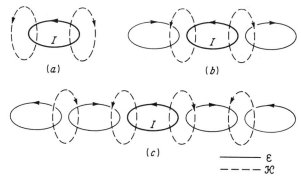

Fig. 2.2 The growth or generation of an electromagnetic wave from a current loop.

Fig. 2.2c, and so forth. The net result is the continual growth and spreading of the electromagnetic field into all space surrounding the current loop. The disturbance moves outward with the velocity of light. A little thought will show that the same characteristic mutual effect between two quantities must always exist for wave motion. That is, quantity A must be generated by quantity B, and vice versa. For example, in an acoustical wave the excess pressure creates a motion of the adjacent air mass. The motion of the air mass by virtue of its inertia in turn creates a condensation, or excess pressure, farther along. The repetition of this process generates the acoustical wave.

For the most part, as at lower frequencies, it is sufficient to consider only the steady-state solution for the electromagnetic field as produced by currents having sinusoidal time dependence. The time derivative may then be eliminated by denoting the time dependence of all quantities as $e^{j\omega t}$ and representing all field vectors as complex-phasor space vectors independent of time. Boldface roman type is used to represent these complex-phasor space vectors. For example, the mathematical representation for the electric field $\mathcal{E}(x, y, z, t)$ will be $\mathbf{E}(x, y, z)e^{j\omega t}$. Each component of $\mathbf{E}$ is in general complex, with a real and imaginary part; thus

$$\mathbf{E} = \mathbf{a}_x(E_{xr} + jE_{xi}) + \mathbf{a}_y(E_{yr} + jE_{yi}) + \mathbf{a}_z(E_{zr} + jE_{zi}) \tag{2.14}$$

where the subscript r refers to the real part and the subscript i refers to the imaginary part. Each component is allowed to be complex in order to provide for an arbitrary time phase for each component. This may be seen by recalling the usual method of obtaining $\mathcal{E}$ from its phasor representation. That is, by definition,

$$\mathcal{E}(x, y, z, t) = \text{Re}\,[\mathbf{E}(x, y, z)e^{j\omega t}] \tag{2.15}$$

Thus

$$\begin{aligned} E_x &= \text{Re}\,[(E_{xr} + jE_{xi})e^{j\omega t}] \\ &= \text{Re}\,(\sqrt{E_{xr}^2 + E_{xi}^2}\,e^{j\omega t + j\phi}) \\ &= \sqrt{E_{xr}^2 + E_{xi}^2}\,\cos(\omega t + \phi) \end{aligned}$$

where $\phi = \tan^{-1}(E_{xi}/E_{xr})$. Unless E_x had both an imaginary part jE_{xi} and a real part E_{xr}, the arbitrary phase angle ϕ would not be present. As a general rule, the time factor $e^{j\omega t}$ will not be written down when the phasor representation is used. However, it is important to remember both the fact that such a time dependence is implied and also the rule (2.15) for obtaining the physical field vector from its phasor representation. The real and imaginary parts of the space components of a vector should not be confused with the space components; for example, E_{xr} and E_{xi} are not two space components of E_x since the component $\mathbf{a}_x E_x$ is always directed along the x axis in space, with the real and imaginary parts simply accounting for an arbitrary time phase or origin.

A further point of interest in connection with the phasor representation is the method used for obtaining the time-average value of a field quantity. For example, if

$$\mathbf{\mathcal{E}} = \mathbf{a}_x E_1 \cos(\omega t + \phi_1) + \mathbf{a}_y E_2 \cos(\omega t + \phi_2) + \mathbf{a}_z E_3 \cos(\omega t + \phi_3)$$

the time-average value of $|\mathbf{\mathcal{E}}|^2$ is

$$\begin{aligned}
|\mathbf{\mathcal{E}}|_{av}^2 &= \frac{1}{T}\int_0^T \mathbf{\mathcal{E}} \cdot \mathbf{\mathcal{E}}\, dt \\
&= \frac{1}{T}\int_0^T [E_1^2 \cos^2(\omega t + \phi_1) + E_2^2 \cos^2(\omega t + \phi_2) \\
&\qquad\qquad + E_3^2 \cos^2(\omega t + \phi_3)]\, dt \\
&= \tfrac{1}{2}(E_1^2 + E_2^2 + E_3^2)
\end{aligned} \quad (2.16)$$

where T is the period, equal to $2\pi/\omega$. The same result is obtained by simply taking one-half of the scalar, or dot, product of $\mathbf{E}$ with the complex conjugate $\mathbf{E}^*$; thus

$$|\mathbf{\mathcal{E}}|_{av}^2 = \tfrac{1}{2}\mathbf{E} \cdot \mathbf{E}^* = \tfrac{1}{2}[(E_{xr}^2 + E_{xi}^2) + (E_{yr}^2 + E_{yi}^2) + (E_{zr}^2 + E_{zi}^2)] \quad (2.17)$$

since $E_x E_x^* = (E_{xr} + jE_{xi})(E_{xr} - jE_{xi}) = E_{xr}^2 + E_{xi}^2$, etc. This is equal to (2.16), since $E_1^2 = E_{xr}^2 + E_{xi}^2$, etc.

By using the phasor representation, the time derivative $\partial/\partial t$ may be replaced by the factor $j\omega$ since $\partial e^{j\omega t}/\partial t = j\omega e^{j\omega t}$. Hence Maxwell's equations, with steady-state sinusoidal time dependence, become

$$\nabla \times \mathbf{E} = -j\omega \mathbf{B} \quad (2.18a)$$
$$\nabla \times \mathbf{H} = j\omega \mathbf{D} + \mathbf{J} \quad (2.18b)$$
$$\nabla \cdot \mathbf{D} = \rho \quad (2.18c)$$
$$\nabla \cdot \mathbf{B} = 0 \quad (2.18d)$$

2.2 Constitutive relations

In material media the auxiliary field vectors $\mathcal{H}$ and $\mathcal{D}$ are defined in terms of the polarization of the material and the fundamental field quantities $\mathcal{B}$ and $\mathcal{E}$. The relationships of $\mathcal{H}$ to $\mathcal{B}$ and of $\mathcal{D}$ to $\mathcal{E}$ are known as constitutive relations, and must be known before solutions for Maxwell's equations can be found.

Consider first the electric case. If an electric field $\mathcal{E}$ is applied to a material body, this force results in a distortion of the atoms or molecules in such a manner as to create effective electric dipoles with a dipole moment $\mathcal{P}$ per unit volume. The total displacement current is the sum

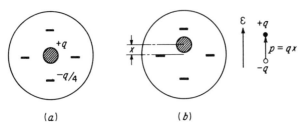

Fig. 2.3 Model for determining the polarization of an atom.

of the vacuum displacement current $\partial \epsilon_0 \mathcal{E}/\partial t$ and the polarization current $\partial \mathcal{P}/\partial t$. To avoid accounting for the polarization current $\partial \mathcal{P}/\partial t$ explicitly, the displacement vector $\mathcal{D}$ is defined as

$$\mathcal{D} = \epsilon_0 \mathcal{E} + \mathcal{P} \tag{2.19}$$

whence the total displacement current density can be written as $\partial \mathcal{D}/\partial t$.

For a great many materials the polarization $\mathcal{P}$ is in the direction of the electric field $\mathcal{E}$, although rarely will $\mathcal{P}$ have the same time phase as $\mathcal{E}$. A simple classical model will serve to illustrate this point. Figure 2.3a shows a model of an atom consisting of a nucleus with charge q surrounded by a spherically symmetrical electron cloud of total charge $-q$. The application of a field $\mathcal{E}$ displaces the electron cloud an effective distance x as in Fig. 2.3b. This displacement is resisted by a restoring force kx proportional to the displacement (Prob. 2.1). In addition, dissipation, or damping, effects are present and result in an additional force, which we shall assume to be proportional to the velocity. If m is the effective mass of the electron cloud, the dynamical equation of motion is obtained by equating the sum of the inertial force $m\,d^2x/dt^2$, viscous force $mv\,dx/dt$, and restoring force kx to the applied force $-q\mathcal{E}$; thus

$$m\frac{d^2x}{dt^2} + mv\frac{dx}{dt} + kx = -q\mathcal{E} \tag{2.20}$$

When $\mathcal{E} = E_x \cos \omega t$, the solution for x is of the form $x = -A \cos(\omega t + \phi)$.

If $E_x \cos \omega t$ is represented by the phasor E_x, and x by the phasor X, the solution for X is readily found to be

$$X = \frac{-qE_x}{-\omega^2 m + j\omega v m + k}$$

and hence

$$x = \operatorname{Re}(Xe^{j\omega t}) = -A \cos(\omega t + \phi)$$

where

$$A = \frac{(q/m)E_x}{[(\omega^2 - \omega_p^2)^2 + \omega^2\nu^2]^{\frac{1}{2}}}$$

$$\phi = \tan^{-1}\frac{\omega\nu}{\omega^2 - \omega_p^2}$$

and we have replaced k/m by ω_p^2.

The dipole moment is p_x, where

$$p_x = -qx = \frac{q^2 E_x}{m[(\omega^2 - \omega_p^2)^2 + \omega^2\nu^2]^{\frac{1}{2}}} \cos(\omega t + \phi) \tag{2.21}$$

For N such atoms per unit volume the polarization per unit volume is $\mathcal{P}_x = Np_x$ and the displacement $\mathcal{D}_x$ is given by

$$\mathcal{D}_x = \epsilon_0 E_x \cos\omega t + \frac{Nq^2 E_x}{m[(\omega^2 - \omega_p^2)^2 + \omega^2\nu^2]^{\frac{1}{2}}} \cos(\omega t + \phi)$$

This equation may be put into the following form also:

$$\mathcal{D}_x = E_x \left\{ \frac{[\epsilon_0(\omega_p^2 - \omega^2) + Nq^2/m]^2 + (\omega\nu\epsilon_0)^2}{(\omega_p^2 - \omega^2)^2 + (\omega\nu)^2} \right\}^{\frac{1}{2}} \cos(\omega t - \theta) \tag{2.22}$$

where

$$\theta = \tan^{-1}\frac{\omega\nu}{\omega_p^2 - \omega^2} - \tan^{-1}\frac{\omega\nu}{\omega_p^2 - \omega^2 + Nq^2/\epsilon_0 m}$$

Two points are of interest in connection with (2.22). One is the linear relationship between $\mathcal{P}$ and $\mathcal{E}$, and hence between $\mathcal{D}$ and $\mathcal{E}$. The second is the time lag in $\mathcal{D}$ relative to $\mathcal{E}$ whenever damping forces are present.

The time-phase difference between $\mathcal{P}$, $\mathcal{E}$, and $\mathcal{D}$ makes it awkward to handle the relations between these quantities unless phasor representation is used. In phasor representation (2.21) and (2.22) become

$$P_x = \frac{q^2 E_x}{(\omega_p^2 - \omega^2 + j\omega\nu)m} \tag{2.23}$$

$$D_x = \frac{\epsilon_0(\omega_p^2 - \omega^2 + j\omega\nu) + Nq^2/m}{\omega_p^2 - \omega^2 + j\omega\nu} E_x \tag{2.24}$$

In general, for linear media, we may write

$$\mathbf{P} = \epsilon_0 \chi_e \mathbf{E} \tag{2.25}$$

where χ_e is a complex constant of proportionality called the electric susceptibility. The equation for $\mathbf{D}$ becomes

$$\mathbf{D} = \epsilon_0 \mathbf{E} + \mathbf{P} = \epsilon_0(1 + \chi_e)\mathbf{E}$$
$$= \epsilon \mathbf{E} = \kappa\epsilon_0 \mathbf{E} = (\epsilon' - j\epsilon'')\mathbf{E} \tag{2.26}$$

20 Foundations for microwave engineering

where $\epsilon = \epsilon_0(1 + \chi_e)$ is called the permittivity, and $\kappa = \epsilon/\epsilon_0$, the dielectric constant of the medium. Note that ϵ is complex whenever damping effects are present and that the imaginary part is always negative. A positive imaginary part would imply energy creation instead of energy loss. [The reader may verify from (2.22) that θ is always positive.]

Loss in a dielectric material may also occur because of a finite conductivity σ. The two mechanisms are indistinguishable as far as external effects related to power dissipation are concerned. The curl equation for **H** may be written as

$$\nabla \times \mathbf{H} = j\omega(\epsilon' - j\epsilon'')\mathbf{E} + \sigma\mathbf{E}$$

where $\mathbf{J} = \sigma\mathbf{E}$ is the conduction current density in the material. We may also write

$$\nabla \times \mathbf{H} = j\omega\left[\epsilon' - j\left(\epsilon'' + \frac{\sigma}{\omega}\right)\right]\mathbf{E} = j\omega\epsilon'\mathbf{E} + (\omega\epsilon'' + \sigma)\mathbf{E} \quad (2.27)$$

whereby $\epsilon'' + \sigma/\omega$ may be considered as the effective imaginary part of the permittivity, or $\omega\epsilon'' + \sigma$ as the total effective conductivity.

The loss tangent of a dielectric medium is defined by

$$\tan \delta_l = \frac{\omega\epsilon'' + \sigma}{\omega\epsilon'} \quad (2.28)$$

Any measurement of $\tan \delta_l$ always includes the effects of finite conductivity σ. At microwave frequencies, however, $\omega\epsilon''$ is usually much larger than σ because of the large value of ω.

Materials for which **P** is linearly related to **E** and in the same direction as **E** are called linear isotropic materials. Nonlinear effects generally occur only for very large applied fields, and as a consequence are rarely encountered in microwave work. However, nonisotropic material is of some importance. If the crystal structure lacks spherical symmetry such as that in a cubic crystal, it may be anticipated that the polarization per unit volume will depend on the direction of the applied field. In Fig. 2.4

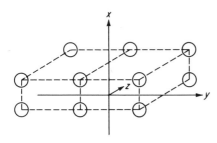

Fig. 2.4 A noncubic crystal exhibiting anisotropic effects.

a two-dimensional sketch of a crystal lacking cubic symmetry is given. The polarization produced when the field is applied along the x axis may be greater than that when the field is applied along the y or z axis because of the greater ease of polarization along the x axis. In this case we must write

$$D_x = \epsilon_{xx} E_x \qquad D_y = \epsilon_{yy} E_y \qquad D_z = \epsilon_{zz} E_z \tag{2.29}$$

where ϵ_{xx}, ϵ_{yy}, and ϵ_{zz} are, in general, all different. The dielectric constants $\kappa_x = \epsilon_{xx}/\epsilon_0$, $\kappa_y = \epsilon_{yy}/\epsilon_0$, $\kappa_z = \epsilon_{zz}/\epsilon_0$ are known as the principal dielectric constants, and the material is said to be anisotropic. If the coordinate system used had a different orientation with respect to the crystal structure, the relation between **D** and **E** would become

$$D_x = \epsilon_{xx} E_x + \epsilon_{xy} E_y + \epsilon_{xz} E_z$$
$$D_y = \epsilon_{yx} E_x + \epsilon_{yy} E_y + \epsilon_{yz} E_z$$
$$D_z = \epsilon_{zx} E_x + \epsilon_{zy} E_y + \epsilon_{zz} E_z$$

or in matrix form,

$$\begin{bmatrix} D_x \\ D_y \\ D_z \end{bmatrix} = \begin{bmatrix} \epsilon_{xx} & \epsilon_{xy} & \epsilon_{xz} \\ \epsilon_{yx} & \epsilon_{yy} & \epsilon_{yz} \\ \epsilon_{zx} & \epsilon_{zy} & \epsilon_{zz} \end{bmatrix} \begin{bmatrix} E_x \\ E_y \\ E_z \end{bmatrix} \tag{2.30}$$

Only for a particular orientation of the coordinate system does (2.30) reduce to (2.29). This particular orientation defines the principal axis of the medium. For anisotropic media the permittivity is referred to as a tensor permittivity (a tensor of rank 2 may be represented by a matrix). For the most part the materials dealt with in this text are isotropic. Nevertheless, an awareness of the existence of anisotropic media and of the nature of the constitutive relations for such media is important.

For the magnetic case, **H** is defined by the constitutive relation

$$\mu_0 \mathbf{H} = \mathbf{B} - \mu_0 \mathbf{M} \tag{2.31}$$

where **M** is the magnetic dipole polarization per unit volume. For most materials (ferromagnetic materials excluded), **M** is linearly related to **B** and hence to **H**. By convention this is expressed by the equation

$$\mathbf{M} = \chi_m \mathbf{H} \tag{2.32}$$

where χ_m is called the magnetic susceptibility. Substituting (2.32) into (2.31) gives

$$\mathbf{B} = \mu_0 (\mathbf{M} + \mathbf{H}) = \mu_0 (1 + \chi_m) \mathbf{H} = \mu \mathbf{H} \tag{2.33}$$

where $\mu = \mu_0 (1 + \chi_m)$ is called the permeability.

As in the electric case, damping forces cause μ to be a complex parameter with a negative imaginary part; that is, $\mu = \mu' - j\mu''$. Also, there are magnetic materials that are anisotropic; in particular, ferrites are anisotropic magnetic materials of great usefulness at microwave frequencies. These exhibit a tensor permeability of the following form,

$$[\mu] = \begin{bmatrix} \mu_1 & j\mu_2 & 0 \\ -j\mu_2 & \mu_1 & 0 \\ 0 & 0 & \mu_3 \end{bmatrix} \qquad (2.34)$$

when a static magnetic field is applied along the axis for which the permeability is μ_3. A discussion of ferrites and their uses is presented later; so further comments on their anisotropic properties is deferred until then.

In Sec. 2.1 care was taken to write Maxwell's equations in a form valid not only in vacuum but also in material media. Thus (2.13) and (2.18) are valid in general, but with the constitutive relations of this section replacing the free-space relations (2.2). Note, however, that it is not possible to write, in general, constitutive relations of the form $\mathfrak{D} = \epsilon \mathcal{E}$, $\mathfrak{B} = \mu \mathcal{H}$, when $\mathfrak{D}$ and $\mathcal{E}$, and likewise $\mathfrak{B}$ and $\mathcal{H}$, are not in time phase. For arbitrary time dependence we must write instead $\mathfrak{D} = \epsilon_0 \mathcal{E} + \mathcal{P}$, $\mathfrak{B} = \mu_0(\mathcal{H} + \mathfrak{M})$ and relate $\mathcal{P}$ and $\mathfrak{M}$ to $\mathcal{E}$ and $\mathcal{H}$ through the dynamical equation of motion governing the polarization mechanism. This difficulty may be circumvented by using the phasor representation for which relations such as $\mathbf{D} = \epsilon \mathbf{E}$ are perfectly valid because the complex nature of ϵ accounts for the difference in time phase.† It should be pointed out, however, that for many materials used at frequencies up to and including microwaves, the losses are so small that $\mathfrak{D}$ and $\mathcal{E}$, and also $\mathcal{H}$ and $\mathfrak{B}$, are very nearly in time phase. In such cases constitutive relations such as $\mathfrak{D} = \epsilon \mathcal{E}$, $\mathfrak{B} = \mu \mathcal{H}$ apply with negligible error. Significant departure in time phase between $\mathfrak{D}$ and $\mathcal{E}$ or $\mathfrak{B}$ and $\mathcal{H}$ occurs only in the vicinity of a natural resonance frequency of the equation of motion for the polarization.

2.3 Static fields

For electric and magnetic fields that are independent of time, the electric and magnetic fields are not coupled, and likewise the current and charge are not coupled. Putting all time derivatives equal to zero in (2.13)

---

† The situation here is like that encountered in a-c circuit analysis, where in phasor notation the voltage V equals the current I multiplied by the impedance Z; that is, $V = IZ$. An Ohm's law of this sort cannot be written for the physical voltage and current, for if $\mathcal{U} = \text{Re}(Ve^{j\omega t}) = V \cos \omega t$, then $\mathcal{I} = \text{Re}(Ie^{j\omega t}) = [V/(R^2 + X^2)^{\frac{1}{2}}] \cos(\omega t - \phi)$, where $\phi = \tan^{-1}(X/R)$. Clearly, $\mathcal{U}$ cannot be equated to $\mathcal{I}$ multiplied by a constant because of the difference in phase.

yields†

$$\nabla \times \mathbf{E} = 0 \tag{2.35a}$$

$$\nabla \cdot \epsilon\mathbf{E} = \rho \tag{2.35b}$$

$$\nabla \times \mathbf{H} = \mathbf{J} \tag{2.36a}$$

$$\nabla \cdot \mathbf{B} = 0 \tag{2.36b}$$

$$\nabla \cdot \mathbf{J} = 0 \tag{2.36c}$$

The last equation is the continuity equation for the special case $\partial\rho/\partial t = 0$.

The static electric field has zero curl, or circulation, and this means that the line integral of $\mathbf{E}$ around any arbitrary closed contour is zero. This property is just the condition that permits $\mathbf{E}$ to be derived from the gradient of a scalar potential function Φ; that is, since $\nabla \times \nabla\Phi$ is identically zero, we may put

$$\mathbf{E} = -\nabla\Phi \tag{2.37}$$

Substituting (2.37) into (2.35b) and assuming that ϵ is a constant independent of the coordinates give

$$-\nabla \cdot \mathbf{E} = \nabla^2\Phi = -\frac{\rho}{\epsilon} \tag{2.38}$$

This equation is known as Poisson's equation. When $\rho = 0$, Laplace's equation

$$\nabla^2\Phi = 0 \tag{2.39}$$

is obtained. The basic field problem in electrostatics is to solve Poisson's or Laplace's equation for a potential function Φ that satisfies specified boundary conditions.

As a simple example consider two infinite conducting planes at $x = 0$, a, as in Fig. 2.5. Let charge be distributed with a density $\rho = \rho_0 x$

† For static fields we are using boldface roman type to represent the physically real vector fields.

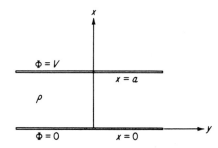

Fig. 2.5 A simple potential problem.

between the two plates.† It is required to find a Φ which is a solution of Poisson's equation and which equals zero on the plane $x = 0$ and V on the plane $x = a$. The potential will depend on x only; so (2.38) becomes

$$\frac{d^2\Phi}{dx^2} = -\rho_0 \frac{x}{\epsilon_0}$$

Integrating this equation twice gives $\Phi = -\rho_0 x^3/6\epsilon_0 + C_1 x + C_2$. Imposing the boundary conditions at $x = 0, a$, yields $0 = C_2$,

$$V = -\frac{\rho_0 a^3}{6\epsilon_0} + C_1 a + C_2$$

and hence $C_2 = 0$, $C_1 = V/a + \rho_0 a^2/6\epsilon_0$. The solution for Φ is thus

$$\Phi = -\frac{\rho_0 x^3}{6\epsilon_0} + \frac{\rho_0 a^2 x}{6\epsilon_0} + \frac{V}{a} x$$

The electric field between the two plates is

$$\mathbf{E} = -\nabla\Phi = -\mathbf{a}_x \frac{\partial \Phi}{\partial x} = \mathbf{a}_x \left(\frac{\rho_0 x^2}{2\epsilon_0} - \frac{\rho_0 a^2}{6\epsilon_0} - \frac{V}{a} \right)$$

The solution for the electrostatic field is greatly facilitated by introduction of the scalar potential Φ. For the same reason it is advantageous to introduce a potential function for the solution of magnetostatic problems. Since $\mathbf{B}$ always has zero divergence, it may be derived from the curl of a vector potential $\mathbf{A}$; that is,

$$\mathbf{B} = \nabla \times \mathbf{A} \tag{2.40}$$

This makes the divergence of $\mathbf{B}$ vanish identically because the divergence of the curl of a vector is identically zero. Using (2.40) in (2.36a) and assuming that μ is constant yield the equation

$$\nabla \times \mu\mathbf{H} = \nabla \times \mathbf{B} = \nabla \times \nabla \times \mathbf{A} = \mu\mathbf{J}$$

A vector identity of use here is $\nabla \times \nabla \times \mathbf{A} = \nabla\nabla \cdot \mathbf{A} - \nabla^2 \mathbf{A}$. The divergence of $\mathbf{A}$ may be placed equal to zero without affecting the value of $\mathbf{B}$ derived from the curl of $\mathbf{A}$, and hence the equation for $\mathbf{A}$ is

$$\nabla^2 \mathbf{A} = -\mu\mathbf{J} \tag{2.41}$$

This equation is a vector Poisson's equation. In rectangular coordinates, (2.41) represents three scalar Poisson's equations, the first being

$$\nabla^2 A_x = -\mu J_x \tag{2.42}$$

In a curvilinear coordinate system, such as a cylindrical coordinate system, (2.41) cannot be written in such a simple component form. The

† The example is somewhat artificial since the assumed charge distribution is not a stable one; i.e., the electric field it produces would cause the charge distribution to change.

reason is that, for example, $\nabla^2 \mathbf{a}_r\, A_r$ does not equal $\mathbf{a}_r\, \nabla^2 A_r$ because, even though the unit vector $\mathbf{a}_r$ is of constant length, its orientation varies from point to point since it is always directed along the radius vector from the origin to the point under consideration. The evaluation of $\nabla^2 \mathbf{A}$ in curvilinear coordinates is made by using the vector identity quoted above to give $\nabla^2 \mathbf{A} = \nabla \nabla \cdot \mathbf{A} - \nabla \times \nabla \times \mathbf{A}$. These latter operations are readily carried out.

The interest in static field solutions at microwave frequencies arises because the field distribution over a cross-sectional plane of a transmission line is a static field distribution and because static field solutions are good approximate solutions to the actual fields in the vicinity of obstacles that are small compared with the wavelength. The potential theory introduced above may be extended to the time-varying case also, and this is done in a following section.

2.4 Wave equation

For convenience, the two curl equations are repeated here:

$$\nabla \times \boldsymbol{\mathcal{E}} = -\frac{\partial \boldsymbol{\mathcal{B}}}{\partial t} \tag{2.43a}$$

$$\nabla \times \boldsymbol{\mathcal{H}} = \frac{\partial \boldsymbol{\mathcal{D}}}{\partial t} \tag{2.43b}$$

where it is assumed for the present that the current density $\boldsymbol{\mathcal{J}}$ is zero in the region of interest. These equations, together with the assumed constitutive relations $\boldsymbol{\mathcal{D}} = \epsilon \boldsymbol{\mathcal{E}}$, $\boldsymbol{\mathcal{B}} = \mu \boldsymbol{\mathcal{H}}$, may be combined to obtain a separate equation for each field. The curl of (2.43a) is

$$\nabla \times \nabla \times \boldsymbol{\mathcal{E}} = -\frac{\partial (\nabla \times \boldsymbol{\mathcal{B}})}{\partial t} = -\mu \frac{\partial (\nabla \times \boldsymbol{\mathcal{H}})}{\partial t}$$

Using (2.43b) and expanding $\nabla \times \nabla \times \boldsymbol{\mathcal{E}}$ now yield

$$\nabla \nabla \cdot \boldsymbol{\mathcal{E}} - \nabla^2 \boldsymbol{\mathcal{E}} = -\mu \epsilon \frac{\partial^2 \boldsymbol{\mathcal{E}}}{\partial t^2}$$

Since ρ is assumed zero and ϵ is taken as a constant, $\nabla \cdot \boldsymbol{\mathcal{E}} = 0$, and we obtain

$$\nabla^2 \boldsymbol{\mathcal{E}} - \mu \epsilon \frac{\partial^2 \boldsymbol{\mathcal{E}}}{\partial t^2} = 0 \tag{2.44}$$

which is a three-dimensional wave equation. The velocity of propagation v is equal to $(\mu \epsilon)^{-\frac{1}{2}}$. In free space v is equal to the velocity of light c. To illustrate the nature of the solutions of (2.44), consider a case where $\boldsymbol{\mathcal{E}}$ has only an x component and depends only on the z coordinate. In this instance

$$\frac{\partial^2 \mathcal{E}_x}{\partial z^2} - \mu\epsilon \frac{\partial^2 \mathcal{E}_x}{\partial t^2} = 0$$

Any function of the form $f(z - vt)$ is a solution of this equation since

$$\frac{\partial^2 f}{\partial z^2} = f'' \qquad \frac{\partial^2 f}{\partial t^2} = v^2 \frac{\partial^2 f}{\partial (vt)^2} = v^2 f''$$

and hence

$$\frac{\partial^2 f}{\partial z^2} - \frac{1}{v^2}\frac{\partial^2 f}{\partial t^2} = 0$$

This solution is illustrated in Fig. 2.6 and clearly represents a disturbance propagating in the positive z direction with velocity v. An equally valid solution is $f(z + vt)$ and represents a disturbance propagating in the negative z direction.

By eliminating the electric field, it is readily found that the magnetic field $\mathcal{H}$ satisfies the wave equation (2.44) also. In practice, however, we solve the wave equation for either $\mathcal{E}$ or $\mathcal{H}$ and then derive the other field by using the appropriate curl equation. When constitutive relations such as $\mathcal{D} = \epsilon \mathcal{E}$ and $\mathcal{B} = \mu \mathcal{H}$ cannot be written, the polarization vectors $\mathcal{P}$ and $\mathcal{M}$ must be exhibited explicitly in Maxwell's equations. Wave equations for $\mathcal{E}$ and $\mathcal{H}$ may still be derived, but $\mathcal{P}$ and $\mathcal{M}$ will now enter as equivalent sources for the field (which they actually are). The derivation is left as a problem at the end of this chapter.

For harmonic time dependence, the equation obtained in place of (2.44) is

$$\nabla^2 \mathbf{E} + k^2 \mathbf{E} = 0 \tag{2.45}$$

where $k^2 = \omega^2 \mu \epsilon$. This equation is referred to as the Helmholtz equation, or reduced wave equation. The constant k is called the wave number

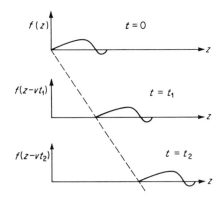

Fig. 2.6 Propagation of a disturbance $f(z - vt)$.

and may be expressed in the form

$$k = \omega\sqrt{\mu\epsilon} = \frac{\omega}{v} = 2\pi\frac{f}{v} = \frac{2\pi}{\lambda} \tag{2.46}$$

where the wavelength λ is equal to v/f. In free space the wave number will be written as k_0, and is equal to $\omega\sqrt{\mu_0\epsilon_0} = 2\pi/\lambda_0$. The magnetic field **H**, as may be surmised, satisfies the same reduced wave equation.

In a medium with finite conductivity σ, a conduction current $\mathcal{J} = \sigma\mathcal{E}$ will exist, and this results in energy loss because of Joule heating. The wave equation in media of this type has a damping term proportional to σ and the first time derivative of the field. In metals, excluding ferromagnetic materials, the permittivity and permeability are essentially equal to their free-space values, at least for frequencies up to and including the microwave range. Thus Maxwell's curl equations become

$$\nabla \times \mathcal{E} = -\mu_0 \frac{\partial \mathcal{H}}{\partial t} \qquad \nabla \times \mathcal{H} = \epsilon_0 \frac{\partial \mathcal{E}}{\partial t} + \sigma\mathcal{E}$$

Elimination of $\mathcal{H}$ in the same manner as before leads to the following wave equation for $\mathcal{E}$:

$$\nabla^2\mathcal{E} - \mu_0\sigma \frac{\partial \mathcal{E}}{\partial t} - \mu_0\epsilon_0 \frac{\partial^2 \mathcal{E}}{\partial t^2} = 0 \tag{2.47}$$

The magnetic field $\mathcal{H}$ also satisfies this equation. For the time-harmonic case damping effects enter in through the complex nature of ϵ and μ, and hence the wave number k. It should be recalled here that, as shown by (2.27), a finite conductivity σ is equivalent to an imaginary term in the permittivity ϵ. In the present case the equivalent permittivity is $\epsilon = \epsilon_0 - j\sigma/\omega$ and the Helmholtz equation is

$$\nabla^2\mathbf{E} + \omega^2\mu_0\epsilon_0\left(1 - j\frac{\sigma}{\omega\epsilon_0}\right)\mathbf{E} = 0 \tag{2.48}$$

In metals the conduction current $\sigma\mathbf{E}$ is generally very much larger than the displacement current $\omega\epsilon_0\mathbf{E}$, so that the latter may be neglected. For example, σ is equal to 5.8×10^7 mhos/m for copper, and at a frequency of 10^{10} cps, $\omega\epsilon_0 = 0.55$, which is much smaller than σ. Only for frequencies in the optical range will the two become comparable. Thus (2.47) may be simplified to

$$\nabla^2\mathcal{E} - \mu_0\sigma \frac{\partial \mathcal{E}}{\partial t} = 0 \tag{2.49}$$

and (2.48) reduces to

$$\nabla^2\mathbf{E} - j\omega\mu_0\sigma\mathbf{E} = 0 \tag{2.50}$$

Equation (2.49) is a diffusion equation similar to that which governs the flow of heat in a thermal conductor.

2.5 Energy and power

When currents exist in conductors as a result of the application of a suitable potential source, energy is expended by the source in maintaining the currents. The energy supplied by the source is stored in the electric and magnetic fields set up by the currents or propagated (radiated) away in the form of an electromagnetic wave. Under steady-state sinusoidal time-varying conditions, the time-average energy stored in the electric field is

$$W_e = \operatorname{Re} \tfrac{1}{4} \int_V \mathbf{E} \cdot \mathbf{D}^* \, dV = \tfrac{1}{4} \int_V \epsilon' \mathbf{E} \cdot \mathbf{E}^* \, dV \qquad (2.51a)$$

If ϵ is a constant and real, (2.51a) becomes

$$W_e = \frac{\epsilon}{4} \int_V \mathbf{E} \cdot \mathbf{E}^* \, dV \qquad (2.51b)$$

The time-average energy stored in the magnetic field is given by

$$W_m = \operatorname{Re} \tfrac{1}{4} \int_V \mathbf{H}^* \cdot \mathbf{B} \, dV = \tfrac{1}{4} \int_V \mu' \mathbf{H} \cdot \mathbf{H}^* \, dV \qquad (2.52a)$$

which, for μ real and constant, becomes

$$W_m = \frac{\mu}{4} \int_V \mathbf{H} \cdot \mathbf{H}^* \, dV \qquad (2.52b)$$

These expressions for W_e and W_m are valid only for nondispersive media, i.e., media for which ϵ and μ can be considered independent of ω in the vicinity of the angular frequency ω with which the fields vary. In general, when the losses are small, so that $\epsilon'' \ll \epsilon'$ and $\mu'' \ll \mu'$, we have

$$W_e = \tfrac{1}{4} \int_V \mathbf{E} \cdot \mathbf{E}^* \frac{\partial \omega \epsilon'}{\partial \omega} \, dV \qquad (2.53a)$$

$$W_m = \tfrac{1}{4} \int_V \mathbf{H} \cdot \mathbf{H}^* \frac{\partial \omega \mu'}{\partial \omega} \, dV \qquad (2.53b)$$

for the time-average stored electric and magnetic energy.

The above equations for the time-average energy in a dispersive medium may be established by considering a classical model of the polarization mechanism similar to that discussed in Sec. 2.2. In a unit volume let the effective oscillating charge of the dipole distribution be $-q$ with an effective mass m. Let the damping force be equal to $m\nu$ times the velocity of the charge. This damping force takes account of collision effects and loss of energy by radiation from the oscillating charge. The equation of motion for the polarization charge displacement u is

$$m \frac{d^2 u}{dt^2} + m\nu \frac{du}{dt} + ku = -q\mathcal{E}$$

where u is in the direction of the field $\mathcal{E}$. In this equation k is the elastic constant giving rise to the restoring force. This constant arises from the Coulomb forces acting on the displaced charge, and hence is of electrical origin. The dipole polarization $\mathcal{P}$ is $-qu$, and the polarization current $\mathcal{J}_p = d\mathcal{P}/dt$. Introducing the polarization current into the equation of motion gives

$$\frac{m}{q^2}\frac{d\mathcal{J}_p}{dt} + \frac{m\nu}{q^2}\mathcal{J}_p + \frac{k}{q^2}\int^t \mathcal{J}_p\, dt = \mathcal{E}$$

This equation is formally the same as that which describes the current in a series LCR circuit with an applied voltage $\mathcal{V}$ equal to $\mathcal{E}$ and with

$$L = \frac{m}{q^2} \qquad R = \frac{m\nu}{q^2} \qquad C = \frac{q^2}{k}$$

An equivalent circuit describing the polarization is illustrated in Fig. 2.7. If a time dependence $e^{j\omega t}$ is assumed and phasor notation is used,

$$J_p = EY = E\frac{R - jX}{R^2 + X^2}$$

where Y is the input admittance and $X = \omega L - 1/\omega C$. Since $P = \epsilon_0 \chi_e E$ and $J_p = j\omega P$, we see that

$$\omega\epsilon_0\chi_e = \omega\epsilon_0(\chi'_e - j\chi''_e) = -jY = \frac{-X - jR}{R^2 + X^2}$$

and hence

$$\omega\epsilon_0\chi'_e = \frac{-X}{R^2 + X^2} \tag{2.54a}$$

$$\omega\epsilon_0\chi''_e = \frac{R}{R^2 + X^2} \tag{2.54b}$$

The time-average power loss associated with the polarization is the same as the power loss in R in the equivalent circuit. This is given by

$$P_l = \tfrac{1}{2}EE^*\frac{R}{R^2 + X^2} = \tfrac{1}{2}EE^*\omega\epsilon_0\chi''_e \tag{2.55}$$

per unit volume. This equation shows that $\omega\epsilon_0\chi''_e = \omega\epsilon''$ is an equivalent conductance. The time-average energy stored in the system is of two forms. First there is the kinetic energy of motion, that is, $\tfrac{1}{2}m(du/dt)^2$ averaged over a cycle, and this is equal to the magnetic energy stored in

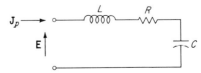

Fig. 2.7 Equivalent circuit for polarization current.

the inductor in the equivalent circuit. This time-average kinetic energy per unit volume is given by

$$U_m = \tfrac{1}{4} L J_p J_p^* = \tfrac{1}{4} EE^* \frac{L}{R^2 + X^2} \tag{2.56a}$$

The second form of stored energy is the potential energy associated with the charge displacement. The time-average value of this energy is equal to the time-average electric energy stored in the capacitor C in the equivalent circuit, and is given by

$$U_e = \tfrac{1}{4} EE^* \frac{1}{\omega^2 C(R^2 + X^2)} \tag{2.56b}$$

The total time-average energy stored per unit volume is $U = U_m + U_e$. Note that U is not given by $\tfrac{1}{4} EE^* \epsilon_0 \chi_e'$. The latter expression gives

$$\tfrac{1}{4} EE^* \epsilon_0 \chi_e' = \tfrac{1}{4} EE^* \frac{-X}{\omega(R^2 + X^2)}$$

$$= \tfrac{1}{4} EE^* \frac{1/\omega^2 C - L}{R^2 + X^2} = U_e - U_m$$

or the difference between the potential and kinetic energy stored.

To obtain an expression for the total stored energy, note that

$$\frac{d}{d\omega}\left(\frac{X}{R^2 + X^2}\right) = \frac{L + 1/\omega^2 C}{R^2 + X^2}\left(1 - \frac{2X^2}{R^2 + X^2}\right)$$

For a low-loss system, $R^2 \ll X^2$, and we then have $1 - 2X^2/(R^2 + X^2) \approx -1$; so

$$\frac{d}{d\omega}\left(\frac{-X}{R^2 + X^2}\right) = \frac{d}{d\omega}(\omega\epsilon_0\chi_e') \approx \frac{L + 1/\omega^2 C}{R^2 + X^2}$$

Multiplying this expression by $\tfrac{1}{4} EE^*$ now gives the total time-average energy stored, as comparison with (2.56a) and (2.56b) shows. Thus the final expression for the time-average electric energy stored in a volume V is given by the volume integral of $U = U_e + U_m$ plus the free-space energy density $\epsilon_0(\mathbf{E} \cdot \mathbf{E}^*)/4$ and is

$$W_e = \int_V \left(U + \frac{\epsilon_0}{4}\mathbf{E} \cdot \mathbf{E}^*\right) dV$$

$$= \int_V \frac{\mathbf{E} \cdot \mathbf{E}^*}{4}\left(\epsilon_0 + \frac{\partial \omega\epsilon_0\chi_e'}{\partial \omega}\right) dV$$

$$= \tfrac{1}{4} \int_V \mathbf{E} \cdot \mathbf{E}^* \frac{\partial \omega\epsilon'}{\partial \omega} dV$$

since $\epsilon' = \epsilon_0(1 + \chi_e')$. This equation is the result given earlier by (2.53a).

A similar type of model may be used to establish (2.53b) for the average stored magnetic energy. It should be pointed out that under time-varying conditions the average stored energy associated with either electric or magnetic polarization includes a kinetic-energy term. This term is negligible at low frequencies and also when ϵ' and μ' are essentially independent of ω for the range of ω of interest. When this energy is not negligible, the modified expressions for stored energy must be used.

Although (2.53) is more general than (2.51) and (2.52), we shall, in the majority of instances, use the latter equations for the stored energy. We thereby tacitly assume that we are dealing with material that is non-dispersive or very nearly so.

The time-average power transmitted across a closed surface S is given by the integral of the real part of one-half of the normal component of the complex Poynting vector $\mathbf{E} \times \mathbf{H}^*$; that is,

$$P = \operatorname{Re} \tfrac{1}{2} \oint_S \mathbf{E} \times \mathbf{H}^* \cdot d\mathbf{S} \qquad (2.57)$$

The above results are obtained from the interpretation of the complex Poynting vector theorem, which may be derived from Maxwell's equations as follows: If the divergence of $\mathbf{E} \times \mathbf{H}^*$, that is, $\nabla \cdot \mathbf{E} \times \mathbf{H}^*$, is expanded, we obtain

$$\nabla \cdot \mathbf{E} \times \mathbf{H}^* = (\nabla \times \mathbf{E}) \cdot \mathbf{H}^* - (\nabla \times \mathbf{H}^*) \cdot \mathbf{E}$$

From Maxwell's equations $\nabla \times \mathbf{E} = -j\omega \mathbf{B}$ and $\nabla \times \mathbf{H}^* = -j\omega \mathbf{D}^* + \mathbf{J}^*$, and hence

$$\nabla \cdot \mathbf{E} \times \mathbf{H}^* = -j\omega \mathbf{B} \cdot \mathbf{H}^* + j\omega \mathbf{D}^* \cdot \mathbf{E} - \mathbf{E} \cdot \mathbf{J}^*$$

The integration of this equation throughout a volume V bounded by a closed surface S gives the complex Poynting vector theorem; i.e.,

$$\tfrac{1}{2} \int_V \nabla \cdot \mathbf{E} \times \mathbf{H}^* \, dV = \tfrac{1}{2} \oint_S \mathbf{E} \times \mathbf{H}^* \cdot d\mathbf{S}$$

$$= -j \frac{\omega}{2} \int_V (\mathbf{B} \cdot \mathbf{H}^* - \mathbf{E} \cdot \mathbf{D}^*) \, dV - \tfrac{1}{2} \int_V \mathbf{E} \cdot \mathbf{J}^* \, dV \qquad (2.58a)$$

where the divergence theorem has been used on the left-hand-side integral. The above result may be rewritten as

$$\tfrac{1}{2} \oint_S \mathbf{E} \times \mathbf{H}^* \cdot (-d\mathbf{S}) = 2j\omega \int_V \left(\frac{\mathbf{B} \cdot \mathbf{H}^*}{4} - \frac{\mathbf{E} \cdot \mathbf{D}^*}{4} \right) dV$$

$$+ \tfrac{1}{2} \int_V \mathbf{E} \cdot \mathbf{J}^* \, dV \qquad (2.58b)$$

where $-d\mathbf{S}$ is a vector element of area directed into the volume V. If the medium in V is characterized by parameters $\epsilon = \epsilon' - j\epsilon''$, $\mu = \mu' - j\mu''$, and conductivity σ, the real and imaginary parts of (2.58) may be equated to give

$$\operatorname{Re} \tfrac{1}{2} \oint_S \mathbf{E} \times \mathbf{H}^* \cdot (-d\mathbf{S}) = \frac{\omega}{2} \int_V (\mu'' \mathbf{H} \cdot \mathbf{H}^* + \epsilon'' \mathbf{E} \cdot \mathbf{E}^*) \, dV$$
$$+ \tfrac{1}{2} \int_V \sigma \mathbf{E} \cdot \mathbf{E}^* \, dV \quad (2.59a)$$

$$\operatorname{Im} \tfrac{1}{2} \oint_S \mathbf{E} \times \mathbf{H}^* \cdot (-d\mathbf{S}) = 2\omega \int_V \left(\mu' \frac{\mathbf{H} \cdot \mathbf{H}^*}{4} - \epsilon' \frac{\mathbf{E} \cdot \mathbf{E}^*}{4} \right) dV \quad (2.59b)$$

Equation (2.59a) is interpreted to state that the real electromagnetic power transmitted through the closed surface S into V is equal to the power loss produced by conduction current $\sigma \mathbf{E}$, resulting in Joule heating plus the power loss resulting from polarization damping forces. Note that $\omega \epsilon''$ could be interpreted as an equivalent conductance, as pointed out in Sec. 2.2. This equation also shows that μ'' and ϵ'' must be positive in order to represent energy loss, and hence the imaginary parts of ϵ and μ must be negative. Equation (2.59b) states that the imaginary part of the complex rate of energy flow into V is equal to 2ω times the net reactive energy $W_m - W_e$ stored in the magnetic and electric fields in V. The complex Poynting vector theorem is essentially an energy-balance equation.

A result analogous to the above may be derived for a conventional network, and serves to demonstrate the validity of the interpretation of (2.58). Consider a simple series RLC circuit as in Fig. 2.7. If the current in the circuit is I and the applied voltage is V, the complex input power is given by

$$\tfrac{1}{2} V I^* = \tfrac{1}{2} Z I I^* = \tfrac{1}{2} I I^* \left(R + j\omega L - \frac{j}{\omega C} \right)$$

The time-average power loss in R, magnetic energy stored in the field around L, and electric energy stored in the field associated with C are given, respectively, by

$$P_l = \tfrac{1}{2} R I I^* \qquad W_m = \tfrac{1}{4} L I I^* \qquad W_e = \frac{1}{4} \frac{I I^*}{\omega^2 C}$$

since the voltage across C is $I/\omega C$. Hence

$$\tfrac{1}{2} V I^* = \tfrac{1}{2} Z I I^* = P_l + 2j\omega (W_m - W_e)$$

which has the same interpretation as (2.58). This equation may also be solved for the impedance Z to give

$$Z = \frac{P_l + 2j\omega (W_m - W_e)}{\tfrac{1}{2} I I^*} \quad (2.60)$$

and provides a general definition of the impedance of a network in terms of the associated power loss and stored reactive energy. The factor $\frac{1}{2}II^*$ in the denominator serves as a normalization factor, and is required in order to make Z independent of the magnitude of the current at the input to the network.

In the case of a general time-varying field, an expansion of $\nabla \cdot \mathcal{E} \times \mathcal{H}$ and substitution from Maxwell's equations (2.13) lead to the following Poynting vector theorem for general time-varying fields:

$$\oint_S \mathcal{E} \times \mathcal{H} \cdot (-d\mathbf{S}) = \int_V \left(\mu_0 \mathcal{H} \cdot \frac{\partial \mathcal{H}}{\partial t} + \mu_0 \mathcal{H} \cdot \frac{\partial \mathcal{M}}{\partial t} + \epsilon_0 \mathcal{E} \cdot \frac{\partial \mathcal{E}}{\partial t} + \mathcal{E} \cdot \frac{\partial \mathcal{P}}{\partial t} + \mathcal{E} \cdot \mathcal{J} \right) dV$$

Since $\mathcal{H} \cdot \partial \mathcal{H}/\partial t = \frac{1}{2} \partial(\mathcal{H} \cdot \mathcal{H})/\partial t$, etc., and the electric and magnetic polarization currents are $\mathcal{J}_p = \partial \mathcal{P}/\partial t$, $\mathcal{J}_m = \mu_0(\partial \mathcal{M}/\partial t)$, we have

$$\oint_S \mathcal{E} \times \mathcal{H} \cdot (-d\mathbf{S}) = \frac{\partial}{\partial t} \int_V \left(\frac{\mu_0 \mathcal{H} \cdot \mathcal{H}}{2} + \frac{\epsilon_0 \mathcal{E} \cdot \mathcal{E}}{2} \right) dV + \int_V [\mathcal{E} \cdot (\mathcal{J} + \mathcal{J}_p) + \mathcal{H} \cdot \mathcal{J}_m] \, dV \quad (2.61)$$

where $-d\mathbf{S}$ is an element of surface area directed into V. This equation states that the rate of energy flow into V is equal to the time rate of change of the free-space field energy stored in V plus the rate of energy dissipation in Joule heating arising from the conduction current $\mathcal{J}$ and, in addition, the instantaneous rate of energy supplied in maintaining the polarization currents. If $\mathcal{M}$ and $\mathcal{H}$, and also $\mathcal{P}$ and $\mathcal{E}$, are in phase, there is no energy loss associated with the polarization currents. If these quantities are not in phase, some energy dissipation takes place, leading to increased heating of the material.

If the susceptibilities χ_e and χ_m can be considered as constants, so that $\partial \mathcal{P}/\partial t = \epsilon_0 \chi_e (\partial \mathcal{E}/\partial t)$ and $\partial \mathcal{M}/\partial t = \chi_m(\partial \mathcal{H}/\partial t)$, then (2.61) becomes

$$\oint_S \mathcal{E} \times \mathcal{H} \cdot (-d\mathbf{S}) = \frac{\partial}{\partial t} \int_V \left(\frac{\mathcal{H} \cdot \mathcal{B}}{2} + \frac{\mathcal{E} \cdot \mathcal{D}}{2} \right) dV + \int_V \mathcal{E} \cdot \mathcal{J} \, dV \quad (2.62)$$

which is the usual form of the Poynting vector theorem. The first term on the right is now interpreted as the instantaneous rate of change of the total electric and magnetic energy stored in the volume V.

The susceptibilities can usually be considered as true constants whenever the inertial and damping forces are small compared with the elastic restoring force in the dynamical equation describing the polarization. For example, with reference to (2.54a), this is the case when k is much greater than $\omega m \nu$ or $\omega^2 m$, that is, when $1/\omega C$ is large compared with ωL and R, so that

$$\epsilon_0 \chi_e' \approx C = \frac{q^2}{k}$$

2.6 Boundary conditions

In order to find the proper and unique solutions to Maxwell's equations for situations of practical interest (these always involve material bodies with boundaries), a knowledge of the behavior of the electromagnetic field at the boundary separating material bodies with different electrical properties is required. From a mathematical point of view, the solution of a partial differential equation, such as a wave equation, in a region V is not unique unless boundary conditions are specified, i.e., the behavior of the field on the boundary of V. Boundary conditions play the same role in the solution of partial differential equations that initial conditions play in the solution of the differential equations that govern the behavior of electric circuits.

As an example, consider the problem of finding a solution to Maxwell's equations inside a cylindrical cavity partially filled with a dielectric medium of permittivity ϵ, as in Fig. 2.8. In practice, the solution is obtained by finding general solutions valid in the two regions labeled R_1 and R_2. These general solutions must satisfy prescribed conditions on the metallic boundaries and in addition contain arbitrary amplitude constants that can be determined only from a knowledge of the boundary conditions to be applied at the air-dielectric boundary separating regions R_1 and R_2.

The integral form of Maxwell's equations provides the most convenient formulation in order to deduce the required boundary conditions. Consider two media with parameters ϵ_1, μ_1 and ϵ_2, μ_2, as in Fig. 2.9a. If there is no surface charge on the boundary, which is the usual case for nonconducting media, the integral of the displacement flux over the surface of the small "coin-shaped" volume centered on the boundary as in Fig. 2.9b gives, in the limit as h tends to zero,

$$\lim_{h \to 0} \oint_S \mathbf{D} \cdot d\mathbf{S} = D_{2n} \Delta S - D_{1n} \Delta S = 0$$

or

$$D_{2n} = D_{1n} = \mathbf{n} \cdot \mathbf{D}_2 = \mathbf{n} \cdot \mathbf{D}_1 \tag{2.63}$$

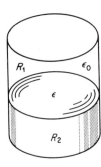

Fig. 2.8 A cylindrical cavity partially filled with a dielectric medium.

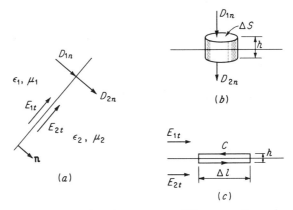

Fig. 2.9 Boundary between two different media.

where n denotes the normal component. The limit $h \to 0$ is taken so that the flux through the sides of the coin-shaped region vanishes. Equation (2.63) simply states that the displacement flux lines are continuous in the direction normal to the boundary. A similar result clearly must hold also for the magnetic flux lines since $\nabla \cdot \mathbf{B} = 0$, and hence, by analogy,

$$\mathbf{n} \cdot \mathbf{B}_2 = \mathbf{n} \cdot \mathbf{B}_1 \tag{2.64}$$

To obtain boundary conditions on the tangential components of the electric field $\mathbf{E}$ and magnetic field $\mathbf{H}$, the circulation integrals for $\mathbf{E}$ and $\mathbf{H}$ are used. If for the contour C in Fig. 2.9c the width h is made to approach zero, the magnetic flux flowing through this contour vanishes and

$$\lim_{h \to 0} \oint_C \mathbf{E} \cdot d\mathbf{l} = \lim_{h \to 0} -j\omega \int_S \mathbf{B} \cdot d\mathbf{S} = 0$$
$$= E_{2t}\,\Delta l - E_{1t}\,\Delta l$$

or

$$E_{1t} = E_{2t} \tag{2.65}$$

For the same contour C the total displacement current directed through the contour vanishes as $h \to 0$, so that

$$\lim_{h \to 0} \oint_C \mathbf{H} \cdot d\mathbf{l} = \lim_{h \to 0} \left(j\omega \int_S \mathbf{D} \cdot d\mathbf{S} \right) = 0$$
$$= (H_{2t} - H_{1t})\,\Delta l$$

or

$$H_{2t} = H_{1t} \tag{2.66}$$

where t denotes the components tangential to the boundary surface. These latter relations state that the components of $\mathbf{E}$ and $\mathbf{H}$ tangent to the boundary are continuous across the boundary; i.e., the tangential components on adjacent sides of the boundary are equal at the boundary surface.

For the boundary conditions at the surface separating a good conductor (any metal) and free space or air, some simplification is possible. As shown in a later section, the electromagnetic field can penetrate into a conductor only a minute distance at microwave frequencies. The field amplitude decays exponentially from its surface value according to e^{-u/δ_s}, where u is the normal distance into the conductor measured from the surface, and δ_s is called the skin depth. The skin depth is given by

$$\delta_s = \left(\frac{2}{\omega\mu\sigma}\right)^{\frac{1}{2}} \tag{2.67}$$

For copper ($\sigma = 5.8 \times 10^7$ mhos/m) at a frequency of 10^{10} cps, the skin depth is 6.6×10^{-5} cm, truly a very small distance. Likewise, the current $\mathbf{J} = \sigma\mathbf{E}$ is concentrated near the surface. As the conductivity is made to approach infinity, δ_s approaches zero and the current is squeezed into a narrower and narrower region and in the limit $\sigma \to \infty$ becomes a true surface current. Since the skin depth is so small at microwave frequencies for metals, the approximation of infinite conductivity may be made with negligible error (an exception is when attenuation is to be calculated, since then infinite conductivity implies no loss). For infinite conductivity the field in the conductor must be zero. Since the flux lines of **B** are continuous and likewise since the tangential component of **E** is continuous across the boundary, it is necessary that

$$\mathbf{n} \cdot \mathbf{B} = 0 \tag{2.68a}$$

$$\mathbf{E}_t = \mathbf{n} \times \mathbf{E} = 0 \tag{2.68b}$$

at the surface of a perfect conductor. This same argument cannot be applied to the normal component of **D** and the tangential component of **H** because, as noted above, a surface current $\mathbf{J}_s$ will exist on the surface in the limit $\sigma \to \infty$. Applying Maxwell's equation

$$\oint_C \mathbf{H} \cdot d\mathbf{l} = j\omega \int \mathbf{D} \cdot d\mathbf{S} + \int \mathbf{J} \cdot d\mathbf{S}$$

to the contour C illustrated in Fig. 2.10 gives

$$\lim_{h \to 0} \oint_C \mathbf{H} \cdot d\mathbf{l} = H_t \, \Delta l = \lim_{h \to 0} \int j\omega \mathbf{D} \cdot d\mathbf{S} + \lim_{h \to 0} \int \mathbf{J} \cdot d\mathbf{S}$$

$$= \lim_{h \to 0} hJ \, \Delta l = J_s \, \Delta l$$

or in vector form,

$$\mathbf{n} \times \mathbf{H} = \mathbf{J}_s \tag{2.68c}$$

Note that the field in the conductor goes to zero, that the total displacement current through C vanishes as $h \to 0$, but that hJ tends to the

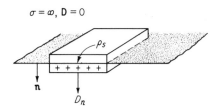 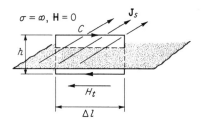

Fig. 2.10 Boundary of a perfect conductor.

limiting value J_s as the conductivity is made infinite and h is made to approach zero. Associated with the surface current is a charge of density ρ_s on which the normal displacement flux lines terminate. Hence, at the surface of a perfect conductor,

$$\mathbf{n} \cdot \mathbf{D} = D_n = \rho_s \tag{2.68d}$$

When it is desired to take into account the large but finite conductivity (as would be the case in attenuation calculations), an impedance boundary condition may be used with little error. The metallic surface exhibits a surface impedance Z_m, with equal resistive and inductive parts, given by

$$Z_m = \frac{1+j}{\sigma \delta_s} \tag{2.69}$$

At the surface a surface current exists, and the relation between this and the electric field tangent to the surface is

$$\mathbf{E}_t = Z_m \mathbf{J}_s \tag{2.70}$$

Note that the tangential electric field cannot be zero for finite conductivity, although it may be very small. Now $\mathbf{n} \times \mathbf{H} = \mathbf{J}_s$, so that

$$\mathbf{E}_t = Z_m \mathbf{J}_s = Z_m \mathbf{n} \times \mathbf{H} \tag{2.71}$$

From (2.69) it is seen that the resistive part of the surface impedance is equal to the d-c resistance per square of a unit square of metal of thickness δ_s. In a later section the above results are verified; so further comments are reserved until then.

In practice, it suffices to make the tangential components of the fields satisfy the proper boundary conditions since, when they do, the normal components of the fields automatically satisfy their appropriate boundary conditions. The reason is that when the fields are a solution of Maxwell's equations, not all the components of the field are independent. For example, when the tangential part of the electric field is continuous across a boundary, the derivatives of the tangential component of electric field with respect to coordinates on the boundary surface are also continuous. Thus the curl of the electric field normal to the surface is continuous, and this implies continuity of the normal component of **B**. More specifically, if the xy plane is the boundary surface and E_x, E_y are continuous, then $\partial E_x/\partial x$, $\partial E_x/\partial y$, $\partial E_y/\partial x$, and $\partial E_y/\partial y$ are also continuous. Hence $-j\omega B_z = \partial E_y/\partial x - \partial E_x/\partial y$ is continuous. For the same reasons continuity of the tangential components of **H** ensures the continuity of the normal component of **D** across a boundary.

In addition to the boundary conditions given above, a boundary condition must be imposed on the field solutions at the edge of a conducting body such as a wedge. The edge condition requires that the energy stored in the field in the vicinity of an edge of a conducting body be finite. This limits the maximum rate at which the field intensities can increase as the edge is approached.† A detailed analysis shows that at the edge of a two-dimensional perfectly conducting wedge with an internal angle ϕ, the field components normal to the edge must not increase any faster than r^α, where r is the perpendicular radial distance away from the edge and

$$\alpha = \frac{n\pi}{2\pi - \phi} - 1$$

where the integer n must be chosen so that α is greater than or equal to $-\tfrac{1}{2}$ at least.

When solving for fields in an infinite region of space, the behavior of the field at infinity must also be specified. This boundary condition is called a radiation condition, and requires that the field at infinity be a wave propagating a finite amount of energy outward, or else that the field vanish so fast that the energy stored in the field and the energy flow at infinity are zero.

2.7 Plane waves

In this section and the two following ones we shall introduce wave solutions by considering plane waves propagating in free space and reflection of a plane wave from a boundary separating free space and a dielectric, or conducting, medium. The latter problem will serve to derive the boundary conditions given by (2.68) to (2.71) in the preceding section.

Plane waves in free space

The electric field is a solution of the Helmholtz equation

$$\nabla^2 \mathbf{E} + k_0^2 \mathbf{E} = \frac{\partial^2 \mathbf{E}}{\partial x^2} + \frac{\partial^2 \mathbf{E}}{\partial y^2} + \frac{\partial^2 \mathbf{E}}{\partial z^2} + k_0^2 \mathbf{E} = 0$$

---

† J. Meixner, The Behavior of Electromagnetic Fields at Edges, *N.Y. Univ. Inst. Math. Sci. Res. Rept.* EM-72, December, 1954. The theory is also discussed in R. E. Collin, "Field Theory of Guided Waves," chap. 1, McGraw-Hill Book Company, New York, 1960.

This vector equation holds for each component of **E**, so that

$$\frac{\partial^2 E_i}{\partial x^2} + \frac{\partial^2 E_i}{\partial y^2} + \frac{\partial^2 E_i}{\partial z^2} + k_0^2 E_i = 0 \qquad i = x, y, z \tag{2.72}$$

The standard procedure for solving a partial differential equation is the method of *separation of variables*. However, this method does not work for all types of partial differential equations in all various coordinate systems, and when it does not work, a solution is very difficult, if not impossible, to obtain. For the Helmholtz equation the method of separation of variables does work in such common coordinate systems as rectangular, cylindrical, and spherical. Hence this method suffices for the class of problems discussed in this text. The basic procedure is to assume for the solution a product of functions each of which is a function of one coordinate variable only. Substitution of this solution into the partial differential equation then separates the partial differential equation into three ordinary differential equations which may be solved by standard means.

In the present case let $E_x = f(x)g(y)h(z)$. Substituting this expression into (2.72) gives

$$ghf'' + fhg'' + fgh'' + k_0^2 fgh = 0$$

where the double prime denotes the second derivative. Dividing this equation by fgh gives

$$\frac{f''}{f} + \frac{g''}{g} + \frac{h''}{h} + k_0^2 = 0 \tag{2.73}$$

Each of the first three terms in (2.73), such as f''/f, is a function of a single independent variable only, and hence the sum of these terms can equal a constant $-k_0^2$ only if each term is constant. Thus (2.73) separates into three equations:

$$\frac{f''}{f} = -k_x^2 \qquad \frac{g''}{g} = -k_y^2 \qquad \frac{h''}{h} = -k_z^2$$

or

$$\frac{d^2 f}{dx^2} + k_x^2 f = 0 \qquad \frac{d^2 g}{dy^2} + k_y^2 g = 0 \qquad \frac{d^2 h}{dz^2} + k_z^2 h = 0 \tag{2.74}$$

where k_x^2, k_y^2, k_z^2 are called separation constants. The only restriction so far on k_x^2, k_y^2, k_z^2 is that their sum must equal k_0^2, that is,

$$k_x^2 + k_y^2 + k_z^2 = k_0^2 \tag{2.75}$$

so that (2.73) will be satisfied.

Equations (2.74) are simple-harmonic differential equations with exponential solutions of the form $e^{\pm jk_x x}$, $e^{\pm jk_y y}$, $e^{\pm jk_z z}$. As one suitable solution for E_x we may therefore choose

$$E_x = Ae^{-jk_x x - jk_y y - jk_z z} \tag{2.76}$$

where A is an amplitude factor. This solution is interpreted as the x component of a wave propagating in the direction specified by the propagation vector

$$\mathbf{k} = \mathbf{a}_x k_x + \mathbf{a}_y k_y + \mathbf{a}_z k_z \tag{2.77}$$

because the scalar product of $\mathbf{k}$ with the position vector

$$\mathbf{r} = \mathbf{a}_x x + \mathbf{a}_y y + \mathbf{a}_z z$$

equals $k_x x + k_y y + k_z z$ and is k_0 times the perpendicular distance from the origin to a plane normal to the vector $\mathbf{k}$, as illustrated in Fig. 2.11. The $\mathbf{k}$ vector may also be written as $\mathbf{k} = \mathbf{n}k_0$, where $\mathbf{n}$ is a unit vector in the direction of $\mathbf{k}$ and k_0 is the magnitude of $\mathbf{k}$ by virtue of (2.75).

Although (2.76) gives a possible solution for E_x, this is not the complete solution for the electric field. Similar solutions for E_y and E_z may be found. The three components of $\mathbf{E}$ are not independent since the divergence relation $\nabla \cdot \mathbf{E} = 0$ must hold in free space. This constraint means that only two components of $\mathbf{E}$ can have arbitrary amplitudes. However, for $\nabla \cdot \mathbf{E}$ to vanish everywhere, all components of $\mathbf{E}$ must have the same spatial dependence, and hence appropriate solutions for E_y and E_z are

$$E_y = Be^{-j\mathbf{k} \cdot \mathbf{r}} \qquad E_z = Ce^{-j\mathbf{k} \cdot \mathbf{r}}$$

with B and C amplitude coefficients. Let $\mathbf{E}_0$ be the vector $\mathbf{a}_x A + \mathbf{a}_y B + \mathbf{a}_z C$; then the total solution for $\mathbf{E}$ may be written in vector form as

$$\mathbf{E} = \mathbf{E}_0 e^{-j\mathbf{k} \cdot \mathbf{r}} \tag{2.78}$$

The divergence condition gives

$$\nabla \cdot \mathbf{E} = \nabla \cdot \mathbf{E}_0 e^{-j\mathbf{k} \cdot \mathbf{r}} = \mathbf{E}_0 \cdot \nabla e^{-j\mathbf{k} \cdot \mathbf{r}} = -j\mathbf{k} \cdot \mathbf{E}_0 e^{-j\mathbf{k} \cdot \mathbf{r}} = 0$$

or

$$\mathbf{k} \cdot \mathbf{E}_0 = 0 \tag{2.79}$$

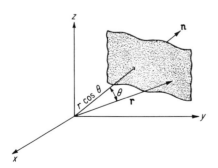

Fig. 2.11 Illustration of plane normal to vector $\mathbf{k}$.

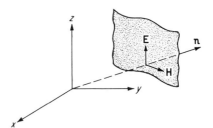

Fig. 2.12 Space relationship between **E**, **H**, and **n** in a TEM wave.

since $\nabla e^{-j\mathbf{k}\cdot\mathbf{r}} = -j\mathbf{k}e^{-j\mathbf{k}\cdot\mathbf{r}}$, as may be verified by expansion in rectangular coordinates. The divergence condition is seen to constrain the amplitudes A, B, C so that the vector $\mathbf{E}_0$ is perpendicular to the direction of propagation as specified by $\mathbf{k}$. The solution (2.78) is called a uniform plane wave since the constant-phase surfaces given by $\mathbf{k}\cdot\mathbf{r} = \text{const}$ are planes and the field $\mathbf{E}$ does not vary on a constant-phase plane.

The solution for **H** is obtained from Maxwell's equation

$$\nabla \times \mathbf{E} = -j\omega\mu_0\mathbf{H}$$

which gives

$$\begin{aligned}\mathbf{H} &= -\frac{1}{j\omega\mu_0} \nabla \times \mathbf{E}_0 e^{-j\mathbf{k}\cdot\mathbf{r}} = \frac{1}{j\omega\mu_0} \mathbf{E}_0 \times \nabla e^{-j\mathbf{k}\cdot\mathbf{r}} \\ &= \frac{1}{\omega\mu_0} \mathbf{k} \times \mathbf{E}_0 e^{-j\mathbf{k}\cdot\mathbf{r}} = \frac{k_0}{\omega\mu_0} \mathbf{n} \times \mathbf{E} \\ &= \sqrt{\frac{\epsilon_0}{\mu_0}} \mathbf{n} \times \mathbf{E} = Y_0 \mathbf{n} \times \mathbf{E} \end{aligned} \quad (2.80)$$

where $Y_0 = \sqrt{\epsilon_0/\mu_0}$ has the dimensions of an admittance and is called the intrinsic admittance of free space. The reciprocal $Z_0 = 1/Y_0$ is called the intrinsic impedance of free space. Note that **H** is perpendicular to **E** and to **n**, and hence both **E** and **H** lie in the constant-phase planes. For this reason this type of wave is called a transverse electromagnetic wave (TEM wave). The spatial relationship between **E**, **H**, and **n** is illustrated in Fig. 2.12.

The physical electric field corresponding to the phasor representation (2.78) is

$$\mathbf{E} = \text{Re}\,(\mathbf{E}_0 e^{-j\mathbf{k}\cdot\mathbf{r}+j\omega t}) = \mathbf{E}_0 \cos(\mathbf{k}\cdot\mathbf{r} - \omega t) \quad (2.81)$$

where, for simplicity, $\mathbf{E}_0$ has been assumed to be real. The wavelength is the distance the wave must propagate to undergo a phase change of 2π. If we let λ_0 denote the wavelength in free space, it follows that

$$|\mathbf{k}|\lambda_0 = k_0\lambda_0 = 2\pi$$

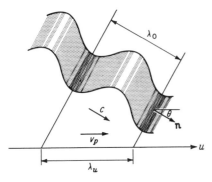

Fig. 2.13 A wave propagating obliquely to the u axis.

so that

$$k_0 = \omega\sqrt{\mu_0\epsilon_0} = \frac{\omega}{c} = \frac{2\pi}{\lambda_0} \qquad (2.82)$$

This result is the familiar relationship between wavelength λ_0, frequency $f = \omega/2\pi$, and velocity c in free space. A wavelength in a direction other than that along the direction of propagation **n** may also be defined. For example, along the direction of the x axis the wavelength is

$$\lambda_x = \frac{2\pi}{k_x} \qquad (2.83)$$

and since k_x is less than k_0, λ_x is greater than λ_0. The phase velocity is the velocity with which an observer would have to move in order to see a constant phase. From (2.81) it is seen that the phase of **E** is constant as long as $\mathbf{k} \cdot \mathbf{r} - \omega t$ is constant. If the angle between **k** and **r** is θ, then $\mathbf{k} \cdot \mathbf{r} - \omega t = k_0 r \cos\theta - \omega t$. Differentiating the relation

$$k_0 r \cos\theta - \omega t = \text{const}$$

gives

$$\frac{dr}{dt} = v_p = \frac{\omega}{k_0 \cos\theta} \qquad (2.84)$$

for the phase velocity v_p in the direction **r**. Along the direction of propagation, $\cos\theta = 1$ and $v_p = \omega/k_0 = c$. In other directions, the phase velocity is greater than c. These results may be understood by reference to Fig. 2.13. When the wave has moved a distance λ_0 along the direction **n**, the constant-phase-plane intersection with the u axis has moved a distance $\lambda_u = \lambda_0 \sec\theta$ along the direction u. For this reason the wavelength and phase velocity along u are greater by a factor $\sec\theta$ than the corresponding quantities measured along the direction of propagation **n**.

The time-average rate of energy flow per unit area in the direction **n** is given by

$$P = \tfrac{1}{2}\operatorname{Re}\mathbf{E} \times \mathbf{H}^* \cdot \mathbf{n} = \tfrac{1}{2}\operatorname{Re}Y_0\mathbf{E} \times (\mathbf{n} \times \mathbf{E}^*) \cdot \mathbf{n} = \tfrac{1}{2}Y_0\mathbf{E}_0 \cdot \mathbf{E}_0^* \qquad (2.85)$$

The time-average energy densities in the electric and magnetic fields of a TEM wave are, respectively,

$$U_e = \frac{\epsilon_0}{4} \mathbf{E} \cdot \mathbf{E}^* = \frac{\epsilon_0}{4} \mathbf{E}_0 \cdot \mathbf{E}_0^*$$

$$U_m = \frac{\mu_0}{4} \mathbf{H} \cdot \mathbf{H}^* = \frac{\mu_0}{4} Y_0^2 (\mathbf{n} \times \mathbf{E}) \cdot (\mathbf{n} \times \mathbf{E}^*) = \frac{\epsilon_0}{4} \mathbf{E}_0 \cdot \mathbf{E}_0^* = U_e$$

and are seen to be equal. Since power is a flow of energy, the velocity v_g of energy propagation is such that

$$(U_e + U_m) v_g = P$$

or

$$v_g = \frac{P}{U_e + U_m} = \frac{\frac{1}{2} Y_0 \mathbf{E}_0 \cdot \mathbf{E}_0^*}{\frac{1}{2} \epsilon_0 \mathbf{E}_0 \cdot \mathbf{E}_0^*} = \frac{Y_0}{\epsilon_0} = c \qquad (2.86)$$

Thus, for a TEM wave in free space, the energy in the field is transported with a velocity $c = 3 \times 10^8$ m/sec, which is also the phase velocity. Since the phase velocity is independent of frequency, a modulated carrier or signal will have all its frequency components propagated with the same velocity c. Hence the signal velocity is also the velocity of light c. Later on, in the study of waveguides, situations arise where the phase velocity is dependent on frequency and consequently is not equal to the velocity of energy propagation or the signal velocity.

2.8 Reflection from a dielectric interface

In Fig. 2.14 the half-space $z \geq 0$ is filled with a dielectric medium with permittivity ϵ (dielectric constant $\kappa = \epsilon/\epsilon_0$; index of refraction $\eta = \sqrt{\kappa}$). A TEM wave is assumed incident from the region $z < 0$. Without loss in generality, the xy axis may be oriented so that the unit vector $\mathbf{n}_1$ specifying the direction of incidence lies in the xz plane. It is convenient to solve this problem as two special cases, namely (1) parallel polarization, where the electric field of the incident wave is coplanar with $\mathbf{n}_1$ and the interface normal, i.e., lies in the xz plane, and (2) perpendicular polariza-

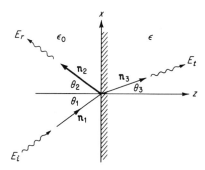

Fig. 2.14 Plane wave incident on a dielectric interface.

tion, where the electric field of the incident wave is perpendicular to the plane of incidence as defined by $\mathbf{n}_1$ and the interface normal, i.e., along the y axis. An incident TEM wave with arbitrary polarization can always be decomposed into a linear sum of perpendicular and parallel polarized waves. The reason for treating the two polarizations separately is that the reflection and transmission coefficients, to be defined, are different for the two cases.

1 Parallel polarization

Let the incident TEM wave be

$$\mathbf{E}_i = \mathbf{E}_1 e^{-jk_0 \mathbf{n}_1 \cdot \mathbf{r}} \qquad \mathbf{H}_i = Y_0 \mathbf{n}_1 \times \mathbf{E}_i \qquad (2.87)$$

where $\mathbf{E}_1$ lies in the xz plane. Part of the incident power will be reflected, and the remainder will be transmitted into the dielectric medium. Let the reflected TEM wave be

$$\mathbf{E}_r = \mathbf{E}_2 e^{-jk_0 \mathbf{n}_2 \cdot \mathbf{r}} \qquad \mathbf{H}_r = Y_0 \mathbf{n}_2 \times \mathbf{E}_r \qquad (2.88)$$

where $\mathbf{n}_2$ and $\mathbf{E}_2$ are to be determined. In the dielectric medium the solution for a TEM wave is the same as that in free space, but with ϵ_0 replaced by ϵ. Thus, in place of $k_0 = \omega\sqrt{\mu_0\epsilon_0}$ and $Y_0 = \sqrt{\epsilon_0/\mu_0}$, the parameters $k = \omega\sqrt{\mu_0\epsilon} = \eta k_0$ and $Y = \sqrt{\epsilon/\mu_0} = \eta Y_0$ are used, where $\eta = \sqrt{\kappa}$ is the index of refraction. The transmitted wave in the dielectric may be expressed by

$$\mathbf{E}_t = \mathbf{E}_3 e^{-jk\mathbf{n}_3 \cdot \mathbf{r}} \qquad \mathbf{H}_t = Y\mathbf{n}_3 \times \mathbf{E}_t \qquad (2.89)$$

with $\mathbf{E}_3$ and $\mathbf{n}_3$ as yet unknown.

The boundary conditions to be applied are the continuity of the tangential components of the electric and magnetic fields at the interface plane $z = 0$. These components must be continuous for all values of x and y on the $z = 0$ plane, and this is possible only if the fields on adjacent sides of the boundary have the same variation with x and y. Hence we must have

$$k_0 n_{1x} = k_0 n_{2x} = k n_{3x} = \eta k_0 n_{3x} \qquad (2.90)$$

i.e., the propagation phase constant along x must be the same for all waves. Since n_{1y} was chosen as zero, it follows that $n_{2y} = n_{3y} = 0$ also. The unit vectors $\mathbf{n}_1$, $\mathbf{n}_2$, $\mathbf{n}_3$ may be expressed as

$$\mathbf{n}_1 = \mathbf{a}_x \sin\theta_1 + \mathbf{a}_z \cos\theta_1$$

$$\mathbf{n}_2 = \mathbf{a}_x \sin\theta_2 - \mathbf{a}_z \cos\theta_2$$

$$\mathbf{n}_3 = \mathbf{a}_x \sin\theta_3 + \mathbf{a}_z \cos\theta_3$$

Equation (2.90) gives

$\sin \theta_1 = \sin \theta_2$

or

$$\theta_1 = \theta_2 \tag{2.91}$$

which is the well-known Snell's law of reflection; in addition, (2.90) gives

$$\sin \theta_1 = \eta \sin \theta_3 \tag{2.92}$$

which is also a well-known result specifying the angle of refraction θ_3 in terms of the angle of incidence θ_1 and the index of refraction η.

The incident electric field $\mathbf{E}_1$ has components $E_{1x} = E_1 \cos \theta_1$,

$E_{1z} = -E_1 \sin \theta_1$

since $\mathbf{n}_1 \cdot \mathbf{E}_1$ must equal zero. Note that E_1 is used to denote the magnitude of the vector $\mathbf{E}_1$. Since the incident electric field has no y component, the reflected and transmitted electric fields also have zero y components.† Expressing all fields in component form, that is,

$E_{2x} = E_2 \cos \theta_2$

$E_{2z} = E_2 \sin \theta_2$, $E_{3x} = E_3 \cos \theta_3$, $E_{3z} = -E_3 \sin \theta_3$, and imposing the boundary condition of continuity of the x component at $z = 0$ yield the relation

$E_1 \cos \theta_1 + E_2 \cos \theta_2 = E_3 \cos \theta_3$

or

$$(E_1 + E_2) \cos \theta_1 = E_3 \sqrt{1 - \sin^2 \theta_3} = E_3 \frac{\sqrt{\kappa - \sin^2 \theta_1}}{\eta} \tag{2.93}$$

by using (2.91) and (2.92). Apart from the propagation factor, the magnetic field is given by

$\mathbf{H}_1 = Y_0 \mathbf{n}_1 \times \mathbf{E}_1 = Y_0 \mathbf{a}_y(-n_{1x} E_{1z} + n_{1z} E_{1x}) = Y_0 \mathbf{a}_y E_1$

$\mathbf{H}_2 = -Y_0 \mathbf{a}_y E_2$

$\mathbf{H}_3 = Y \mathbf{a}_y E_3$

and has only a y component. Continuity of this magnetic field at the boundary requires that

$$Y_0(E_1 - E_2) = Y E_3 = \eta Y_0 E_3 \tag{2.94}$$

† If the reflected and transmitted electric fields were assumed to have a y component, the boundary conditions which must apply would show that these are, indeed, zero.

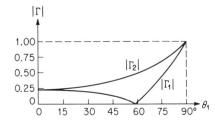

Fig. 2.15 Modulus of reflection coefficient at a dielectric interface for $\kappa = 2.56$, $|\Gamma_1|$ parallel polarization, $|\Gamma_2|$ perpendicular polarization.

If a reflection coefficient Γ_1 and a transmission coefficient T_1 are introduced according to the following relations,

$$\Gamma_1 = \frac{\text{amplitude of reflected electric field}}{\text{amplitude of incident electric field}} = \frac{E_2}{E_1} \quad (2.95a)$$

$$T_1 = \frac{\text{amplitude of transmitted electric field}}{\text{amplitude of incident electric field}} = \frac{E_3}{E_1} \quad (2.95b)$$

then the boundary conditions (2.93) and (2.94) may be expressed as

$$1 + \Gamma_1 = T_1 \frac{(\kappa - \sin^2 \theta_1)^{\frac{1}{2}}}{\eta \cos \theta_1} \quad (2.96a)$$

$$1 - \Gamma_1 = \eta T_1 \quad (2.96b)$$

These equations may be solved to give the Fresnel reflection and transmission coefficients for the case of parallel polarization, namely,

$$\Gamma_1 = \frac{(\kappa - \sin^2 \theta_1)^{\frac{1}{2}} - \kappa \cos \theta_1}{(\kappa - \sin^2 \theta_1)^{\frac{1}{2}} + \kappa \cos \theta_1} \quad (2.97a)$$

$$T_1 = \frac{2\eta \cos \theta_1}{(\kappa - \sin^2 \theta_1)^{\frac{1}{2}} + \kappa \cos \theta_1} \quad (2.97b)$$

An interesting feature of Γ_1 is that it vanishes for an angle of incidence $\theta_1 = \theta_b$, called the Brewster angle, where, from (2.97a),

$$\kappa - \sin^2 \theta_b = \kappa^2 \cos^2 \theta_b$$

or

$$\sin \theta_b = \left(\frac{\kappa}{\kappa + 1}\right)^{\frac{1}{2}} \quad (2.98)$$

At this particular angle of incidence all the incident power is transmitted into the dielectric medium. In Fig. 2.15 the reflection coefficient Γ_1 is plotted as a function of θ_1 for polystyrene, for which $\kappa = 2.56$.

2 Perpendicular polarization

For perpendicular polarization the roles of electric and magnetic fields are interchanged so that the electric field has only a y component. The fields may, however, still be expressed in the form given by (2.87) to (2.89), but with $\mathbf{E}_1$, $\mathbf{E}_2$, and $\mathbf{E}_3$ having y components only. As in the previous case, the boundary conditions must hold for all values of x and y on the $z = 0$ plane. Therefore Snell's laws of reflection and refraction again result; i.e., (2.91) and (2.92) must be satisfied. In place of the boundary conditions (2.93) and (2.94), we have

$$E_1 + E_2 = E_3 \qquad (2.99a)$$

$$Y_0(E_1 - E_2) \cos \theta_1 = Y E_3 \cos \theta_3 \qquad (2.99b)$$

Introducing the following reflection and transmission coefficients,

$$\Gamma_2 = \frac{E_2}{E_1} \qquad T_2 = \frac{E_3}{E_1}$$

into (2.99) yields

$$1 + \Gamma_2 = T_2 \qquad (2.100a)$$

$$1 - \Gamma_2 = T_2 \frac{(\kappa - \sin^2 \theta_1)^{\frac{1}{2}}}{\cos \theta_1} \qquad (2.100b)$$

The Fresnel reflection and transmission coefficients for the case of perpendicular polarization thus are

$$\Gamma_2 = \frac{\cos \theta_1 - (\kappa - \sin^2 \theta_1)^{\frac{1}{2}}}{(\kappa - \sin^2 \theta_1)^{\frac{1}{2}} + \cos \theta_1} \qquad (2.101a)$$

$$T_2 = \frac{2 \cos \theta_1}{(\kappa - \sin^2 \theta_1)^{\frac{1}{2}} + \cos \theta_1} \qquad (2.101b)$$

A notable difference for this case is the nonexistence of a Brewster angle for which Γ_2 vanishes. For comparison with the case of parallel polarization, Γ_2 is plotted in Fig. 2.15 for $\kappa = 2.56$.

2.9 Reflection from a conducting plane

The essential features of the behavior of the electromagnetic field at the surface of a good conductor may be deduced from an analysis of the simple problem of a TEM wave incident normally onto a conducting plane. The problem is illustrated in Fig. 2.16, which shows a medium with parameters ϵ, μ, σ filling the half-space $z \geq 0$. Let the electric field be polarized along the x axis so that the incident and reflected fields may

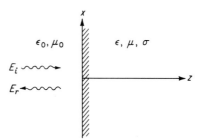

Fig. 2.16 A TEM wave incident normally on a conducting plane.

be expressed as

$$\mathbf{E}_i = E_1 \mathbf{a}_x e^{-jk_0 z}$$
$$\mathbf{H}_i = Y_0 E_1 \mathbf{a}_y e^{-jk_0 z} \quad (2.102a)$$
$$\mathbf{E}_r = \Gamma E_1 \mathbf{a}_x e^{+jk_0 z}$$
$$\mathbf{H}_r = -Y_0 \Gamma E_1 \mathbf{a}_y e^{+jk_0 z} \quad (2.102b)$$

where Γ is the reflection coefficient.

In the conducting medium the conduction current $\sigma \mathbf{E}$ is much greater than the displacement current $j\omega\epsilon\mathbf{E}$, so that Helmholtz's equation reduces to (2.50); i.e.

$$\nabla^2 \mathbf{E} - j\omega\mu\sigma \mathbf{E} = 0$$

The transmitted field is a solution of

$$\left(\frac{\partial^2}{\partial z^2} - j\omega\mu\sigma \right) \mathbf{E}_t = 0$$

since no x or y variation is assumed. The solution for a wave with an x component only and propagating in the z direction is

$$\mathbf{E}_t = E_3 \mathbf{a}_x e^{-\gamma z} \quad (2.103a)$$

with a corresponding magnetic field

$$\mathbf{H}_t = -\frac{1}{j\omega\mu} \nabla \times \mathbf{E}_t = \frac{\gamma}{j\omega\mu} \mathbf{a}_y E_3 e^{-\gamma z} \quad (2.103b)$$

where

$$\gamma = (j\omega\mu\sigma)^{\frac{1}{2}} = \frac{1+j}{\delta_s} \quad (2.104)$$

and the skin depth $\delta_s = (\omega\mu\sigma/2)^{-\frac{1}{2}}$. The propagation constant $\gamma = \alpha + j\beta$ has equal phase and attenuation constants. In the conductor the fields decay by an amount e^{-1} in a distance of one skin depth δ_s, which is a very small distance for metals at microwave frequencies (about 10^{-5} cm). The intrinsic impedance of the metal is Z_m, where

$$Z_m = \frac{j\omega\mu}{\gamma} = \frac{j\omega\mu}{(j\omega\mu\sigma)^{\frac{1}{2}}} = \frac{1+j}{\sigma\delta_s} \quad (2.105)$$

and is very small compared with the intrinsic impedance $Z_0 = (\mu_0/\epsilon_0)^{\frac{1}{2}}$ of free space. For example, for copper at 10^4 Mc, $Z_m = 0.026(1+j)$ ohms as compared with 377 ohms for Z_0. Note that (2.103b) may be written as

$$\mathbf{H}_t = \frac{1}{Z_m} \mathbf{a}_y E_3 e^{-\gamma z} = Y_m \mathbf{a}_y E_3 e^{-\gamma z}$$

which shows that the ratio of the magnitudes of the electric field to the magnetic field for a TEM wave in a conductor is the intrinsic impedance Z_m.

Returning to the boundary-value problem and imposing the boundary conditions of continuity of tangential fields at the boundary plane $z = 0$ give

$$(1 + \Gamma)E_1 = E_3 = TE_1 \tag{2.106a}$$

$$(1 - \Gamma)Y_0 E_1 = H_3 = Y_m E_3 = Y_m TE_1 \tag{2.106b}$$

where $E_3/E_1 = T$, the transmission coefficient. Solving (2.106) for the reflection coefficient Γ and T yields

$$\Gamma = \frac{Z_m - Z_0}{Z_m + Z_0} \tag{2.107a}$$

$$T = 1 + \Gamma = \frac{2Z_m}{Z_m + Z_0} \tag{2.107b}$$

Since $|Z_m|$ is small compared with Z_0, the reflection coefficient Γ is almost equal to -1 and the transmission coefficient T is very small. Almost all the incident power is reflected from the metallic boundary. As the conductivity σ is made to approach infinity, the impedance Z_m approaches zero and in the limit $\Gamma = -1$ and $T = 0$. Hence, for a perfect conductor, the tangential electric field at the surface is zero and the tangential magnetic field has a value equal to twice that of the incident wave.

The current density in the conductor is $\mathbf{J} = \sigma \mathbf{E}_t = \sigma TE_1 \mathbf{a}_x e^{-\gamma z}$. The total current per unit width of conductor along y is

$$\mathbf{J}_s = \int_0^\infty \mathbf{J}\, dz = \sigma TE_1 \mathbf{a}_x \int_0^\infty e^{-\gamma z}\, dz = \frac{\sigma TE_1 \mathbf{a}_x}{\gamma} \quad \text{amp/m}$$

This result may also be expressed in the following form:

$$\mathbf{J}_s = \frac{2\sigma Z_m^2 E_1}{(Z_m + Z_0)j\omega\mu} \mathbf{a}_x \tag{2.108}$$

by substituting for T from (2.107b) and replacing γ by $j\omega\mu/Z_m$ from (2.105). As $\sigma \to \infty$, the limiting value of $\mathbf{J}_s$ becomes

$$\mathbf{J}_s = \frac{2E_1}{Z_0}\mathbf{a}_x = 2Y_0 E_1 \mathbf{a}_x \tag{2.109}$$

since $Z_m \to 0$ and $\sigma Z_m^2 \to j\omega\mu$. This current exists only on the surface of the conductor since, as $\sigma \to \infty$, the skin depth $\delta_s \to 0$; i.e., the field decays infinitely fast with distance into the conductor. When σ is infinite, $\Gamma = -1$ and the total tangential magnetic field at the surface is $2Y_0 E_1 \mathbf{a}_y$ and equal in magnitude to $\mathbf{J}_s$. In vector form the boundary conditions at the surface of a perfect conductor are thus seen to be

$$\mathbf{n} \times \mathbf{E} = 0 \tag{2.110a}$$

$$\mathbf{n} \times \mathbf{H} = \mathbf{J}_s \tag{2.110b}$$

where $\mathbf{n}$ is a unit outward normal at the conductor surface.

For finite conductivity the current density at the surface is $\sigma T E_1$ and the magnetic field at the surface is $Y_m T E_1$. In terms of these quantities the total current per unit width may be expressed as

$$\mathbf{J}_s = \frac{\sigma T E_1}{\gamma} \mathbf{a}_x = \frac{\sigma Z_m}{\gamma}(Y_m T E_1)\mathbf{a}_x = Y_m T E_1 \mathbf{a}_x$$

In other words, the total current per unit width is equal to the tangential magnetic field at the surface.

The time-average power transmitted into the conductor per unit area is given by the real part of one-half of the complex Poynting vector at the surface, and is

$$P_t = \tfrac{1}{2} \operatorname{Re} \mathbf{E} \times \mathbf{H}^* \cdot \mathbf{a}_z = \tfrac{1}{2} T T^* E_1 E_1^* \operatorname{Re} Y_m = \tfrac{1}{4} T T^* E_1 E_1^* \sigma \delta_s \quad (2.111)$$

The reader may readily verify that this is equal to the result obtained from a volume integral of $\mathbf{J} \cdot \mathbf{J}^*$; that is,

$$P_t = \frac{1}{2\sigma} \int_0^\infty \mathbf{J} \cdot \mathbf{J}^* \, dz$$

Equation (2.111) may be simplified with little error by making the following approximation:

$$\sigma T T^* = \frac{4\sigma Z_m Z_m^*}{(Z_m + Z_0)(Z_m + Z_0)^*}$$
$$\approx \frac{4\sigma Z_m Z_m^*}{Z_0^2} = \frac{8}{\sigma \delta_s^2 Z_0^2}$$

whence (2.111) becomes

$$P_t \approx \frac{1}{2} \frac{(2Y_0 E_1)(2Y_0 E_1^*)}{\sigma \delta_s} \quad (2.112)$$

Note that $2Y_0 E_1$ is the value of the magnetic field, tangent to the surface, that would exist if σ were infinite. Hence an excellent approximate technique for evaluating power loss in a conductor is to find the tangential magnetic field, say H_t, that would exist for a perfect conductor, and then compute the power loss according to the relation

$$P_t = \tfrac{1}{2} \operatorname{Re}(H_t H_t^* Z_m) = \tfrac{1}{2} \operatorname{Re}(J_s J_s^* Z_m) \quad (2.113)$$

This procedure is equivalent to assuming that the metal exhibits a surface impedance Z_m and the current is essentially the same as that which would exist for infinite conductivity.

The procedure outlined above for power-loss calculations is widely used in microwave work. Although the derivation was based on a consideration of a very special boundary-value problem, the same conclusions result for more complex structures such as conducting spheres and cylinders. In general, the technique of characterizing the metal by a surface impedance Z_m and assuming that the current $\mathbf{J}_s$ is the same as that for infinite conductivity is valid as long as the conductor surface has a radius of curvature at least a few skin depths in magnitude.

2.10 Potential theory

The wave solutions presented in the previous sections have all been source-free solutions; i.e., the nature of the sources giving rise to the field was not considered. When it is necessary to consider the specific field generated by a given source, as in antenna problems, waveguide and cavity coupling, etc., this is greatly facilitated by introducing an auxiliary vector potential function **A**. As will be seen, the vector potential **A** is determined by the current source, and the total electromagnetic field may be derived from **A**.

Since $\nabla \cdot \mathbf{B} = 0$ always, this condition will hold identically if **B** is expressed as the curl of a vector potential **A** since the divergence of the curl of a vector is identically zero. Thus let

$$\mathbf{B} = \nabla \times \mathbf{A} \tag{2.114}$$

The assumed time dependence $e^{j\omega t}$ is not written out explicitly in (2.114) since this is a phasor representation. The curl equation for **E** gives

$$\nabla \times \mathbf{E} = -j\omega \mathbf{B} = -j\omega \nabla \times \mathbf{A}$$

or

$$\nabla \times (\mathbf{E} + j\omega \mathbf{A}) = 0$$

Now the curl of the gradient of a scalar function Φ is identically zero; so the general integral of the above equation is

$$\mathbf{E} + j\omega \mathbf{A} = -\nabla \Phi$$

or

$$\mathbf{E} = -j\omega \mathbf{A} - \nabla \Phi \tag{2.115}$$

Substituting this expression into the $\nabla \times \mathbf{H}$ equation gives

$$\nabla \times \mathbf{H} = \frac{1}{\mu} \nabla \times \nabla \times \mathbf{A} = j\omega \epsilon \mathbf{E} + \mathbf{J} = \omega^2 \epsilon \mathbf{A} - j\omega \epsilon \nabla \Phi + \mathbf{J} \tag{2.116}$$

Up to this point the divergences of **A** and $\nabla \Phi$ have not been specified [note that (2.114) specifies the curl of **A** only]. Therefore a relation between $\nabla \cdot \mathbf{A}$ and Φ may be chosen so as to simplify (2.116). Expanding $\nabla \times \nabla \times \mathbf{A}$ to give $\nabla \nabla \cdot \mathbf{A} - \nabla^2 \mathbf{A}$ enables us to write (2.116) as

$$\nabla \nabla \cdot \mathbf{A} - \nabla^2 \mathbf{A} = k^2 \mathbf{A} - j\omega \epsilon \mu \nabla \Phi + \mu \mathbf{J}$$

where $k^2 = \omega^2 \mu \epsilon$. If now the following condition is specified,

$$\nabla \nabla \cdot \mathbf{A} = -j\omega \epsilon \mu \nabla \Phi$$

or

$$\nabla \cdot \mathbf{A} = -j\omega\mu\epsilon\Phi \tag{2.117}$$

this equation simplifies to

$$\nabla^2 \mathbf{A} + k^2 \mathbf{A} = -\mu \mathbf{J} \tag{2.118}$$

Thus **A** is a solution of the inhomogeneous Helmholtz equation, the current **J** being the source term. The condition imposed on $\nabla \cdot \mathbf{A}$ and Φ in (2.117) is called the Lorentz condition in honor of the man first to propose its use.

In the preceding derivation three of Maxwell's equations have been used and are therefore satisfied. The fourth equation, $\nabla \cdot \mathbf{D} = \rho$, must also hold, and this will be shown to be the case provided the Lorentz condition is obeyed. Hence the three equations (2.114), (2.115), and (2.118), together with the Lorentz condition (2.117), are fully equivalent to Maxwell's equations. To verify the equation $\nabla \cdot \mathbf{D} = \rho$, take the divergence of (2.115) to obtain

$$\nabla \cdot \epsilon \mathbf{E} = -j\omega\epsilon \nabla \cdot \mathbf{A} - \epsilon \nabla^2 \Phi \tag{2.119}$$

where ϵ is a constant. Using the Lorentz condition yields

$$\nabla \cdot \mathbf{D} = -j\omega\epsilon \nabla \cdot \mathbf{A} - \nabla^2 \frac{\nabla \cdot \mathbf{A}}{-j\omega\mu} = \frac{1}{j\omega\mu} \nabla \cdot (\nabla^2 \mathbf{A} + k^2 \mathbf{A}) = -\frac{1}{j\omega} \nabla \cdot \mathbf{J}$$

by using (2.118) and noting that $\nabla^2 \nabla \cdot \mathbf{A} = \nabla \cdot \nabla^2 \mathbf{A}$; that is, these differential operators commute. Now $\nabla \cdot \mathbf{J} = -j\omega\rho$ from the continuity equation; so we obtain

$$\nabla \cdot \mathbf{D} = -\frac{1}{j\omega}(-j\omega\rho) = \rho$$

If, instead of eliminating Φ in (2.119), $\nabla \cdot \mathbf{A}$ is eliminated by use of the Lorentz condition, we get

$$\nabla \cdot \mathbf{D} = \rho = -j\omega\epsilon(-j\omega\mu\epsilon\Phi) - \epsilon \nabla^2 \Phi$$

or

$$\nabla^2 \Phi + k^2 \Phi = -\frac{\rho}{\epsilon} \tag{2.120}$$

Hence the scalar potential Φ is a solution of the inhomogeneous scalar Helmholtz equation, with the charge density ρ as a source term.

For the time-varying field, **J** and ρ are not independent, and hence the field can be determined in terms of **A** and **J** alone. The scalar potential can always be found from the Lorentz relation, and ρ from the continuity equation, but explicit knowledge of these is not required in order to solve radiation problems. For convenience, the pertinent equations are summarized here:

$$\mathbf{B} = \nabla \times \mathbf{A} \tag{2.121a}$$

$$\mathbf{E} = -j\omega\mathbf{A} - \nabla\Phi = -j\omega\mathbf{A} + \frac{\nabla\nabla \cdot \mathbf{A}}{j\omega\mu\epsilon} = \frac{k^2\mathbf{A} + \nabla\nabla \cdot \mathbf{A}}{j\omega\mu\epsilon} \tag{2.121b}$$

$$\nabla^2\mathbf{A} + k^2\mathbf{A} = -\mu\mathbf{J} \tag{2.121c}$$

where the Lorentz condition was used to eliminate $\nabla\Phi$ in (2.121b). Note that, in rectangular coordinates, (2.121c) is three scalar equations of the form

$$\nabla^2 A_x + k^2 A_x = -\mu J_x$$

but that, in other coordinate systems, $\nabla^2\mathbf{A}$ must be expanded according to the relation $\nabla^2\mathbf{A} = \nabla\nabla \cdot \mathbf{A} - \nabla \times \nabla \times \mathbf{A}$.

The simplest solution to (2.121c) is that for an infinitesimal current element $\mathbf{J}(x', y', z') = \mathbf{J}(\mathbf{r}')$ located at the point x', y', z', as specified by the position vector $\mathbf{r}' = \mathbf{a}_x x' + \mathbf{a}_y y' + \mathbf{a}_z z'$, as in Fig. 2.17. This solution is

$$\mathbf{A}(x, y, z) = \mathbf{A}(\mathbf{r}) = \frac{\mu}{4\pi} \mathbf{J}(\mathbf{r}') \frac{e^{-jkR}}{R} \tag{2.122}$$

where $R = |\mathbf{r} - \mathbf{r}'|$ is the magnitude of the distance from the source point to the field point at which $\mathbf{A}$ is evaluated; i.e.,

$$R = [(x - x')^2 + (y - y')^2 + (z - z')^2]^{\frac{1}{2}}$$

In terms of this fundamental solution, the vector potential from a general current distribution may be obtained by superposition. Thus, adding up all the contributions from each infinitesimal current element gives

$$\mathbf{A}(\mathbf{r}) = \frac{\mu}{4\pi} \int_V \mathbf{J}(x', y', z') \frac{e^{-jkR}}{R} dx' \, dy' \, dz' = \frac{\mu}{4\pi} \int_V \mathbf{J}(\mathbf{r}') \frac{e^{-jk|\mathbf{r}-\mathbf{r}'|}}{|\mathbf{r} - \mathbf{r}'|} dV' \tag{2.123}$$

where the integration is over the total volume occupied by the current. Note that the solution for $\mathbf{A}$ as given by (2.122) is a spherical wave propagating radially outward from $\mathbf{J}$ and with an amplitude falling off as $1/R$. The solution (2.123) is a superposition of such elementary spherical waves.

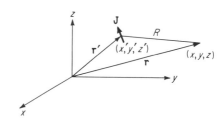

Fig. 2.17 Coordinates used to describe vector potential from a current element.

★2.11 Derivation of solution for vector potential

In this section a detailed derivation of the solution to the inhomogeneous Helmholtz equation for a unit current source is given. A unit source is a source of unit strength, localized at a point in space (a familiar example is a point charge). Such a unit source in a three-dimensional space is a generalization of a unit current impulse localized at a time t' along the time coordinate. A current pulse is represented by the Dirac delta function $\delta(t - t')$ in circuit theory, where $\delta(t - t')$ has the property

$$\delta(t - t') = 0 \qquad t \neq t' \tag{2.124a}$$

and at $t = t'$ it becomes infinite but is integrable to give

$$\int_{t'-\tau}^{t'+\tau} \delta(t - t')\, dt = 1 \tag{2.124b}$$

A further property is that, for any function $f(t)$ which is continuous at t',

$$\int_{t'-\tau}^{t'+\tau} f(t)\delta(t - t')\, dt = f(t') \tag{2.124c}$$

This result follows since τ can be chosen so small that, in the interval $t' - \tau < t < t' + \tau$, the function $f(t)$ differs by a vanishing amount from $f(t')$ since $f(t)$ is continuous at t'. Hence (2.124c) may be written as

$$f(t') \int_{t'-\tau}^{t'+\tau} \delta(t - t')\, dt = f(t')$$

by virtue of (2.124b).

As the preceding discussion has shown, the delta function is a convenient mathematical way to represent a source of unit strength localized at a point along a coordinate axis, in the above example along the time axis. In an N-dimensional space a product of N delta functions, one for each coordinate, may be used to represent a unit source. Thus, in three dimensions, a unit source is represented by

$$\delta(x - x')\delta(y - y')\delta(z - z') = \delta(\mathbf{r} - \mathbf{r}') \tag{2.125}$$

where $\delta(\mathbf{r} - \mathbf{r}')$ is an abbreviated notation for the product of the three one-dimensional delta functions. The source function $\delta(\mathbf{r} - \mathbf{r}')$ has the following properties:

$$\delta(\mathbf{r} - \mathbf{r}') = 0 \qquad \mathbf{r} \neq \mathbf{r}' \tag{2.126a}$$

$$\int_V \delta(\mathbf{r} - \mathbf{r}')\, dV = \begin{cases} 1 & \mathbf{r}' \text{ in } V \\ 0 & \mathbf{r}' \text{ not in } V \end{cases} \tag{2.126b}$$

$$\int_V \mathbf{F}(\mathbf{r})\delta(\mathbf{r} - \mathbf{r}')\, dV = \begin{cases} \mathbf{F}(\mathbf{r}') & \mathbf{r}' \text{ in } V \\ 0 & \mathbf{r}' \text{ not in } V \end{cases} \tag{2.126c}$$

where $\mathbf{F}$ is an arbitrary vector (or scalar) function that is continuous at

Sec. 2.11   **Electromagnetic theory**   **55**

$\mathbf{r}'$, that is, at x', y', z'. These properties follow from the properties of the one-dimensional delta functions that make up $\delta(\mathbf{r} - \mathbf{r}')$.

A unit current source directed along the unit vector $\mathbf{a}$ at $\mathbf{r}'$ may be expressed as $\mathbf{J} = \mathbf{a}\delta(\mathbf{r} - \mathbf{r}')$. The vector potential is a solution of

$$\nabla^2 \mathbf{A} + k^2 \mathbf{A} = -\mu \mathbf{a}\delta(\mathbf{r} - \mathbf{r}') \tag{2.127}$$

Since the current is in the direction $\mathbf{a}$, the vector potential must also be in this direction, and hence $\mathbf{A} = A\mathbf{a}$. Equation (2.127) may therefore be written as a scalar equation:

$$\nabla^2 A + k^2 A = -\mu \delta(\mathbf{r} - \mathbf{r}') \tag{2.128}$$

At all points $\mathbf{r} \neq \mathbf{r}'$, A is a solution of

$$\nabla^2 A + k^2 A = 0 \tag{2.129}$$

If the source point $\mathbf{r}'$ is considered as the origin in a spherical coordinate system, then, since no angle variables occur in the source term in (2.128), the solution for A must have spherical symmetry about the source point $\mathbf{r}'$. Thus, in terms of the spherical radial coordinate $R = |\mathbf{r} - \mathbf{r}'|$, which is the radial distance from the origin at $\mathbf{r}'$, (2.129) is a function of R only and may be written as

$$\frac{1}{R^2}\frac{\partial}{\partial R}\left(R^2 \frac{\partial A}{\partial R}\right) + k^2 A = 0$$

or

$$\frac{d^2 A}{dR^2} + \frac{2}{R}\frac{dA}{dR} + k^2 A = 0 \tag{2.130}$$

after expressing the R-dependent part of ∇^2 in spherical coordinates. In anticipation of a spherical-wave solution, let $A = f(R)e^{-jkR}$. Substitution in (2.130) leads to the following equation for $f(R)$,

$$\frac{d^2 f}{dR^2} + \left(\frac{2}{R} - 2jk\right)\frac{df}{dR} - \frac{2jk}{R}f = 0$$

which is readily verified to have the solution $f = C/R$, where C is an arbitrary constant. Consequently, the solution to (2.129) is $A = Ce^{-jkR}/R$. This solution is singular at $R = 0$, and the singularity must correspond to that of the source term at this point.

To determine the constant C, integrate (2.128) throughout a small sphere of radius r_0 centered on $\mathbf{r}'$ and use the delta-function property (2.126b) to obtain

$$\int_0^{2\pi}\int_0^{\pi}\int_0^{r_0}(\nabla^2 A + k^2 A)R^2 \sin\theta \, d\theta \, d\phi \, dR$$
$$= \int_V (\nabla^2 A + k^2 A)\, dV = -\mu\int_V \delta(\mathbf{r} - \mathbf{r}')\, dV = -\mu$$

Now the integral of the term $k^2 R^2 A$, which is proportional to R^2, will vanish as r_0 tends to zero. Hence, for sufficiently small r_0,

$$\int_V \nabla^2 A \, dV = -\mu$$

Since $\nabla^2 A = \nabla \cdot \nabla A$, the divergence theorem may be used to give

$$\int_V \nabla^2 A \, dV = \oint_S \nabla A \cdot d\mathbf{S} = \oint_S \nabla A \cdot \mathbf{a}_r r_0^2 \, d\Omega$$

since $d\mathbf{S} = \mathbf{a}_r r_0^2 \, d\Omega$, where $d\Omega$ is an element of solid angle. Since A is a function of R only, $\nabla A = \mathbf{a}_r (\partial A / \partial R)$, and hence

$$r_0^2 \oint_S \nabla A \cdot \mathbf{a}_r \, d\Omega = r_0^2 \oint_S \frac{\partial A}{\partial R} \, d\Omega = 4\pi r_0^2 \frac{\partial A}{\partial R} = -\mu$$

Evaluating $\partial A / \partial R$ for $R = r_0$ shows that

$$4\pi r_0^2 \frac{\partial A}{\partial R} = -4\pi C r_0^2 \left(\frac{jk}{r_0} e^{-jkr_0} + \frac{e^{-jkr_0}}{r_0^2} \right) = -4\pi C$$

in the limit as r_0 tends to zero. Hence $4\pi C = \mu$, or $C = \mu/4\pi$, in order for the singularity in the solution for A to correspond to that for a unit source.

The above solution for the vector potential from a unit source, namely,

$$\mathbf{A}(\mathbf{r}) = \frac{\mu}{4\pi} \frac{e^{-jk|\mathbf{r}-\mathbf{r}'|}}{|\mathbf{r}-\mathbf{r}'|} \mathbf{a} \tag{2.131}$$

is clearly a function of both the source point and field point. Since (2.131) is the solution for a unit source, it is often called a Green's function and denoted by the symbol $\mathbf{G}$ as

$$\mathbf{G}(\mathbf{r}|\mathbf{r}') = G(\mathbf{r}|\mathbf{r}')\mathbf{a} = G(x,y,z|x',y',z')\mathbf{a} = \frac{\mu}{4\pi} \frac{e^{-jk|\mathbf{r}-\mathbf{r}'|}}{|\mathbf{r}-\mathbf{r}'|} \mathbf{a} \tag{2.132}$$

because, by definition, a Green's function is the solution of a differential equation for a unit source.

The vector potential from a general current distribution may now be expressed in the form

$$\mathbf{A}(\mathbf{r}) = \frac{\mu}{4\pi} \int_V \mathbf{J}(\mathbf{r}') \frac{e^{-jk|\mathbf{r}-\mathbf{r}'|}}{|\mathbf{r}-\mathbf{r}'|} \, dV' = \int_V \mathbf{J}(\mathbf{r}') G(\mathbf{r}|\mathbf{r}') \, dV' \tag{2.133}$$

since any current distribution $\mathbf{J}$ may be considered as a sum of weighted unit sources.

2.12 Lorentz reciprocity theorem

The Lorentz reciprocity theorem is one of the most useful theorems in the solution of electromagnetic problems, since it may be used to deduce a number of fundamental properties of practical devices. It provides the

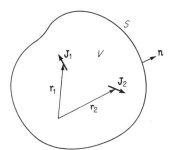

Fig. 2.18 Illustration for Lorentz reciprocity theorem.

basis for demonstrating the reciprocal properties of microwave circuits and for showing that the receiving and transmitting characteristics of antennas are the same. It also may be used to establish the orthogonality properties of the modes that may exist in waveguides and cavities.† Another important use is in deriving suitable field expansions (analogous to a Fourier series expansion) for the fields radiated or coupled into waveguides and cavities by probes, loops, or coupling apertures.

To derive the theorem, consider a volume V bounded by a closed surface S as in Fig. 2.18. Let a current source $\mathbf{J}_1$ in V produce a field $\mathbf{E}_1$, $\mathbf{H}_1$, while a second source $\mathbf{J}_2$ produces a field $\mathbf{E}_2$, $\mathbf{H}_2$. These fields satisfy Maxwell's equations; so

$$\nabla \times \mathbf{E}_1 = -j\omega\mu\mathbf{H}_1 \qquad \nabla \times \mathbf{H}_1 = j\omega\epsilon\mathbf{E}_1 + \mathbf{J}_1$$
$$\nabla \times \mathbf{E}_2 = -j\omega\mu\mathbf{H}_2 \qquad \nabla \times \mathbf{H}_2 = j\omega\epsilon\mathbf{E}_2 + \mathbf{J}_2$$

Expanding the relation $\nabla \cdot (\mathbf{E}_1 \times \mathbf{H}_2 - \mathbf{E}_2 \times \mathbf{H}_1)$ and using Maxwell's equation show that

$$\begin{aligned}\nabla \cdot (\mathbf{E}_1 \times \mathbf{H}_2 &- \mathbf{E}_2 \times \mathbf{H}_1) \\ &= (\nabla \times \mathbf{E}_1) \cdot \mathbf{H}_2 - (\nabla \times \mathbf{H}_2) \cdot \mathbf{E}_1 - (\nabla \times \mathbf{E}_2) \cdot \mathbf{H}_1 + (\nabla \times \mathbf{H}_1) \cdot \mathbf{E}_2 \\ &= -\mathbf{J}_2 \cdot \mathbf{E}_1 + \mathbf{J}_1 \cdot \mathbf{E}_2 \end{aligned} \qquad (2.134)$$

Integrating both sides over the volume V and using the divergence theorem give

$$\int_V \nabla \cdot (\mathbf{E}_1 \times \mathbf{H}_2 - \mathbf{E}_2 \times \mathbf{H}_1)\, dV = \oint_S (\mathbf{E}_1 \times \mathbf{H}_2 - \mathbf{E}_2 \times \mathbf{H}_1) \cdot \mathbf{n}\, dS$$
$$= \int_V (\mathbf{E}_2 \cdot \mathbf{J}_1 - \mathbf{E}_1 \cdot \mathbf{J}_2)\, dV \qquad (2.135)$$

where $\mathbf{n}$ is the unit outward normal to S.

† In any waveguide or cavity an infinite number of field solutions are possible. Any one solution is called a mode for the same reason that the various solutions for vibrating strings and membranes are called modes. Orthogonality of modes is discussed in Sec. 3.9.

Equation (2.135) is the basic form of the Lorentz reciprocity theorem.† For a number of typical situations that occur, the surface integral vanishes. If S is a perfectly conducting surface, then $\mathbf{n} \times \mathbf{E}_1 = \mathbf{n} \times \mathbf{E}_2 = 0$ on S. Since $\mathbf{E}_1 \times \mathbf{H}_2 \cdot \mathbf{n} = (\mathbf{n} \times \mathbf{E}_1) \cdot \mathbf{H}_2$, etc., the surface integral vanishes in this case. If the surface S is characterized by a surface impedance Z_m, then, according to (2.71),

$$\mathbf{E}_t = -Z_m \mathbf{n} \times \mathbf{H}, \quad \text{or} \quad \mathbf{n} \times \mathbf{E} = -Z_m \mathbf{n} \times (\mathbf{n} \times \mathbf{H})$$

[note that in (2.71) $\mathbf{n}$ points into the region occupied by the field, and hence the minus sign is used here, since $\mathbf{n}$ is directed out of V]. Consequently,

$$(\mathbf{n} \times \mathbf{E}_1) \cdot \mathbf{H}_2 - (\mathbf{n} \times \mathbf{E}_2) \cdot \mathbf{H}_1$$
$$= -Z_m [\mathbf{n} \times (\mathbf{n} \times \mathbf{H}_1)] \cdot \mathbf{H}_2 + Z_m [\mathbf{n} \times (\mathbf{n} \times \mathbf{H}_2)] \cdot \mathbf{H}_1$$
$$= Z_m (\mathbf{n} \times \mathbf{H}_2) \cdot (\mathbf{n} \times \mathbf{H}_1) - Z_m (\mathbf{n} \times \mathbf{H}_1) \cdot (\mathbf{n} \times \mathbf{H}_2) = 0$$

and the surface integral vanishes again.

Another example where the surface integral vanishes is when S is chosen as a spherical surface at infinity for which $\mathbf{n} = \mathbf{a}_r$. The radiated field at infinity is a spherical TEM wave for which

$$\mathbf{H} = Y \mathbf{a}_r \times \mathbf{E} = \left(\frac{\epsilon}{\mu}\right)^{\frac{1}{2}} \mathbf{a}_r \times \mathbf{E}$$

Therefore

$$(\mathbf{n} \times \mathbf{E}_1) \cdot \mathbf{H}_2 - (\mathbf{n} \times \mathbf{E}_2) \cdot \mathbf{H}_1 = Y(\mathbf{a}_r \times \mathbf{E}_1) \cdot (\mathbf{a}_r \times \mathbf{E}_2)$$
$$- Y(\mathbf{a}_r \times \mathbf{E}_2) \cdot (\mathbf{a}_r \times \mathbf{E}_1) = 0$$

and the surface integral vanishes.

Actually, for any surface S which encloses all the sources for the field, the surface integral will vanish. This result may be seen by applying (2.135) to the volume V bounded by S and the surface of a sphere of infinite radius. There are no sources in this volume, and since the surface integral over the surface of the sphere with infinite radius is zero, we must have, from (2.135),

$$\oint_S (\mathbf{E}_1 \times \mathbf{H}_2 - \mathbf{E}_2 \times \mathbf{H}_1) \cdot (-\mathbf{n}) \, dS = 0$$
$$= \oint_S (\mathbf{E}_1 \times \mathbf{H}_2 - \mathbf{E}_2 \times \mathbf{H}_1) \cdot \mathbf{n} \, dS$$

Hence the surface integral taken over any closed surface S surrounding all the sources vanishes.

---

† In anisotropic media with nonsymmetrical permittivity or permeability tensors, a modified form must be used. See, for example, R. F. Harrington and A. T. Villeneuve, Reciprocity Relations for Gyrotropic Media, *IRE Trans.*, vol. MTT-6, pp. 308–310, July, 1958.

Sec. 2.12 Electromagnetic theory

When the surface integral vanishes, (2.135) reduces to

$$\int_V \mathbf{E}_1 \cdot \mathbf{J}_2 \, dV = \int_V \mathbf{E}_2 \cdot \mathbf{J}_1 \, dV \tag{2.136}$$

If $\mathbf{J}_1$ and $\mathbf{J}_2$ are infinitesimal current elements, then

$$\mathbf{E}_1(\mathbf{r}_2) \cdot \mathbf{J}_2(\mathbf{r}_2) = \mathbf{E}_2(\mathbf{r}_1) \cdot \mathbf{J}_1(\mathbf{r}_1) \tag{2.137}$$

which states that the field $\mathbf{E}_1$ produced by $\mathbf{J}_1$ has a component along $\mathbf{J}_2$ that is equal to the component along $\mathbf{J}_1$ of the field radiated by $\mathbf{J}_2$ when $\mathbf{J}_1$ and $\mathbf{J}_2$ have unit magnitude. The form (2.137) is essentially the reciprocity principle used in circuit analysis except that $\mathbf{E}$ and $\mathbf{J}$ are replaced by the voltage V and current I. The applications of the reciprocity theorem are illustrated at various points throughout the text and hence are not discussed further at this time.

Problems

2.1 An atom of atomic number Z has a nuclear charge Ze and Z electrons revolving around it. As a model of this atom, consider the nucleus as a point charge and treat the electron cloud as a total charge $-Ze$ distributed uniformly throughout a sphere of radius r_0. When an external field E is applied, the nucleus is displaced an amount x. Show that a restoring force $x(Ze)^2/4\pi r_0^3 \epsilon_0$ is produced and must be equal to ZeE. Thus show that the induced dipole moment is $p = 4\pi\epsilon_0 r_0^3 E$ and is linearly related to E.

2.2 In a certain material the equation of motion for the polarization is

$$\frac{d^2 \mathcal{P}}{dt^2} + \nu \frac{d\mathcal{P}}{dt} + \omega_0^2 \mathcal{P} = 2\epsilon_0 \omega_0^2 \mathcal{E}$$

where $\mathcal{E}$ is the total field in the dielectric. Find the relation between $\mathcal{P}$ and $\mathcal{E}$ when $\mathcal{E} = \mathrm{Re}\,(Ee^{i\omega t})$ and E is real. If $\omega_0 = 10^{11}$ and $\nu = 10^{10}$, over what frequency range can a relationship such as $\mathcal{D} = \epsilon \mathcal{E} = \epsilon_0 \mathcal{E} + \mathcal{P}$ be written if it is assumed that the criterion to be used is that the phase difference between $\mathcal{D}$ and $\mathcal{E}$ should not exceed 5°? Plot the magnitude and phase angle of the dielectric constant $\kappa = \epsilon/\epsilon_0 = (\epsilon' - j\epsilon'')/\epsilon_0$ as a function of ω.

2.3 A dielectric material is characterized by a matrix (tensor) permittivity

$$\begin{bmatrix} \epsilon_{xx} & \epsilon_{xy} & \epsilon_{xz} \\ \epsilon_{xy} & \epsilon_{yy} & \epsilon_{yz} \\ \epsilon_{xz} & \epsilon_{yz} & \epsilon_{zz} \end{bmatrix} = \frac{\epsilon_0}{4} \begin{bmatrix} 7 & 3 & -2\sqrt{0.5} \\ 3 & 7 & -2\sqrt{0.5} \\ -2\sqrt{0.5} & -2\sqrt{0.5} & 10 \end{bmatrix}$$

when referred to the xyz coordinate frame. If the coordinate axis is rotated into the principal axis u, v, w, the permittivity is exhibited in diagonal form:

$$[\epsilon] = \begin{bmatrix} \epsilon_{uu} & 0 & 0 \\ 0 & \epsilon_{vv} & 0 \\ 0 & 0 & \epsilon_{ww} \end{bmatrix}$$

Find the principal axis and the values of the principal dielectric constants ϵ_{uu}/ϵ_0, etc.
Hint: By definition, along a principal axis a scalar equation such as $D_u = \epsilon_{uu} E_u$

holds. In general, if **D** is directed along a principal axis, then

$$\begin{bmatrix} D_x \\ D_y \\ D_z \end{bmatrix} = \frac{\epsilon_0}{4} \begin{bmatrix} 7 & 3 & -2\sqrt{0.5} \\ 3 & 7 & -2\sqrt{0.5} \\ -2\sqrt{0.5} & -2\sqrt{0.5} & 10 \end{bmatrix} \begin{bmatrix} E_x \\ E_y \\ E_z \end{bmatrix} = \lambda \begin{bmatrix} E_x \\ E_y \\ E_z \end{bmatrix}$$

or in words, when **D** is directed along a principal axis, it is related to **E** by a scalar constant λ. The above constitutes a set of three homogeneous equations, of which the first is

$$\left(\frac{7\epsilon_0}{4} - \lambda\right) E_x + \frac{3\epsilon_0}{4} E_y - \frac{2\epsilon_0 \sqrt{0.5}}{4} E_z = 0$$

Verify that, for a solution, the following determinant must vanish:

$$\begin{vmatrix} 7 - 4\lambda/\epsilon_0 & 3 & -2\sqrt{0.5} \\ 3 & 7 - 4\lambda/\epsilon_0 & -2\sqrt{0.5} \\ -2\sqrt{0.5} & -2\sqrt{0.5} & 10 - 4\lambda/\epsilon_0 \end{vmatrix} = 0$$

This cubic equation gives three roots for λ, which may be identified as ϵ_{uu}, ϵ_{vv}, ϵ_{ww}. For any one root, say ϵ_{uu}, the components of a vector directed along the corresponding principal axis are proportional to the cofactors of the above determinant. This type of problem is called a matrix eigenvalue problem. The λ's are the eigenvalues.

Answer: $\epsilon_{uu} = 3\epsilon_0$, $\epsilon_{vv} = 2\epsilon_0$, $\epsilon_{ww} = \epsilon_0$.

Unit vectors along the principal axis are

$\mathbf{a}_u = 0.5\mathbf{a}_x + 0.5\mathbf{a}_y - \sqrt{0.5}\,\mathbf{a}_z$

$\mathbf{a}_v = 0.5\mathbf{a}_x + 0.5\mathbf{a}_y + \sqrt{0.5}\,\mathbf{a}_z$

$\mathbf{a}_w = \sqrt{0.5}\,\mathbf{a}_x - \sqrt{0.5}\,\mathbf{a}_y$

2.4 In the interior of a medium with conductivity σ and permittivity ϵ, free charge is distributed with a density $\rho_0(x, y, z)$ at time $t = 0$. Show that the charge decays according to

$$\rho = \rho_0 e^{-t/\tau} \qquad \tau = \frac{\epsilon}{\sigma}$$

Evaluate the relaxation time τ for copper for which $\sigma = 5.8 \times 10^7$ mhos/m, $\epsilon = \epsilon_0$. Find τ for sea water also for which $\sigma = 4$ mhos/m and $\epsilon = 80\epsilon_0$. If the relaxation time is short compared with the period of an applied time-harmonic field, there will be negligible accumulation of free charge and $\nabla \cdot \mathbf{D}$ may be assumed to be zero. What is the upper frequency limit for which this is true in the case of copper and sea water, i.e., the frequency for which τ is equal to the period?

Hint: Use the continuity equation, Ohm's law, and the divergence equation for **D**.

2.5 Show that, when the relaxation time for a material is small compared with the period of the time-harmonic field, the displacement current may be neglected in comparison with the conduction current.

2.6 Consider two concentric spheres of radii a and b. The outer sphere is kept at a potential V, and the inner sphere at zero potential. Solve Laplace's equation in spherical coordinates to find the potential and electric field between the spheres. Take $b > a$.

2.7 Solve Laplace's equation to find the potential and electric field between two coaxial cylinders of radii a and b if the center cylinder is kept at a potential V and the outer cylinder at zero potential. Take $b > a$.

2.8 Derive (2.45) from (2.18).

2.9 Derive (2.47).

2.10 Express the scalar Helmholtz equation $\nabla^2\psi + k^2\psi = 0$ in cylindrical coordinates. If $\psi = f(\phi)g(r)h(z)$, find the differential equations satisfied by f, g, and h.

2.11 When material polarization $\mathcal{P}$ and $\mathcal{M}$ are explicitly taken into account, show that the wave equations satisfied by $\mathcal{E}$ and $\mathcal{H}$ are

$$\nabla^2\mathcal{H} - \mu_0\epsilon_0\frac{\partial^2\mathcal{H}}{\partial t^2} = -\nabla\nabla\cdot\mathcal{M} + \mu_0\epsilon_0\frac{\partial^2\mathcal{M}}{\partial t^2} - \frac{\partial}{\partial t}\nabla\times\mathcal{P} - \nabla\times\mathcal{J}$$

$$\nabla^2\mathcal{E} - \mu_0\epsilon_0\frac{\partial^2\mathcal{E}}{\partial t^2} = \mu_0\frac{\partial^2\mathcal{P}}{\partial t^2} + \mu_0\frac{\partial\mathcal{J}}{\partial t} + \mu_0\frac{\partial}{\partial t}\nabla\times\mathcal{M} + \frac{1}{\epsilon_0}\nabla\rho - \frac{\nabla\nabla\cdot\mathcal{P}}{\epsilon_0}$$

Note that $\nabla\cdot\mathcal{B} = 0$; so $\nabla\cdot\mathcal{H} = -\nabla\cdot\mathcal{M}$ and $\nabla\cdot\mathcal{D} = \rho$; so $\nabla\cdot\epsilon_0\mathcal{E} = \rho - \nabla\cdot\mathcal{P}$. Examination of the source terms in the above equations shows that $\partial\mathcal{P}/\partial t$ is a polarization current analogous to conduction current $\mathcal{J}$.

2.12 Derive (2.62).

2.13 Between two perfectly conducting coaxial cylinders of radii a and b, $b > a$, the electromagnetic field is given by

$$\mathbf{E} = \mathbf{a}_r E_0 r^{-1} e^{-jk_0 z} \qquad \mathbf{H} = \mathbf{a}_\phi Y_0 E_0 r^{-1} e^{-jk_0 z}$$

where $k_0 = \omega(\mu_0\epsilon_0)^{1/2}$, $Y_0 = (\epsilon_0/\mu_0)^{1/2}$. Find the potential difference between the cylinders and the total current on the inner and outer cylinders. Express the power in terms of the voltage and current, and show that it is equal to that computed from an integration of the complex Poynting vector over the coaxial-line cross section. Show that the characteristic impedance of the line is $V/I = (Z_0/2\pi)\ln(b/a) = 60\ln(b/a)$, where V is the voltage and I is the total current on one cylinder.

2.14 A round wire of radius r_0 much greater than the skin depth δ_s has a uniform electric field E applied in the axial direction at its surface. Use the surface-impedance concept to find the total current on the wire. Show that the ratio of the a-c impedance of the wire to the d-c resistance is

$$\frac{Z_{ac}}{R_{dc}} = \frac{r_0\sigma}{2}Z_m$$

Evaluate this ratio for copper at $f = 10^6$ cycles for $\sigma = 5.8 \times 10^7$ mhos/m, $r_0 = 0.1$ cm, $\mu = \mu_0$.

2.15 The half-space $z \geq 0$ is filled with a material with permittivity ϵ_0 and permeability $\mu \neq \mu_0$. A parallel polarized plane TEM wave is incident at an angle θ_1, as in Fig. 2.14. Find the reflection and transmission coefficients for the electric field. Does a Brewster angle exist for which the reflection coefficient vanishes?

2.16 Repeat Prob. 2.15 for the case of a perpendicular polarized incident wave. Does a Brewster angle exist? If so, obtain an expression for it.

2.17 The half-space $z \geq 0$ is filled with a material with permeability μ and permittivity ϵ. When a plane wave is incident normally on this material, show that the reflection and transmission coefficients are

$$\Gamma = \frac{Z - Z_0}{Z + Z_0} \qquad T = 1 + \Gamma = \frac{2Z}{Z + Z_0}$$

where $Z = (\mu/\epsilon)^{1/2}$, $Z_0 = (\mu_0/\epsilon_0)^{1/2}$. Choose an electric field with an x component only.

2.18 The half-space $z \geq 0$ is filled with a material of permittivity ϵ_2 and with $\mu = \mu_0$. A second sheet with permittivity ϵ_1 is placed in front. A plane wave is incident normally on the structure from the left, as illustrated. Verify that the reflection coefficient at the first interface vanishes if $\epsilon_1 = (\epsilon_2 \epsilon_0)^{\frac{1}{2}}$ and the thickness $d = \frac{1}{4}\lambda_0(\epsilon_0/\epsilon_1)^{\frac{1}{2}}$. The electric field may be assumed to have an x component only. The matching layer is known as a quarter-wave transformer (actually an impedance transformer). This matching technique is used to reduce reflections from optical lenses and is called *lens blooming*, or *coated lenses*.

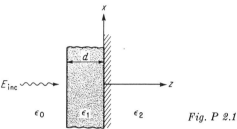

Fig. P 2.18

2.19 In terms of the vector potential **A** from a short current element $\Delta z I_0 \mathbf{a}_z$ located at the origin, show that the radiated electric and magnetic fields are

$$\mathbf{H} = \frac{I_0 \Delta z}{4\pi} \left(\frac{jk_0}{r} + \frac{1}{r^2} \right) \mathbf{a}_\phi \sin \theta \, e^{-jk_0 r}$$

$$\mathbf{E} = -\frac{I_0 \Delta z}{2\pi} \frac{jZ_0}{k_0} \left(\frac{jk_0}{r^2} + \frac{1}{r^3} \right) \mathbf{a}_r \cos \theta \, e^{-jk_0 r} - \frac{I_0 \Delta z}{4\pi} \frac{jZ_0}{k_0} \left(\frac{-k_0^2}{r} + \frac{jk_0}{r^2} + \frac{1}{r^3} \right) \mathbf{a}_\theta \sin \theta \, e^{-jk_0 r}$$

Hint: Use (2.122) and (2.121), and express **A** as components in a spherical coordinate system r, θ, ϕ. Note that $\mathbf{a}_z = \mathbf{a}_r \cos \theta - \mathbf{a}_\theta \sin \theta$.

2.20 A dielectric may be characterized by its dipole polarization **P** per unit volume. If $\rho = \mathbf{J} = 0$ and **P** is taken into account explicitly, show that, if a vector potential **A** is introduced according to $\mathbf{B} = \nabla \times \mathbf{A}$, then **A** is a solution of

$$\nabla^2 \mathbf{A} + k_0^2 \mathbf{A} = -j\omega\mu_0 \mathbf{P}$$

and that the fields are given by

$$\mathbf{B} = \nabla \times \mathbf{A} \qquad \mathbf{E} = \frac{\nabla \nabla \cdot \mathbf{A} + k_0^2 \mathbf{A}}{j\omega\mu_0\epsilon_0}$$

Note that a Lorentz condition is used. Thus an electric dipole **P** is equivalent to a current element $j\omega\mathbf{P}$, or alternatively, a current element **J** may be considered as an electric dipole $\mathbf{P} = \mathbf{J}/j\omega$.

2.21 A small current loop constitutes a magnetic dipole $\mathbf{M} = IS\mathbf{a}$, where I is the current, S the area of the loop, and **a** a vector normal to the plane of the loop and pointing in the direction that a right-hand screw, rotating in the direction of the current, would advance. The field radiated by such a current loop, with linear dimensions much smaller than a wavelength, may be obtained by a potential theory analogous to that given in Prob. 2.20 by treating the loop as a magnetic dipole **M**. Thus replace **B** by $\mu_0 \mathbf{H} + \mu_0 \mathbf{M}$ in Maxwell's equation and treat **M** as a source term. Since ρ is zero, $\nabla \cdot \mathbf{D} = 0$, and this permits **D** to be expressed as $\mathbf{D} = -\nabla \times \mathbf{A}_m$, where $\mathbf{A}_m$ is a magnetic-type vector potential. By paralleling the development in the text for the

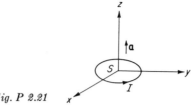

Fig. P 2.21

potential **A**, show that the following relations are obtained:

$\nabla^2 \mathbf{A}_m + k_0{}^2 \mathbf{A}_m = -j\omega\mu_0\epsilon_0 \mathbf{M}$

$\mathbf{D} = -\nabla \times \mathbf{A}_m$

$\mathbf{H} = \dfrac{k_0{}^2 \mathbf{A}_m + \nabla\nabla \cdot \mathbf{A}_m}{j\omega\mu_0\epsilon_0}$

Hence, for a z-directed magnetic dipole at the origin,

$\mathbf{A}_m = \dfrac{j\omega\mu_0\epsilon_0 \mathbf{M}}{4\pi r} e^{-jk_0 r}$

from which the fields are readily found.

2.22 Consider an arbitrary current element $\mathbf{J}_1$ in front of a perfectly conducting plane. This current radiates a field $\mathbf{E}_1$ having zero tangential components on the conducting plane. Use the Lorentz reciprocity theorem to show that a current $\mathbf{J}_2$ parallel to the conducting plane and an infinitesimal distance in front of it does not radiate.

References

1. Fano, R. M., L. J. Chu, and R. B. Adler: "Electromagnetic Fields, Energy, and Forces," John Wiley & Sons, Inc., New York, 1960.
2. Kraus, J. D.: "Electromagnetics," McGraw-Hill Book Company, New York, 1953.
3. Plonsey, R., and R. E. Collin: "Principles and Applications of Electromagnetic Fields," McGraw-Hill Book Company, New York, 1961.
4. Reitz, J. R., and F. J. Milford: "Foundations of Electromagnetic Theory," Addison-Wesley Publishing Company, Inc., Reading, Mass., 1960.
5. Stratton, J. A.: "Electromagnetic Theory," McGraw-Hill Book Company, New York, 1941.

3

Transmission lines and waveguides

Transmission lines and waveguides are used to transport electromagnetic energy at microwave frequencies from one point in a system to another without radiation of energy taking place. The two main characteristics desired in a transmission line or waveguide are single-mode propagation over a wide band of frequencies and small attenuation. A great variety of transmission lines and waveguides having these two essential features have been investigated. Most of the structures considered fall into one of the following three categories: (1) transmission lines on which the dominant mode of propagation is a transverse electromagnetic wave, (2) closed cylindrical conducting tubes, and (3) open-boundary structures that support a surface-wave mode of propagation. It will not be possible to examine in detail all the different structures that have been introduced for waveguiding. We shall restrict the discussion to examining the basic theory of transmission lines and empty cylindrical waveguides, with specific reference to the commonly used coaxial transmission line and the rectangular and circular waveguides, and to presenting an introduction to surface waveguides. The extensions and modifications of the theory, necessary for analyzing other specific structures, are not difficult. The student should find little difficulty in reading the literature devoted to various types of waveguides, once familiarity with the theory given here is acquired.

Transmission lines consist of two or more parallel conductors and will guide a transverse electromagnetic (TEM) wave. The common forms of transmission lines are the two-conductor line, the shielded two-conductor line, and the coaxial line. Another form of transmission line that has come into prominence in the last few years is the microwave strip line, which consists of a thin conducting ribbon separated from a wider ground plane by a dielectric sheet or placed between two ground planes to form a shielded structure. The two main advantages obtained with strip lines are the reduction in size and weight and the ability to use printed-circuit techniques for the construction of the strip lines and associated components such as bends, junctions, filters, etc. A good introduction to strip lines and associated components may be found in a special issue of *IRE Transactions* devoted entirely to this subject.†

The common forms of waveguides are the rectangular and circular guides, which will be analyzed in detail in this chapter. A number of other structures have also been proposed that offer some distinct advantages for certain applications, but the theory of many of them is not sufficiently different from that of the common rectangular and circular guides to warrant detailed treatment.‡

3.1 Classification of wave solutions

The transmission lines and waveguides to be analyzed in this chapter are all characterized by having axial uniformity. Their cross-sectional shape and electrical properties do not vary along the axis, which is chosen as the z axis. Since sources are not considered, the electric and magnetic fields are solutions of the homogeneous vector Helmholtz equation, i.e.,

$$\nabla^2 \mathbf{E} + k_0^2 \mathbf{E} = 0 \qquad \nabla^2 \mathbf{H} + k_0^2 \mathbf{H} = 0$$

The type of solution sought is that corresponding to a wave that propagates along the z axis. Since the Helmholtz equation is separable, it is possible to find solutions of the form $f(z)g(x, y)$, where f is a function of z only and g is a function of x and y or other suitable transverse coordinates only. The second derivative with respect to z enters into the wave equation in a manner similar to the second derivative with respect to time. By analogy with $e^{j\omega t}$ as the time dependence, the z dependence can be assumed to be $e^{\pm j\beta z}$. This assumption will lead to wave solutions of the

---

† Special Issue on Microwave Strip Circuits, *IRE Trans.*, vol. MTT-3, March, 1955.

‡ As an introduction to some of these other structures, the following references may be consulted. For a complete survey with an extensive bibliography the book by Harvey should be consulted.

S. B. Cohn, Properties of Ridged Waveguides, *Proc. IRE*, vol. 35, pp. 783–788, August, 1947.

G. Goubau and J. R. Christian, Some Aspects of Beam Waveguides for Long Distance Transmission at Optical Frequencies, *IEEE Trans.*, vol. MTT-12, pp. 212–220, March, 1964.

A. F. Harvey, "Microwave Engineering," chaps. 1, 9, 10, and 22, Academic Press Inc., New York, 1963.

S. Hopfer, Design of Ridged Waveguides, *IRE Trans.*, vol. MTT-3, pp. 20–29, October, 1955.

M. Sugi and T. Nakahara, O-guide and X-guide: An Advanced Surface Wave Transmission Concept, *IRE Trans.*, vol. MTT-7, pp. 366–369, July, 1959.

F. J. Tischer, Properties of the H-guide at Microwave and Millimetre Wave Regions, *Proc. IEE*, vol. 106B, suppl. 13, 1959.

F. J. Tischer, The Groove Guide: A Low Loss Waveguide for Millimetre Waves, *IEEE Trans.*, vol. MTT-11, pp. 201–296, September, 1963.

form $\cos(\omega t \pm \beta z)$ and $\sin(\omega t \pm \beta z)$, which are appropriate for describing wave propagation along the z axis. A wave propagating in the positive z direction is represented by $e^{-j\beta z}$, and $e^{j\beta z}$ corresponds to a wave propagating in the negative z direction. With an assumed z dependence $e^{-j\beta z}$, the del operator becomes $\nabla = \nabla_t + \nabla_z = \nabla_t - j\beta \mathbf{a}_z$, since $\nabla_z = \mathbf{a}_z\, \partial/\partial z$. Note that ∇_t is the transverse part and equals $\mathbf{a}_x\, \partial/\partial x + \mathbf{a}_y\, \partial/\partial y$ in rectangular coordinates. The propagation phase constant β will turn out to depend on the waveguide configuration.

Considerable simplification of Maxwell's equations is obtained by decomposing all fields into transverse and axial components and separating out the z dependence. Thus, let (the time dependence $e^{j\omega t}$ is suppressed)

$$\mathbf{E}(x, y, z) = \mathbf{E}_t(x, y, z) + \mathbf{E}_z(x, y, z)$$
$$= \mathbf{e}(x, y)e^{-j\beta z} + \mathbf{e}_z(x, y)e^{-j\beta z} \qquad (3.1)$$
$$\mathbf{H}(x, y, z) = \mathbf{H}_t(x, y, z) + \mathbf{H}_z(x, y, z)$$
$$= \mathbf{h}(x, y)e^{-j\beta z} + \mathbf{h}_z(x, y)e^{-j\beta z} \qquad (3.2)$$

where $\mathbf{E}_t$, $\mathbf{H}_t$ are the transverse (x and y) components, and $\mathbf{E}_z$, $\mathbf{H}_z$ are the axial components. Note also that $\mathbf{e}(x, y)$, $\mathbf{h}(x, y)$ are transverse vector functions of the transverse coordinates only, and $\mathbf{e}_z(x, y)$, $\mathbf{h}_z(x, y)$ are axial vector functions of the transverse coordinates.

Consider the $\nabla \times \mathbf{E}$ equation, which may be expanded to give

$$\nabla \times \mathbf{E} = (\nabla_t - j\beta \mathbf{a}_z) \times (\mathbf{e} + \mathbf{e}_z)e^{-j\beta z} = -j\omega\mu_0 (\mathbf{h} + \mathbf{h}_z)e^{-j\beta z}$$

or

$$\nabla_t \times \mathbf{e} - j\beta \mathbf{a}_z \times \mathbf{e} + \nabla_t \times \mathbf{e}_z - j\beta \mathbf{a}_z \times \mathbf{e}_z = -j\omega\mu_0 \mathbf{h} - j\omega\mu_0 \mathbf{h}_z$$

The term $\mathbf{a}_z \times \mathbf{e}_z = 0$, and $\nabla_t \times \mathbf{e}_z = \nabla_t \times \mathbf{a}_z e_z = -\mathbf{a}_z \times \nabla_t e_z$. Note also that $\nabla_t \times \mathbf{e}$ is directed along the z axis only, since it involves factors such as $\mathbf{a}_x \times \mathbf{a}_y$, $\mathbf{a}_x \times \mathbf{a}_x$, $\mathbf{a}_y \times \mathbf{a}_x$, and $\mathbf{a}_y \times \mathbf{a}_y$, whereas $\mathbf{a}_z \times \mathbf{e}$ and $\nabla_t \times \mathbf{e}_z$ have transverse components only. Consequently, when the transverse and axial components of the above equation are equated, there results

$$\nabla_t \times \mathbf{e} = -j\omega\mu_0 \mathbf{h}_z \qquad (3.3a)$$
$$\nabla_t \times \mathbf{e}_z - j\beta \mathbf{a}_z \times \mathbf{e} = -\mathbf{a}_z \times \nabla_t e_z - j\beta \mathbf{a}_z \times \mathbf{e} = -j\omega\mu_0 \mathbf{h} \qquad (3.3b)$$

In a similar manner the $\nabla \times \mathbf{H}$ equation yields

$$\nabla_t \times \mathbf{h} = j\omega\epsilon_0 \mathbf{e}_z \qquad (3.3c)$$
$$\mathbf{a}_z \times \nabla_t h_z + j\beta \mathbf{a}_z \times \mathbf{h} = -j\omega\epsilon_0 \mathbf{e} \qquad (3.3d)$$

The divergence equation $\nabla \cdot \mathbf{B} = 0$ becomes

$$\nabla \cdot \mathbf{B} = \nabla \cdot \mu_0 \mathbf{H} = (\nabla_t - j\beta \mathbf{a}_z) \cdot (\mathbf{h} + \mathbf{h}_z)\mu_0 e^{-j\beta z}$$
$$= (\nabla_t \cdot \mathbf{h} - j\beta \mathbf{a}_z \cdot \mathbf{h}_z)\mu_0 e^{-j\beta z} = 0$$

or

$$\nabla_t \cdot \mathbf{h} = j\beta h_z \tag{3.3e}$$

Similarly, $\nabla \cdot \mathbf{D} = 0$ gives

$$\nabla_t \cdot \mathbf{e} = j\beta e_z \tag{3.3f}$$

This reduced form of Maxwell's equations will prove to be very useful in formulating the solutions for waveguiding systems.

For a large variety of waveguides of practical interest it turns out that all the boundary conditions can be satisfied by fields that do not have all components present. Specifically, for transmission lines, the solution of interest is a transverse electromagnetic wave with transverse components only, that is, $E_z = H_z = 0$, whereas for waveguides, solutions with $E_z = 0$ or $H_z = 0$ are possible. Because of the widespread occurrence of such field solutions, the following classification of solutions is of particular interest.

1. Transverse electromagnetic (TEM) waves. For TEM waves, $E_z = H_z = 0$. The electric field may be found from the transverse gradient of a scalar function $\Phi(x, y)$, which is a function of the transverse coordinates only and is a solution of the two-dimensional Laplace's equation.
2. Transverse electric (TE), or H, modes. These solutions have $E_z = 0$, but $H_z \neq 0$. All the field components may be derived from the axial component H_z of magnetic field.
3. Transverse magnetic (TM), or E, modes. These solutions have $H_z = 0$, but $E_z \neq 0$. The field components may be derived from E_z.

In some cases it will be found that a TE or TM mode by itself will not satisfy all the boundary conditions. However, in such cases linear combinations of TE and TM modes may be used, since such linear combinations always provide a complete and general solution. Although other possible types of wave solutions may be constructed, the above three types are the most useful in practice and by far the most commonly used ones.

The appropriate equations to be solved to obtain TEM, TE, or TM modes will be derived below by placing E_z and H_z, E_z, and H_z, respectively, equal to zero in Maxwell's equations.

TEM waves

For TEM waves $e_z = h_z = 0$; so (3.3) reduces to

$$\nabla_t \times \mathbf{e} = 0 \tag{3.4a}$$

$$\beta \mathbf{a}_z \times \mathbf{e} = \omega\mu_0 \mathbf{h} \tag{3.4b}$$

$$\nabla_t \times \mathbf{h} = 0 \tag{3.4c}$$

$$\beta \mathbf{a}_z \times \mathbf{h} = -\omega\epsilon_0 \mathbf{e} \tag{3.4d}$$

$$\nabla_t \cdot \mathbf{h} = 0 \tag{3.4e}$$

$$\nabla_t \cdot \mathbf{e} = 0 \tag{3.4f}$$

The vanishing of the transverse curl of **e** means that the line integral of **e** around any closed path in the xy plane is zero. This must clearly be so since there is no axial magnetic flux passing through such a contour. Although $\nabla_t \times \mathbf{h} = 0$ when there are no volume currents present, the line integral of **h** will not vanish for a transmission line with conductors on which axial currents may exist. This point will be considered again later when transmission lines are analyzed. Equation (3.4a) is just the condition that permits **e** to be expressed as the gradient of a scalar potential. Hence, let

$$\mathbf{e}(x, y) = -\nabla_t \Phi(x, y) \tag{3.5}$$

Using (3.4f) shows that Φ is a solution of the two-dimensional Laplace equation

$$\nabla_t^2 \Phi(x, y) = 0 \tag{3.6}$$

The electric field is thus given by

$$\mathbf{E}_t(x, y, z) = -\nabla_t \Phi(x, y) e^{-j\beta z}$$

But this field must also satisfy the Helmholtz equation

$$\nabla^2 \mathbf{E}_t + k_0^2 \mathbf{E}_t = 0$$

Since $\nabla = \nabla_t - j\beta \mathbf{a}_z$, $\nabla^2 = \nabla_t^2 - \beta^2$, that is, the second derivative with respect to z gives a factor $-\beta^2$, this reduces to

$$\nabla_t^2 \mathbf{E}_t + (k_0^2 - \beta^2)\mathbf{E}_t = 0$$

or

$$\nabla_t[\nabla_t^2 \Phi + (k_0^2 - \beta^2)\Phi] = 0$$

This shows that $\beta = \pm k_0$ for TEM waves, a result to be anticipated from the wave solutions discussed in Chap. 2. The magnetic field may be found from the $\nabla \times \mathbf{E}$ equation, i.e., from (3.4b); thus

$$\frac{\omega\mu_0}{k_0}\mathbf{h} = \mathbf{a}_z \times \mathbf{e} = Z_0 \mathbf{h} \tag{3.7}$$

In summary, for TEM waves, first find a scalar potential Φ which is a solution of

$$\nabla_t^2 \Phi(x, y) = 0 \tag{3.8a}$$

and satisfies the proper boundary conditions. The fields are then given by

$$\mathbf{E} = \mathbf{E}_t = \mathbf{e}e^{\mp jk_0 z} = -\nabla_t \Phi e^{\mp jk_0 z} \tag{3.8b}$$

$$\mathbf{H} = \mathbf{H}_t = \pm \mathbf{h} e^{\mp jk_0 z} = \pm Y_0 \mathbf{a}_z \times \mathbf{e} e^{\mp jk_0 z} \tag{3.8c}$$

where $k_0 = \omega(\mu_0 \epsilon_0)^{1/2}$, $Y_0 = (\epsilon_0/\mu_0)^{1/2}$, and $e^{-jk_0 z}$ represents a wave propagating in the $+z$ direction and $e^{jk_0 z}$ corresponds to wave propagation in the $-z$ direction. For TEM waves, Z_0 is the wave impedance, and from (3.8c) it is seen that, for wave propagation in the $+z$ direction,

$$\frac{E_x}{H_y} = -\frac{E_y}{H_x} = Z_0 \tag{3.9a}$$

whereas for propagation in the $-z$ direction,

$$\frac{E_x}{H_y} = -\frac{E_y}{H_x} = -Z_0 \tag{3.9b}$$

TE waves

For transverse electric (TE) waves, h_z plays the role of a potential function from which the rest of the field components may be obtained. The magnetic field $\mathbf{H}$ is a solution of

$$\nabla^2 \mathbf{H} + k_0^2 \mathbf{H} = 0$$

Separating the above equation into transverse and axial parts and replacing ∇^2 by $\nabla_t^2 - \beta^2$ yield

$$\nabla_t^2 h_z(x, y) + k_c^2 h_z(x, y) = 0 \tag{3.10a}$$

$$\nabla_t^2 \mathbf{h} + k_c^2 \mathbf{h} = 0 \tag{3.10b}$$

where $k_c^2 = k_0^2 - \beta^2$ and a z dependence $e^{-j\beta z}$ is assumed. Unlike the case of TEM waves, β^2 will not equal k_0^2 for TE waves. Instead, β is determined by the parameter k_c^2 in (3.10a). When this equation is solved, subject to appropriate boundary conditions, the eigenvalue k_c^2 will be found to be a function of the waveguide configuration.

The Maxwell equations (3.3) with $e_z = 0$ become

$$\nabla_t \times \mathbf{e} = -j\omega\mu_0 \mathbf{h}_z \tag{3.11a}$$

$$\beta \mathbf{a}_z \times \mathbf{e} = \omega\mu_0 \mathbf{h} \tag{3.11b}$$

$$\nabla_t \times \mathbf{h} = 0 \tag{3.11c}$$

$$\mathbf{a}_z \times \nabla_t h_z + j\beta \mathbf{a}_z \times \mathbf{h} = -j\omega\epsilon_0 \mathbf{e} \tag{3.11d}$$

$$\nabla_t \cdot \mathbf{h} = j\beta h_z \tag{3.11e}$$

$$\nabla_t \cdot \mathbf{e} = 0 \tag{3.11f}$$

The transverse curl of (3.11c) gives

$$\nabla_t \times (\nabla_t \times \mathbf{h}) = \nabla_t \nabla_t \cdot \mathbf{h} - \nabla_t^2 \mathbf{h} = 0$$

Replacing $\nabla_t \cdot \mathbf{h}$ by $j\beta h_z$ from (3.3e) and $\nabla_t^2 \mathbf{h}$ by $-k_c^2 \mathbf{h}$ from (3.10b) leads to the solution for $\mathbf{h}$ in terms of h_z; namely,

$$\mathbf{h} = -\frac{j\beta}{k_c^2} \nabla_t h_z \tag{3.12}$$

To find $\mathbf{e}$ in terms of $\mathbf{h}$, take the vector product of (3.11b) with $\mathbf{a}_z$ to obtain

$$\beta \mathbf{a}_z \times (\mathbf{a}_z \times \mathbf{e}) = \beta[(\mathbf{a}_z \cdot \mathbf{e})\mathbf{a}_z - (\mathbf{a}_z \cdot \mathbf{a}_z)\mathbf{e}] = -\beta \mathbf{e} = \omega \mu_0 \mathbf{a}_z \times \mathbf{h}$$

or

$$\mathbf{e} = -\frac{\omega \mu_0}{\beta} \mathbf{a}_z \times \mathbf{h} = -\frac{k_0}{\beta} Z_0 \mathbf{a}_z \times \mathbf{h} \tag{3.13}$$

The factor $k_0 Z_0/\beta$ has the dimensions of an impedance, and is called the wave impedance of TE, or H, modes. It will be designated by the symbol Z_h, so that

$$Z_h = \frac{k_0}{\beta} Z_0 \tag{3.14}$$

Thus, in component form, (3.13) gives

$$\frac{e_x}{h_y} = -\frac{e_y}{h_x} = Z_h \tag{3.15}$$

for a wave with z dependence $e^{-j\beta z}$.

The remaining equations in the set (3.11) do not yield any new results; so the solution for TE waves may be summarized as follows: First find a solution for h_z, where

$$\nabla_t^2 h_z + k_c^2 h_z = 0 \tag{3.16a}$$

Then

$$\mathbf{h} = -\frac{j\beta}{k_c^2} \nabla_t h_z \tag{3.16b}$$

and

$$\mathbf{e} = -Z_h \mathbf{a}_z \times \mathbf{h} \tag{3.16c}$$

where

$$\beta = (k_0^2 - k_c^2)^{\frac{1}{2}} \quad \text{and} \quad Z_h = \frac{k_0 Z_0}{\beta}$$

Complete expressions for the fields are

$$\mathbf{H} = \pm \mathbf{h} e^{\mp j\beta z} + \mathbf{h}_z e^{\mp j\beta z} \tag{3.16d}$$

$$\mathbf{E} = \mathbf{E}_t = \mathbf{e} e^{\mp j\beta z} \tag{3.16e}$$

Note that in (3.16d) the sign in front of $\mathbf{h}$ is reversed for a wave propagating in the $-z$ direction since $\mathbf{h}$ will be defined by (3.16b), with β positive regardless of whether propagation is in the $+z$ or $-z$ direction. The sign in front of **e** does not change since it involves the factor β twice, once in the expression for **h** and again in Z_h. Only the sign of one of **e** or **h** can change if a reversal in the direction of energy flow is to occur. That is, the solution for a wave propagating in the $-z$ direction can be chosen as $\mathbf{E} = -\mathbf{e}e^{j\beta z}$, $\mathbf{H} = (\mathbf{h} - \mathbf{h}_z)e^{j\beta z}$ or as $\mathbf{E} = \mathbf{e}e^{j\beta z}$, $\mathbf{H} = (-\mathbf{h} + \mathbf{h}_z)e^{j\beta z}$. One solution is the negative of the other. The latter solution is arbitrarily chosen as the standard in this text.

TM waves

The TM, or E, waves have $h_z = 0$, but the axial electric field e_z is not zero. These modes may be considered the dual of the TE modes in that the roles of electric and magnetic fields are interchanged. The derivation of the equations to be solved parallels that for TE waves, and hence only the final results will be given.

First obtain a solution for e_z, where

$$\nabla_t^2 e_z + k_c^2 e_z = 0 \tag{3.17a}$$

subject to the boundary conditions imposed. This will serve to determine the eigenvalue k_c^2. The transverse fields are then given by

$$\mathbf{E}_t = \mathbf{e}e^{\mp j\beta z} = -\frac{j\beta}{k_c^2}\nabla_t e_z e^{\mp j\beta z} \tag{3.17b}$$

$$\mathbf{H}_t = \pm \mathbf{h}e^{\mp j\beta z} = \pm Y_e \mathbf{a}_z \times \mathbf{e}e^{\mp j\beta z} \tag{3.17c}$$

where $\beta = (k_0^2 - k_c^2)^{\frac{1}{2}}$ and the wave admittance Y_e for TM waves is given by

$$Y_e = Z_e^{-1} = \frac{k_0}{\beta} Y_0 \tag{3.17d}$$

The dual nature of TE and TM waves is exhibited by the relation

$$Z_e Z_h = Z_0^2 \tag{3.18}$$

which holds when both types of waves have the same value of β and is derivable from (3.14) and (3.17d). The complete expression for the electric field is

$$\mathbf{E} = \mathbf{E}_t + \mathbf{E}_z = \mathbf{e}e^{\mp j\beta z} \pm \mathbf{e}_z e^{\mp j\beta z}$$

$$= \left(-\frac{j\beta}{k_c^2}\nabla_t e_z \pm \mathbf{e}_z\right)e^{\mp j\beta z} \tag{3.19}$$

It is convenient to keep the sign of **e** the same for propagation in either the $+z$ or $-z$ direction. Since $\nabla \cdot \mathbf{E} = 0$, that is, $\nabla_t \cdot \mathbf{E}_t + \partial E_z/\partial z = 0$, this requires that the z component of electric field be $-\mathbf{e}_z e^{j\beta z}$ for a wave propagating in the $-z$ direction, because $\nabla_t \cdot \mathbf{E}_t$ does not change sign, whereas $\partial E_z/\partial z$ does, in view of the change in sign in front of β. The transverse magnetic field must also change sign upon reversal of the direction of propagation in order to obtain a change in the direction of energy flow. For reference, this sign convention is summarized below. The transverse variations of the fields are represented by the functions $\mathbf{e}$, $\mathbf{h}$, $\mathbf{e}_z$, and $\mathbf{h}_z$, independent of the direction of propagation. Waves propagating in the $+z$ direction are then given by

$$\mathbf{E} = \mathbf{E}^+ = (\mathbf{e} + \mathbf{e}_z)e^{-j\beta z} \qquad (3.20a)$$

$$\mathbf{H} = \mathbf{H}^+ = (\mathbf{h} + \mathbf{h}_z)e^{-j\beta z} \qquad (3.20b)$$

For propagation in the $-z$ directions the fields are

$$\mathbf{E} = \mathbf{E}^- = (\mathbf{e} - \mathbf{e}_z)e^{j\beta z} \qquad (3.21a)$$

$$\mathbf{H} = \mathbf{H}^- = (-\mathbf{h} + \mathbf{h}_z)e^{j\beta z} \qquad (3.21b)$$

Additional superscripts $(+)$ or $(-)$ will be used when it is necessary to indicate the direction of propagation. The previously derived equations for TEM, TE, and TM modes are valid in a medium with electrical constants ϵ, μ, provided these are used to replace ϵ_0, μ_0. A finite conductivity can also be taken into account by making ϵ complex, i.e., replacing ϵ by $\epsilon - j\sigma/\omega$.

The wave impedance introduced in the solutions is an extremely useful concept in practice. The wave impedance is always chosen to relate the transverse components of the electric and magnetic fields. The sign is always such that if i, j, k is a cyclic labeling of the coordinates and propagation is along the positive direction of coordinate k, the ratio $E_i/H_j = (Z_w)_k$ is positive. Here $(Z_w)_k$ is the wave impedance referred to the k axis as the direction of propagation. If i, j, k form an odd permutation of the coordinates, then E_i/H_j is negative. The usefulness of the wave-impedance concept stems from the fact that the power is given in terms of the transverse fields alone. For example, for TE waves,

$$\begin{aligned} P &= \tfrac{1}{2}\operatorname{Re}\int_S \mathbf{E} \times \mathbf{H}^* \cdot \mathbf{a}_z\, dx\, dy \\ &= \tfrac{1}{2}\operatorname{Re}\int_S \mathbf{e} \times \mathbf{h}^* \cdot \mathbf{a}_z\, dx\, dy \\ &= -\tfrac{1}{2}\operatorname{Re}\int_S Z_h(\mathbf{a}_z \times \mathbf{h}) \times \mathbf{h}^* \cdot \mathbf{a}_z\, dx\, dy \\ &= \frac{Z_h}{2}\int_S \mathbf{h}\cdot\mathbf{h}^*\, dx\, dy = \frac{Y_h}{2}\int_S \mathbf{e}\cdot\mathbf{e}^*\, dx\, dy \end{aligned}$$

upon expanding the integrand. Thus the wave impedance enables the power transmitted to be expressed in terms of one of the transverse fields alone. A further property of the wave impedance, which will be dealt with later, is that it provides a basis for an analogy between conventional multiconductor transmission lines and waveguides.

3.2 Transmission lines (field analysis)

Lossless transmission line

A transmission line consists of two or more parallel conductors. Typical examples are the two-conductor line, shielded two-conductor line, and coaxial line with cross sections, as illustrated in Fig. 3.1. Initially, it will be assumed that the conductors are perfectly conducting and that the medium surrounding the conductors is air, with $\epsilon \approx \epsilon_0$, $\mu \approx \mu_0$. The effect of small losses will be considered later.

With reference to Fig. 3.2, let the one conductor be at a potential $V_0/2$ and the other conductor at $-V_0/2$. To determine the field of a TEM wave, a suitable potential $\Phi(x, y)$ must be found first. It is necessary that Φ be a solution of

$$\nabla_t^2 \Phi = 0$$

and satisfy the boundary conditions

$$\Phi = \begin{cases} \dfrac{V_0}{2} & \text{on } S_2 \\ -\dfrac{V_0}{2} & \text{on } S_1 \end{cases}$$

Since Φ is unique only to within an additive constant, we could equally

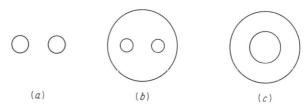

Fig. 3.1 Cross sections of typical transmission lines. (a) Two-conductor line; (b) shielded two-conductor line; (c) coaxial line.

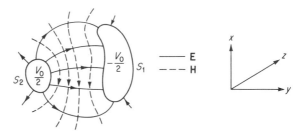

Fig. 3.2 Cross section of a general two-conductor line showing transverse field patterns.

well choose $\Phi = V_0$ on S_2 and $\Phi = 0$ on S_1. If a solution for Φ is possible, a TEM mode or field solution is also possible. When two or more conductors are present, this is always the case. The solution for Φ is an electrostatic problem that can be solved when the line configuration is simple enough, as exemplified in Fig. 3.1.

The fields are given by (3.8), and for propagation in the $+z$ direction are

$$\mathbf{E} = \mathbf{E}_t = \mathbf{e}e^{-jk_0z} = -\nabla_t\Phi e^{-jk_0z} \tag{3.22a}$$

$$\mathbf{H} = \mathbf{H}_t = Y_0\mathbf{a}_z \times \mathbf{e}e^{-jk_0z} \tag{3.22b}$$

The line integral of $\mathbf{e}$ between the two conductors is

$$\int_{S_1}^{S_2} \mathbf{e} \cdot d\mathbf{l} = \int_{S_1}^{S_2} -\nabla_t\Phi \cdot d\mathbf{l}$$

$$= -\int_{S_1}^{S_2} \frac{d\Phi}{dl} dl = -[\Phi(S_2) - \Phi(S_1)] = -V_0$$

Associated with the electric field is a unique voltage wave

$$V = V_0 e^{-jk_0 z} \tag{3.23}$$

since the line integral of $\mathbf{e}$ between S_1 and S_2 is independent of the path chosen because $\mathbf{e}$ is the gradient of a scalar potential.

The line integral of $\mathbf{h}$ around one conductor, say S_2, gives

$$\oint_{S_2} \mathbf{h} \cdot d\mathbf{l} = \oint_{S_2} J_s \, dl = I_0$$

by application of Ampère's law, $\nabla \times \mathbf{H} = j\omega\mathbf{D} + \mathbf{J}$, and noting that there is no axial displacement flux D_z for a TEM mode. On the conductors the boundary conditions require $\mathbf{n} \times \mathbf{e} = 0$ and $\mathbf{n} \times \mathbf{h} = \mathbf{J}_s$, where $\mathbf{n}$ is a unit outward normal and $\mathbf{J}_s$ is the surface current density. Since $\mathbf{n}$ and $\mathbf{h}$ lie in a transverse plane, the current $\mathbf{J}_s$ is in the axial direction. In the region remote from the conductors, $\nabla_t \times \mathbf{h} = 0$, but the line integral around a conductor is not zero because of the current that exists. The current on the two conductors is oppositely directed, as may be verified from the expression $\mathbf{n} \times \mathbf{h} = \mathbf{J}_s$. Associated with the magnetic field there is a unique current wave

$$I = I_0 e^{-jk_0 z} \tag{3.24}$$

Since the potential Φ is independent of frequency, it follows that the transverse fields $\mathbf{e}$ and $\mathbf{h}$ are also independent of frequency and are, in actual fact, the static field distributions which exist between the conductors if the potential difference is V_0 and currents I_0, $-I_0$ exist on S_2, S_1, respectively. The magnetic lines of flux coincide with the equipotential lines, since $\mathbf{e}$ and $\mathbf{h}$ are orthogonal, as seen from (3.22b).

Example 3.1 *Coaxial line* Figure 3.3 illustrates a coaxial transmission line for which the solution for a TEM mode will be constructed. In

Sec. 3.2   Transmission lines and waveguides   75

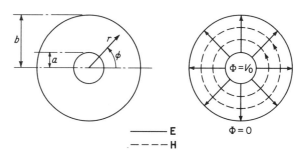

Fig. 3.3 Coaxial transmission line.

cylindrical coordinates r, ϕ, z, the two-dimensional Laplace equation is

$$\frac{1}{r}\frac{\partial}{\partial r}\left(r\frac{\partial \Phi}{\partial r}\right) + \frac{1}{r^2}\frac{\partial^2 \Phi}{\partial \phi^2} = 0$$

or for a potential function independent of the angular coordinate ϕ,

$$\frac{1}{r}\frac{\partial}{\partial r}\left(r\frac{\partial \Phi}{\partial r}\right) = 0$$

Integrating this equation twice gives

$$\Phi = C_1 \ln r + C_2$$

Imposing the boundary conditions $\Phi = V_0$ at $r = a$, $\Phi = 0$ at $r = b$, gives

$$V_0 = C_1 \ln a + C_2 \qquad 0 = C_1 \ln b + C_2$$

and hence $C_2 = -C_1 \ln b$, $C_1 = V_0/[\ln (a/b)]$,

$$\Phi = V_0 \frac{\ln (r/b)}{\ln (a/b)} \tag{3.25}$$

The electric and magnetic fields of a TEM mode propagating in the $+z$ direction are given by (3.22) and are

$$\mathbf{E} = -\mathbf{a}_r \frac{\partial \Phi}{\partial r} e^{-jk_0 z} = -\frac{V_0}{\ln (a/b)} \frac{\mathbf{a}_r}{r} e^{-jk_0 z}$$

$$= \frac{V_0}{\ln (b/a)} \frac{\mathbf{a}_r}{r} e^{-jk_0 z} \tag{3.26a}$$

$$\mathbf{H} = Y_0 \mathbf{a}_z \times \mathbf{e} e^{-jk_0 z} = \frac{Y_0 V_0}{\ln (b/a)} \frac{\mathbf{a}_\phi}{r} e^{-jk_0 z} \tag{3.26b}$$

The potential difference between the two conductors is obviously V_0; so the voltage wave associated with the electric field is

$$V = V_0 e^{-jk_0 z} \tag{3.27}$$

The current density on the inner conductor is

$$\mathbf{J}_s = \mathbf{n} \times \mathbf{H} = \mathbf{a}_r \times \mathbf{H} = \frac{Y_0 V_0}{\ln (b/a)} \frac{\mathbf{a}_z}{a} e^{-jk_0 z}$$

The total current, apart from the factor $e^{-jk_0 z}$, is

$$I_0 = \frac{Y_0 V_0}{a \ln (b/a)} \int_0^{2\pi} a \, d\phi = \frac{Y_0 V_0 2\pi}{\ln (b/a)} \qquad (3.28)$$

The current on the inner surface of the outer conductor is readily shown to be equal to I_0 also, but directed in the $-z$ direction. The current wave associated with the magnetic field is therefore

$$I = I_0 e^{-jk_0 z} \qquad (3.29)$$

The power, or rate of energy flow, along the line is given by

$$P = \frac{1}{2} \operatorname{Re} \int_a^b \int_0^{2\pi} \mathbf{E} \times \mathbf{H}^* \cdot \mathbf{a}_z r \, dr \, d\phi = \frac{1}{2} \frac{Y_0 V_0^2}{[\ln (b/a)]^2} \int_a^b \int_0^{2\pi} \frac{d\phi \, dr}{r}$$

$$= \frac{\pi Y_0 V_0^2}{\ln (b/a)} \qquad (3.30)$$

The power transmitted is seen to be also given, as anticipated, by the expression

$$\tfrac{1}{2} \operatorname{Re} (VI^*) = \tfrac{1}{2} V_0 I_0 = \tfrac{1}{2} V_0^2 \frac{2\pi Y_0}{\ln (b/a)}$$

The characteristic impedance of the line is defined by the ratio

$$\frac{V_0}{I_0} = Z_c \qquad (3.31)$$

in terms of which the power may be expressed as $P = \tfrac{1}{2} Z_c I_0^2 = \tfrac{1}{2} Y_c V_0^2$, where Y_c is the characteristic admittance of the line and equal to Z_c^{-1}. The characteristic impedance is a function of the cross-sectional shape of the transmission line.

Transmission line with small losses

Practical transmission lines always have some loss caused by the finite conductivity of the conductors and also loss that may be present in the dielectric material surrounding the conductors. Consider first the case when the conductors are surrounded by a dielectric with permittivity $\epsilon = \epsilon' - j\epsilon''$ but the conductors are still considered to be perfect. The presence of a lossy dielectric does not affect the solution for the scalar potential Φ. Consequently, the field solution is formally the same as for the ideal line, except that k_0 and Y_0 are replaced by $k = k_0(\kappa' - j\kappa'')^{\frac{1}{2}}$ and $Y = Y_0(\kappa' - j\kappa'')^{\frac{1}{2}}$, where the dielectric constant $\kappa = \kappa' - j\kappa'' = \epsilon/\epsilon_0$.

For small losses such that $\kappa'' \ll \kappa'$, the propagation constant is

$$jk = \alpha + j\beta = j(\kappa')^{\frac{1}{2}}k_0\left(1 - j\frac{\kappa''}{\kappa'}\right)^{\frac{1}{2}} \approx j(\kappa')^{\frac{1}{2}}k_0 + \frac{\kappa'' k_0}{2(\kappa')^{\frac{1}{2}}}$$

Thus

$$\alpha = \frac{\kappa'' k_0}{2(\kappa')^{\frac{1}{2}}} \qquad (3.32a)$$

$$\beta = (\kappa')^{\frac{1}{2}}k_0 \qquad (3.32b)$$

where α is the attenuation constant and β is the phase constant. The wave consequently attenuates according to $e^{-\alpha z}$ as it propagates in the $+z$ direction.

It will be instructive to derive the above expression for α by means of a perturbation method that is widely used in the evaluation of the attenuation, or damping, factor for a low-loss physical system. This method is based on the assumption that the introduction of a small loss does not substantially perturb the field from its loss-free value. The known field distribution for the loss-free case is then used to evaluate the loss in the system, and from this the attenuation constant can be calculated. In the present case, if $\kappa'' = 0$, the loss-free solution is

$$\mathbf{E} = -\nabla_t \Phi e^{-jkz} \qquad \mathbf{H} = Y\mathbf{a}_z \times \mathbf{E}$$

where $k = (\kappa')^{\frac{1}{2}}k_0$ and $Y = (\kappa')^{\frac{1}{2}}Y_0$. When κ'' is small but not zero, the imaginary part of ϵ, that is, ϵ'', is equivalent to a conductivity

$$\sigma = \omega\epsilon'' = \omega\epsilon_0 \kappa''$$

A conductivity σ results in a shunt current $\mathbf{J} = \sigma \mathbf{E}$ between the two conductors. The power loss per unit length of line is

$$P_l = \frac{1}{2\sigma}\int_S \mathbf{J} \cdot \mathbf{J}^* \, dS = \frac{\omega\epsilon''}{2}\int_S \mathbf{E} \cdot \mathbf{E}^* \, dS \qquad (3.33)$$

where the integration is over the cross section of the line, and the loss-free solution for $\mathbf{E}$ is used to carry out the evaluation of P_l. Since loss is present, the power propagated along the line must decrease according to a factor $e^{-2\alpha z}$. The rate of decrease of power propagated along the line equals the power loss. If the power at $z = 0$ is P_0, then at z it is $P = P_0 e^{-2\alpha z}$. Consequently,

$$-\frac{\partial P}{\partial z} = P_l = 2\alpha P_0 e^{-2\alpha z} = 2\alpha P \qquad (3.34)$$

which states that the power loss at any plane z is directly proportional to the total power P present at this plane. The power propagated along the line is given by

$$P = \tfrac{1}{2} \operatorname{Re} \int_S \mathbf{E} \times \mathbf{H}^* \cdot \mathbf{a}_z \, dS$$
$$= \frac{Y}{2} \operatorname{Re} \int_S \mathbf{E} \times (\mathbf{a}_z \times \mathbf{E}^*) \cdot \mathbf{a}_z \, dS = \frac{Y}{2} \int_S \mathbf{E} \cdot \mathbf{E}^* \, dS$$

Hence the attenuation α is given by

$$\alpha = \frac{P_l}{2P} = \frac{\sigma}{2Y} = \frac{\omega \epsilon''}{2Y_0(\kappa')^{\frac{1}{2}}} = k_0 \frac{\kappa''}{2(\kappa')^{\frac{1}{2}}}$$

which is the same as the expression (3.32a). For this example the perturbation method does not offer any advantage. However, often the field solution for the lossy case is very difficult to find, in which case the perturbation method is extremely useful and simple to carry out by comparison with other methods. The case of transmission lines with conductors having finite conductivity is an important example of this, and is discussed below.

If the conductors of a transmission line have a finite conductivity, they exhibit a surface impedance

$$Z_m = \frac{1+j}{\sigma \delta_s} \tag{3.35}$$

where $\delta_s = (2/\omega \mu \sigma)^{\frac{1}{2}}$ is the skin depth (Sec. 2.9). At the surface the electric field must have a tangential component equal to $Z_m \mathbf{J}_s$, where $\mathbf{J}_s$ is the surface current density. Therefore it is apparent that an axial component of electric field must be present, and consequently the field is no longer that of a TEM wave. The axial component of electric field gives rise to a component of the Poynting vector directed into the conductor, and this accounts for the power loss in the conductor. Generally, it is very difficult to find the exact solution for the fields when the conductors have finite conductivity. However, since $|Z_m|$ is very small compared with Z_0, the axial component of electric field is also very small relative to the transverse components. Thus the field is very nearly that of the TEM mode in the loss-free case. The perturbation method outlined earlier may be used to evaluate the attenuation caused by finite conductivity.

The current density $\mathbf{J}_s$ is taken equal to $\mathbf{n} \times \mathbf{H}$, where $\mathbf{n}$ is the unit outward normal to the conductor surface and $\mathbf{H}$ is the *loss-free* magnetic field. The power loss in the surface impedance per unit length of line is

$$P_l = \tfrac{1}{2} \operatorname{Re} Z_m \oint_{S_1 + S_2} \mathbf{J}_s \cdot \mathbf{J}_s^* \, dl$$
$$= \frac{R_m}{2} \oint_{S_1 + S_2} (\mathbf{n} \times \mathbf{H}) \cdot (\mathbf{n} \times \mathbf{H}^*) \, dl$$
$$= \frac{R_m}{2} \oint_{S_1 + S_2} \mathbf{H} \cdot \mathbf{H}^* \, dl \tag{3.36}$$

where $R_m = 1/\sigma\delta_s$ is the high-frequency surface resistance, and

$$(\mathbf{n} \times \mathbf{H}) \cdot (\mathbf{n} \times \mathbf{H}^*) = \mathbf{n} \cdot \mathbf{H} \times (\mathbf{n} \times \mathbf{H}^*)$$
$$= \mathbf{n} \cdot [(\mathbf{H} \cdot \mathbf{H}^*)\mathbf{n} - (\mathbf{H} \cdot \mathbf{n})\mathbf{H}^*] = \mathbf{H} \cdot \mathbf{H}^*$$

since $\mathbf{n} \cdot \mathbf{H} = 0$ for the infinite-conductivity case. The integration is taken around the periphery $S_1 + S_2$ of the two conductors. The attenuation constant arising from conductor loss is thus

$$\alpha = \frac{P_l}{2P} = \frac{R_m \oint_{S_1+S_2} \mathbf{H} \cdot \mathbf{H}^* \, dl}{2Z \int \mathbf{H} \cdot \mathbf{H}^* \, dS} \tag{3.37}$$

where the power propagated along the line is given by

$$\operatorname{Re} \tfrac{1}{2} \int \mathbf{E} \times \mathbf{H}^* \cdot \mathbf{a}_z \, dS = \tfrac{1}{2} Z \int \mathbf{H} \cdot \mathbf{H}^* \, dS$$

and Z is the intrinsic impedance of the medium; that is, $Z = (\mu/\epsilon)^{\frac{1}{2}}$.

When both dielectric and conductor losses are present, the attenuation constant is the sum of the attenuation constants arising from each cause, provided both attenuation constants are small.

Example 3.2 Lossy coaxial line Let the coaxial line in Fig. 3.3 be filled with a lossy dielectric ($\epsilon = \epsilon' - j\epsilon''$), and let the conductors have finite conductivity σ. For the loss-free case ($\epsilon'' = 0, \sigma = \infty$) the fields are given by (3.26), with k_0 and Y_0 replaced by $k = (\epsilon'/\epsilon_0)^{\frac{1}{2}} k_0$,

$$Y = \left(\frac{\epsilon'}{\epsilon_0}\right)^{\frac{1}{2}} Y_0$$

Thus

$$\mathbf{E} = \frac{V_0}{\ln(b/a)} \frac{\mathbf{a}_r}{r} e^{-jkz} \tag{3.38a}$$

$$\mathbf{H} = \frac{YV_0}{\ln(b/a)} \frac{\mathbf{a}_\phi}{r} e^{-jkz} \tag{3.38b}$$

The power propagated along the line is

$$P = \tfrac{1}{2} \operatorname{Re} \int_0^{2\pi} \int_a^b \mathbf{E} \times \mathbf{H}^* \cdot \mathbf{a}_z r \, d\phi \, dr = \frac{\pi Y V_0^2}{\ln(b/a)} \tag{3.39}$$

The power loss P_{l1} from the lossy dielectric is, from (3.33),

$$P_{l1} = \frac{\omega\epsilon''}{2} \int_0^{2\pi} \int_a^b \mathbf{E} \cdot \mathbf{E}^* r \, d\phi \, dr = \frac{\omega\epsilon'' V_0^2 \pi}{\ln(b/a)} \tag{3.40a}$$

The power loss from finite conductivity is given by (3.36), and is

$$P_{l2} = \frac{R_m}{2} \frac{Y^2 V_0^2}{[\ln(b/a)]^2} \int_0^{2\pi} \left(\frac{1}{a} + \frac{1}{b}\right) d\phi$$

$$= \frac{R_m \pi Y^2 V_0^2}{[\ln(b/a)]^2} \frac{b+a}{ab} \tag{3.40b}$$

Hence the attenuation constant α for the coaxial line is given by

$$\alpha = \frac{P_{l1} + P_{l2}}{2P} = \frac{\omega \epsilon''}{2Y} + \frac{R_m Y}{2 \ln (b/a)} \frac{b+a}{ab}$$

$$= \frac{k_0 \kappa''}{2(\kappa')^{\frac{1}{2}}} + \frac{R_m}{2Z \ln (b/a)} \frac{b+a}{ab} \tag{3.41}$$

For the lossy case the propagation constant is consequently taken as

$$\alpha + j\beta = \alpha + jk$$

with α given by (3.41).

3.3 Transmission lines (distributed-circuit analysis)

In the previous section it was shown that a unique voltage and current wave was associated with the electric and magnetic fields of a TEM mode on a transmission line. Also, the transverse fields of a TEM mode have a transverse variation with the coordinates that is the same as for static fields. For these reasons the transmission line can be described in a unique manner as a distributed-parameter electric network. Energy storage in the magnetic field is accounted for by the series inductance L per unit length, whereas energy storage in the electric field is accounted for by the distributed shunt capacitance C per unit length. Power loss in the conductors is taken into account by a series resistance R per unit length. Finally, the power loss in the dielectric may be included by introducing a shunt conductance G per unit length. Suitable definitions for the parameters L, C, R, and G based on the above concepts are

$$L = \frac{\mu}{I_0 I_0^*} \int_S \mathbf{H} \cdot \mathbf{H}^* \, dS \tag{3.42a}$$

$$C = \frac{\epsilon'}{V_0 V_0^*} \int_S \mathbf{E} \cdot \mathbf{E}^* \, dS \tag{3.42b}$$

$$R = \frac{R_m}{I_0 I_0^*} \oint_{S_1 + S_2} \mathbf{H} \cdot \mathbf{H}^* \, dl \tag{3.42c}$$

$$G = \frac{\omega \epsilon''}{V_0 V_0^*} \int \mathbf{E} \cdot \mathbf{E}^* \, dS \tag{3.42d}$$

where I_0 is the total current on the line, and V_0 the potential difference. These expressions are obtained, for example, by equating the magnetic energy $\frac{1}{4} I_0 I_0^* L = W_m$ stored in the equivalent series inductance L to the expression for W_m in terms of the field. The above definitions are readily

Table 3.1 Parameters of common transmission lines†

	Z_c	R
(two parallel wires, spacing D, diameter d)	$\dfrac{1}{\pi}\left(\dfrac{\mu_0}{\epsilon'}\right)^{\frac{1}{2}} \cosh^{-1}\dfrac{D}{d}$	$\dfrac{2R_m}{\pi d} \dfrac{D/d}{[(D/d)^2 - 1]^{\frac{1}{2}}}$
(coaxial, inner radius a, outer b)	$\dfrac{1}{2\pi}\left(\dfrac{\mu_0}{\epsilon'}\right)^{\frac{1}{2}} \ln\dfrac{b}{a}$	$\dfrac{R_m}{2\pi}\left(\dfrac{1}{a} + \dfrac{1}{b}\right)$
(shielded pair) $p = \dfrac{D}{d}$, $q = \dfrac{D}{a}$	$\dfrac{1}{\pi}\left(\dfrac{\mu_0}{\epsilon'}\right)^{\frac{1}{2}}\left[\ln\left(2p\dfrac{1-q^2}{1+q^2}\right) - \dfrac{1+4p^2}{16p^4}(1-4q^2)\right]$	$\dfrac{2R_m}{\pi d}\left[1 + \dfrac{1+2p^2}{4p^4}(1-4q^2)\right] + \dfrac{8R_m}{\pi a} q^2\left[(1+q^2) - \dfrac{1+4p^2}{8p^4}\right]$

† For all TEM transmission lines

$$C = \frac{(\mu_0\epsilon')^{\frac{1}{2}}}{Z_c} \qquad L = (\mu_0\epsilon')^{\frac{1}{2}} Z_c \qquad G = \frac{\omega\epsilon''C}{\epsilon'}$$

$$\alpha_d = \frac{GZ_c}{2} \qquad \alpha_c = \frac{RY_c}{2} \qquad R_m = \frac{1}{\sigma\delta_s} = \left(\frac{\omega\mu}{2\sigma}\right)^{\frac{1}{2}}$$

shown to be equivalent to the other commonly used definitions such as†

$$L = \frac{\text{magnetic flux linkage}}{\text{total current}} \qquad (3.43a)$$

$$C = \frac{\text{total charge per unit length}}{\text{voltage difference between conductors}} \qquad (3.43b)$$

$$G = \frac{\text{total shunt current}}{\text{voltage difference between conductors}} \qquad (3.43c)$$

The series resistance R is most conveniently defined on an energy basis as in (3.42c) since the current density $\mathbf{J}_s = \mathbf{n} \times \mathbf{H}$ is not always uniformly distributed around the conductor periphery. Parameters of some common transmission lines are given in Table 3.1.

The equivalent circuit of a section of transmission line of differential length dz is illustrated in Fig. 3.4. If the voltage and current at the input are $\mathcal{U}(z, t)$, $\mathcal{I}(z, t)$ and at the output are

$$\mathcal{U} + \frac{\partial \mathcal{U}}{\partial z} dz \qquad \mathcal{I} + \frac{\partial \mathcal{I}}{\partial z} dz$$

† See, for example, R. Plonsey and R. E. Collin, "Principles and Applications of Electromagnetic Fields," sec. 10.5, McGraw-Hill Book Company, New York, 1961.

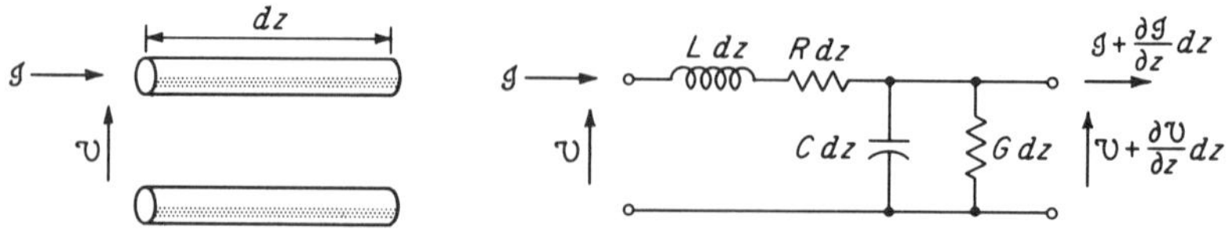

Fig. 3.4 Equivalent circuit of a differential length of transmission line.

then Kirchhoff's laws give

$$\mathcal{V} - \left(\mathcal{V} + \frac{\partial \mathcal{V}}{\partial z} dz\right) = \mathcal{I} R\, dz + L\, dz\, \frac{\partial \mathcal{I}}{\partial t}$$

or

$$\frac{\partial \mathcal{V}}{\partial z} = -\mathcal{I} R - L \frac{\partial \mathcal{I}}{\partial t} \tag{3.44a}$$

Similarly,

$$\mathcal{I} - \left(\mathcal{I} + \frac{\partial \mathcal{I}}{\partial z} dz\right) = \mathcal{V} G\, dz + C\, dz\, \frac{\partial \mathcal{V}}{\partial t}$$

or

$$\frac{\partial \mathcal{I}}{\partial z} = -\mathcal{V} G - C \frac{\partial \mathcal{V}}{\partial t} \tag{3.44b}$$

The first equation states that the potential difference between the input and output is equal to the potential drop across R and L. The second equation states that the output current is less than the input current by an amount equal to the shunt current flowing through C and G. Differentiating (3.44a) with respect to z and (3.44b) with respect to time t gives

$$\frac{\partial^2 \mathcal{V}}{\partial z^2} = -R \frac{\partial \mathcal{I}}{\partial z} - L \frac{\partial^2 \mathcal{I}}{\partial t\, \partial z} \tag{3.45a}$$

$$\frac{\partial^2 \mathcal{I}}{\partial t\, \partial z} = -G \frac{\partial \mathcal{V}}{\partial t} - C \frac{\partial^2 \mathcal{V}}{\partial t^2} \tag{3.45b}$$

Using (3.44b) and (3.45b) in (3.45a) now gives the following equation for the line voltage $\mathcal{V}$:

$$\frac{\partial^2 \mathcal{V}}{\partial z^2} = R\left(G\mathcal{V} + C \frac{\partial \mathcal{V}}{\partial t}\right) + L\left(G \frac{\partial \mathcal{V}}{\partial t} + C \frac{\partial^2 \mathcal{V}}{\partial t^2}\right)$$

or

$$\frac{\partial^2 \mathcal{V}}{\partial z^2} - (RC + LG) \frac{\partial \mathcal{V}}{\partial t} - LC \frac{\partial^2 \mathcal{V}}{\partial t^2} - RG\mathcal{V} = 0 \tag{3.46}$$

The current $\mathcal{I}$ satisfies this one-dimensional wave equation also. If a solution in the form of a propagating wave

$$\mathcal{V} = \text{Re}\,(Ve^{-\gamma z + j\omega t})$$

is assumed, substitution into (3.46) shows that the propagation constant γ must be a solution of

$$\gamma^2 - j\omega(RC + LG) + \omega^2 LC - RG = 0 \qquad (3.47)$$

If only the steady-state sinusoidally time-varying solution is desired, phasor notation may be used. If we let V and I represent the voltage and current without the time dependence $e^{j\omega t}$, the basic equations (3.44) may be written as

$$\frac{\partial V}{\partial z} = -(R + j\omega L)I \qquad (3.48a)$$

$$\frac{\partial I}{\partial z} = -(G + j\omega C)V \qquad (3.48b)$$

The wave equation (3.46) becomes

$$\frac{\partial^2 V}{\partial z^2} - (RG - \omega^2 LC)V - j\omega(RC + LG)V = 0 \qquad (3.49)$$

The general solution to (3.49) is

$$V = V^+ e^{-\gamma z} + V^- e^{\gamma z} \qquad (3.50)$$

where $\gamma = \alpha + j\beta$ is given by

$$\gamma = [-\omega^2 LC + RG + j\omega(RC + LG)]^{\frac{1}{2}} \qquad (3.51)$$

from (3.47). The constants V^+ and V^- are arbitrary amplitude constants for waves propagating in the $+z$ and $-z$ directions, respectively. The solution for the current I may be found from (3.48a), and is

$$I = I^+ e^{-\gamma z} - I^- e^{+\gamma z} = \frac{\gamma}{R + j\omega L}(V^+ e^{-\gamma z} - V^- e^{\gamma z}) \qquad (3.52)$$

The parameter

$$Z_c = \frac{R + j\omega L}{\gamma} = \left(\frac{R + j\omega L}{G + j\omega C}\right)^{\frac{1}{2}} \qquad (3.53)$$

is called the characteristic impedance of the line since it is equal to the ratio V^+/I^+ and V^-/I^-. Note that $\gamma = [(R + j\omega L)(G + j\omega C)]^{\frac{1}{2}}$.

Loss-free transmission line

For a line without loss, i.e., for which $R = G = 0$, the propagation constant is

$$\gamma = j\beta = j\omega\sqrt{LC} \qquad (3.54)$$

and the characteristic impedance is pure real and given by

$$Z_c = \sqrt{\frac{L}{C}} \tag{3.55}$$

According to the field analysis, β is also equal to $\omega(\mu\epsilon)^{1/2}$, and hence

$$LC = \mu\epsilon \tag{3.56}$$

for a transmission line. This result may also be verified from the solutions for L and C, as shown below in the section on transmission-line parameters. Using (3.56) in (3.55) shows that the characteristic impedance is also given by

$$Z_c = \sqrt{\frac{L}{C}} = \sqrt{\frac{\mu\epsilon}{C^2}} = \frac{\epsilon}{C}\sqrt{\frac{\mu}{\epsilon}} = Z\frac{\epsilon}{C} \tag{3.57}$$

where Z is the intrinsic impedance of the medium. The characteristic impedance differs from the intrinsic impedance Z by a factor ϵ/C, which is a function of the line configuration only.

Low-loss transmission line

For most microwave transmission lines the losses are very small; that is, $R \ll \omega L$ and $G \ll \omega C$. When this is the case, the term RG in the expression (3.51) for γ may be neglected. A binomial expansion then gives

$$\gamma \approx j\omega\sqrt{LC} + \tfrac{1}{2}\sqrt{LC}\left(\frac{R}{L} + \frac{G}{C}\right) = \alpha + j\beta \tag{3.58}$$

To first order the characteristic impedance is still given by (3.55) or (3.57). Thus the phase constant for a low-loss line is

$$\beta = \omega\sqrt{LC} \tag{3.59a}$$

and the attenuation constant α is

$$\alpha = \tfrac{1}{2}\sqrt{LC}\left(\frac{R}{L} + \frac{G}{C}\right) = \tfrac{1}{2}(RY_c + GZ_c) \tag{3.59b}$$

where $Y_c = Z_c^{-1} = \sqrt{C/L}$ is the characteristic admittance of the transmission line.

★3.4 Transmission-line parameters

In this section the field analysis to determine the circuit parameters L, R, C, and G for a transmission line is examined in greater detail. This will serve further to correlate the field analysis and circuit analysis of transmission lines.

Consider first the case of a loss-free line such as that illustrated in Fig. 3.2. When the scalar potential Φ has been determined, the charge density on the conductors may be found from the normal component of electric field at the surface; that is, $\rho_s = \epsilon \mathbf{n} \cdot \mathbf{e} = -\epsilon \mathbf{n} \cdot \nabla \Phi = -\epsilon \, \partial \Phi / \partial n$, where ϵ is the permittivity of the medium surrounding the conductors. The total charge Q per unit length on conductor S_2 is

$$Q = \oint_{S_2} \epsilon \mathbf{n} \cdot \mathbf{e} \, dl$$

The total charge on the conductor S_1 is $-Q$ per meter. The potential of S_2 is V_0, and hence the capacitance C per unit length is

$$C = \frac{Q}{V_0} = \frac{\epsilon \int_{S_2} \mathbf{n} \cdot \mathbf{e} \, dl}{\int_{S_2}^{S_1} \mathbf{e} \cdot d\mathbf{l}} \tag{3.60}$$

The total current on S_2 is

$$I_0 = \oint_{S_2} \mathbf{h} \cdot d\mathbf{l} = \oint_{S_2} Y \mathbf{n} \cdot \mathbf{e} \, dl = \frac{YQ}{\epsilon}$$

since $|\mathbf{h}| = Y|\mathbf{e}| = Y \mathbf{n} \cdot \mathbf{e}$ at the surface of S_2 because the normal component of $\mathbf{h}$ and the tangential component of $\mathbf{e}$ are zero at the perfectly conducting surface S_2. The characteristic impedance of the line is given by

$$Z_c = \frac{V_0}{I_0} = \frac{V_0 \epsilon}{YQ} = \frac{\epsilon Z}{C} \tag{3.61}$$

A knowledge of the capacitance per unit length suffices to determine the characteristic impedance.

To determine the inductance L per unit length, refer to Fig. 3.5, which illustrates the magnetic flux lines around the conductors. Since $\mathbf{h}$ is orthogonal to $\mathbf{e}$, these coincide with the equipotential lines. All the flux lines from the $\Phi = 0$ to the $\Phi = V_0/2$ line link the current on S_2. The flux linkage is the total flux cutting any path joining the $\Phi = 0$ line to the surface S_2. If a path such as $P_1 S_2$ or $P_2 S_2$ is chosen, which is orthogonal to the flux lines, this path coincides with a line of electric force. The

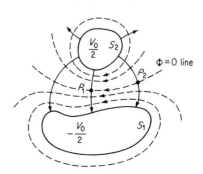

Fig. 3.5 Magnetic flux lines in a transmission line.

flux cutting such a path is

$$\psi = \int_{P_1}^{S_2} \mu h \, dl = \mu Y \int_{P_1}^{S_2} -\mathbf{e} \cdot d\mathbf{l} = \mu Y \frac{V_0}{2}$$

since $|\mathbf{h}| = Y|\mathbf{e}|$ for a TEM wave. The inductance of one conductor of the line is

$$L_1 = \frac{\psi}{I_0} = \mu Y \frac{V_0}{2I_0}$$

The inductance of both conductors per unit length is twice this value; so

$$L = \mu Y \frac{V_0}{I_0} = \mu Y Z_c \tag{3.62}$$

From this relation and (3.61) it is seen that $Z = \mu Z_c/L = CZ_c/\epsilon$, and hence

$$Z^2 = \frac{\mu}{\epsilon} = \frac{\mu Z_c}{L} \frac{CZ_c}{\epsilon}$$

which gives

$$Z_c = \sqrt{\frac{L}{C}} \tag{3.63}$$

Equations (3.61) and (3.62) also show that

$$\mu\epsilon = LC \tag{3.64}$$

for a transmission line. The above expressions for C and L can also be obtained from the definitions based on stored energy. The derivation is left as a problem.

If the dielectric has a complex permittivity $\epsilon = \epsilon' - j\epsilon''$, where ϵ'' includes the conductivity of the dielectric if it is not zero, the total shunt current consists of a displacement current I_D and a conduction current I_S. The current leaving conductor S_2 per unit length is

$$I = I_D + I_S = j\omega\epsilon \oint_{S_2} \mathbf{e} \cdot \mathbf{n} \, dl = j\omega\epsilon' \oint_{S_2} \mathbf{e} \cdot \mathbf{n} \, dl + \omega\epsilon'' \oint_{S_2} \mathbf{e} \cdot \mathbf{n} \, dl$$

where the first integral on the right gives the displacement current and the second integral gives the conduction current. The total shunt admittance is given by $Y = j\omega C + G = (I_S + I_D)/V_0$, and hence it is seen that

$$G = \frac{I_S}{V_0} = \frac{I_S}{I_D} \frac{I_D}{V_0} = \frac{\omega\epsilon''}{\epsilon'} C \tag{3.65}$$

since $j\omega C = I_D/V_0$ and $j\omega C/j\omega\epsilon' = C/\epsilon'$. This relation shows that G differs from C by the factor $\omega\epsilon''/\epsilon'$ only.

The transmission-line loss from finite conductivity may be accounted for by a series resistance R per unit length provided R is chosen so that

$$\tfrac{1}{2}RI_0^2 = \frac{R_m}{2}\oint_{S_1+S_2}|\mathbf{h}|^2\,dl \tag{3.66}$$

The right-hand side gives the total power loss per unit length arising from the high-frequency resistance of the conductors. In terms of this quantity, the resistance R is thus defined as

$$R = R_m\frac{\oint_{S_1+S_2}|\mathbf{h}|^2\,dl}{\left(\oint_{S_2}|\mathbf{h}|\,dl\right)^2} \tag{3.67}$$

where $R_m = 1/\sigma\delta_s$ and δ_s is the skin depth.

A further effect of the finite conductivity is to increase the series inductance of the line by a small amount because of the penetration of the magnetic field into the conductor. This skin-effect inductance L_s is readily evaluated on an energy basis. The surface impedance Z_m has an inductive part $jX_m = j/\sigma\delta_s$ equal in magnitude to R_m. The magnetic energy stored in X_m is (note that X_m is equivalent to a surface inductance $X_m/\omega = L_m$)

$$W_m = \frac{X_m}{4\omega}\oint_{S_1+S_2}|\mathbf{J}_s|^2\,dl$$

$$= \frac{X_m}{4\omega}\oint_{S_1+S_2}|\mathbf{h}|^2\,dl = \frac{X_m}{4\omega}\frac{RI_0^2}{R_m} = \frac{RI_0^2}{4\omega}$$

by using (3.66) to replace the integral. Defining L_s by the relation

$$\tfrac{1}{4}L_sI_0^2 = W_m$$

gives

$$\omega L_s = R \tag{3.68}$$

The series inductive reactance of the line is increased by an amount equal to the series resistance. However, for low-loss lines, $R \ll \omega L$, so that $L_s \ll L$, and the correction is not significant for most practical lines. The inductance L_s is called the internal inductance since it arises from flux linkage internal to the conductor surfaces.

It should not come as a surprise to find that $\omega L_s = R$ since both the inductive reactance and resistance arise from the penetration of the current and fields into the conductor. The effect of this penetration into the conductor by an effective distance equal to the skin depth δ_s is correctly accounted for in a simplified manner by introduction of the surface impedance $Z_m = (1+j)/\sigma\delta_s$.

Example 3.3 Coaxial-line parameters For the coaxial line of Fig. 3.3 the potential Φ is given by

$$\Phi = V_0\frac{\ln(r/b)}{\ln(a/b)}$$

The charge on the inner conductor is

$$Q = \epsilon \int_0^{2\pi} \mathbf{a}_r \cdot \mathbf{e} a \, d\phi = \epsilon \int_0^{2\pi} -\frac{\partial \Phi}{\partial r} a \, d\phi$$

$$= \frac{-\epsilon V_0}{\ln (a/b)} \int_0^{2\pi} d\phi = \frac{2\pi \epsilon V_0}{\ln (b/a)}$$

Hence the capacitance per unit length is

$$C = \frac{\epsilon'}{\epsilon} \frac{Q}{V_0} = \frac{2\pi \epsilon'}{\ln (b/a)} \tag{3.69}$$

since the capacitance arises only from that part of the charge associated with ϵ' whereas ϵ'' gives rise to the shunt conductance.

The magnetic field is given by (3.38b) as

$$\mathbf{H} = \mathbf{h} e^{-jk_0 z} = \frac{YV_0}{\ln (b/a)} \frac{\mathbf{a}_\phi}{r} e^{-jk_0 z}$$

The current I_0 is

$$I_0 = \int_0^{2\pi} \mathbf{h} \cdot \mathbf{a}_\phi a \, d\phi = \frac{2\pi Y V_0}{\ln (b/a)}$$

Thus the characteristic impedance is

$$Z_c = \frac{V_0}{I_0} = \frac{Z \ln (b/a)}{2\pi} \tag{3.70}$$

The flux linking the center conductor is

$$\psi = \mu \int_a^b \mathbf{h} \cdot \mathbf{a}_\phi \, dr = \frac{\mu Y V_0}{\ln (b/a)} \int_a^b \frac{dr}{r} = \mu Y V_0$$

Consequently, the inductance per unit length is

$$L = \frac{\psi}{I_0} = \frac{\mu Y V_0}{2\pi Y V_0} \ln \frac{b}{a} = \frac{\mu}{2\pi} \ln \frac{b}{a} \tag{3.71}$$

from which it is seen that $LC = \mu \epsilon'$ and $Z_c = (L/C)^{1/2}$.

The shunt conductance G is given by $\omega \epsilon'' C / \epsilon'$, and is

$$G = \frac{\omega \epsilon''}{\epsilon'} \frac{2\pi \epsilon'}{\ln (b/a)} = \frac{2\pi \omega \epsilon''}{\ln (b/a)} \tag{3.72}$$

To find the series resistance the power loss in the inner and outer conductors must be evaluated. This was done in Example 3.2, with the result [Eq. (3.40b)]

$$\tfrac{1}{2} R I_0^2 = P_{l2} = \frac{R_m \pi Y^2 V_0^2}{[\ln (b/a)]^2} \frac{b+a}{ab}$$

Solving for R gives

$$R = \frac{R_m}{2\pi} \frac{b+a}{ab} \tag{3.73}$$

The internal inductance L_s is equal to R/ω; so the total series line inductance per unit length is

$$L + L_s = \frac{\mu}{2\pi} \ln \frac{b}{a} + \frac{b+a}{2\pi\omega a b \delta_s \sigma} \tag{3.74}$$

3.5 Terminated transmission line

In this section the properties of a transmission line terminated in an arbitrary load impedance Z_L are examined. This will serve to illustrate how the forward and backward propagating waves can be combined to satisfy the boundary conditions at a termination. Figure 3.6 illustrates schematically a transmission line terminated in a load impedance Z_L. The line is assumed lossless and with a characteristic impedance Z_c and a propagation constant $\gamma = j\beta$. It should be noted that at microwave frequencies conventional low-frequency resistors, inductors, or capacitors, when connected across the two conductors of a transmission line, may behave as impedance elements with quite different characteristics from the low-frequency behavior.

If a voltage wave $V^+ e^{-j\beta z}$ with an associated current $I^+ e^{-j\beta z}$ is incident on the termination, a reflected voltage wave $V^- e^{j\beta z}$ with a current $-I^- e^{j\beta z}$ will, in general, be created. The ratio of the reflected and incident wave amplitudes is determined by the load impedance only. At the load the total line voltage must equal the impressed voltage across the load and the line current must be continuous through the load. Hence, if Z_L is located at $z = 0$,

$$V = V^+ + V^- = V_L \tag{3.75a}$$
$$I = I^+ - I^- = I_L \tag{3.75b}$$

But $I^+ = Y_c V^+$, $I^- = Y_c V^-$, and $V_L/I_L = Z_L$ by definition of load impedance. Therefore

$$V^+ + V^- = V_L \tag{3.76a}$$
$$V^+ - V^- = \frac{Z_c}{Z_L} V_L \tag{3.76b}$$

The ratio of V^- to V^+ is usually described by a voltage reflection coeffi-

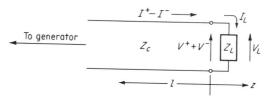

Fig. 3.6 Terminated transmission line.

cient Γ defined as

$$\Gamma = \frac{V^-}{V^+} \tag{3.77}$$

In place of (3.76) we may write

$$V^+(1 + \Gamma) = V_L$$

$$V^+(1 - \Gamma) = \frac{Z_c}{Z_L} V_L$$

Dividing one equation by the other yields

$$\frac{1 + \Gamma}{1 - \Gamma} = \frac{Z_L}{Z_c} \tag{3.78}$$

The quantity Z_L/Z_c is called the normalized load impedance (load impedance measured in units of Z_c), and $(1 + \Gamma)/(1 - \Gamma)$ is then the normalized input impedance seen looking toward the load at $z = 0$. The normalized load impedance will be expressed as $\bar{Z}_L$, with the bar on top signifying a normalized impedance in general. Solving for the voltage reflection coefficient Γ gives

$$\Gamma = \frac{Z_L - Z_c}{Z_L + Z_c} = \frac{Z_L/Z_c - 1}{Z_L/Z_c + 1} = \frac{\bar{Z}_L - 1}{\bar{Z}_L + 1} \tag{3.79}$$

Analogous to a voltage reflection coefficient, a current reflection coefficient Γ_I could also be introduced. In the present case

$$\Gamma_I = \frac{-I^-}{I^+} = -\frac{Y_c V^-}{Y_c V^+} = -\Gamma$$

In this text, however, only the voltage reflection coefficient will be used; so the adjective "voltage" can be dropped without confusion.

The incident voltage wave can be considered as transmitting a voltage V_L across the load, and a voltage transmission coefficient T can be defined as giving V_L in terms of V^+; thus

$$V_L = TV^+ = (1 + \Gamma)V^+$$

So

$$T = 1 + \Gamma \tag{3.80}$$

A corresponding current transmission coefficient is not used in this book.

Returning to (3.79), it is seen that if $Z_L = Z_c$, the reflection coefficient is zero. In this case all the power in the incident wave is transmitted to the load and none of it is reflected back toward the generator. The power delivered to the load in this case is

$$P = \tfrac{1}{2} \text{Re}\,(VI^*) = \tfrac{1}{2}|V^+|^2 Y_c = \tfrac{1}{2}|V_L|^2 Y_L \tag{3.81}$$

The load is said to be matched to the transmission line when $\Gamma = 0$.

If Z_L does not equal Z_c, the load is mismatched to the line and a reflected wave is produced. The power delivered to the load is now given by

$$P = \tfrac{1}{2} \operatorname{Re}(V_L I_L^*) = \tfrac{1}{2} \operatorname{Re}[(V^+ + V^-)(I^+ - I^-)^*]$$
$$= \tfrac{1}{2} \operatorname{Re}[Y_c(V^+ + V^-)(V^+ - V^-)^*] = \tfrac{1}{2} \operatorname{Re}[Y_c|V^+|^2(1 + \Gamma)(1 - \Gamma)^*]$$
$$= \tfrac{1}{2} Y_c |V^+|^2 (1 - |\Gamma|^2) \tag{3.82}$$

The final result states the physically obvious result that the power delivered to the load is the incident power minus that reflected from the load.

In the absence of reflection, the magnitude of the voltage along the line is a constant equal to $|V^+|$. When a reflected wave also exists, the incident and reflected waves interfere to produce a standing-wave pattern along the line. The voltage at any point on the line ($z < 0$) is given by

$$V = V^+ e^{-j\beta z} + \Gamma V^+ e^{j\beta z}$$

and has a magnitude given by

$$|V| = |V^+| \, |1 + \Gamma e^{2j\beta z}| = |V^+| \, |1 + \Gamma e^{-2j\beta l}|$$

where $l = -z$ is the positive distance measured from the load toward the generator, as in Fig. 3.6. Let Γ be equal to $\rho e^{j\theta}$, where $\rho = |\Gamma|$; then†

$$|V| = |V^+| \, |1 + \rho e^{j(\theta - 2\beta l)}| = |V^+|\{[1 + \rho \cos(\theta - 2\beta l)]^2$$
$$+ \rho^2 \sin^2(\theta - 2\beta l)\}^{\frac{1}{2}}$$
$$= |V^+|\{(1 + \rho)^2 - 2\rho[1 - \cos(\theta - 2\beta l)]\}^{\frac{1}{2}}$$
$$= |V^+|\left[(1 + \rho)^2 - 4\rho \sin^2\left(\beta l - \frac{\theta}{2}\right)\right]^{\frac{1}{2}} \tag{3.83}$$

This result shows that $|V|$ oscillates back and forth between maximum values of $|V^+|(1 + \rho)$ when $\beta l - \theta/2 = n\pi$ and minimum values $|V^+|(1 - \rho)$ when $\beta l - \theta/2 = n\pi + \pi/2$, where n is an integer. These results also agree with physical intuition since they state that voltage maxima occur when the incident and reflected waves add in phase and that voltage minima occur when they add 180° out of phase. Successive maxima and minima are spaced a distance $d = \pi/\beta = \lambda\pi/2\pi = \lambda/2$ apart, where λ is the wavelength for TEM waves in the medium surrounding the conductors. The distance between a maximum and the nearest minimum is $\lambda/4$.

Since the current reflection coefficient is equal to $-\Gamma$, the current waves subtract whenever the voltage waves add up in phase. Hence

---

† The symbol ρ denotes both charge density and the modulus of the reflection coefficient. The context makes it clear which quantity is under discussion; so confusion should not occur.

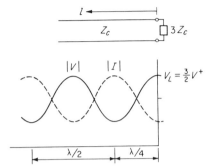

Fig. 3.7 Voltage and current standing-wave patterns on a line terminated in a load impedance equal to $3Z_c$.

current maxima and minima are displaced $\lambda/4$ from the corresponding voltage maxima and minima. Figure 3.7 illustrates the voltage and current standing-wave patterns that result when Z_L is a pure resistance equal to $3Z_c$.

The ratio of the maximum line voltage to the minimum line voltage is called the voltage standing-wave ratio S; thus

$$S = \frac{|V^+|(1 + \rho)}{|V^+|(1 - \rho)} = \frac{1 + \rho}{1 - \rho} \tag{3.84}$$

This is a parameter of considerable importance in practice for the following reasons: At microwave frequencies instruments for the direct absolute measurement of voltage or current are difficult to construct and use. On the other hand, devices to measure relative voltage or current (or electric or magnetic field) amplitudes are easy to construct. A typical device is a small probe inserted into the region of the electric field around a line. The output of the probe is connected to a crystal rectifier, and produces an output current which is a measure of the relative electric field or voltage at the probe position. By moving the probe along the line, the standing-wave ratio can be measured directly in terms of the maximum and minimum probe currents. The location of a voltage minimum can also be measured, and this permits the phase angle θ of Γ to be calculated. Since ρ is known from the measured value of S, Γ is specified, and the normalized load impedance may be calculated from (3.78).

Although the reflection coefficient was introduced as a measure of the ratio of reflected- to incident-wave amplitudes at the load, the definition may be extended to give the corresponding voltage ratio at any point on the line. Thus, at $z = -l$, the reflection coefficient is

$$\Gamma(l) = \frac{V^- e^{-j\beta l}}{V^+ e^{j\beta l}} = \frac{V^-}{V^+} e^{-2j\beta l} = \Gamma_L e^{-2j\beta l} \tag{3.85}$$

where $\Gamma_L = V^-/V^+$ now denotes the reflection coefficient of the load.

The normalized impedance, seen looking toward the load, at $z = -l$, is

$$\bar{Z}_{in} = \frac{Z_{in}}{Z_c} = \frac{V}{IZ_c} = \frac{V^+e^{j\beta l} + V^-e^{-j\beta l}}{V^+e^{j\beta l} - V^-e^{-j\beta l}}$$

$$= \frac{1 + \Gamma(l)}{1 - \Gamma(l)} = \frac{1 + \Gamma_L e^{-2j\beta l}}{1 - \Gamma_L e^{-2j\beta l}} \quad (3.86)$$

By replacing Γ_L by $(Z_L - Z_c)/(Z_L + Z_c)$ and $e^{\pm j\beta l}$ by $\cos \beta l \pm j \sin \beta l$, this result may be expressed as

$$\bar{Z}_{in} = \frac{Z_{in}}{Z_c} = \frac{Z_L + jZ_c \tan \beta l}{Z_c + jZ_L \tan \beta l} \quad (3.87)$$

A similar result holds for the normalized input admittance; so

$$\bar{Y}_{in} = \frac{Y_{in}}{Y_c} = \frac{Y_L + jY_c \tan \beta l}{Y_c + jY_L \tan \beta l} = \frac{\bar{Y}_L + j \tan \beta l}{1 + j\bar{Y}_L \tan \beta l} \quad (3.88)$$

Of particular interest are two special cases, namely, $\beta l = \pi$ or $l = \lambda/2$ and $\beta l = \pi/2$ or $l = \lambda/4$, for which

$$Z_{in}\left(l = \frac{\lambda}{2}\right) = Z_L \quad (3.89a)$$

$$Z_{in}\left(l = \frac{\lambda}{4}\right) = \frac{Z_c^2}{Z_L} \quad (3.89b)$$

The first is equivalent to an ideal one-to-one impedance transformer, whereas in the second case the impedance has been inverted with respect to Z_c^2.

Terminated lossy line

In the case of a lossy line with propagation constant $\gamma = j\beta + \alpha$, the previous equations hold except that $j\beta$ must be replaced by $j\beta + \alpha$, where α is usually so small that, for the short lengths of line used in most experimental setups, the neglect of α is justified. Nevertheless, it is of some interest to examine the behavior of a lossy transmission line terminated in a load Z_L. One simplifying assumption will be made, and this is that the characteristic impedance Z_c can still be considered real. This assumption is certainly valid for low-loss lines of the type used at microwave frequencies. A detailed calculation justifying this assumption for a typical case is called for in Prob. 3.6.

Clearly, the presence of an attenuation constant α does not affect the definition of the voltage reflection coefficient Γ_L for the load. However, at any other point a distance l toward the generator, the reflection coefficient is now given by

$$\Gamma(l) = \Gamma_L e^{-2j\beta l - 2\alpha l} \quad (3.90)$$

94 Foundations for microwave engineering

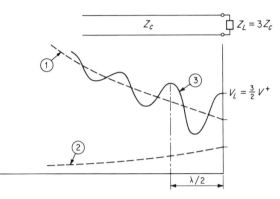

Fig. 3.8 Voltage-standing-wave pattern on a lossy transmission line. (1) Envelope of incident-wave amplitude; (2) envelope of reflected-wave amplitude; (3) standing-wave pattern.

As l is increased, Γ decreases exponentially until, for large l, it essentially vanishes. Thus, whenever a load Z_L is viewed through a long section of lossy line, it appears to be matched to the line since Γ is negligible at the point considered. This effect may also be seen from the expression for the input impedance, namely,

$$Z_{\text{in}} = Z_c \frac{1 + \Gamma_L e^{-2j\beta l - 2\alpha l}}{1 - \Gamma_L e^{-2j\beta l - 2\alpha l}} = Z_c \frac{Z_L + Z_c \tanh(j\beta l + \alpha l)}{Z_c + Z_L \tanh(j\beta l + \alpha l)} \quad (3.91)$$

which approaches Z_c for l large since $\tanh x$ approaches 1 for x large and not a pure imaginary quantity.

The losses also have the effect of reducing the standing-wave ratio S toward unity as the point of observation is moved away from the load toward the generator. As the generator is approached, the incident-wave amplitude increases exponentially whereas the reflected-wave amplitude decreases exponentially. The result is a standing-wave pattern of the type illustrated in Fig. 3.8. For illustrative purposes a relatively large value of α has been assumed here.

The power delivered to the load is given by

$$P_L = \tfrac{1}{2} \operatorname{Re}(V_L I_L^*) = \tfrac{1}{2}|V_L|^2 Y_L = \frac{Y_c}{2}|V^+|^2(1 - |\Gamma_L|^2) \quad (3.92)$$

as before. At some point $z = -l$, the power directed toward the load is

$$P(l) = \tfrac{1}{2} \operatorname{Re}(VI^*) = \tfrac{1}{2}|V|^2 Y_c = \frac{Y_c}{2}|V^+ e^{\alpha l}|^2[1 - |\Gamma(l)|^2]$$

$$= \frac{Y_c}{2}|V^+|^2(e^{2\alpha l} - |\Gamma_L|^2) \quad (3.93)$$

where $|\Gamma|e^{\alpha l}$ has been replaced by $|\Gamma_L|$. Of the power given by (3.93), only that portion corresponding to P_L as given by (3.92) is delivered to the load. The remainder is dissipated in the lossy line, this remainder being given by

$$P(l) - P_L = \frac{Y_c}{2}|V^+|^2(e^{2\alpha l} - 1)$$

3.6 Rectangular waveguide

The rectangular waveguide with a cross section as illustrated in Fig. 3.9 is an example of a waveguiding device that will not support a TEM wave. Consequently, it turns out that unique voltage and current waves do not exist, and the analysis of the waveguide properties has to be carried out as a field problem rather than as a distributed-parameter-circuit problem.

In a hollow cylindrical waveguide a transverse electric field can exist only if a time-varying axial magnetic field is present. Similarly, a transverse magnetic field can exist only if either an axial displacement current or an axial conduction current is present, as Maxwell's equations show. Since a TEM wave does not have any axial field components and there is no center conductor on which a conduction current can exist, a TEM wave cannot be propagated in a cylindrical waveguide.

The types of waves that can be supported (propagated) in a hollow empty waveguide are the TE and TM modes discussed in Sec. 3.1. The essential properties of all hollow cylindrical waveguides are the same, so that an understanding of the rectangular guide provides insight into the behavior of other types as well. As for the case of the transmission line, the effect of losses is initially neglected. The attenuation is computed later by using the perturbation method given earlier, together with the loss-free solution for the currents on the walls.

The essential properties of empty loss-free waveguides, which the detailed analysis to follow will establish, are that there is a double infinity of possible solutions for both TE and TM waves. These waves, or modes, may be labeled by two identifying integer subscripts n and m, for example, TE_{nm}. The integers n and m pertain to the number of standing-wave interference maxima occurring in the field solutions that describe

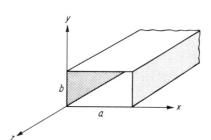

Fig. 3.9 Rectangular waveguide.

the variation of the fields along the two transverse coordinates. It will be found that each mode has associated with it a characteristic cutoff frequency $f_{c,nm}$ below which the mode does not propagate and above which the mode does propagate. The cutoff frequency is a geometrical parameter dependent on the waveguide cross-sectional configuration. When f_c has been determined, it is found that the propagation factor β is given by

$$\beta = (k_0^2 - k_c^2)^{\frac{1}{2}} \tag{3.94}$$

where $k_0 = \omega \sqrt{\mu_0 \epsilon_0}$ and $k_c = 2\pi f_c \sqrt{\mu_0 \epsilon_0}$. The guide wavelength is readily seen to be given by

$$\lambda_g = \frac{2\pi}{\beta} = \frac{\lambda_0}{(1 - \lambda_0^2/\lambda_c^2)^{\frac{1}{2}}} = \frac{\lambda_0}{\sqrt{1 - f_c^2/f^2}} \tag{3.95}$$

where λ_0 is the free-space wavelength of plane waves at the frequency $f = \omega/2\pi$. Since k_c differs for different modes, there is always a lower band of frequencies for which only one mode propagates (except when k_c may be the same for two or more modes). In practice, waveguides are almost universally restricted to operation over this lower-frequency band for which only the dominant mode propagates, because of the difficulties associated with coupling signal energy into and out of a waveguide when more than one mode propagates. This latter difficulty arises because of the different values of the propagation phase constant β for different modes, since this means that the signal carried by the two or more modes does not remain in phase as the modes propagate along the guide. This necessitates the use of separate coupling probes for each mode at both the input and output and thus leads to increased system complexity and cost.

Another feature common to all empty uniform waveguides is that the phase velocity v_p is greater than the velocity of light c by the factor λ_g/λ_0. On the other hand, the velocity at which energy and a signal are propagated is the group velocity v_g and is smaller than c by the factor λ_0/λ_g. Also, since β, and hence λ_g, v_p, and v_g, are functions of frequency, any signal consisting of several frequencies is dispersed, or spread out, in both time and space as it propagates along the guide. This dispersion results from the different velocities at which the different frequency components propagate. If the guide is very long, considerable signal distortion may take place. Group and signal velocities are discussed in detail in Sec. 3.11.

With some of the general properties of waveguides considered, it is now necessary to consider the detailed analysis that will establish the above properties and that, in addition, will provide the relation between k_c and the guide configuration, the expressions for power and attenuation, etc. The case of TE modes in a loss-free empty rectangular guide is considered first.

TE waves

For TE, or H, modes, $e_z = 0$ and all the remaining field components can be determined from the axial magnetic field h_z by means of (3.16). The axial field h_z is a solution of

$$\nabla_t^2 h_z + k_c^2 h_z = 0$$

or

$$\frac{\partial^2 h_z}{\partial x^2} + \frac{\partial^2 h_z}{\partial y^2} + k_c^2 h_z = 0 \tag{3.96}$$

If a product solution $h_z = f(x)g(y)$ is assumed, (3.96) becomes

$$\frac{1}{f}\frac{d^2 f}{dx^2} + \frac{1}{g}\frac{d^2 g}{dy^2} + k_c^2 = 0$$

after substituting fg for h_z and dividing the equation by fg. The term $\frac{1}{f}\frac{d^2 f}{dx^2}$ is a function of x only, $\frac{1}{g}\frac{d^2 g}{dy^2}$ is a function of y only, and k_c^2 is a constant, and hence this equation can hold for all values of x and y only if each term is constant. Thus we may write

$$\frac{1}{f}\frac{d^2 f}{dx^2} = -k_x^2 \quad \text{or} \quad \frac{d^2 f}{dx^2} + k_x^2 f = 0$$

$$\frac{1}{g}\frac{d^2 g}{dy^2} = -k_y^2 \quad \text{or} \quad \frac{d^2 g}{dy^2} + k_y^2 g = 0$$

where $k_x^2 + k_y^2 = k_c^2$ in order that the sum of the three terms may vanish. The use of the *separation-of-variables* technique has reduced the partial differential equation (3.96) to two ordinary simple-harmonic second-order equations. The solutions for f and g are easily found to be

$$f = A_1 \cos k_x x + A_2 \sin k_x x$$
$$g = B_1 \cos k_y y + B_2 \sin k_y y$$

where A_1, A_2, B_1, B_2 are arbitrary constants. These constants, as well as the separation constants k_x, k_y, can be further specified by considering the boundary conditions that h_z must satisfy. Since the normal component of the transverse magnetic field $\mathbf{h}$ must vanish at the perfectly conducting waveguide wall, (3.16b) shows that $\mathbf{n} \cdot \nabla_t h_z = 0$ at the walls, where $\mathbf{n}$ is a unit normal vector at the walls. When this condition holds, tangential $\mathbf{e}$ will also vanish on the guide walls, as (3.16c) shows. The requirements on h_z are thus

$$\frac{\partial h_z}{\partial x} = 0 \quad \text{at } x = 0, a$$

$$\frac{\partial h_z}{\partial y} = 0 \quad \text{at } y = 0, b$$

where the guide cross section is taken to be that in Fig. 3.9. In the solution for f, the boundary conditions give

$$-k_x A_1 \sin k_x x + k_x A_2 \cos k_x x = 0 \qquad \text{at } x = 0, a$$

Hence, from the condition at $x = 0$, it is found that $A_2 = 0$. At $x = a$, it is necessary for $\sin k_x a = 0$, and this specifies k_x to have the values

$$k_x = \frac{n\pi}{a} \qquad n = 0, 1, 2, \ldots$$

In a similar manner it is found that $B_2 = 0$ and

$$k_y = \frac{m\pi}{b} \qquad m = 0, 1, 2, \ldots$$

Both n and m equal to zero yields a constant for the solution for h_z and no other field components; so this trivial solution is of no interest.

If we use the above relations and put $A_1 B_1 = A_{nm}$, the solutions for h_z are seen to be

$$h_z = A_{nm} \cos \frac{n\pi x}{a} \cos \frac{m\pi y}{b} \tag{3.97}$$

for $n = 0, 1, 2, \ldots$; $m = 0, 1, 2, \ldots$; $n = m \neq 0$. The constant A_{nm} is an arbitrary amplitude constant associated with the nmth mode. For the nmth mode the cutoff wave number is designated $k_{c,nm}$, given by

$$k_{c,nm} = \left[\left(\frac{n\pi}{a} \right)^2 + \left(\frac{m\pi}{b} \right)^2 \right]^{\frac{1}{2}} \tag{3.98}$$

and is clearly a function of the guide dimensions only. The propagation constant for the nmth mode is given by

$$\gamma_{nm} = j\beta_{nm} = j(k_0^2 - k_{c,nm}^2)^{\frac{1}{2}}$$

$$= j\left[\left(\frac{2\pi}{\lambda_0} \right)^2 - \left(\frac{n\pi}{a} \right)^2 - \left(\frac{m\pi}{b} \right)^2 \right]^{\frac{1}{2}} \tag{3.99}$$

When $k_0 > k_{c,nm}$, β_{nm} is pure real and the mode propagates; when $k_0 < k_{c,nm}$, then γ_{nm} is real but β_{nm} is imaginary and the propagation factor is $e^{-\gamma_{nm}|z|}$, which shows that the mode decays rapidly with distance $|z|$ from the point at which it is excited. This decay is not associated with energy loss, but is a characteristic feature of the solution. Such decaying, or evanescent, modes may be used to represent the local diffraction, or fringing, fields that exist in the vicinity of coupling probes and obstacles in waveguides. The frequency separating the propagation and no-propagation bands is designated the cutoff frequency $f_{c,nm}$. This is given by the solution of $k_0 = k_{c,nm}$; that is,

$$f_{c,nm} = \frac{c}{\lambda_{c,nm}} = \frac{c}{2\pi} k_{c,nm} = \frac{c}{2\pi} \left[\left(\frac{n\pi}{a} \right)^2 + \left(\frac{m\pi}{b} \right)^2 \right]^{\frac{1}{2}} \tag{3.100}$$

where c is the velocity of light. The cutoff wavelength is given by

$$\lambda_{c,nm} = \frac{2ab}{(n^2 b^2 + m^2 a^2)^{\frac{1}{2}}} \tag{3.101}$$

A typical guide may have $a = 2b$, in which case

$$\lambda_{c,nm} = \frac{2a}{(n^2 + 4m^2)^{\frac{1}{2}}}$$

and $\lambda_{c,10} = 2a$, $\lambda_{c,01} = a$, $\lambda_{c,11} = 2a/\sqrt{5}$, etc. In this example there is a band of wavelengths from a to $2a$, that is, a frequency band

$$\frac{c}{2a} < f < \frac{c}{a}$$

for which only the H_{10} mode propagates. This is the dominant mode in a rectangular guide and the one most commonly used in practice. Above the frequency c/a, other modes may propagate; so the useful frequency band in the present case is a one-octave band from $c/2a$ to c/a.

The remainder of the field components for the TE$_{nm}$, or H_{nm}, mode are readily found from (3.97) by using (3.16). The results for the complete nmth solution are

$$H_z = A_{nm} \cos \frac{n\pi x}{a} \cos \frac{m\pi y}{b} e^{\mp j\beta_{nm} z} \tag{3.102a}$$

$$H_x = \pm j \frac{\beta_{nm}}{k_{c,nm}^2} A_{nm} \frac{n\pi}{a} \sin \frac{n\pi x}{a} \cos \frac{m\pi y}{b} e^{\mp j\beta_{nm} z} \tag{3.102b}$$

$$H_y = \pm j \frac{\beta_{nm}}{k_{c,nm}^2} A_{nm} \frac{m\pi}{b} \cos \frac{n\pi x}{a} \sin \frac{m\pi y}{b} e^{\mp j\beta_{nm} z} \tag{3.102c}$$

$$E_x = Z_{h,nm} A_{nm} j \frac{\beta_{nm}}{k_{c,nm}^2} \frac{m\pi}{b} \cos \frac{n\pi x}{a} \sin \frac{m\pi y}{b} e^{\mp j\beta_{nm} z} \tag{3.102d}$$

$$E_y = -Z_{h,nm} A_{nm} j \frac{\beta_{nm}}{k_{c,nm}^2} \frac{n\pi}{a} \sin \frac{n\pi x}{a} \cos \frac{m\pi y}{b} e^{\mp j\beta_{nm} z} \tag{3.102e}$$

where the wave impedance for the nmth H mode is given by

$$Z_{h,nm} = \frac{k_0}{\beta_{nm}} Z_0 \tag{3.103}$$

When the mode does not propagate, $Z_{h,nm}$ is imaginary, indicating that there is no net energy flow associated with the evanescent mode. A general field with $E_z = 0$ can be described in a complete manner by a linear superposition of all the H_{nm} modes.

Power

For a propagating H_{nm} mode the power, or rate of energy flow, in the positive z direction is given by

$$P_{nm} = \tfrac{1}{2} \operatorname{Re} \int_0^a \int_0^b \mathbf{E} \times \mathbf{H}^* \cdot \mathbf{a}_z \, dx \, dy$$

$$= \tfrac{1}{2} \operatorname{Re} \int_0^a \int_0^b (E_x H_y^* - E_y H_x^*) \, dx \, dy$$

$$= \tfrac{1}{2} \operatorname{Re} Z_{h,nm} \int_0^a \int_0^b (H_y H_y^* + H_x H_x^*) \, dx \, dy \qquad (3.104)$$

If we substitute from (3.102b) and (3.102c) and note that

$$\int_0^a \int_0^b \sin^2 \frac{n\pi x}{a} \cos^2 \frac{m\pi y}{b} \, dx \, dy = \int_0^a \int_0^b \cos^2 \frac{n\pi x}{a} \sin^2 \frac{m\pi y}{b} \, dx \, dy$$

$$= \begin{cases} \dfrac{ab}{4} & n \neq 0,\, m \neq 0 \\ \dfrac{ab}{2} & n \text{ or } m = 0 \end{cases}$$

we find that

$$P_{nm} = |A_{nm}|^2 \frac{\tfrac{1}{2}ab}{\epsilon_{0n}\epsilon_{0m}} \frac{\beta_{nm}^2}{k_{c,nm}^4} Z_{h,nm} \left[\left(\frac{m\pi}{b}\right)^2 + \left(\frac{n\pi}{a}\right)^2 \right]$$

$$= \frac{|A_{nm}|^2 ab}{2\epsilon_{0n}\epsilon_{0m}} \left(\frac{\beta_{nm}}{k_{c,nm}}\right)^2 Z_{h,nm} \qquad (3.105)$$

where ϵ_{0m} is the Neumann factor and equal to 1 for $m = 0$ and equal to 2 for $m > 0$.

If two modes, say the H_{nm} and H_{rs} modes, were present simultaneously, it would be found that the power is the sum of that contributed by each individual mode, that is, $P_{nm} + P_{rs}$. This is a general property of loss-free waveguides, and is discussed in detail in a later section. This power orthogonality arises because of the orthogonality of the functions (eigenfunctions) that describe the transverse variation of the fields when integrated over the guide cross section; e.g.,

$$\int_0^a \sin \frac{n\pi x}{a} \sin \frac{r\pi x}{a} \, dx = 0 \qquad n \neq r$$

Even when small losses are present the energy flow may be taken to be that contributed by each individual mode, with negligible error in all cases except when two or more degenerate modes are present. Degenerate modes are modes which have the same propagation constant γ, and for these the presence of even small losses may result in strong coupling between the modes. This phenomenon is explained more fully in Sec. 3.10.

Attenuation

If the waveguide walls have finite conductivity, there will be a continuous loss of power to the walls as the modes propagate along the guide. Consequently, the phase constant $j\beta$ is perturbed and becomes $\gamma = \alpha + j\beta$, where α is an attenuation constant that gives the rate at which the mode amplitude must decay as the mode progresses along the guide. For practical waveguides the losses caused by finite conductivity are so small that the attenuation constant may be calculated using the perturbation method outlined in Sec. 3.2 in connection with lossy transmission lines.† The method will be illustrated for the dominant H_{10} mode only. For the H_{nm} and also the E_{nm} modes, the calculation differs only in that somewhat greater algebraic manipulation is required.

For the H_{10} mode, the fields are given by (apart from the factor $e^{-j\beta_{10}z}$)

$$h_z = A_{10} \cos \frac{\pi x}{a}$$

$$h_x = j \frac{\beta_{10}}{k_{c,10}^2} A_{10} \frac{\pi}{a} \sin \frac{\pi x}{a}$$

$$e_y = -Z_{h,10} A_{10} \frac{j\beta_{10}}{k_{c,10}^2} \frac{\pi}{a} \sin \frac{\pi x}{a}$$

as reference to (3.102) shows. From (3.105) the rate of energy flow along the guide is

$$P_{10} = |A_{10}|^2 \frac{ab}{4} \left(\frac{\beta_{10}}{k_{c,10}}\right)^2 Z_{h,10}$$

The currents on the lossy walls are assumed to be the same as the loss-free currents, and hence are given by

$$\mathbf{J}_s = \mathbf{n} \times \mathbf{H}$$

where $\mathbf{n}$ is a unit inward directed normal at the guide wall. Thus, on the walls at $x = 0, a$, the surface currents are

$$\mathbf{J}_s = \begin{cases} \mathbf{a}_x \times \mathbf{H} = -\mathbf{a}_y A_{10} & x = 0 \\ -\mathbf{a}_x \times \mathbf{H} = -\mathbf{a}_y A_{10} & x = a \end{cases}$$

whereas on the upper and lower walls the currents are

$$\mathbf{J}_s = \begin{cases} \mathbf{a}_y \times \mathbf{H} = -\mathbf{a}_z \dfrac{j\beta_{10}}{k_{c,10}^2} A_{10} \dfrac{\pi}{a} \sin \dfrac{\pi x}{a} + \mathbf{a}_x A_{10} \cos \dfrac{\pi x}{a} & y = 0 \\ -\mathbf{a}_y \times \mathbf{H} = \mathbf{a}_z \dfrac{j\beta_{10}}{k_{c,10}^2} A_{10} \dfrac{\pi}{a} \sin \dfrac{\pi x}{a} - \mathbf{a}_x A_{10} \cos \dfrac{\pi x}{a} & y = b \end{cases}$$

With a finite conductivity σ, the waveguide walls may be characterized

† The case for degenerate modes may require a modified analysis, and this is covered in Sec. 3.10.

Table 3.2 Properties of modes in a rectangular guide†

	TE modes	TM modes
H_z	$\cos\dfrac{n\pi x}{a}\cos\dfrac{m\pi y}{b}e^{-j\beta_{nm}z}$	0
E_z	0	$\sin\dfrac{n\pi x}{a}\sin\dfrac{m\pi y}{b}e^{-j\beta_{nm}z}$
E_x	$Z_{h,nm}H_y$	$\dfrac{-j\beta_{nm}n\pi}{ak_{c,nm}^2}\cos\dfrac{n\pi x}{a}\sin\dfrac{m\pi y}{b}e^{-j\beta_{nm}z}$
E_y	$-Z_{h,nm}H_x$	$\dfrac{-j\beta_{nm}m\pi}{bk_{c,nm}^2}\sin\dfrac{n\pi x}{a}\cos\dfrac{m\pi y}{b}e^{-j\beta_{nm}z}$
H_x	$\dfrac{j\beta_{nm}n\pi}{ak_{c,nm}^2}\sin\dfrac{n\pi x}{a}\cos\dfrac{m\pi y}{b}e^{-j\beta_{nm}z}$	$-\dfrac{E_y}{Z_{e,nm}}$
H_y	$\dfrac{j\beta_{nm}m\pi}{bk_{c,nm}^2}\cos\dfrac{n\pi x}{a}\sin\dfrac{m\pi y}{b}e^{-j\beta_{nm}z}$	$\dfrac{E_x}{Z_{e,nm}}$
$Z_{h,nm}$	$\dfrac{k_0}{\beta_{nm}}Z_0$	
$Z_{e,nm}$		$\dfrac{\beta_{nm}}{k_0}Z_0$
$k_{c,nm}$	$\left[\left(\dfrac{n\pi}{a}\right)^2+\left(\dfrac{m\pi}{b}\right)^2\right]^{1/2}$	
β_{nm}	$(k_0^2-k_{c,nm}^2)^{1/2}$	
$\lambda_{c,nm}$	$\dfrac{2ab}{(n^2b^2+m^2a^2)^{1/2}}$	
α	$\dfrac{2R_m}{bZ_0(1-k_{c,nm}^2/k_0^2)^{1/2}}\left[\left(1+\dfrac{b}{a}\right)\dfrac{k_{c,nm}^2}{k_0^2}\right.$ $\left.+\dfrac{b}{a}\left(\dfrac{\epsilon_{0m}}{2}-\dfrac{k_{c,nm}^2}{k_0^2}\right)\dfrac{n^2ab+m^2a^2}{n^2b^2+m^2a^2}\right]$	$\dfrac{2R_m}{bZ_0(1-k_{c,nm}^2/k_0^2)^{1/2}}\dfrac{n^2b^3+m^2a^3}{n^2b^2a+m^2a^3}$

† $R_m=(\omega\mu_0/2\sigma)^{1/2}$, $\epsilon_{0m}=1$ for $m=0$ and 2 for $m>0$. The expression for α is not valid for degenerate modes (Sec. 3.10).

as exhibiting a surface impedance given by

$$Z_m=\frac{1+j}{\sigma\delta_s}=(1+j)R_m$$

where δ_s is the skin depth. The power loss in the resistive part R_m of Z_m per unit length of guide is

$$P_l=\frac{R_m}{2}\oint_{\substack{\text{guide}\\\text{walls}}}\mathbf{J}_s\cdot\mathbf{J}_s^*\,dl$$

$$=\frac{R_m|A_{10}|^2}{2}\left(2\int_0^b dy+2\int_0^a\frac{\beta_{10}^2\pi^2}{k_{c,10}^4 a^2}\sin^2\frac{\pi x}{a}dx+2\int_0^a\cos^2\frac{\pi x}{a}dx\right)$$

Since $k_{c,10} = \pi/a$, the above gives

$$P_l = R_m |A_{10}|^2 \left[b + \frac{a}{2}\left(\frac{\beta_{10}}{k_{c,10}}\right)^2 + \frac{a}{2} \right]$$

If P_0 is the power at $z = 0$, then $P_{10} = P_0 e^{-2\alpha z}$ is the power in the guide at any z. The rate of decrease of power propagated is

$$-\frac{dP_{10}}{dz} = 2\alpha P_{10} = P_l$$

and equals the power loss, as indicated in the above equation. The attenuation constant α for the H_{10} mode is thus seen to be

$$\alpha = \frac{P_l}{2P_{10}} = \frac{R_m \left[b + \frac{a}{2}\left(\frac{\beta_{10}}{k_{c,10}}\right)^2 + \frac{a}{2} \right]}{\frac{ab}{2}\left(\frac{\beta_{10}}{k_{c,10}}\right)^2 Z_{h,10}}$$

$$= \frac{R_m}{ab\beta_{10}k_0 Z_0}(2bk_{c,10}^2 + ak_0^2) \qquad \text{nepers/m} \tag{3.106}$$

The attenuation for other TE$_{nm}$ modes is given by the formula in Table 3.2, which summarizes the solutions for TE$_{nm}$ and also TM$_{nm}$ modes. In Fig. 3.10 the attenuation for the H_{10} mode in a copper rectangular guide is given as a function of frequency. To convert attenuation given in nepers to decibels, multiply by 8.7.

The theoretical formulas for attenuation give results in good agreement with experimental values for frequencies below about 5,000 Mc. For higher frequencies, measured values of α may be considerably higher, depending on the smoothness of the waveguide surface. If surface imperfections of the order of magnitude of the skin depth δ_s are present, it is readily appreciated that the effective surface area is much greater, resulting in greater loss. By suitably polishing the surface, the experi-

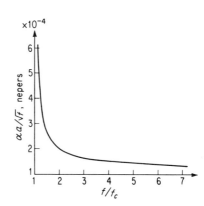

Fig. 3.10 Attenuation of H_{10} mode in a copper rectangular waveguide, $a = 2b$, f in units of 10^{10} cps.

mental values of attenuation are found to be in substantial agreement with the theoretical values.†

Dominant TE₁₀ mode

Since the TE_{10} mode is the dominant mode in a rectangular guide, and also the most commonly used mode, it seems appropriate to examine this mode in more detail. From the results given earlier, the field components for this mode are described by the following (propagation in the $+z$ direction assumed):

$$H_z = A \cos \frac{\pi x}{a} e^{-j\beta z} \tag{3.107a}$$

$$H_x = \frac{j\beta}{k_c} A \sin \frac{\pi x}{a} e^{-j\beta z} \tag{3.107b}$$

$$E_y = -jAZ_h \frac{\beta}{k_c} \sin \frac{\pi x}{a} e^{-j\beta z} \tag{3.107c}$$

where the subscript 10 has been dropped for convenience since this discussion pertains only to the TE_{10} mode. The parameters β, k_c, and Z_h are given by

$$k_c = \frac{\pi}{a} \tag{3.108a}$$

$$\beta = \left[k_0{}^2 - \left(\frac{\pi}{a}\right)^2 \right]^{\frac{1}{2}} \tag{3.108b}$$

$$Z_h = -\frac{E_y}{H_x} = \frac{k_0}{\beta} Z_0 \tag{3.108c}$$

The guide wavelength λ_g is

$$\lambda_g = \frac{2\pi}{\beta} = \frac{\lambda_0}{[1 - (\lambda_0/2a)^2]^{\frac{1}{2}}} \tag{3.108d}$$

since the cutoff wavelength $\lambda_c = 2a$. The phase and group velocities are

$$v_p = \frac{\lambda_g}{\lambda_0} c \tag{3.108e}$$

$$v_g = \frac{\lambda_0}{\lambda_g} c \tag{3.108f}$$

and are discussed in detail in Sec. 3.11.

In Fig. 3.11 the magnetic and electric field lines associated with the TE_{10} mode are illustrated. Note that the magnetic flux lines encircle the electric field lines; so these can be considered to be the source (displacement current) for the magnetic field. On the other hand, the

† See J. Allison and F. A. Benson, Surface Roughness and Attenuation of Precision Drawn, Chemically Polished, Electropolished, Electroplated and Electroformed Waveguides, *Proc. IEE* (*London*), vol. 102, pt. B, pp. 251–259, 1955.

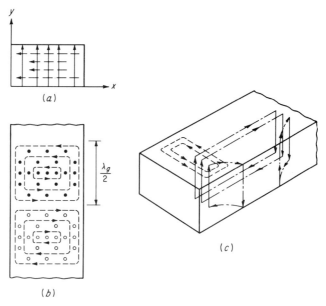

Fig. 3.11 Magnetic and electric field lines for the TE_{10} mode. (a) Transverse plane; (b) top view; (c) mutual total current and magnetic field linkages.

Fig. 3.12 Decomposition of TE_{10} mode into two plane waves.

electric field lines terminate in an electric charge distribution on the inner surface of the upper and lower waveguide walls. This charge oscillates back and forth in the axial and transverse directions and thus constitutes an axial and transverse conduction current that forms the continuation of the displacement current. The total current, displacement plus conduction, forms a closed linkage of the magnetic field lines, and as such may be regarded as being generated by the changing magnetic flux they enclose. This completes the required mutual-support action between the electric and magnetic fields which is required for wave propagation.

The fields for a TE_{10} mode may be decomposed into the sum of two plane TEM waves propagating along zigzag paths between the two waveguide walls at $x = 0$ and $x = a$, as in Fig. 3.12. For the electric field we have

$$E_y = -\frac{Z_h}{2}\frac{\beta}{k_c}\left(e^{j\pi x/a - j\beta z} - e^{-j\pi x/a - j\beta z}\right)$$

If π/a and β are expressed as

$$\frac{\pi}{a} = k_0 \sin \theta \qquad \beta = k_0 \cos \theta$$

the relation $(\pi/a)^2 + \beta^2 = k_0^2$ still holds. The electric field is now given by

$$E_y = \frac{Z_h}{2} \frac{\beta}{k_c} \left(e^{-jk_0(x \sin \theta + z \cos \theta)} - e^{-jk_0(-x \sin \theta + z \cos \theta)} \right)$$

which is clearly two plane waves propagating at angles $\pm \theta$ with respect to the z axis, as illustrated. Alternatively, the field may be pictured as a plane wave reflecting back and forth between the two guide walls. As shown in Sec. 2.7, the constant phase planes associated with these obliquely propagating plane waves move in the z direction at the phase velocity $c/\cos \theta = \beta c/k_0$, and this is the reason why the phase velocity of the TE_{10} mode exceeds the velocity of light. Since the energy in a TEM wave propagates with the velocity c in the direction in which the plane wave propagates, this energy will propagate down the guide at a velocity equal to the component of c along the z axis. This component is $v_g = c \cos \theta = (k_0/\beta)c$ and is the group velocity for the TE_{10} mode. When $\theta = \pi/2$, the plane waves reflect back and forth, but do not progress down the guide; so the mode is cutoff.

The above decomposition of the TE_{10} mode into two plane waves may be extended to the TE_{nm} modes also. When n and m are both different from zero, four plane waves result. Although such superpositions of plane waves may be used to construct the field solutions for rectangular guides, this is a rather cumbersome approach. However, it does lend insight into why the phase velocity exceeds that of light, as well as other properties of the modes.

TM modes

For TM modes, h_z equals zero and e_z plays the role of a potential function from which the remaining field components may be derived. This axial electric field satisfies the reduced Helmholtz equation

$$\nabla_t^2 e_z + k_c^2 e_z = 0 \qquad (3.109)$$

of the same type encountered earlier for h_z, that is, (3.96). The solution may be found by using the separation-of-variables method. In the present case the boundary conditions require that e_z vanish at $x = 0$, a and $y = 0, b$. This condition requires that the solution for e_z be

$$e_z = A_{nm} \sin \frac{n\pi x}{a} \sin \frac{m\pi y}{b} \qquad (3.110)$$

instead of a product of cosine functions which was suitable for describing h_z. Again, there are a doubly infinite number of solutions corresponding to various integers n and m. However, unlike the situation for TE modes, $n = 0$ and $m = 0$ are not solutions. The cutoff wave number is given by the same expression as for TE modes; that is,

$$k_{c,nm} = \left[\left(\frac{n\pi}{a}\right)^2 + \left(\frac{m\pi}{b}\right)^2\right]^{\frac{1}{2}} \tag{3.111}$$

and the propagation factor β_{nm} by

$$\beta_{nm} = (k_0^2 - k_{c,nm}^2)^{\frac{1}{2}} \tag{3.112}$$

The lowest-order propagating mode is the $n = m = 1$ mode, and this has a cutoff wavelength equal to $2ab/(a^2 + b^2)^{\frac{1}{2}}$. Note that the TE_{10} mode can propagate at a lower frequency (longer wavelength), thus verifying that this is the dominant mode.† It should also be noted that for the same values of n and m, the TE_{nm} and TM_{nm} modes are degenerate since they have the same propagation factor. Another degeneracy occurs when $a = b$, for in this case the four modes TE_{nm}, TE_{mn}, TM_{nm}, and TM_{mn} all have the same propagation constant. Still further degeneracies exist if a is an integer multiple of b, or vice versa.

The rest of the solution for TM modes is readily constructed using the general equations (3.17) given in Sec. 3.1. A summary of this solution is given in Table 3.2. The TM modes are the dual of the TE modes and apart from minor differences have essentially the same properties. For this reason it does not seem necessary to repeat the preceding discussion.

3.7 Circular waveguides

Figure 3.13 illustrates a cylindrical waveguide with a circular cross section of radius a. In view of the cylindrical geometry involved, cylindrical coordinates are most appropriate for the analysis to be carried out. Since the general properties of the modes that may exist are similar to those for the rectangular guide, this section is not as detailed.

† In any hollow waveguide the dominant mode is a TE mode because the boundary conditions $e_z = 0$ for TM modes always require e_z to have a greater spatial variation in the transverse plane than that for h_z for the lowest-order TE mode, and hence the smallest value of k_c occurs for TE modes. Hence a TE mode has the lowest cutoff frequency, i.e., is the dominant mode.

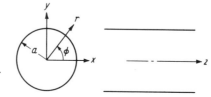

Fig. 3.13 The circular cylindrical waveguide.

TM modes

For the TM modes a solution of

$$\nabla_t^2 e_z + k_c^2 e_z = 0$$

is required such that e_z will vanish at $r = a$. When we express the transverse laplacian Δ_t^2 in cylindrical coordinates (Appendix I), this equation becomes

$$\frac{\partial^2 e_z}{\partial r^2} + \frac{1}{r}\frac{\partial e_z}{\partial r} + \frac{1}{r^2}\frac{\partial^2 e_z}{\partial \phi^2} + k_c^2 e_z = 0 \qquad (3.113)$$

The separation-of-variables method may be used to reduce the above to two ordinary differential equations. Consequently, it is assumed that a product solution $f(r)g(\phi)$ exists for e_z. Substituting for e_z into (3.113) and dividing the equation by fg yield

$$\frac{1}{f}\frac{d^2 f}{dr^2} + \frac{1}{rf}\frac{df}{dr} + \frac{1}{r^2 g}\frac{d^2 g}{d\phi^2} + k_c^2 = 0$$

Multiplying this result by r^2 gives

$$\frac{r^2}{f}\frac{d^2 f}{dr^2} + \frac{r}{f}\frac{df}{dr} + r^2 k_c^2 = -\frac{1}{g}\frac{d^2 g}{d\phi^2}$$

The left-hand side is a function of r only, whereas the right-hand side depends on ϕ only. Therefore this equation can hold for all values of the variables only if both sides are equal to some constant ν^2. As a result, (3.113) is seen to separate into the following two equations:

$$\frac{d^2 f}{dr^2} + \frac{1}{r}\frac{df}{dr} + \left(k_c^2 - \frac{\nu^2}{r^2}\right)f = 0 \qquad (3.114a)$$

$$\frac{d^2 g}{d\phi^2} + \nu^2 g = 0 \qquad (3.114b)$$

In this case the field inside the waveguide must be periodic in ϕ with period 2π, that is, single-valued. It is therefore necessary to choose ν equal to an integer n, in which case the general solution to (3.114b) is

$$g(\phi) = A_1 \cos n\phi + A_2 \sin n\phi$$

where A_1 and A_2 are arbitrary constants.

Equation (3.114a) is Bessel's differential equation and has two solutions (a second-order differential equation always has two independent solutions) $J_\nu(k_c r)$ and $Y_\nu(k_c r)$, called Bessel functions of the first and second kind, respectively, and of order ν.† For the problem under investigation here, only $J_n(k_c r)$ is a physically acceptable solution since $Y_n(k_c r)$ becomes infinite at $r = 0$. The final solution for e_z may thus

† Y_ν is also called a Neumann function.

be expressed as

$$e_z(r, \phi) = (A_1 \cos n\phi + A_2 \sin n\phi) J_n(k_c r) \qquad (3.115)$$

Reference to Appendix II shows that $J_n(x)$ behaves like a damped sinusoidal function and passes through zero in a quasi-periodic fashion. Since e_z must vanish when $r = a$, it is necessary to choose $k_c a$ in such a manner that $J_n(k_c a) = 0$. If the mth root of the equation $J_n(x) = 0$ is designated p_{nm}, the allowed values (eigenvalues) of k_c are

$$k_{c,nm} = \frac{p_{nm}}{a} \qquad (3.116)$$

The values of p_{nm} for the first three modes for $n = 0, 1, 2$ are given in Table 3.3. As in the case of the rectangular guide, there are a doubly infinite number of solutions.

Each choice of n and m specifies a particular TM_{nm} mode (eigenfunction). The integer n is related to the number of circumferential variations in the field, whereas m relates to the number of radial variations. The propagation constant for the nmth mode is given by

$$\beta_{nm} = \left(k_0^2 - \frac{p_{nm}^2}{a^2}\right)^{\frac{1}{2}} \qquad (3.117)$$

the cutoff wavelength by

$$\lambda_{c,nm} = \frac{2\pi a}{p_{nm}} \qquad (3.118)$$

and the wave impedance by

$$Z_{e,nm} = \frac{\beta_{nm}}{k_0} Z_0 \qquad (3.119)$$

A cutoff phenomenon similar to that for the rectangular guide exists. For the dominant TM mode, $\lambda_c = 2\pi a/p_{01} = 2.61a$, a value 30 percent greater than the waveguide diameter.

Expressions for the remaining field components may be derived by using the general equations (3.17). Energy flow and attenuation may

Table 3.3 Values of p_{nm} for TM modes

n	p_{n1}	p_{n2}	p_{n3}
0	2.405	5.520	8.654
1	3.832	7.016	10.174
2	5.135	8.417	11.620

Table 3.4 Properties of modes in circular waveguides

	TE modes	TM modes
H_z	$J_n\left(\dfrac{p'_{nm}r}{a}\right)e^{-j\beta_{nm}z}\begin{Bmatrix}\cos n\phi \\ \sin n\phi\end{Bmatrix}$	0
E_z	0	$J_n\left(\dfrac{p_{nm}r}{a}\right)e^{-j\beta_{nm}z}\begin{Bmatrix}\cos n\phi \\ \sin n\phi\end{Bmatrix}$
H_r	$-\dfrac{j\beta_{nm}p'_{nm}}{ak_{c,nm}^2}J'_n\left(\dfrac{p'_{nm}r}{a}\right)e^{-j\beta_{nm}z}\begin{Bmatrix}\cos n\phi \\ \sin n\phi\end{Bmatrix}$	$-\dfrac{E_\phi}{Z_{e,nm}}$
H_ϕ	$-\dfrac{jn\beta_{nm}}{rk_{c,nm}^2}J_n\left(\dfrac{p'_{nm}r}{a}\right)e^{-j\beta_{nm}z}\begin{Bmatrix}-\sin n\phi \\ \cos n\phi\end{Bmatrix}$	$\dfrac{E_r}{Z_{e,nm}}$
E_r	$Z_{h,nm}H_\phi$	$-\dfrac{j\beta_{nm}p_{nm}}{ak_{c,nm}^2}J'_n\left(\dfrac{p_{nm}r}{a}\right)e^{-j\beta_{nm}z}\begin{Bmatrix}\cos n\phi \\ \sin n\phi\end{Bmatrix}$
E_ϕ	$-Z_{h,nm}H_r$	$-\dfrac{jn\beta_{nm}}{rk_{c,nm}^2}J_n\left(\dfrac{p_{nm}r}{a}\right)e^{-j\beta_{nm}z}\begin{Bmatrix}-\sin n\phi \\ \cos n\phi\end{Bmatrix}$
β_{nm}	$\left[k_0^2 - \left(\dfrac{p'_{nm}}{a}\right)^2\right]^{1/2}$	$\left[k_0^2 - \left(\dfrac{p_{nm}}{a}\right)^2\right]^{1/2}$
$Z_{h,nm}$	$\dfrac{k_0}{\beta_{nm}}Z_0$	
$Z_{e,nm}$		$\dfrac{\beta_{nm}}{k_0}Z_0$
$k_{c,nm}$	$\dfrac{p'_{nm}}{a}$	$\dfrac{p_{nm}}{a}$
$\lambda_{c,nm}$	$\dfrac{2\pi a}{p'_{nm}}$	$\dfrac{2\pi a}{p_{nm}}$
Power	$\dfrac{Z_0 k_0 \beta_{nm}\pi}{4k_{c,nm}^4}(p'^2_{nm}-n^2)J_n^2(p'_{nm})\epsilon_{0n}$	$\dfrac{Y_0 k_0 \beta_{nm}\pi}{4k_{c,nm}^4}p_{nm}^2[J'_n(k_{c,nm}a)]^2\epsilon_{0n}$
α	$\dfrac{R_m}{aZ_0}\left(1-\dfrac{k_{c,nm}^2}{k_0^2}\right)^{-1/2}\times\left[\dfrac{k_{c,nm}^2}{k_0^2}+\dfrac{n^2}{(p'_{nm})^2-n^2}\right]$	$\dfrac{R_m}{aZ_0}\left(1-\dfrac{k_{c,nm}^2}{k_0^2}\right)^{-1/2}$

be determined by methods similar to those used for the rectangular guide. A summary of the results is given in Table 3.4.

TE modes

The solution for TE modes parallels that for the TM modes with the exception that the boundary conditions require that $\partial h_z/\partial r$ vanish at

$r = a$. An appropriate solution for h_z is

$$h_z(r, \phi) = (B_1 \cos n\phi + B_2 \sin n\phi) J_n(k_c r) \tag{3.120}$$

with the requirement that

$$\frac{dJ_n(k_c r)}{dr} = 0 \quad \text{at } r = a \tag{3.121}$$

The roots of (3.121) will be designated by p'_{nm}; so the eigenvalues $k_{c,nm}$ are given by

$$k_{c,nm} = \frac{p'_{nm}}{a} \tag{3.122}$$

Table 3.5 lists the values of the roots for the first few modes. Note

Table 3.5 Values of p'_{nm} for TE modes

n	p'_{n1}	p'_{n2}	p'_{n3}
0	3.832	7.016	10.174
1	1.841	5.331	8.536
2	3.054	6.706	9.970

that $p'_{0m} = p_{1m}$ since $dJ_0(x)/dx = -J_1(x)$, and hence the TE$_{0m}$ and TM$_{1m}$ modes are degenerate.

The first TE mode to propagate is the TE$_{11}$ mode, having a cutoff wavelength $\lambda_{c,11} = 3.41a$. This mode is seen to be the dominant mode for the circular waveguide, and is normally the one used. A sketch of the field lines in the transverse plane for this mode is given in Fig. 3.14.

If the expression for the attenuation constant for TE modes is examined, it will be seen that, for the TE$_{0m}$ modes, the attenuation is

$$\alpha = \frac{R_m}{aZ_0} \frac{f_{c,0m}^2}{f(f^2 - f_{c,0m}^2)^{\frac{1}{2}}} \tag{3.123}$$

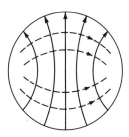

Fig. 3.14 Field lines for TE$_{11}$ mode in a circular guide.

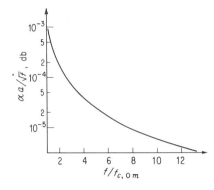

Fig. 3.15 Attenuation of low-loss TE_{0m} modes in a circular copper waveguide, f in units of 10^{10} cps, $f_{c,10} = 1.83 \times 10^{10}/a$ cps, where a = radius in centimeters.

and falls off as $f^{-\frac{3}{2}}$ for high frequencies since R_m increases as $f^{\frac{1}{2}}$. The rapid decrease in attenuation with increasing frequency is a unique property of the TE_{0m} modes in circular waveguides and makes possible the construction of very long low-loss waveguide communication links.[†] In Fig. 3.15 the attenuation in decibels for the TE_{0m} modes as a function of frequency for a copper waveguide is plotted. Although very low attenuations are achieved for frequencies well above the cutoff frequency $f_{c,01}$, certain practical difficulties are encountered which limit the overall performance of such guides to less than the theoretical predictions. These practical difficulties stem from operating the guide at a frequency well above the dominant-TE_{11}-mode cutoff frequency, i.e., in a frequency region where many modes can propagate. Any small irregularity in the guide causes some of the power in the TE_{01} mode to be converted into power in other modes (mode conversion). Two serious effects arise from mode conversion. The most obvious effect is the loss of power in the desired TE_{01} mode when some of this power is converted into other more rapidly attenuating modes. The more serious effect arises when the power in the TE_{01} mode is converted into power in other modes, with different propagation phase constants and at a position farther along the guide converted back into a TE_{01} mode, since this leads to signal distortion. To avoid signal distortion arising from this mode conversion and reconversion, it is desirable that the waveguide have a high attenuation for the undesired modes, so that these will be rapidly attenuated and not converted back into a TE_{01} mode. The currents associated with the TE_{0m} modes are in the circumferential direction only. This property may be utilized to construct mode filters that will suppress modes having

† S. E. Miller, Waveguide as a Communication Medium, *Bell System Tech. J.*, vol. 33, pp. 1209–1265, November, 1954; also Millimeter Waves in Communications, *Proc. Symp. Millimeter Waves*, Polytechnic Institute of Brooklyn, New York, 1959, pp. 25–43.

currents directed along the waveguide axis. Special waveguide linings exhibiting high conductivity in the circumferential direction and low conductivity in the axial direction have been used as mode suppressors. Another technique that shows considerable promise is the use of a small-pitch helical winding with a suitable supporting jacket for the waveguide. The helix guide has a very low conductivity in the axial direction, whereas in the circumferential direction the conductivity is essentially the same as for a solid cylindrical waveguide wall.†

★ 3.8 Surface waveguides

The rectangular and circular waveguides discussed in the two previous sections are examples of closed-boundary waveguides. Another class of waveguides, generally called open-boundary waveguides, is characterized by having the fields outside the boundary. These guides support modes of propagation called surface waves. Examples of surface-wave guides are a dielectric rod, dielectric-coated conducting wire, dielectric sheet placed on a conducting plane, a corrugated conducting cylinder, etc.‡ The field of a surface wave decays exponentially in the direction normal to and away from the guiding structure. At the shorter wavelengths this results in a field confinement sufficiently great to make the effective guide size small enough to have practical utility. Surface waveguides are therefore of considerable interest for potential applications in the millimeter wavelength region and, of course, the submillimeter and optical regions, where the construction of a conventional hollow-pipe waveguide is difficult or impossible. In this section a brief intro-

---

† S. P. Morgan and J. A. Young, Helix Waveguide, *Bell System Tech. J.*, vol. 35, pp. 1347–1384, November, 1956.
  J. A. Morrison, Heat Loss of Circular Electric Waves in Helix Waveguides, *IRE Trans.*, vol. MTT-6, pp. 173–177, April, 1958.
  T. Hosono and S. Kohno, The Transmission of TE_{01} Waves in Helix Waveguides, *IRE Trans.*, vol. MTT-7, pp. 370–373, July, 1959.
  G. W. Luderer and H. G. Unger, Circular Electric Wave Propagation in Periodic Structures, *Bell System Tech. J.*, vol. 43, pp. 755–783, March, 1964.

‡ For example, see:
  C. H. Chandler, An Investigation of Dielectric Rod as a Wave Guide, *J. Appl. Phys.*, vol. 20, pp. 1188–1192, December, 1949.
  S. P. Schlesinger and D. D. King, Dielectric Image Lines, *IRE Trans.*, vol. MTT-6, pp. 291–299, July, 1958.
  R. S. Elliott, On the Theory of Corrugated Plane Surfaces, *IRE Trans.*, vol. AP-2, pp. 71–81, April, 1954.
  H. E. M. Barlow and A. E. Karbowiak, An Experimental Investigation of the Properties of Corrugated Cylindrical Surface Waveguides, *Proc. IEE*, vol. 101, pt. III, pp. 182–188, May, 1954.
  M. Cohn, TE Modes of the Dielectric Loaded Trough Line, *IRE Trans.*, vol. MTT-8, pp. 449–454, July, 1960.

114 Foundations for microwave engineering

duction to surface waveguides is presented. A more extensive account may be found elsewhere.†

Surface waves on a dielectric-coated plane‡

One of the simplest surface-wave structures to analyze is that consisting of a dielectric sheet of thickness t placed on an infinite conducting plane, as in Fig. 3.16. A solution for TM modes having $H_z = 0$ and no variation with respect to y is determined below. In the dielectric denote k_c by k_d and in the air region $x > t$, let k_c equal jh. The axial electric field e_z is a solution of

$$\nabla_t^2 e_z + k_c^2 e_z = 0$$

or

$$\frac{\partial^2 e_z}{\partial x^2} + k_d^2 e_z = 0 \qquad 0 < x < t \tag{3.124a}$$

$$\frac{\partial^2 e_z}{\partial x^2} - h^2 e_z = 0 \qquad x > t \tag{3.124b}$$

The cutoff wave number k_c must be different in the two regions since $k_c^2 = k^2 - \beta^2$, where $k = \omega(\mu\epsilon)^{\frac{1}{2}}$, the latter quantity depending on the medium. Appropriate solutions of these equations in the two regions such that e_z vanishes on the conducting plane at $x = 0$ and at x approach-

---

† R. E. Collin, "Field Theory of Guided Waves," chap. 11, McGraw-Hill Book Company, New York, 1960.
  H. E. M. Barlow and J. Brown, "Radio Surface Waves," Oxford University Press, Fair Lawn, N.J., 1962.
‡ S. S. Attwood, Surface Wave Propagation over a Coated Plane Conductor, *J. Appl. Phys.*, vol. 22, pp. 504–509, April, 1954.

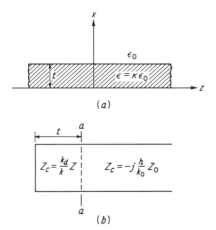

Fig. 3.16 (a) Dielectric-coated plane; (b) equivalent circuit of transverse section.

ing infinity are

$$e_z = A \sin k_d x \qquad 0 < x < t \qquad (3.125a)$$

$$e_z = Be^{-hx} \qquad x > t \qquad (3.125b)$$

where A and B are constants. The reason for choosing $k_c = jh$ in the air region was in anticipation of obtaining a decaying field for large x as in (3.125b). At the air-dielectric boundary $x = t$, the tangential electric and magnetic fields must be continuous for all values of z. This requires that the propagation constant β be the same in both regions, which will be the case if

$$\beta^2 = \kappa k_0^2 - k_d^2 = k_0^2 + h^2$$

or

$$k_d^2 + h^2 = (\kappa - 1)k_0^2 \qquad (3.126)$$

Also, it is necessary that

$$A \sin k_d t = Be^{-ht} \qquad (3.127)$$

in order that e_z be continuous at $x = t$. One further relation is necessary in order to determine the three quantities A/B, k_d, and h. This relation is obtained from the requirement that h_y be continuous at $x = t$. Since h_y is given by (from 3.17b and 3.17c)

$$h_y = -\frac{j\beta}{k_c^2}\frac{k}{\beta} Y(\mathbf{a}_z \times \nabla_t e_z) \cdot \mathbf{a}_y$$

where kY equals $k_0 Y_0$ in air and $\kappa k_0 Y_0$ in the dielectric, this third relation is

$$\frac{\kappa}{k_d^2} \frac{\partial e_z}{\partial x}\bigg|_{x=t_-} = \frac{-1}{h^2} \frac{\partial e_z}{\partial x}\bigg|_{x=t_+}$$

or

$$\kappa A k_d^{-1} \cos k_d t = h^{-1} B e^{-ht} \qquad (3.128)$$

Dividing (3.128) by (3.127) and multiplying both sides by t give

$$\kappa h t = k_d t \tan k_d t \qquad (3.129)$$

A simultaneous solution of (3.126) and (3.129) yields the allowed values of k_d and h. If a solution for h real exists, this solution is a surface wave, since its field decays exponentially in the x direction and propagates in the z direction.

The eigenvalue equation (3.129) may be derived in a more direct manner by using the *transverse-resonance technique*. This technique is an application of transmission-line theory to the equivalent circuit of a transverse section of the guiding structure and is applicable to any waveguide that is uniform in the direction of propagation, i.e., along the z axis. In the present case the mode being considered is an E mode with respect to the x axis. The propagation constant in the x direction has been denoted as jk_d in the dielectric region and as h in the free-space region. Using (3.17d) and interpreting $j\beta$ in that equation as the propagation constant in the x direction, the wave impedance for the E mode in the x direction is seen to be given by $k_d Z/k$ in the dielectric region and by hZ_0/jk_0 in the air region. The equivalent circuit of the transverse section is thus a short-circuited transmission line of length t and impedance $k_d Z/k$ connected to an infinite line with imaginary impedance $-jhZ_0/k_0$, as in Fig. 3.16b. At the plane aa, the transverse resonance condition requires that the sum of the impedances seen looking toward the short circuit and that at the input to the infinite line vanish. Thus we have

$$j\frac{k_d Z}{k} \tan k_d t - j\frac{hZ_0}{k_0} = 0$$

or

$$k_d \tan k_d t = \kappa h$$

since $Z/k = Z_0/\kappa k_0$. This equation is the same as (3.129).

Although the transverse-resonance technique leads to the required eigenvalue equation in a direct manner, it does require a priori knowledge of the type of field the structure will support so that an appropriate equivalent circuit of the transverse section may be constructed. Further examples of the application of the method are given in the problems at the end of this chapter.

By using graphical methods the solutions for k_d and h are readily found. Equation (3.129) is plotted in the ht-$k_d t$ plane in Fig. 3.17. Superimposed on these curves are the circles $(ht)^2 + (k_d t)^2 = (\kappa - 1)(k_0 t)^2$. Common points of intersection yield the allowed solutions, or eigen-

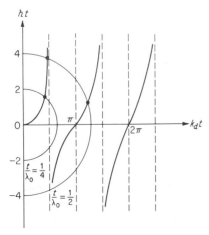

Fig. 3.17 Graphical solution for k_d and h.

values. The curves in Fig. 3.17 are plotted for $\kappa = 2.56$ (polystyrene), $t/\lambda_0 = 0.25, 0.5$. Note that no matter how small t or how large λ_0, there is always one solution at least. This occurs from an intersection with the first branch of the tangent function, and is called the TM_0 mode. For larger values of t or smaller values of λ_0, the circles intersect with other branches of the tangent function, yielding higher-order TM_n modes. The points of intersection in the region $ht < 0$ are not physically valid solutions since the corresponding field would be exponentially increasing for large positive x.

Unlike the rectangular guide, the TM_0 surface wave has no low-frequency cutoff. However, this is by no means a general property of surface-wave guides. A feature of surface-wave guides is that, at any particular frequency, only a finite number of modes exist. Also, when a mode does exist, its phase velocity is less than that of light in free space since

$$v_p = \frac{k_0}{\beta} c = \frac{k_0 c}{(k_0^2 + h^2)^{\frac{1}{2}}} < c \tag{3.130}$$

For this reason a surface wave is called a *slow wave*. This property is not too surprising since it could be expected that the phase velocity should be between that of light in free space and that in the dielectric, which is $c\kappa^{-\frac{1}{2}}$. This is actually the case since

$$k_0 < \beta = (k_0^2 + h^2)^{\frac{1}{2}} = (\kappa k_0^2 - k_d^2)^{\frac{1}{2}} < \kappa^{\frac{1}{2}} k_0$$

Using the general equations (3.17), the complete solution for the TM surface-wave mode is found to be

$$E_z = A \sin k_d x \, e^{-j\beta z}$$

$$E_x = -\frac{j\beta}{k_d} A \cos k_d x \, e^{-j\beta z} \tag{3.131a}$$

$$H_y = -j\kappa \frac{k_0 Y_0}{k_d} A \cos k_d x \, e^{-j\beta z}$$

for $0 < x < t$, and

$$E_z = A \sin k_d t \, e^{-h(x-t)-j\beta z}$$

$$E_x = -\frac{j\beta}{h} A \sin k_d t \, e^{-h(x-t)-j\beta z} \tag{3.131b}$$

$$H_y = -j \frac{k_0 Y_0}{h} A \sin k_d t \, e^{-h(x-t)-j\beta z}$$

for $x > t$. The constant B has been replaced by $Ae^{ht} \sin k_d t$ as obtained from (3.127). Expressions for power and attenuation caused by dielectric and conductor losses may be obtained by methods analogous to those used earlier. The details are left as problems.

Dielectric-coated wire guide†

As a second example of a surface-wave guide, the case of a circular wire coated with dielectric is considered. This structure is illustrated in Fig. 3.18, and consists of a conducting wire of radius a surrounded by a thin dielectric sleeve of thickness t and outer radius b.

A solution for a circularly symmetric E mode, that is, TM mode, will be constructed. Again, let $k_c = k_d$ in the dielectric and let $k_c = jh$ in the surrounding air region $r > b$. With no ϕ variation, e_z is a solution of Bessel's equation,

$$\frac{d^2 e_z}{dr^2} + \frac{1}{r}\frac{d e_z}{dr} + k_c^2 e_z = 0 \tag{3.132}$$

as reference to (3.114) shows. In the region $a < r < b$, which does not include the origin, the general solution for e_z is a linear combination of the zeroth-order Bessel functions of the first and second kinds; i.e.,

$$e_z = C_1 J_0(k_d r) + C_2 Y_0(k_d r) \qquad a < r < b \tag{3.133}$$

In the region $r > b$ the field must decay for large r in order to correspond to a surface wave. A suitable linear combination of $J_0(jhr)$ and $Y_0(jhr)$ will provide such a decay. For hr large, these functions have the following asymptotic forms:

$$J_0(jhr) \sim \sqrt{\frac{2}{jhr\pi}} \cos\left(jhr - \frac{\pi}{4}\right)$$

$$Y_0(jhr) \sim \sqrt{\frac{2}{jhr\pi}} \sin\left(jhr - \frac{\pi}{4}\right)$$

from which it is seen that

$$J_0(jhr) + jY_0(jhr) \sim \sqrt{\frac{2}{jhr\pi}}\, e^{-hr - j\pi/4}$$

Instead of using the functions J_n and Y_n when the arguments are imaginary, new functions with real arguments are introduced, according to

† G. Goubau, Surface Waves and Their Applications to Transmission Lines, *J. Appl. Phys.*, vol. 21, pp. 1119–1128, November, 1950; also Single Conductor Surface Wave Transmission Lines, *Proc. IRE*, vol. 39, pp. 619–624, June, 1951.

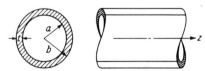

Fig. 3.18 Dielectric-coated wire.

the relations

$$I_n(x) = j^{-n}J_n(jx) \tag{3.134a}$$

$$K_n(x) = \frac{\pi}{2}j^{n+1}[J_n(jx) + jY_n(jx)] \tag{3.134b}$$

Clearly, K_n decays exponentially; so this function with $n = 0$ is appropriate for describing e_z in the region $r > b$. The function I_n cannot be used because it behaves like e^{hr} at infinity. Thus

$$e_z = C_3 K_0(hr) \qquad r > b \tag{3.135}$$

At $r = a$, e_z must vanish; so it is necessary that

$$C_1 J_0(k_d a) + C_2 Y_0(k_d a) = 0$$

or

$$C_2 = -C_1 \frac{J_0(k_d a)}{Y_0(k_d a)} \tag{3.136}$$

At $r = b$, e_z must be continuous, and hence

$$C_3 K_0(hb) = C_1 \left[J_0(k_d b) - \frac{Y_0(k_d b) J_0(k_d a)}{Y_0(k_d a)} \right] \tag{3.137}$$

This equation contains essentially three unknown constants, h, k_d, and C_3/C_1. Note that only the ratio C_3/C_1 is considered as unknown since the relative amplitude of the field is arbitrary. A second equation connecting these parameters is

$$\beta^2 = \kappa k_0^2 - k_d^2 = k_0^2 + h^2$$

or

$$k_d^2 + h^2 = (\kappa - 1)k_0^2 \tag{3.138}$$

A third relation is required before a solution can be found. This relation will be obtained from the boundary condition that the tangential magnetic field be continuous at $r = b$.

In terms of e_z, the transverse electric field is given by [see (3.17)]

$$\mathbf{e} = -\frac{j\beta}{k_c^2} \nabla_t e_z$$

In cylindrical coordinates

$$\nabla_t = \mathbf{a}_r \frac{\partial}{\partial r} + \frac{\mathbf{a}_\phi}{r} \frac{\partial}{\partial \phi}$$

and hence

$$\mathbf{e} = e_r \mathbf{a}_r = -\frac{j\beta}{k_c^2} \frac{\partial e_z}{\partial r} \mathbf{a}_r$$

The magnetic field is given by

$$\mathbf{h} = Y_e \mathbf{a}_z \times \mathbf{e}$$

where $Y_e = kY/\beta$. On noting that $k_c = k_d$, $kY = \kappa k_0 Y_0$ in the dielectric and that $k_c = jh$, $kY = k_0 Y_0$ in air, the transverse fields are found to be

$$e_r = -\frac{j\beta}{k_d^2}[C_1 k_d J_0'(k_d r) + C_2 k_d Y_0'(k_d r)] \tag{3.139a}$$

$$h_\phi = \frac{\kappa k_0 Y_0}{\beta} e_r \tag{3.139b}$$

for $a < r < b$, and

$$e_r = \frac{j\beta}{h^2} h C_3 K_0'(hr) \tag{3.140a}$$

$$h_\phi = \frac{k_0 Y_0}{\beta} e_r \tag{3.140b}$$

for $r > b$, where the prime denotes differentiation with respect to the argument $k_d r$ or hr; for example, $\partial K_0/\partial r = h\, \partial K_0/\partial hr = hK_0'$.

Continuity of h_ϕ at $r = b$ yields the equation

$$\frac{jk_0 Y_0}{h} C_3 K_0'(hb) = \frac{-j\kappa k_0 Y_0}{k_d}[C_1 J_0'(k_d b) + C_2 Y_0'(k_d b)]$$

or

$$C_3 k_d K_0'(hb) = -\kappa h C_1 \left[J_0'(k_d b) - \frac{J_0(k_d a) Y_0'(k_d b)}{Y_0(k_d a)} \right] \tag{3.141}$$

If we now divide (3.141) by (3.137), we obtain

$$\frac{k_d K_0'(hb)}{K_0(hb)} = -\kappa h \frac{J_0'(k_d b) Y_0(k_d a) - Y_0'(k_d b) J_0(k_d a)}{J_0(k_d b) Y_0(k_d a) - J_0(k_d a) Y_0(k_d b)} \tag{3.142}$$

This transcendental equation must be solved simultaneously with (3.138) in order to determine h and k_d. Graphical methods similar to that employed for the previous dielectric-sheet problem may be used. However, for the case when $t = b - a \ll a$, an approximate solution is readily obtained by expanding the Bessel functions that involve $k_d b$ in a Taylor series about the point $k_d a$. Thus, for example,

$$J_0(k_d b) \approx J_0(k_d a) + \left.\frac{dJ_0(k_d b)}{dk_d b}\right|_{k_d a} k_d(b - a)$$

The parameter h is of the same order of magnitude as k_0, and since $a \ll \lambda_0$ for a typical surface-wave guide, hb is small for small t. In this case

$$K_0(hb) \approx -\ln 0.89 hb$$

$$K_0'(hb) = K_1(hb) \approx \frac{1}{hb}$$

Using the above Taylor series expansion and the small-argument approximations for K_0 and K_1 reduces (3.142) to the following simple form:

$$\kappa h^2 b \ln 0.89 hb = -(\kappa - 1)k_0^2 t + h^2 t \approx -(\kappa - 1)k_0^2 t \quad (3.143)$$

since, as this equation shows, $h^2 \ll k_0^2$.

As a typical example consider a copper wire with $a = 0.09$ cm, $b = 0.1$ cm, $t = 0.01$ cm, $\kappa = 2.56$, and $\lambda_0 = 3.14$ cm. The solution for h gives $h = 0.258$ neper/cm. Since $k_0 = 2$, $h^2 \ll k_0^2$, which justifies dropping the last term in (3.143). The propagation factor β comes out equal to 2.02 rad/cm. For hr large,

$$K_0(hr) \sim \left(\frac{\pi}{2hr}\right)^{\frac{1}{2}} e^{-hr}$$

so in a distance of two or three wavelengths the field has decayed to a relatively small value; e.g., for $r = 2\lambda_0$ the field is reduced by a factor 0.052 from its value at $r = b$. The effective guide diameter is about four wavelengths, or 12.5 cm in this case. Although this might be considered inconveniently large at this wavelength, at much higher frequencies, say $\lambda_0 = 3$ mm, an effective diameter of four wavelengths is quite small. For this reason surface-wave guides are normally not used except for high frequencies. By increasing the dielectric constant κ or the thickness t, the decay constant h can be increased, resulting in a field more tightly bound to the surface. Although this is an advantage, the greater concentration of the field near the boundary increases the losses, which results in a greater attenuation in the direction of propagation z.

3.9 Power orthogonality

In a hollow cylindrical metal waveguide that is free of all losses (perfectly conducting walls), each possible mode of propagation carries energy independently of all other modes that may be present. This result is readily proved for any particular guide by using the expressions for the fields appropriate to that guide. Since it can also be proved in general without any great difficulty, this is done in the following discussion.

Let $\mathbf{E}_1$, $\mathbf{H}_1$ and $\mathbf{E}_2$, $\mathbf{H}_2$ be two linearly independent solutions (modes) of Maxwell's equations in a given waveguide. These fields may be expressed as (for TE modes place $\mathbf{e}_z = 0$ and for TM modes put $\mathbf{h}_z = 0$)

$$\mathbf{E}_1 = \mathbf{e}_1 e^{-j\beta_1 z} + \mathbf{e}_{z1} e^{-j\beta_1 z} \quad (3.144a)$$

$$\mathbf{H}_1 = \mathbf{h}_1 e^{-j\beta_1 z} + \mathbf{h}_{z1} e^{-j\beta_1 z} \quad (3.144b)$$

$$\mathbf{E}_2 = \mathbf{e}_2 e^{-j\beta_2 z} + \mathbf{e}_{z2} e^{-j\beta_2 z} \quad (3.145a)$$

$$\mathbf{H}_2 = \mathbf{h}_2 e^{-j\beta_2 z} + \mathbf{h}_{z2} e^{-j\beta_2 z} \quad (3.145b)$$

In order to prove that these modes carry energy independently (power orthogonality), it must be shown that

$$\text{Re} \int_{S_0} (\mathbf{E}_1 + \mathbf{E}_2) \times (\mathbf{H}_1^* + \mathbf{H}_2^*) \cdot \mathbf{a}_z \, dS$$
$$= \text{Re} \int_{S_0} (\mathbf{E}_1 \times \mathbf{H}_1^* + \mathbf{E}_2 \times \mathbf{H}_2^*) \cdot \mathbf{a}_z \, dS$$
$$= \int_{S_0} (\mathbf{e}_1 \times \mathbf{h}_1 + \mathbf{e}_2 \times \mathbf{h}_2) \cdot \mathbf{a}_z \, dS$$

where $\mathbf{e}_1$ and $\mathbf{h}_1$ have been assumed to be real, which is permissible in a loss-free guide since then β is real, hence $k_c^2 = k_0^2 - \beta^2$ is real, and thus the solutions to $(\nabla_t^2 + k_c^2)(\mathbf{e}_1, \mathbf{h}_1) = 0$ may be chosen to be real. In the above integral S_0 is an arbitrary cross-sectional plane in the guide. In other words, the interaction terms

$$\int_{S_0} (\mathbf{E}_1 \times \mathbf{H}_2^* + \mathbf{E}_2 \times \mathbf{H}_1^*) \cdot \mathbf{a}_z \, dS$$
$$= \int_{S_0} (\mathbf{e}_1 \times \mathbf{h}_2 e^{-j\beta_1 z + j\beta_2 z} + \mathbf{e}_2 \times \mathbf{h}_1 e^{j\beta_1 z - j\beta_2 z}) \cdot \mathbf{a}_z \, dS$$

must vanish so that the rate of energy flow (power) is the sum of those due to each mode separately.

Consider a section of waveguide bounded by planes at z_1 and z_2, as in Fig. 3.19. The closed surface S bounding the volume V consists of the perfectly conducting walls S_w and the transverse surfaces S_1 and S_2. Since V contains no sources, the fields in V satisfy the equations $\nabla \times \mathbf{E}_i = -j\omega\mu\mathbf{H}_i$ and $\nabla \times \mathbf{H}_i = j\omega\epsilon\mathbf{E}_i$, where $i = 1$ or 2. Expanding $\nabla \cdot (\mathbf{E}_1 \times \mathbf{H}_2 - \mathbf{E}_2 \times \mathbf{H}_1)$ gives

$$j\omega\mu(\mathbf{H}_1 \cdot \mathbf{H}_2 - \mathbf{H}_1 \cdot \mathbf{H}_2) + j\omega\epsilon(\mathbf{E}_1 \cdot \mathbf{E}_2 - \mathbf{E}_1 \cdot \mathbf{E}_2) = 0$$

upon using the above equations. Thus, since the divergence is zero, the total flux of $\mathbf{E}_1 \times \mathbf{H}_2 - \mathbf{E}_2 \times \mathbf{H}_1$ through any closed surface is zero. Hence

$$\oint_S (\mathbf{E}_1 \times \mathbf{H}_2 - \mathbf{E}_2 \times \mathbf{H}_1) \cdot \mathbf{n} \, dS = 0 \quad (3.146)$$

On S_w, $\mathbf{E}_1 \times \mathbf{H}_2 \cdot \mathbf{n} = \mathbf{n} \times \mathbf{E}_1 \cdot \mathbf{H}_2 = 0$, and similarly for the other term. Also, $\mathbf{n} = \mathbf{a}_z$ on S_1 and $-\mathbf{a}_z$ on S_2. Hence (3.146) becomes

$$\int_{S_1} (\mathbf{E}_1 \times \mathbf{H}_2 - \mathbf{E}_2 \times \mathbf{H}_1) \cdot \mathbf{a}_z \, dS - \int_{S_2} (\mathbf{E}_1 \times \mathbf{H}_2 - \mathbf{E}_2 \times \mathbf{H}_1) \cdot \mathbf{a}_z \, dS = 0$$
$$(3.147)$$

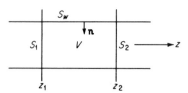

Fig. 3.19 Section of waveguide bounded by surfaces S_1, S_2, and S_w.

Any term in (3.147) involves only the transverse fields since only $(\mathbf{e} \times \mathbf{h})$ is directed in the z direction; i.e., the other parts of $(\mathbf{e} + \mathbf{e}_z) \times (\mathbf{h} + \mathbf{h}_z)$ are normal to $\mathbf{a}_z$. If the expressions (3.144) and (3.145) are substituted into (3.147), then the integral over S_1 depends on z_1 and that over S_2 on z_2 only. Since z_1 and z_2 can be arbitrarily chosen, it is necessary for each integral to be zero (an exception is when $\beta_1 = -\beta_2$, in which case the integrals are independent of z_1 and z_2 and cancel identically). Thus, when $\beta_1 \neq -\beta_2$,

$$\int_{S_0} (\mathbf{E}_1 \times \mathbf{H}_2 - \mathbf{E}_2 \times \mathbf{H}_1) \cdot \mathbf{a}_z \, dS = 0$$

where S_0 is any transverse plane. Furthermore, this integral equals

$$e^{-j\beta_1 z_0 - j\beta_2 z_0} \int_{S_0} (\mathbf{e}_1 \times \mathbf{h}_2 - \mathbf{e}_2 \times \mathbf{h}_1) \cdot \mathbf{a}_z \, dS = 0$$

or

$$\int_{S_0} (\mathbf{e}_1 \times \mathbf{h}_2 - \mathbf{e}_2 \times \mathbf{h}_1) \cdot \mathbf{a}_z \, dS = 0 \qquad (3.148)$$

In place of the field $\mathbf{E}_2, \mathbf{H}_2$ given by (3.145), an equally valid field to use in (3.146) and the subsequent equations is the same mode propagating in the $-z$ direction. In this case

$$\mathbf{E}_2 = \mathbf{e}_2 e^{j\beta_2 z} - \mathbf{e}_{z2} e^{j\beta_2 z} \qquad (3.149a)$$

$$\mathbf{H}_2 = -\mathbf{h}_2 e^{j\beta_2 z} + \mathbf{h}_{z2} e^{j\beta_2 z} \qquad (3.149b)$$

In place of (3.148) it is now found that (replace $\mathbf{h}_2$ by $-\mathbf{h}_2$)

$$\int_{S_0} (-\mathbf{e}_1 \times \mathbf{h}_2 - \mathbf{e}_2 \times \mathbf{h}_1) \cdot \mathbf{a}_z \, dS = 0 \qquad (3.150)$$

Addition and subtraction of (3.148) and (3.150) now show that

$$\int_{S_0} \mathbf{e}_1 \times \mathbf{h}_2 \cdot \mathbf{a}_z \, dS = \int_{S_0} \mathbf{e}_2 \times \mathbf{h}_1 \cdot \mathbf{a}_z \, dS = 0 \qquad (3.151)$$

which completes the proof that the power-interaction terms between two nondegenerate ($\beta_1 \neq \beta_2$) modes vanish.

When $\beta_1 = \beta_2$, the modes are degenerate and the foregoing proof breaks down. If the modes are considered to propagate in the $+z$ and $-z$ directions, the two integrals in (3.147) become independent of z_1 and z_2 and hence are identically the same and cancel. The result (3.150) therefore does not necessarily follow. On the other hand, if both modes are considered to propagate in the $+z$ direction or the $-z$ direction, the two integrals in (3.147) involve factors $e^{\mp j\beta_1 z} e^{\mp j\beta_2 z}$ and depend on z. In this case each integral must vanish separately and the result (3.148) may be obtained. However, without the companion relationship (3.150), it is not possible to show that each term vanishes separately.

When degenerate modes do not satisfy a power-orthogonality relation, it is possible to define new modes which are linear combinations of the old such that power orthogonality holds. Thus define new modes

$$\mathbf{E}_1' = \mathbf{E}_1 \tag{3.152a}$$
$$\mathbf{H}_1' = \mathbf{H}_1 \tag{3.152b}$$
$$\mathbf{E}_2' = \mathbf{E}_1 + C\mathbf{E}_2 \tag{3.153a}$$
$$\mathbf{H}_2' = \mathbf{H}_1 + C\mathbf{H}_2 \tag{3.153b}$$

The new modes will be orthogonal if the constant C is chosen so that the interaction terms vanish. The rate of energy flow is given by

$$\tfrac{1}{2} \operatorname{Re} \int_{S_0} \operatorname{Re} (\mathbf{E}_1' + \mathbf{E}_2') \times (\mathbf{H}_1' + \mathbf{H}_2')^* \cdot \mathbf{a}_z \, dS$$
$$= \tfrac{1}{2} \operatorname{Re} \int_{S_0} [\mathbf{E}_1' \times (\mathbf{H}_1')^* + \mathbf{E}_2' \times (\mathbf{H}_2')^*] \cdot \mathbf{a}_z \, dS$$
$$+ \tfrac{1}{2} \operatorname{Re} \int_{S_0} [\mathbf{E}_1' \times (\mathbf{H}_2')^* + \mathbf{E}_2' \times (\mathbf{H}_1')^*] \cdot \mathbf{a}_z \, dS$$

The interaction term vanishes if

$$\int_{S_0} [\mathbf{E}_1 \times (\mathbf{H}_1 + C\mathbf{H}_2)^* + (\mathbf{E}_1 + C\mathbf{E}_2) \times \mathbf{H}_1^*] \cdot \mathbf{a}_z \, dS = 0$$

Let $P_{ij} = \tfrac{1}{2} \int_{S_0} \mathbf{E}_i \times \mathbf{H}_j^* \cdot \mathbf{a}_z \, dS$, where $i = 1, 2$. Then the preceding equation becomes

$$P_{11} + CP_{12} + P_{11} + CP_{21} = 0$$

or

$$C = -\frac{2P_{11}}{P_{12} + P_{21}} \tag{3.154}$$

With the constant C chosen in this manner, power orthogonality for the new (primed) modes is ensured. This procedure may be extended in an obvious way when more than two modes are degenerate.

In the case of lossy waveguides, power orthogonality does not hold exactly, although for nondegenerate modes the assumption of power orthogonality is a good approximation for low-loss guides. The nature of the coupling of modes by finite conducting walls is discussed in the next section.

★3.10 Attenuation for degenerate modes

The basic principle employed in the perturbation method of evaluating the attenuation constant for a given waveguide mode is the following: It is assumed that the presence of small losses does not change the field distribution or wall currents to a first approximation. The loss-free field and current solutions are thus used to evaluate the power loss in the medium filling the guide (if present) and the power loss in the finite conducting walls. An attenuation constant is then introduced such that the rate of decrease of power transmission along the guide just equals the power dissipation in the lossy walls and media filling the guide.

If two or more modes are present simultaneously, the method may be applied to each mode individually, provided the power transmitted and the power loss are the sum of those contributed by each mode. This requires that the cross-interaction terms in the expressions for power transmitted and power loss vanish. If this is not the case, the modes are said to be coupled together by the finite losses. In general, the coupling of modes is very small except for degenerate modes. It is also important to note that in many cases even degenerate modes are not coupled together.

Consider two modes with fields $\mathbf{E}_1, \mathbf{H}_1$ and $\mathbf{E}_2, \mathbf{H}_2$. Let the corresponding surface currents on the guide walls be $\mathbf{J}_1$ and $\mathbf{J}_2$, respectively. When the walls have finite conductivity, the electric fields of the modes have tangential components equal to $Z_m \mathbf{J}_1$ and $Z_m \mathbf{J}_2$ instead of zero values at the guide wall, where Z_m is the surface impedance of the wall. In general, $\mathbf{E}_2$ will have a component parallel to $\mathbf{J}_1^*$; that is, $\mathbf{E}_2 \cdot \mathbf{J}_1^* \neq 0$, and similarly, $\mathbf{E}_1 \cdot \mathbf{J}_2^* \neq 0$. This means that the current $\mathbf{J}_1$ can supply power to the field $\mathbf{E}_2$, and $\mathbf{J}_2$ can supply power to $\mathbf{E}_1$. However, if the integral of these products around the guide boundary is zero, there will be no net power transfer. Thus the modes are coupled by the wall losses only if the following integrals do not vanish:

$$\oint_C \mathbf{E}_2 \cdot \mathbf{J}_1^* \, dl \qquad \oint_C \mathbf{E}_1 \cdot \mathbf{J}_2^* \, dl$$

where the integration is around the guide boundary. Since

$$\mathbf{E}_2 \cdot \mathbf{J}_1^* = Z_m \mathbf{J}_2 \cdot \mathbf{J}_1^*$$

the vanishing of these integrals occurs only when the currents of the two modes are orthogonal with respect to integration around the guide boundary.

It was shown in the previous section that when the modes are degenerate, the energy flow along the guide is not necessarily the sum of those from each mode. However, it was shown that, by defining new modes which were linear combinations of the old modes, the power-interaction terms for the new modes could be made to vanish. This technique may be extended not only so that the new modes have zero power-interaction terms, but also so that the surface currents associated with these new modes are orthogonal. These new modes are then completely uncoupled, and their attenuation may be determined from a consideration of each mode individually, following the usual perturbation method (often called the power-loss method). The mathematical formulation of the foregoing ideas is now presented.

★Mathematical description of mode coupling

Let there be N degenerate modes in a loss-free waveguide. Let the fields of these modes be given by

$$\mathbf{E}_n = (\mathbf{e}_n + \mathbf{e}_{zn})e^{-j\beta_0 z} \tag{3.155a}$$

$$\mathbf{H}_n = (\mathbf{h}_n + \mathbf{h}_{zn})e^{-j\beta_0 z} \tag{3.155b}$$

where $n = 1, 2, \ldots, N$. All modes have the same propagation constant β_0. If some of the modes are TM, then $\mathbf{h}_{zn} = 0$, whereas for TE modes, $\mathbf{e}_{zn} = 0$. Let the surface currents associated with these modes be

$$\mathbf{J}_n = \mathbf{n} \times \mathbf{H}_n \qquad n = 1, 2, \ldots, N \tag{3.156}$$

Define the following interaction terms for the above modes:

$$P_{nm} = \tfrac{1}{2} \int_S \mathbf{E}_n \times \mathbf{H}_m^* \cdot \mathbf{a}_z \, dS \tag{3.157}$$

$$W_{nm} = \frac{R_m}{2} \oint_C \mathbf{J}_n \cdot \mathbf{J}_m^* \, dl \tag{3.158}$$

where R_m is the resistive part of Z_m.

A new set of modes will now be constructed such that the cross-interaction terms are zero. Let the new modes be defined as

$$\mathbf{E}'_s = \sum_{n=1}^{N} C_n{}^s \mathbf{E}_n \tag{3.159a}$$

$$\mathbf{H}'_s = \sum_{n=1}^{N} C_n{}^s \mathbf{H}_n \tag{3.159b}$$

where the $C_n{}^s$ are amplitude constants and $s = 1, 2, \ldots, N$, so that there will be N new modes which are linear combinations of the N old modes. Define interaction terms for the new modes as

$$P'_{sr} = \tfrac{1}{2} \int_S \mathbf{E}'_s \times (\mathbf{H}'_r)^* \cdot \mathbf{a}_z \, dS \tag{3.160}$$

$$W'_{sr} = \frac{R_m}{2} \oint_C \mathbf{J}'_s \cdot (\mathbf{J}'_r)^* \, dl \tag{3.161}$$

The constants $C_n{}^s$ will be determined so that all the cross-interaction terms P'_{sr}, W'_{sr}, $s \neq r$, for the new modes vanish. Note that we may choose all $\mathbf{e}_n$ and $\mathbf{h}_n$ real and then h_{zn} is pure imaginary, as reference to (3.3e) shows. If we use (3.155) in (3.157) and the result (3.148), which holds for any two degenerate modes, we see that $P_{nm} = P_{mn}$ and may be chosen real. Since $\mathbf{J}_n \cdot \mathbf{J}_m^* = \mathbf{H}_n \cdot \mathbf{H}_m^*$, the W_{nm} are also real and $W_{nm} = W_{mn}$.

When (3.159) is substituted into (3.160) and (3.161), we obtain

$$P'_{sr} = \sum_{n=1}^{N} \sum_{m=1}^{N} C_n{}^s (C_m{}^r)^* P_{nm} = P'_{sr} \delta_{sr} \tag{3.162}$$

$$W'_{sr} = \sum_{n=1}^{N} \sum_{m=1}^{N} C_n{}^s (C_m{}^r)^* W_{nm} = W'_{sr} \delta_{sr} \tag{3.163}$$

where $\delta_{sr} = 0$ for $s \neq r$ and equals unity for $r = s$. Since s and r range from 1 to N, there are N^2 equations in each set above. The right-hand sides of (3.162) and (3.163) express the conditions imposed on the $C_n{}^s$.

To simplify the solution of the above, it is convenient to introduce the following matrices:

$$[P] = \begin{bmatrix} P_{11} & P_{12} & \cdots & P_{1N} \\ P_{21} & P_{22} & \cdots & P_{2N} \\ \cdots\cdots\cdots\cdots\cdots\cdots \\ P_{N1} & P_{N2} & \cdots & P_{NN} \end{bmatrix} \tag{3.164a}$$

$$[W] = \begin{bmatrix} W_{11} & W_{12} & \cdots & W_{1N} \\ W_{21} & W_{22} & \cdots & W_{2N} \\ \cdots\cdots\cdots\cdots\cdots\cdots \\ W_{N1} & W_{N2} & \cdots & W_{NN} \end{bmatrix} \tag{3.164b}$$

$$[C^s] = \begin{bmatrix} C_1{}^s \\ C_2{}^s \\ \cdot \\ \cdot \\ \cdot \\ C_N{}^s \end{bmatrix} \tag{3.164c}$$

$$[C^s]_t = [C_1{}^s \quad C_2{}^s \quad \cdots \quad C_N{}^s] \tag{3.164d}$$

where the subscript t denotes the transposed matrix. The matrix forms of (3.162) and (3.163) are

$$[C^s]_t [P][C^r]^* = P'_{sr} \delta_{sr}$$
$$[C^s]_t [W][C^r]^* = W'_{sr} \delta_{sr}$$

Multiplying the first equation by W'_{rr} and the second by P'_{rr} for $s = r$ gives

$$W'_{rr}[C^r]_t [P][C^r]^* = W'_{rr} P'_{rr}$$
$$P'_{rr}[C^r]_t [W][C^r]^* = W'_{rr} P'_{rr}$$

Subtracting these equations now gives

$$[C^r]_t (W'_{rr}[P] - P'_{rr}[W])[C^r]^* = 0$$

since W'_{rr} and P'_{rr} are scalar constants. Since the $[C^r]_t$ matrix is not zero, it is necessary that

$$\left([W] - \frac{W'_{rr}}{P'_{rr}}[P]\right)[C^r]^* = 0 \tag{3.165}$$

This expression is a set of homogeneous equations for $[C^r]^*$, and will have a solution only if the determinant vanishes. A solution exists only if

$$\left|[W] - \frac{W'_{rr}}{P'_{rr}}[P]\right| = \begin{vmatrix} W_{11} - \lambda_r P_{11} & \cdots & W_{1N} - \lambda_r P_{1N} \\ \cdots & \cdots & \cdots \\ W_{N1} - \lambda_r P_{N1} & \cdots & W_{NN} - \lambda_r P_{NN} \end{vmatrix} = 0 \tag{3.166}$$

where $\lambda_r = W'_{rr}/P'_{rr}$ is the eigenvalue for this matrix-eigenvalue problem. Since (3.166) is an $N \times N$ determinant, expansion yields an Nth-order polynomial in λ_r. That is, there will be N roots for λ_r, and these can be labeled $\lambda_1, \lambda_2, \lambda_3, \ldots, \lambda_N$. For any one root, say λ_r, the corresponding coefficients C_n^r that define the new rth mode are proportional to the cofactors of the determinant. Only the relative values of the C_n^r are determined since the absolute amplitudes of the new modes are arbitrary. The solution obtained from (3.165) and (3.166) will satisfy all the relations in (3.162) and (3.163). To show this, consider two solutions which satisfy the equations

$$(P'_{rr}[W] - W'_{rr}[P])[C^r]^* = 0 \qquad (P'_{ss}[W] - W'_{ss}[P])[C^s]^* = 0$$

Since $[P]$ and $[W]$ are real and symmetric, the eigenvalues $\lambda_r = W'_{rr}/P'_{rr}$ are real. Thus the transposed complex conjugate of the second equation is

$$[C^s]_t(P'_{ss}[W] - W'_{ss}[P]) = 0$$

If we premultiply the first equation by $[C^s]_t$ and postmultiply the last one by $[C^r]^*$, we obtain

$$[C^s]_t(P'_{rr}[W] - W'_{rr}[P])[C^r]^* = 0 \qquad [C^s]_t(P'_{ss}[W] - W'_{ss}[P])[C^r]^* = 0$$

If we now subtract these two equations, we find that

$$[C^s]_t\{(P'_{rr} - P'_{ss})[W] - (W'_{rr} - W'_{ss})[P]\}[C^r]^* = 0$$

We may split the eigenvalues λ_r arbitrarily into factors W'_{rr} and P'_{rr} so that $P'_{rr} - P'_{ss}$ and $W'_{rr} - W'_{ss}$ are arbitrary factors. Thus we conclude that

$$[C^s]_t[W][C^r]^* = [C^s]_t[P][C^r]^* = 0$$

and hence (3.162) and (3.163) hold not only for $s = r$, as guaranteed by (3.166), but also for all $s \ne r$.

The new modes are uncoupled; consequently the power transmitted along the guide for the rth mode is

$$\tfrac{1}{2}\int_S \mathbf{E}'_r \times (\mathbf{H}'_r)^* \cdot \mathbf{a}_z \, dS = P'_{rr}$$

and the power loss in the walls is

$$\frac{R_m}{2} \oint_C \mathbf{J}'_r \cdot (\mathbf{J}'_r)^* \, dl = W'_{rr}$$

Hence the attenuation constant for the rth mode is

$$\alpha_r = \frac{W'_{rr}}{2P'_{rr}} = \frac{\lambda_r}{2} \qquad (3.167)$$

Thus the roots of the determinant give the attenuation constants for the new uncoupled modes (apart from a factor $\frac{1}{2}$).

The propagation constants for the new modes are

$$\gamma_r = j\beta_0 + \alpha_r \qquad (3.168)$$

The new modes are no longer degenerate. The presence of a wall impedance $Z_m = (1 + j)R_m$ actually results in a small perturbation of the phase constant β_0 also. A more careful analysis shows that β_0 is perturbed so that

$$\gamma_r = j(\beta_0 + \alpha_r) + \alpha_r \qquad (3.169)$$

However, since $\alpha_r \ll \beta_0$, the change in β_0 is negligible. This result may be obtained by replacing R_m by Z_m in the above equations.†

If the original modes were nondegenerate, then, as shown in Sec. 3.9, $P_{nm} = 0$ for $n \neq m$. Likewise, $W_{nm} = 0$ for $n \neq m$, and hence to a first approximation these modes are not coupled when their fields are perturbed by the introduction of small losses. Although there are many examples of degenerate modes in waveguides in the majority of cases, it will be found that these modes are not coupled together. For example, the TE_{0m} and TM_{1m} modes in a circular guide are degenerate, but they nevertheless remain uncoupled because the fields and currents of the TE_{0m} modes have no variation with the angle ϕ, whereas the fields and currents of the TM_{1m} modes vary either as $\cos \phi$ or as $\sin \phi$. Thus all factors P_{nm}, W_{nm}, $n \neq m$, are zero. An example where mode coupling does take place is that of TE_{nm} and TM_{nm} modes in a rectangular guide. Another case of considerable practical importance is that of multiconductor TEM-wave transmission lines.

Coupling of TE_{11} and TM_{11} modes in a rectangular guide

The theory discussed above will be applied to the case of degenerate TE_{11} and TM_{11} modes in a rectangular guide of dimensions $a \times b$. The TE_{11} mode will be designated as the field $\mathbf{E}_1$, $\mathbf{H}_1$. From Table 3.2 the fields

† J. J. Gustincic, A General Power Loss Method for Attenuation of Cavities and Waveguides, *IEEE Trans.*, vol. MTT-11, pp. 83–87, January, 1963.

are (apart from the propagation factor $e^{-j\beta_0 z}$)

$$\mathbf{H}_1 = \frac{j\beta_0\pi}{ak_c{}^2}\left(\sin\frac{\pi x}{a}\cos\frac{\pi y}{b}\right)\mathbf{a}_x + \frac{j\beta_0\pi}{bk_c{}^2}\left(\cos\frac{\pi x}{a}\sin\frac{\pi y}{b}\right)\mathbf{a}_y$$
$$+ \left(\cos\frac{\pi x}{a}\cos\frac{\pi y}{b}\right)\mathbf{a}_z$$

$$\mathbf{E}_1 = \frac{k_0 Z_0}{\beta_0}(\mathbf{a}_z H_{1y} - \mathbf{a}_y H_{1x})$$

$$\mathbf{E}_2 = \frac{-j\beta_0\pi}{ak_c{}^2}\left(\cos\frac{\pi x}{a}\sin\frac{\pi y}{b}\right)\mathbf{a}_x - \frac{j\beta_0\pi}{bk_c{}^2}\left(\sin\frac{\pi x}{a}\cos\frac{\pi y}{b}\right)\mathbf{a}_y$$
$$+ \left(\sin\frac{\pi x}{a}\sin\frac{\pi y}{b}\right)\mathbf{a}_z$$

$$\mathbf{H}_2 = \frac{k_0}{\beta_0 Z_0}(-\mathbf{a}_x E_{2y} + \mathbf{a}_y E_{2x})$$

where $\beta_0 = \beta_{11}$ and $k_c{}^2 = k_{c,11}^2 = (\pi/a)^2 + (\pi/b)^2$.

The factors P_{nm} are readily determined and are found to be given by

$$P_{11} = \tfrac{1}{2}\int_0^a\int_0^b \mathbf{E}_1 \times \mathbf{H}_1^* \cdot \mathbf{a}_z\, dx\, dy = \frac{abk_0 Z_0 \beta_0}{8k_c{}^2}$$

$$P_{12} = \tfrac{1}{2}\int_0^a\int_0^b \mathbf{E}_1 \times \mathbf{H}_2^* \cdot \mathbf{a}_z\, dx\, dy = 0$$

$$P_{21} = 0$$

$$P_{22} = \frac{k_0 \beta_0 ab}{8Z_0 k_c{}^2}$$

The coupling between the wall currents arises from the z-directed currents only because the TM modes have only axial currents. The axial currents for fields 1 and 2 are

$$J_{1z} = \begin{cases} \dfrac{-j\beta_0\pi}{k_c{}^2 a}\sin\dfrac{\pi x}{a} & y = 0, b \\[6pt] \dfrac{j\beta_0\pi}{k_c{}^2 b}\sin\dfrac{\pi y}{b} & x = 0, a \end{cases}$$

$$J_{2z} = \begin{cases} \dfrac{-jk_0\pi}{k_c{}^2 b Z_0}\sin\dfrac{\pi x}{a} & y = 0, b \\[6pt] \dfrac{-jk_0\pi}{k_c{}^2 a Z_0}\sin\dfrac{\pi y}{b} & x = 0, a \end{cases}$$

Since W_{11} involves the transverse currents also, these are given:

$$J_{1x} = \cos\frac{\pi x}{a} \qquad y = 0, b$$

$$J_{1y} = -\cos\frac{\pi y}{b} \qquad x = 0, a$$

The factors W_{nm} are now readily found to be

$$W_{11} = \frac{R_m}{2} \oint \mathbf{J}_1 \cdot \mathbf{J}_1^* \, dl = \frac{R_m}{2} \left[2 \int_0^a \left(\frac{\beta_0^2 \pi^2}{k_c^4 a^2} \sin^2 \frac{\pi x}{a} + \cos^2 \frac{\pi x}{a} \right) dx \right.$$
$$\left. + 2 \int_0^b \left(\frac{\beta_0^2 \pi^2}{k_c^4 b^2} \sin^2 \frac{\pi y}{b} + \cos^2 \frac{\pi y}{b} \right) dy \right]$$
$$= \frac{R_m}{2} \left(\frac{\beta_0^2 \pi^2}{k_c^4 ab} + 1 \right)(a + b)$$

$$W_{12} = \frac{R_m}{2} \frac{\beta_0 k_0 \pi^2}{k_c^4 ab Z_0} (a - b) = W_{21}$$

$$W_{22} = \frac{R_m}{2} \frac{k_0^2 \pi^2}{k_c^4 a^2 b^2 Z_0^2} (a^3 + b^3)$$

The eigenvalue equation (3.166) becomes

$$\begin{vmatrix} \frac{R_m}{2}\left(\frac{\beta_0^2\pi^2}{k_c^4 ab}+1\right)(a+b) - \lambda_r \frac{abk_0 Z_0 \beta_0}{8k_c^2} & \frac{R_m}{2}\frac{\beta_0 k_0 \pi^2}{k_c^4 abZ_0}(a-b) \\ \frac{R_m}{2}\frac{\beta_0 k_0 \pi^2}{k_c^4 abZ_0}(a-b) & \frac{R_m}{2}\frac{k_0^2\pi^2}{k_c^4 a^2 b^2 Z_0^2}(a^3+b^3) - \lambda_r \frac{k_0 \beta_0 ab}{8Z_0 k_c^2} \end{vmatrix} = 0$$

(3.170)

In view of the complexity of this equation, further reduction is not carried out. At this point it seems just as convenient to substitute numerical values into (3.170).

As an example, consider a rectangular guide with $a = 2b = 2.54$ cm and $R_m = 2.6 \times 10^{-7} \sqrt{f}$ ohms for copper. The computed values of α are plotted as a function of f/f_c in Fig. 3.20. The broken curves are values of α obtained when mode coupling is neglected [obtained by equating each diagonal element in (3.170) equal to zero]. Although the

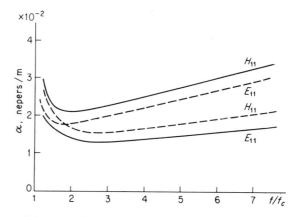

Fig. 3.20 Attenuation in a copper guide with $a = 2b = 2.54$ cm. Solid line: attenuation taking mode coupling into account; broken line: attenuation neglecting mode coupling.

132 Foundations for microwave engineering

two curves of α for the new modes (coupled TE_{11} and TM_{11} modes) are labeled H_{11} and E_{11}, not too much significance should be attached to this nomenclature since the coupling is so strong that neither of the new modes can be considered as being approximately an E or H mode.

3.11 Wave velocities

In any system capable of supporting propagating waves, a number of wave velocities occur that pertain to signal propagation, energy propagation, wavefront propagation, etc. These various velocities are examined below in the context of propagation in waveguides.

Phase velocity

The phase velocity of a wave in a waveguide was introduced earlier and shown to be equal to

$$v_p = \frac{\lambda_g}{\lambda_0} c \qquad (3.171)$$

for an air-filled guide. Here λ_g is the guide wavelength, λ_0 the free-space wavelength, and c the velocity of light. The phase velocity is the velocity an observer must move with in order to see a constant phase for the wave propagating along the guide. It is noted that the phase velocity is greater than the velocity of light, and since the principle of relativity states that no signal or energy can be propagated at a velocity exceeding that of light, the physical significance of the phase velocity might very well be questioned.

The clue to the significance of the phase velocity comes from the recognition that this velocity entered into wave solutions that had a steady-state time dependence of $e^{j\omega t}$. A pure monochromatic (single-frequency) wave of this type exists only if the source was turned on at $t = -\infty$ and is kept on for all future time as well. This is clearly not a physically realizable situation. In actual fact the source must be turned on at some finite time, which can be chosen as $t = 0$. The generated signal is then of the form illustrated in Fig. 3.21. Associated with the sudden step-like beginning of the signal is a broad frequency band, as a Fourier analysis shows. If this signal is injected into the guide at $z = 0$, an observer a distance l farther along the guide will, in actual fact, see no

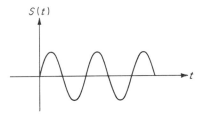

Fig. 3.21 Sinusoidal signal applied at time $t = 0$.

signal until a time l/c has elapsed. In other words, the wavefront will propagate along the guide with a velocity c. At the time l/c, the observer will begin to see the arrival of the transient associated with the switching on of the signal. After a suitable period of time has elapsed, the transient will have died out, and the observer will then see the steady-state sinusoidally varying wave. Once steady-state conditions prevail, the phase velocity can be introduced to describe the velocity at which a constant phase point appears to move along the guide. Note, however, that there is no information being transmitted along the guide once steady-state conditions have been established. Thus the phase velocity is not associated with any physical entity such as a signal, wavefront, or energy flow velocity. The term signal is used here to denote a time function that can convey information to the observer. Thus the step change at $t = 0$ is a signal, but once steady-state conditions are achieved, the observer does not receive any more information. A better understanding of the above features will be obtained after the group velocity, discussed below, has been examined.

Group velocity

The physical definition of group velocity is the velocity with which a signal consisting of a very narrow band of frequency components propagates. The appropriate tool for the analysis of this situation is the Fourier transform. If a time function is denoted by $f(t)$, this function of time has associated with it a frequency spectrum $F(\omega)$ given by the Fourier transform of $f(t)$,

$$F(\omega) = \int_{-\infty}^{\infty} f(t) e^{-j\omega t} \, dt \qquad (3.172a)$$

Conversely, if the spectrum $F(\omega)$ is known, the time function may be found from the inverse Fourier transform relation

$$f(t) = \frac{1}{2\pi} \int_{-\infty}^{\infty} F(\omega) e^{j\omega t} \, d\omega \qquad (3.172b)$$

From Eq. (3.172b) it is seen that the Fourier transform represents $f(t)$ as a superposition of steady-state sinusoidal functions of infinite duration. These relations are a generalization of the Fourier series relations. If the time function is passed through a device having a response $Z(\omega)$ that is a function of frequency, e.g., filter, the output time function $f_o(t)$ will have a frequency spectrum $F(\omega)Z(\omega)$, and hence, by (3.172b), is given by

$$f_o(t) = \frac{1}{2\pi} \int_{-\infty}^{\infty} F(\omega) Z(\omega) e^{j\omega t} \, d\omega$$

In general, $Z(\omega) = |Z(\omega)| e^{-j\psi(\omega)}$; so

$$f_o(t) = \frac{1}{2\pi} \int_{-\infty}^{\infty} |Z(\omega)| F(\omega) e^{j(\omega t - \psi)} \, d\omega \qquad (3.173)$$

If the output f_o is to be an exact reproduction of the input, then in (3.173) it is necessary for $|Z|$ to equal a constant A, and ψ must be a linear function of ω, say $a\omega + b$. In this case

$$f_o(t) = \frac{A}{2\pi} e^{-jb} \int_{-\infty}^{\infty} F(\omega) e^{j\omega(t-a)} \, d\omega \tag{3.174a}$$

If we put $t - a = t'$, the above becomes

$$f_o(t' + a) = \frac{Ae^{-jb}}{2\pi} \int_{-\infty}^{\infty} F(\omega) e^{j\omega t'} \, d\omega = Ae^{-jb} f(t') \tag{3.174b}$$

as comparison with (3.172b) shows. Thus the output time function is

$$f_o(t' + a) = f_o(t) = Ae^{-jb} f(t') = Ae^{-jb} f(t - a) \tag{3.175}$$

i.e., an exact duplicate of the input, apart from a constant multiplier and a time delay a. Thus the conditions given on $|Z|$ and ψ are those sufficient for a distortion-free system.

Now, in a waveguide, the transverse variations of the field are independent of frequency. The only essential frequency-dependent part of the field solution is the propagation factor $e^{-j\beta z}$ since

$$\beta = (k_0^2 - k_c^2)^{\frac{1}{2}} = \left(\frac{\omega^2}{c^2} - k_c^2\right)^{\frac{1}{2}}$$

is a function of frequency. Thus a waveguide of length l, in which the field has a time dependence $e^{j\omega t}$, $\omega > 0$, can be considered as a frequency filter with a response $e^{-j\beta l}$. Since β is not a linear function of ω, it may be anticipated that some signal distortion will occur for propagation in a waveguide. For an ideal TEM-wave transmission line, $\beta = k_0 = \omega/c$ and distortion-free transmission is possible. However, practical lines have an attenuation which depends on frequency ($R_m \propto \sqrt{f}$), and this will produce distortion. Fortunately, for narrowband signals neither waveguides nor transmission lines produce significant distortion unless very long lines are used.

Consider now a time function $f(t)$ having a band of frequencies between $-f_m$ and f_m. This signal is used to modulate a carrier of frequency $f_c \gg f_m$. The resultant is

$$S(t) = f(t) \cos \omega_c t = \text{Re}\,[f(t) e^{j\omega_c t}] \tag{3.176}$$

If $F(\omega)$ is the spectrum of $f(t)$, the spectrum of $S(t)$ is

$$F_s(\omega) = \int_{-\infty}^{\infty} e^{-j\omega t} f(t) \frac{e^{j\omega_c t} + e^{-j\omega_c t}}{2} \, d\omega$$
$$= \tfrac{1}{2}[F(\omega - \omega_c) + F(\omega + \omega_c)] \tag{3.177}$$

These spectra are illustrated in Fig. 3.22.

For positive ω the waveguide response is $e^{-j\beta(\omega)l}$. For negative ω the response must be chosen as $e^{j\beta(\omega)l}$ since, if the time variation is $e^{-j\omega t}$, the

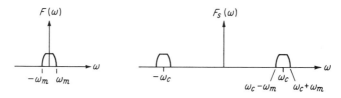

Fig. 3.22 Spectrum of $f(t)$ and $S(t)$.

sign in front of β must be positive for propagation in the positive z direction. In other words, all physical systems will have a response such that $|Z(\omega)|$ is an even function of ω and $\psi(\omega)$ is an odd function of ω. Since β is an even function, the sign must be changed for $\omega < 0$. These even and odd symmetry properties are required simply so that the output time function is real, a physical requirement. The output spectrum for the waveguide is thus

$$F_o(\omega) = \tfrac{1}{2}[F(\omega - \omega_c)e^{-j\beta(\omega)l} + F(\omega + \omega_c)e^{j\beta(\omega)l}]$$

and the output signal is

$$S_o(t) = \frac{1}{2\pi}\int_{-\infty}^{\infty} F_o(\omega)e^{j\omega t}\,d\omega \tag{3.178}$$

The analysis that follows is simplified if the signal is represented in complex form as $f(t)e^{j\omega_c t}$ with a spectrum $F(\omega - \omega_c)$. In this case only the positive half of the spectrum needs to be considered, and the output signal is given by

$$S_o(t) = \operatorname{Re}\frac{1}{2\pi}\int_{\omega_c-\omega_m}^{\omega_c+\omega_m} F(\omega - \omega_c)e^{j\omega t - j\beta(\omega)l}\,d\omega \tag{3.179}$$

since $F(\omega - \omega_c)$ is zero outside the band $\omega_c - \omega_m \leq \omega \leq \omega_c + \omega_m$. If the band is very narrow, $\omega_m \ll \omega_c$, then $\beta(\omega)$ may be approximated by the first few terms in a Taylor series expansion about ω_c. Thus

$$\beta(\omega) = \beta(\omega_c) + \frac{d\beta}{d\omega}\bigg|_{\omega_c}(\omega - \omega_c) + \frac{1}{2}\frac{d^2\beta}{d\omega^2}\bigg|_{\omega_c}(\omega - \omega_c)^2 + \cdots \tag{3.180}$$

Retaining the first two terms only and letting $\beta(\omega_c) = \beta_0$ and $d\beta/d\omega_c = \beta_0'$ at ω_c, (3.179) gives

$$S_o(t) = \operatorname{Re}\frac{1}{2\pi}\int_{\omega_c-\omega_m}^{\omega_c+\omega_m} F(\omega - \omega_c)e^{j\omega(t-\beta_0'l)}e^{-j\beta_0 l + j\beta_0'l\omega_c}\,d\omega$$

If this is compared with (3.174) and (3.175), it is seen that

$$S_o(t) = \operatorname{Re}\,[e^{-j\beta_0 l + j\beta_0'l\omega_c}f(t - \beta_0'l)e^{j\omega_c(t-\beta_0'l)}]$$
$$= f(t - \beta_0'l)\cos(\omega_c t - \beta_0 l) \tag{3.181}$$

To the order of approximation used here, the input modulating signal $f(t)$ is reproduced without distortion but with a time delay $\beta_0'l$. This is to be anticipated since β was approximated by a linear function of ω in the band $\omega_c - \omega_m$ to $\omega_c + \omega_m$ (distortion-free condition). The signal delay defines the group velocity v_g, which is equal to the distance l divided by the delay time; thus

$$v_g = \frac{l}{l\beta_0'} = \left(\frac{d\beta}{d\omega}\right)^{-1} \tag{3.182}$$

This is also the signal velocity. Note, however, that this velocity has significance only if the band, or "group," of frequencies making up the signal is so narrow that β may be approximated by a linear function throughout the frequency band of interest. If this is not the case, more terms in the expansion (3.180) must be retained and signal distortion will occur. In this case the group velocity as given by (3.182) is no longer the signal velocity. In fact, because of signal distortion, no unique signal velocity exists any longer. Different portions of the signal will travel with different velocities, and the resultant signal becomes dispersed in both time and space.

In the case of a waveguide,

$$\begin{aligned} v_g &= \left(\frac{d\beta}{d\omega}\right)^{-1} = c\frac{dk_0}{d\beta} = \left[\frac{d(\omega^2/c^2 - k_c^2)^{\frac{1}{2}}}{d\omega}\right]^{-1} \\ &= \frac{\beta c^2}{\omega} = \frac{\beta}{k_0}c = \frac{\lambda_0}{\lambda_g}c \end{aligned} \tag{3.183}$$

It is seen that $v_g < c$ and that $v_g v_p = c^2$ for a waveguide.

A typical plot of k_0 versus β for a waveguide is given in Fig. 3.23. From this plot it can be seen that for a narrow band of frequencies a linear approximation for β is good. Also note that for high frequencies (large k_0) β becomes equal to k_0. Thus frequencies well above the cutoff frequency f_c suffer very little dispersion and propagate essentially with the velocity of light c. No frequency components below the cutoff frequency f_c can propagate along the guide.

The equality of the wavefront velocity and the velocity of light can be readily explained by means of Fig. 3.23. The switching on of a signal results in an initial transient that has a spectrum of frequencies extend-

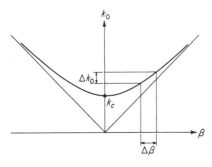

Fig. 3.23 Plot of k_0 versus β for a waveguide.

ing out to infinity. Any small group of frequencies at the high end of the spectrum will have a group velocity equal to c since $dk_0/d\beta$ equals unity for k_0 near infinity. Thus the high-frequency part of the transient will propagate along the guide with a velocity c. The lower-frequency components will propagate with smaller group velocities and arrive later.

Energy-flow velocity

Power is a flow of energy, and consequently there is a velocity of energy flow such that the average energy density in the guide multiplied by this velocity is equal to the power. In a waveguide it turns out that this velocity of energy flow is equal to the group velocity. A proof for the case of E modes will be given below, that for H modes being very similar.

For E modes the field is given by [see (3.17)]

$$\mathbf{E}_t = -\frac{j\beta}{k_c^2} \nabla_t e_z e^{-j\beta z} \qquad \mathbf{H}_t = \frac{k_0 Y_0}{\beta} \mathbf{a}_z \times \mathbf{E}_t$$

The average rate of energy flow, or power, is given by

$$P = \frac{1}{2} \int_S Y_e |\mathbf{E}_t|^2 \, dS = \frac{1}{2} \frac{k_0 Y_0}{\beta} \int_S |\mathbf{E}_t|^2 \, dS \tag{3.184}$$

where the integration is over the guide cross section.

The energy density in the magnetic field per unit length of guide is

$$U_m = \frac{\mu_0}{4} \int_S |\mathbf{H}_t|^2 \, dS = \frac{\mu_0}{4} \frac{k_0^2 Y_0^2}{\beta^2} \int_S |\mathbf{E}_t|^2 \, dS \tag{3.185}$$

The energy density in the electric field per unit length of guide is equal to that in the magnetic field. This is readily shown to be the case by using the complex Poynting vector theorem, which states that (Sec. 2.5)

$$\tfrac{1}{2} \int_S \mathbf{E} \times \mathbf{H}^* \cdot \mathbf{a}_z \, dS = P + 2j\omega(W_m - W_e)$$

where the integration is over the guide cross section, and the term on the right gives the power transmitted past the plane S plus $2j\omega$ times the net reactive energy stored in the guide beyond the plane S. Since the integral of the complex Poynting vector over a cross section S for a propagating mode in a loss-free guide is real, it follows that $W_m = W_e$. In addition, since the location of the transverse plane S is arbitrary, it also follows that the energy densities U_m and U_e per unit length of guide are equal.

The velocity of energy flow may now be found from the relation

$$v = \frac{P}{U_e + U_m} = \frac{P}{2U_m} = \frac{k_0 Y_0}{\beta} \frac{\beta^2}{\mu_0 k_0^2 Y_0^2}$$

$$= \frac{\beta}{\mu_0 k_0 Y_0} = \frac{\beta}{k_0 \sqrt{\mu_0 \epsilon_0}} = \frac{\beta}{k_0} c = v_g \tag{3.186}$$

and comes out equal to the group velocity as stated earlier.

Problems

3.1 Figure P 3.1 illustrates a three-conductor transmission line. Since potential is arbitrary to within an additive constant, the shield S_0 can be chosen to be at zero potential. Show that there are two linearly independent solutions for TEM waves in this transmission line. If S_0 encloses N conductors, how many TEM-wave solutions are possible?

Hint: Note that the potential can be arbitrarily specified on S_1 and S_2.

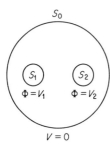

Fig. P 3.1

3.2 Show that power transmitted along a transmission line is given by

$$P = \frac{Y_0}{2} \int |\nabla_t \Phi|^2 \, dx \, dy$$

For Prob. 3.1 show that this equals $\frac{1}{2}(V_1 I_1 + V_2 I_2)$ by using Green's first identity (Appendix I) to convert the surface integral to a contour integral around the conductor boundaries.

3.3 For Prob. 3.1 choose two new potential solutions Φ'_1, Φ'_2 which are linear combinations of potentials Φ_1 and Φ_2 that satisfy the boundary conditions $\Phi'_1 = V_1$ on S_1 and zero on S_0, S_2; $\Phi'_2 = V_2$ on S_2 and zero on S_0, S_1. That is, choose

$$\Phi'_1 = C_{11}\Phi_1 + C_{12}\Phi_2$$
$$\Phi'_2 = C_{21}\Phi_1 + C_{22}\Phi_2$$

Determine expressions for the constants C_{ij} such that the rates of energy flow in the TEM waves derived from the new potentials are independent.

Hint: For no power interaction verify that the interaction term is

$$\int \nabla_t \Phi'_1 \cdot \nabla_t \Phi'_2 \, dx \, dy = 0$$

3.4 Derive Eqs. (3.17) for TM waves.

3.5 Show that, for an air-filled coaxial line, minimum attenuation occurs when $x \ln x = 1 + x$, $x = b/a$. What is the corresponding characteristic impedance?

Hint: Hold outer radius b constant and find $d\alpha/da$.

3.6 Evaluate Z_c for a lossy coaxial line using (3.53) and computed values of R, G, L, and C. Assume $b = 3a = 1$ cm, $f = 10^9$ cps, and $\epsilon = (2.56 - j0.001)\epsilon_0$. Verify that

$$\text{Im } Z_c \ll \text{Re } Z_c \quad \text{and} \quad Z_c \approx \left(\frac{L}{C}\right)^{\frac{1}{2}}$$

3.7 Obtain expressions for the voltage and current standing-wave patterns on a lossless open-circuited transmission line. Sketch these patterns.

3.8 A transmission line with $Z_c = 50$ ohms is terminated in an impedance $25 + j25$ ohms. Find the reflection coefficient, standing-wave ratio, and fraction of the incident power delivered to the load.

3.9 Verify (3.87) and compute Z_{in} at a distance $\lambda_0/4$ from the termination given in Prob. 3.8.

3.10 On a transmission line with $Z_c = 50$ ohms, the voltage at a distance $0.4\lambda_0$ from the load is $4 + j2$. The corresponding current is -2 amp. Determine the normalized load impedance.

★3.11 Use the energy definitions of L and C [Eqs. (3.42)] to derive the results given by (3.60) and (3.62). Evaluate L and C for a coaxial transmission line by this method.

★3.12 Figure P 3.12 illustrates a strip transmission line consisting of a conducting strip embedded in a dielectric slab between two wide conducting planes. Assume that there is no fringing field and that the potential varies linearly between the center and outer conductors. Obtain expressions for the parameters R, G, L, C, and Z_c for this transmission line.

Fig. P 3.12

3.13 Derive the solution for a TE_{10} mode in a rectangular guide of wide dimension a and height b when the guide is filled with dielectric of permittivity ϵ. Show that the cutoff frequency is given by $f_c = c/2a\kappa^{\frac{1}{2}}$, where c is the free-space velocity of light and κ is the dielectric constant. Show that the guide wavelength is smaller for a dielectric-filled guide than for an air-filled guide.

3.14 Obtain an expression for the attenuation of a TE_{10} mode in a dielectric-filled guide when $\epsilon = \epsilon_1 - j\epsilon_2$ but the walls are perfectly conducting. Obtain an exact expression and compare with the results deduced by an application of the perturbation method.

3.15 Obtain a solution for an H wave in the parallel-plate transmission line with centered dielectric slab as illustrated. Assume that the plates are perfectly conducting and infinitely wide. Can a TEM wave propagate in this structure? Why?

Hint: Assume $h_z = \cos k_d x$ for $|x| \leq a/2$ and $h_z = Ae^{-h|x|}$ for $|x| > a/2$. Verify that $k_d^2 + h^2 = (\kappa - 1)k_0^2$. Match the tangential fields at $x = a/2$ to obtain an equation for A and one relating the parameters h and k_d.

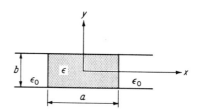

Fig. P 3.15

3.16 Obtain solutions for TE$_{n0}$ modes in the partially filled waveguide illustrated in Fig. P 3.16.

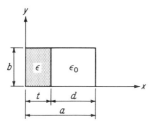

Fig. P 3.16

Hint: Assume that

$$h_z = \begin{cases} \cos k_d x & 0 < x < t \\ A \cos h(a - x) & t < x < a \end{cases}$$

and match the tangential fields at $x = t$. Thus show that

$$\beta^2 = k_0^2 - h^2 = \kappa k_0^2 - k_d^2$$

and that

$$h \tan k_d t = -k_d \tan hd$$

Note that there are an infinite number of solutions for h and k_d corresponding to various TE$_{n0}$ modes.

Obtain numerical values for β, h, and k_d when $k_0 = 2$, $t = 1$ cm, $d = 1.5$ cm, and $\kappa = 4$. Note that there is a lowest-order solution for h pure imaginary.

3.17 Show that, in a coaxial line with inner radius a and outer radius b, there are solutions for TE$_{nm}$ and TM$_{nm}$ modes. A suitable solution for e_z and h_z is

$$[A J_n(k_c r) + Y_n(k_c r)] \cos n\phi$$

Obtain equations (transcendental in nature) for determining the cutoff wave number k_c by imposing proper boundary conditions at $r = a, b$.

3.18 Obtain expressions for the surface currents of a TE$_{10}$ mode in a rectangular guide. A narrow slot may be cut in a waveguide along a current flow line without appreciably disturbing the field. Show that, for the TE$_{10}$ mode, narrow centered axial slots may be cut in the broad face of a rectangular guide. This principle is used in standing-wave detectors to provide suitable points of entry for a probe used to sample the interior waveguide field.

3.19 Find the surface currents for the H_{01} mode in a circular guide.

3.20 For TE modes in a waveguide, write $\mathbf{H}_t = -I(z) \nabla_t h_z$, $\mathbf{E}_t = V(z) \mathbf{a}_z \times \nabla_t h_z$. Use Maxwell's equations to show that $V(z)$ and $I(z)$ satisfy the transmission-line equations

$$\frac{dV}{dz} = -j\omega\mu_0 I \qquad \frac{dI}{dz} = -\left(j\omega\epsilon_0 + \frac{k_c^2}{j\omega\mu_0}\right) V$$

Construct an equivalent distributed-parameter circuit for these modes. For TM modes put $\mathbf{E}_t = -V(z)\nabla_t e_z$, $\mathbf{H}_t = -I(z)\mathbf{a}_z \times \nabla_t e_z$, and show that

$$\frac{dV}{dz} = -\left(j\omega\mu_0 + \frac{k_c{}^2}{j\omega\epsilon_0}\right)I \qquad \frac{dI}{dz} = -j\omega\epsilon_0 V$$

Construct an equivalent distributed-parameter circuit for these modes.†

3.21 Obtain an expression for power in a TE_{11} mode in a circular guide. (See Appendix II for Bessel-function integrals.)

3.22 Derive the expression for attenuation for TE_{0m} modes in a circular waveguide, i.e., Eq. (3.123).

3.23 Find the attenuation in decibels per mile for an H_{01} mode in a circular copper guide of 1 in. diameter when operated at a frequency of 10 times the cutoff frequency.

3.24 Consider an infinitely long rectangular guide. The guide is filled with dielectric for $z \geq 0$, having a dielectric constant κ. An H_{10} mode is incident from $z < 0$. At $z = 0$, a reflected H_{10} mode and a transmitted H_{10} mode are produced because of the discontinuity. Show that the reflection coefficient is given by $(Z_2 - Z_1)/(Z_2 + Z_1)$, where Z_1 is the wave impedance in the empty guide and Z_2 is the wave impedance in the dielectric-filled guide. Show that the ratio of the wave impedances equals the ratio of the guide wavelengths.

★3.25 Obtain an expression for the power in a TM surface-wave mode on a grounded dielectric slab.

★3.26 Obtain solutions for TE surface-wave modes on a dielectric slab placed on a conducting plane. From a graphical solution of the eigenvalue equation, verify that all TE surface-wave modes have lower cutoff frequencies.

★3.27 Consider a plane surface which presents a surface reactance jX_s to the electromagnetic field. Show that, when the surface reactance is inductive, a TM surface-wave mode with a decay constant $k_0 X_s/Z_0$ can exist.

★3.28 Use the transverse-resonance technique to derive the eigenvalue equation for TE_{n0} modes in the partially filled rectangular guide of Prob. 3.16. Verify that the wave impedances in the x direction in the two regions are $k_0 Z_0/h$ and $kZ/k_d = k_0 Z_0/k_d$.

★3.29 Use the transverse-resonance technique to derive the eigenvalue equations called for in Probs. 3.26 and 3.27.

3.30 Consider an ideal loss-free transmission line of length l. The far end is short-circuited. At the input end a battery of voltage V_0 is switched across the line at time $t = 0$. Sketch the voltage wave on the line for times $0 < t < l/c$, $l/c < t < 2l/c$, $2l/c < t < 3l/c$.

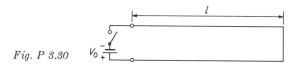

Fig. P 3.30

† S. A. Schelkunoff, *Bell System Tech. J.*, vol. 34, p. 995, September, 1955.

3.31 Consider the transmission-line circuit illustrated. At time $t = 0$ a battery of voltage V_0 is switched across the line at the input. Determine the output load voltage V as a function of time.

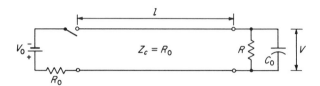

Fig. P 3.31

Hint: This transient problem may be solved in a manner similar to that used in low-frequency circuit theory. The governing equations for the transmission line are

$$\frac{\partial \mathbb{U}}{\partial z} = -L \frac{\partial \mathcal{I}}{\partial t} \qquad \frac{\partial \mathcal{I}}{\partial z} = -C \frac{\partial \mathbb{U}}{\partial t}$$

The time derivative may be eliminated by taking the Laplace transform. The transformed solutions for $\mathbb{U}$ and $\mathcal{I}$ are

$$V^+ e^{-sz/c} + V^- e^{sz/c} \qquad I^+ e^{-sz/c} + I^- e^{sz/c}$$

By transforming the circuit equations for the load termination and the input voltage, the resultant equations may be solved for the Laplace transform of the load voltage. The inverse transform then gives the load voltage as a function of time.

The foregoing procedure may be simplified by first replacing the battery by a source $V_0 e^{j\omega t}$ and obtaining the transfer function $V/V_0 = Z_t(j\omega)$ as a function of ω for this steady-state problem. Replacing $j\omega$ by s then gives the transfer function in the s domain. The Laplace transform of the output voltage is then

$$V(s) = \frac{V_0}{s} Z_t(j\omega = s)$$

since the Laplace transform of the input step voltage is V_0/s. The output voltage is obtained from the inverse Laplace transform of $V(s)$.

3.32 The permittivity ϵ is generally a function $\epsilon(\omega)$ of ω. Obtain an expression for the group velocity of a coaxial line filled with dielectric. Neglect frequency dependence of the attenuation due to conductor loss.

3.33 A rectangular guide of dimensions $a = 2b = 2.5$ cm is operated at a frequency of 10^{10} cps. A pulse-modulated carrier of the above frequency is transmitted through the guide. How much pulse delay time is introduced by a guide 100 m long?

3.34 An amplitude-modulated wave $(1 + m \cos \omega_m t) \cos \omega_c t$ is transmitted through the guide in Prob. 3.33. If $f_m = 20$ kc and $f_c = 10{,}000$ Mc, how long must the guide be before the upper and lower sidebands will undergo a relative phase shift of 180°?

Hint: Compute the phase velocity for each frequency component.

3.35 A signal $(2 + \cos 2\pi f_1 t + \cos 2\pi f_2 t) \cos 2\pi f_c t$ is transmitted through a rectangular guide 1,000 m long. If $a = 2.5$ cm, $f_c = 10{,}000$ Mc, $f_1 = 10$ Mc, and $f_2 = 12$ Mc and a TE_{10} mode is used, determine the exact form of the output signal.

Hint: Find the phase shift for each frequency component of the input signal, and recombine these at the output into sine and cosine terms. Sketch the input and output low-frequency modulating signals, and note the distortion produced.

★3.36 A plane surface exhibits a surface reactance $jX_s = j(a\omega^2 - b)/\omega$. As in Prob. 3.27, find the solution for a TM surface wave. Derive an expression for the group velocity for this wave.

References

1. Atwater, H. A.: "Introduction to Microwave Theory," McGraw-Hill Book Company, New York, 1962.
2. Bronwell, A. B., and R. E. Beam: "Theory and Application of Microwaves," McGraw-Hill Book Company, New York, 1947.
3. Collin, R. E.: "Field Theory of Guided Waves," McGraw-Hill Book Company, New York, 1960.
4. Ghose, R. N.: "Microwave Circuit Theory and Analysis," McGraw-Hill Book Company, New York, 1963.
5. Harrington, R. F.: "Time-harmonic Electromagnetic Fields," McGraw-Hill Book Company, New York, 1961.
6. Jordan, E. C.: "Electromagnetic Waves and Radiating Systems," Prentice-Hall, Inc., Englewood Cliffs, N.J., 1950.
7. Ramo, S., and J. R. Whinnery: "Fields and Waves in Modern Radio," 2d ed., John Wiley & Sons, Inc., New York, 1953.
8. Reich, H. J., P. F. Ordung, H. L. Krauss, and J. K. Skalnik: "Microwave Theory and Techniques," D. Van Nostrand Company, Inc., Princeton, N.J., 1953.
9. Schelkunoff, S. A.: "Electromagnetic Waves," D. Van Nostrand Company, Inc., Princeton, N.J., 1943.
10. Slater, J. C.: "Microwave Transmission," McGraw-Hill Book Company, New York, 1942.
11. Southworth, G. C.: "Principles and Applications of Waveguide Transmission," D. Van Nostrand Company, Inc., Princeton, N.J., 1950.
12. Toraldo di Francia, G.: "Electromagnetic Waves," Interscience Publishers, Inc., New York, 1956.

4
Circuit theory for waveguiding systems[†]

At low frequencies the interconnection of resistors, capacitors, and inductors results in a circuit. Such circuits are normally linear, so that the superposition principle may be used to find the response when more than one exciting source is present. Kirchhoff's laws form the basis for the analysis, whether in terms of loop currents or node voltages. In these low-frequency circuits the various elements are connected by conducting wires, and generally the length of these connecting wires is not critical or important.

At microwave frequencies equivalent reactive and resistive elements may also be connected to form a microwave circuit. In place of connecting wires, transmission lines and waveguides are used. The length of the connecting link is often several wavelengths, and hence propagation effects become very important. The analysis of microwave circuits is therefore by necessity somewhat more involved than that for the low-frequency case. The circuit theory of transmission-line circuits has been well developed for several decades, and, as shown below, the circuit theory for waveguide systems is formally the same.

Many of the circuit-analysis techniques and circuit properties that are valid at low frequencies are also valid for microwave circuits. In actual fact, low-frequency circuit analysis is a special case of microwave circuit analysis. As a consequence, a study of microwave circuits provides a deeper physical insight into conventional circuit theory. In this chapter the physical basis for a circuit theory for waveguiding systems is developed. In later chapters we shall utilize this foundation in the study of impedance matching, waveguide devices, resonators, filters, etc.

[†] The basic theory of microwave circuits is developed in C. G. Montgomery, R. H. Dicke, and E. M. Purcell, "Principles of Microwave Circuits," McGraw-Hill Book Company, New York, 1948. Much of the material presented here in Secs. 4.1 to 4.9 must of necessity be similar, in view of its basic nature.

4.1 Equivalent voltages and currents

At microwave frequencies voltmeters and ammeters for the direct measurement of voltages and currents do not exist. For this reason voltage and current, as a measure of the level of electrical excitation of a circuit, do not play a primary role at microwave frequencies. On the other hand, it is useful to be able to describe the operation of a microwave circuit in terms of voltages, currents, and impedances in order to make optimum use of low-frequency circuit concepts. For the most part this can be done. There is, however, a notable difference, namely, the nonuniqueness of the voltages and currents in most instances. It was noted in the preceding chapter that for the TEM wave on a transmission line there existed a voltage and a current wave uniquely related to the transverse electric and magnetic fields, respectively. In the case of TE and TM modes in a waveguide, no unique voltage or current waves exist that have the same physical significance as those associated with the TEM wave on a transmission line. This result might have been anticipated since the guide boundary is a closed conducting boundary, and one is at a loss as to the two points on the boundary between which the voltage should be measured. Furthermore, if voltage is defined as the line integral of the transverse electric field between two chosen points on the boundary, it is found that for TM waves the line integral is zero (Probs. 4.1 and 4.2), whereas for TE waves the value of the line integral depends on the path of integration that is chosen. For these reasons the introduction of voltage and current waves, to be associated with waveguide modes, is done on an equivalent basis and has formal significance only. The basis for the introduction of equivalent voltages and currents is discussed below.

In the previous chapter it was shown that propagating waveguide modes have the following properties:

1. Power transmitted is given by an integral involving the transverse electric and transverse magnetic fields only.

2. In a loss-free guide supporting several modes of propagation, the power transmitted is the sum of that contributed by each mode individually.

3. The transverse fields vary with distance along the guide according to a propagation factor $e^{\pm j\beta z}$ only.

4. The transverse magnetic field is related to the transverse electric field by a simple constant, the wave impedance of the mode; i.e.,

$$Z_w \mathbf{h} = \mathbf{a}_z \times \mathbf{e}$$

for a mode propagating in the $+z$ direction.

These properties suggest letting equivalent voltage and current waves be introduced proportional to the transverse electric and magnetic fields, respectively, since the transverse fields have properties similar to those of the voltage and current waves on a transmission line. This is, in actual fact, what is done.

A propagating waveguide mode may be expressed in general as

$$\mathbf{E} = C^+\mathbf{e}e^{-j\beta z} + C^+\mathbf{e}_z e^{-j\beta z} \tag{4.1a}$$

$$\mathbf{H} = C^+\mathbf{h}e^{-j\beta z} + C^+\mathbf{h}_z e^{-j\beta z} \tag{4.1b}$$

for propagation in the $+z$ direction, and

$$\mathbf{E} = C^-\mathbf{e}e^{j\beta z} - C^-\mathbf{e}_z e^{j\beta z} \tag{4.2a}$$

$$\mathbf{H} = -C^-\mathbf{h}e^{j\beta z} + C^-\mathbf{h}_z e^{j\beta z} \tag{4.2b}$$

for propagation in the $-z$ direction. In (4.1) and (4.2), C^+ and C^- are arbitrary amplitude constants. Note also that if the mode is a TE or a TM mode, then $\mathbf{e}_z$ or $\mathbf{h}_z$ is zero accordingly. Let the following equivalent voltage and current waves be introduced:

$$V = V^+ e^{-j\beta z} + V^- e^{j\beta z} \tag{4.3a}$$

$$I = I^+ e^{-j\beta z} - I^- e^{j\beta z} \tag{4.3b}$$

where $V^+ = K_1 C^+$, $V^- = K_1 C^-$, and $I^+ = K_2 C^+$, $I^- = K_2 C^-$. K_1 and K_2 are constants of proportionality that will establish the relation between voltages and the transverse electric field and currents and the transverse magnetic field. In order to conserve power, it is necessary that

$$\tfrac{1}{2} V^+ (I^+)^* = \frac{|C^+|^2}{2} \int_S \mathbf{e} \times \mathbf{h}^* \cdot \mathbf{a}_z \, dS$$

or

$$K_1 K_2^* = \int \mathbf{e} \times \mathbf{h}^* \cdot \mathbf{a}_z \, dS \tag{4.4}$$

By proper normalization of the functions $\mathbf{e}$ and $\mathbf{h}$, the product $K_1 K_2^*$ can be made equal to unity. Although (4.4) provides one relationship between K_1 and K_2, a second relation is required before they are determined. This second relationship can be chosen in a variety of ways. For example, the voltage and current waves given by (4.3) may be thought of as existing on a fictitious transmission line that is equivalent to the waveguide. As such, it may be desirable to choose the characteristic impedance of this transmission line equal to unity, in which case

$$Z_c = \frac{V^+}{I^+} = \frac{V^-}{I^-} = \frac{K_1}{K_2} = 1 \tag{4.5}$$

As an alternative, it might be desirable to choose the characteristic impedance equal to the wave impedance, in which case

$$Z_c = \frac{K_1}{K_2} = Z_w \qquad (4.6)$$

Other possibilities are obvious and equally valid. In this text either the definition (4.5) or (4.6) is used. The one that is used will be stated, or else it will be clear from the discussion which definition is being utilized. When the equivalent voltages and currents are chosen so that the equivalent transmission-line characteristic impedance is unity, we shall refer to them as *normalized* voltages and currents. Note that even though equivalent transmission lines may be used to represent a waveguide, the propagation constant of this line must be taken as that for the waveguide.

A waveguide supporting N propagating modes may now be formally represented as N fictitious transmission lines supporting equivalent voltage and current waves (from property 2 listed above for waveguide modes). Thus we have

$$V = \sum_{n=1}^{N} (V_n^+ e^{-j\beta_n z} + V_n^- e^{j\beta_n z}) \qquad (4.7a)$$

$$I = \sum_{n=1}^{N} (I_n^+ e^{-j\beta_n z} - I_n^- e^{j\beta_n z})$$

$$= \sum_{n=1}^{N} (V_n^+ Y_n e^{-j\beta_n z} - V_n^- Y_n e^{j\beta_n z}) \qquad (4.7b)$$

where the Y_n are arbitrarily chosen characteristic admittances for the equivalent transmission lines. When an obstacle is inserted into a waveguide supporting N modes of propagation, these modes are in general coupled together by the obstacle. This coupling can be described in terms of an equivalent circuit made up of impedance elements. This impedance description of obstacles in waveguides is developed in the next section. Once the equivalent voltage and current amplitudes have been determined, the waveguide fields are known from the relations

$$\mathbf{E}_t = \sum_{n=1}^{N} (V_n^+ K_{1n}^{-1} e^{-j\beta_n z} + V_n^- K_{1n}^{-1} e^{j\beta_n z}) \mathbf{e}_n \qquad (4.8a)$$

$$\mathbf{H}_t = \sum_{n=1}^{N} (I_n^+ K_{2n}^{-1} e^{-j\beta_n z} - I_n^- K_{2n}^{-1} e^{j\beta_n z}) \mathbf{h}_n \qquad (4.8b)$$

and the specified proportionality constants K_{1n}, K_{2n} for each mode. The axial field components may be found from (4.8) by the use of Maxwell's equations. Note that the equivalent current wave amplitude for propagation in the $-z$ direction is expressed by $-I_n^-$, and hence the corresponding transverse magnetic field is proportional to $-K_{2n}^{-1} I_n^-$. When a waveguide supports several modes of propagation simultaneously at the same frequency, the number of electrical ports will exceed the number of physical ports. That is, power can be fed to a given load by means of any of the propagating modes, and all these modes may be common to a single physical waveguide input port.†

4.2 Impedance description of waveguide elements and circuits

One-port circuits

A one-port circuit (equivalent to a two-terminal network) is a circuit for which power can enter or leave through a single waveguide or transmission line. A short-circuited transmission line and a short-circuited waveguide containing a metallic post as illustrated in Fig. 4.1 are examples of one-port circuits.

For one-port devices of the above type, a knowledge of any two of the four quantities V^+, V^-, $V = V^+ + V^-$, $I = I^+ - I^-$ will serve to describe the effect of the one-port device on an incident wave (it is assumed that the waveguide supports only one propagating mode). These quantities must, of course, be referred to a terminal plane such as t in Fig. 4.1 in order to be unambiguously specified. A terminal plane, or reference plane, is the equivalent of a terminal pair in a low-frequency network. In the present instance an impedance description is desired. If the total voltage and current at the terminal plane are

$$V = V^+ + V^-$$
$$I = I^+ - I^- = Y_c(V^+ - V^-)$$

where Y_c is the equivalent characteristic admittance (actual character-

---

† The microwave equivalent-circuit theory presented in this chapter may be extended to include nonpropagating modes. However, when nonpropagating modes are included, the impedance and scattering matrices do not have the same properties as when only propagating modes are present at the terminal planes. See H. Haskal, Matrix Description of Waveguide Discontinuities in the Presence of Evanescent Modes, *IEEE Trans.*, vol. MTT-12, pp. 184–188, March, 1964.

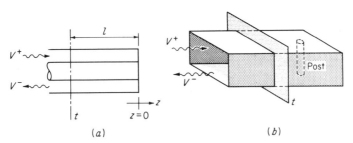

Fig. 4.1 One-port circuits. (*a*) Short-circuited coaxial line; (*b*) short-circuited waveguide with post.

istic admittance for the transmission line), the input impedance is given by

$$Z_{in} = \frac{V}{I} = \frac{V^+ + V^-}{V^+ - V^-} Z_c \tag{4.9}$$

The complex Poynting vector may be used to establish the physical nature of one-port impedance functions. From (2.59) and (2.60) we have

$$\tfrac{1}{2} \oint_S \mathbf{E} \times \mathbf{H}^* \cdot \mathbf{n} \, dS = P_l + 2j\omega(W_m - W_e) \tag{4.10}$$

where $\mathbf{n}$ is a unit inward normal to the closed surface S, P_l is the power dissipated in the volume bounded by S, and $W_m - W_e$ is the net reactive energy stored within S. For the surface S choose the terminal plane, the guide walls, and the short-circuiting plane. For perfectly conducting walls and short circuit, $\mathbf{n} \times \mathbf{E} = 0$; so the integral reduces to that over the terminal plane only. Thus

$$\tfrac{1}{2} \oint_t \mathbf{E} \times \mathbf{H}^* \cdot \mathbf{a}_z \, dS = P_l + 2j\omega(W_m - W_e) \tag{4.11}$$

Now at the terminal plane the transverse fields are [see (4.8)]

$$\mathbf{E}_t = K_1^{-1}(V^+ + V^-)\mathbf{e} = K_1^{-1} V \mathbf{e} \tag{4.12a}$$
$$\mathbf{H}_t = K_2^{-1}(I^+ - I^-)\mathbf{h} = K_2^{-1} I \mathbf{h} \tag{4.12b}$$

Hence (4.11) becomes

$$\tfrac{1}{2}(K_1 K_2^*)^{-1} V I^* \int_t \mathbf{e} \times \mathbf{h}^* \cdot \mathbf{a}_z \, dS = \tfrac{1}{2} V I^* = P_l + 2j\omega(W_m - W_e) \tag{4.13}$$

If now V is replaced by IZ_{in}, we find that

$$Z_{in} = \frac{P_l + 2j\omega(W_m - W_e)}{\tfrac{1}{2} I I^*} = R + jX \tag{4.14}$$

This relates the input impedance to the power loss and net reactive energy stored in the volume beyond the terminal plane. Inasmuch as the current I may be an equivalent current, the corresponding impedance Z_{in} is an equivalent one also. Since P_l, W_m, and W_e are all proportional to $|I^+|^2$, and hence also proportional to $|I|^2$ in view of the linearity of the field equations, the equivalent resistance R and reactance X in (4.14) are independent of the amplitude of the incident wave.

By replacing I^* by $Y_{in}^* V^*$ in (4.13), we obtain, after taking the complex conjugate,

$$Y_{in} = \frac{P_l - 2j\omega(W_m - W_e)}{\tfrac{1}{2} V V^*} = G + jB \tag{4.15}$$

for the input admittance of the one-port device. The susceptance B is positive (capacitive in nature) only if $W_e > W_m$.

The evaluation of an input impedance by means of the general definition (4.14) will be carried out for the simplest case, that of a short-circuited coaxial line. In the short-circuited coaxial line of Fig. 4.1, the fields in the one-port device are given by

$$\mathbf{E} = \frac{V^+}{\ln(b/a)} \frac{\mathbf{a}_r}{r} (e^{-jk_0z} - e^{jk_0z})$$

$$\mathbf{H} = \frac{Y_0 V^+}{\ln(b/a)} \frac{\mathbf{a}_\phi}{r} (e^{-jk_0z} + e^{jk_0z})$$

since the electric field must vanish at the short-circuited position $z = 0$. If the terminal plane is located at $z = -l$, then

$$W_e = \frac{\epsilon_0}{4} \int_0^{2\pi} \int_a^b \int_{-l}^0 |\mathbf{E}|^2 r \, d\phi \, dr \, dz$$

$$= \frac{2\pi\epsilon_0 |V^+|^2}{\ln(b/a)} \int_{-l}^0 \sin^2 k_0 z \, dz$$

$$= \frac{\pi\epsilon_0 |V^+|^2}{\ln(b/a)} \left(l - \frac{\sin 2k_0 l}{2k_0} \right)$$

Similarly, it is found that

$$W_m = \frac{\pi\mu_0 Y_0^2 |V^+|^2}{\ln(b/a)} \left(l + \frac{\sin 2k_0 l}{2k_0} \right)$$

The total current at the terminal plane at $z = -l$ is

$$I = Y_c V^+ (e^{jk_0l} + e^{-jk_0l}) = 2Y_c V^+ \cos k_0 l$$

Using (4.14) now gives

$$Z_{\text{in}} = \frac{4j\omega\pi\epsilon_0 |V^+|^2}{\ln(b/a)} \frac{\sin 2k_0 l}{k_0(4Y_c^2 |V^+|^2) \cos^2 k_0 l}$$

$$= \frac{j\omega\pi\epsilon_0}{\ln(b/a)} \frac{\sin 2k_0 l}{k_0 Y_c^2 \cos^2 k_0 l} = jZ_c \tan k_0 l \quad (4.16)$$

on using the relations $Y_c = 2\pi Y_0/[\ln(b/a)]$, $\sin 2k_0 l = 2 \sin k_0 l \cos k_0 l$, and $k_0 = \omega(\mu_0\epsilon_0)^{1/2}$. This result for the input impedance of a short-circuited coaxial line, as obtained from the general definition (4.14), agrees with the simple computation based directly on expressions for the total voltage and current at the terminal plane. However, the purpose of introducing (4.14) was not as a computational tool, but rather for the physical insight it provides into the nature of the impedance function for a one-port circuit.

The second example of a one-port circuit as illustrated in Fig. 4.1 cannot be evaluated in as straightforward a manner as for the coaxial line because it does not consist of a uniform unperturbed waveguide. The presence of a conducting post within the termination results in induced currents on the post that will excite a multitude of waveguide modes. However, since it is assumed that only one mode propagates (the TE_{10} mode), all the other modes decay exponentially in both directions away from the post. By choosing the terminal plane sufficiently far away from the post, the fields at this plane are essentially just those of the incident and reflected dominant modes. The evanescent modes excited by the post will store reactive energy, and this will contribute to the input reactance as viewed from the terminal plane, as reference to (4.14) shows. The presence of the post within the termination modifies the input reactance by changing the amplitude of the reflected dominant wave in just the right amount to account for the additional reactive energy stored.

As seen from the preceding discussion, it is important when dealing with waveguide structures to choose terminal planes sufficiently far away from obstacles that excite evanescent modes, so that only dominant-mode fields have significant amplitudes at these reference planes. This will ensure that all the reactive energy associated with the nonpropagating modes that make up the fringing field around the obstacle is taken into account in the expression for the input reactance. This precaution is particularly important in any experimental setup used to measure the impedance function for a particular obstacle. Once the impedance has been properly determined at a given terminal plane, it may be referred to any other terminal plane by using the impedance-transformation formula

$$Z(l_2) = Z_c \frac{Z(l_1) + jZ_c \tan \beta(l_2 - l_1)}{Z_c + jZ(l_1) \tan \beta(l_2 - l_1)} \tag{4.17}$$

where l_1 is the location of terminal plane 1, and l_2 specifies the location of the new terminal plane. In particular, shift in the terminal-plane position by a multiple of $\lambda_g/2$ leaves the impedance invariant. Thus an impedance may be referred to terminal planes located in the near vicinity of an obstacle where now it is understood that this impedance describes the effect of the obstacle on the dominant mode only, and does not imply that the total field at this particular terminal plane is that of the dominant mode only. In other words, the impedance description of a waveguide element or obstacle gives information on the effect this element has on the dominant propagating mode but does not give any information on the detailed field structure near the obstacle. Fortunately, the latter information is rarely required.

Lossless one-port termination

If there are no losses present in a one-port circuit, the input impedance is a pure reactance given by

$$jX = \frac{4j\omega(W_m - W_e)}{II^*} \tag{4.18}$$

The assumption of a lossless structure is often a very good approximation for microwave circuits. If $W_m = W_e$, the input reactance vanishes and a condition of resonance exists. There are actually two possibilities, namely, $W_m = W_e$ but $I \neq 0$, and $W_m = W_e$ but $V \neq 0$. The first corresponds to a zero in the input reactance (series resonance), whereas the second corresponds to a zero in the input susceptance (parallel resonance) as given by

$$jB = \frac{4j\omega(W_e - W_m)}{VV^*} \tag{4.19}$$

When the input reactance is zero, the input susceptance must be infinite, which implies that $V = 0$ at the terminal plane. This latter condition is possible since, for a pure reactive termination, all the incident power is reflected, so that the total voltage along the waveguide is a standing wave of the form $\sin \beta l$. In the case of a zero for the susceptance function B, the reactance X must be infinite (have a pole) and I must vanish at the terminal plane. It may be anticipated, then, that the reactance and susceptance functions will have a number of zeros and poles, that is, frequencies at which they vanish or become infinite. This behavior is clearly evident in the expression for the reactance of a short-circuited coaxial line, which is

$$jX = jZ_c \tan k_0 l = jZ_c \tan \frac{\omega l}{c} \tag{4.20}$$

A plot of X against frequency is given in Fig. 4.2. In particular, note that the slope of the reactance is always positive; that is, $\partial X/\partial \omega > 0$. This positive-slope condition means that the poles and zeros of X must alternate as ω is increased from zero to infinity. We shall show below that this is a general property of any reactive one-port circuit, a result known as Foster's reactance theorem. First, however, it will be instructive to rewrite (4.20) by using the infinite-product representation and also

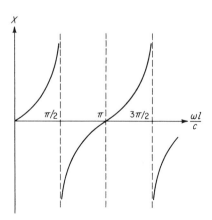

Fig. 4.2 Input reactance of a short-circuited coaxial line.

the partial-fraction expansion of the tangent function:†

$$X = Z_c \tan k_0 l = \frac{Z_c \dfrac{\omega l}{c} \prod_{n=1}^{\infty} \left[1 - \left(\dfrac{\omega l}{n\pi c}\right)^2\right]}{\prod_{n=0}^{\infty} \left[1 - \dfrac{(\omega l)^2}{(n+\frac{1}{2})^2 (\pi c)^2}\right]}$$

$$= Z_c \frac{2\omega c}{l} \sum_{n=1,3,\ldots}^{\infty} \frac{1}{(n\pi c/2l)^2 - \omega^2} \qquad (4.21)$$

The first form contains the product of an infinite number of factors in both the numerator and denominator and clearly exhibits both the zeros and poles and their alternating occurrence. The second form exhibits the poles very clearly, but information on the zero locations is lost. In the vicinity of a pole, say that at $\omega = \omega_n = n\pi c/2l$, all terms in the partial-fraction expansion are small except the nth term, so that

$$X \approx \frac{2\omega c}{l} \frac{Z_c}{(\omega_n - \omega)(\omega_n + \omega)} \approx \frac{cZ_c}{l(\omega_n - \omega)} \qquad (4.22)$$

since $\omega \approx \omega_n$. This behavior near a pole is similar to that for a simple LC parallel network for which

$$X = \frac{-\omega L}{\omega^2 LC - 1} \approx \frac{\omega_0}{2}\sqrt{\frac{L}{C}} \frac{1}{\omega_0 - \omega} \qquad (4.23)$$

where $\omega_0 = (LC)^{-\frac{1}{2}}$. However, the microwave network is a good deal more complicated, for it has an infinite number of poles and zeros, and not just a double zero and a single pole as a simple parallel LC circuit has [the zeros occur at $\omega = 0$, where ωL vanishes, and at infinity, where $(\omega C)^{-1}$ vanishes]. These similarities and differences are important to note since they are characteristic of microwave networks in general, even though we have demonstrated some of these properties for a short-circuited coaxial line only.

★4.3 Foster's reactance theorem

The theorem that will now be proved is that the rate of change of the reactance X and susceptance B with ω is positive. Once this result is established, it follows that the poles and zeros of a reactance function

† E. A. Guillemin, "The Mathematics of Circuit Analysis," chap. 6, John Wiley & Sons, Inc., New York, 1949.
  E. T. Copson, "Theory of Functions of a Complex Variable," Oxford University Press, Fair Lawn, N.J., 1935.
  J. Pierpont, "Functions of a Complex Variable," Dover Publications, Inc., New York, 1959.

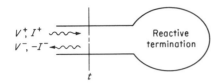

Fig. 4.3 A one-port reactive termination.

must alternate in position along the ω axis. Figure 4.3 illustrates a general one-port reactive termination. The field within the termination satisfies Maxwell's equations

$$\nabla \times \mathbf{E} = -j\omega\mu\mathbf{H} \qquad \nabla \times \mathbf{H} = j\omega\epsilon\mathbf{E}$$

The derivative with respect to ω of the complex conjugate of these equations gives

$$\nabla \times \frac{\partial \mathbf{E}^*}{\partial \omega} = j\omega\mu\frac{\partial \mathbf{H}^*}{\partial \omega} + j\mathbf{H}^*\frac{\partial \omega\mu}{\partial \omega} \qquad \nabla \times \frac{\partial \mathbf{H}^*}{\partial \omega} = -j\omega\epsilon\frac{\partial \mathbf{E}^*}{\partial \omega} - j\mathbf{E}^*\frac{\partial \omega\epsilon}{\partial \omega}$$

Consider next the quantity

$$\nabla \cdot \left(\mathbf{E} \times \frac{\partial \mathbf{H}^*}{\partial \omega} + \frac{\partial \mathbf{E}^*}{\partial \omega} \times \mathbf{H} \right) = \nabla \times \mathbf{E} \cdot \frac{\partial \mathbf{H}^*}{\partial \omega} - \mathbf{E} \cdot \nabla \times \frac{\partial \mathbf{H}^*}{\partial \omega}$$
$$+ \nabla \times \frac{\partial \mathbf{E}^*}{\partial \omega} \cdot \mathbf{H} - \frac{\partial \mathbf{E}^*}{\partial \omega} \cdot \nabla \times \mathbf{H}$$

Substituting from above gives

$$\nabla \cdot \left(\mathbf{E} \times \frac{\partial \mathbf{H}^*}{\partial \omega} + \frac{\partial \mathbf{E}^*}{\partial \omega} \times \mathbf{H} \right) = j\left(\mathbf{H} \cdot \mathbf{H}^* \frac{\partial \omega\mu}{\partial \omega} + \mathbf{E} \cdot \mathbf{E}^* \frac{\partial \omega\epsilon}{\partial \omega} \right)$$
$$+ j\omega\left(\mu\mathbf{H} \cdot \frac{\partial \mathbf{H}^*}{\partial \omega} - \mu\mathbf{H} \cdot \frac{\partial \mathbf{H}^*}{\partial \omega} + \epsilon\mathbf{E} \cdot \frac{\partial \mathbf{E}^*}{\partial \omega} - \epsilon\mathbf{E} \cdot \frac{\partial \mathbf{E}^*}{\partial \omega} \right)$$

The second term on the right-hand side vanishes; so we have

$$\nabla \cdot \left(\mathbf{E} \times \frac{\partial \mathbf{H}^*}{\partial \omega} + \frac{\partial \mathbf{E}^*}{\partial \omega} \times \mathbf{H} \right) = j\left(\mathbf{H} \cdot \mathbf{H}^* \frac{\partial \omega\mu}{\partial \omega} + \mathbf{E} \cdot \mathbf{E}^* \frac{\partial \omega\epsilon}{\partial \omega} \right)$$

If we integrate throughout the volume of the termination and use the divergence theorem on the left-hand side, we obtain

$$\oint_S \left(\mathbf{E} \times \frac{\partial \mathbf{H}^*}{\partial \omega} + \frac{\partial \mathbf{E}^*}{\partial \omega} \times \mathbf{H} \right) \cdot d\mathbf{S} = -j\int_V \left(\mathbf{H} \cdot \mathbf{H}^* \frac{\partial \omega\mu}{\partial \omega} + \mathbf{E} \cdot \mathbf{E}^* \frac{\partial \omega\epsilon}{\partial \omega} \right) dV$$
$$= -4j(W_m + W_e) \qquad (4.24a)$$

where $W_m + W_e$ is the total time-average energy stored in the lossless termination, as reference to (2.53) in Sec. 2.5 shows, and $d\mathbf{S}$ is chosen directed into the volume.

Since both $\mathbf{n} \times \mathbf{E}$ and $\mathbf{n} \times \partial\mathbf{E}/\partial\omega$, where $\mathbf{n}$ is a unit inward normal, vanish on the perfectly conducting waveguide walls, the surface integral reduces to an integral over the terminal plane t only. On the terminal plane we have

$$\int_{t} \left(\mathbf{E} \times \frac{\partial \mathbf{H}^*}{\partial \omega} + \frac{\partial \mathbf{E}^*}{\partial \omega} \times \mathbf{H} \right) \cdot \mathbf{n} \, dS = V \frac{\partial I^*}{\partial \omega} + \frac{\partial V^*}{\partial \omega} I \quad (4.24b)$$

where V and I are the equivalent terminal voltage and current. Now $V = jIX$ for a lossless reactive termination; so

$$\frac{\partial V^*}{\partial \omega} = -jX \frac{\partial I^*}{\partial \omega} - jI^* \frac{\partial X}{\partial \omega}$$

Thus

$$V \frac{\partial I^*}{\partial \omega} + \frac{\partial V^*}{\partial \omega} I = jXI \frac{\partial I^*}{\partial \omega} - jXI \frac{\partial I^*}{\partial \omega} - jII^* \frac{\partial X}{\partial \omega}$$

and hence we find that (4.24b) yields

$$II^* \frac{\partial X}{\partial \omega} = 4(W_m + W_e)$$

or

$$\frac{\partial X}{\partial \omega} = \frac{4(W_m + W_e)}{II^*} \quad (4.25)$$

The right-hand side is proportional to the total energy stored in the termination and can never be negative. Consequently, the slope of the reactance function must always be positive. If I is replaced by jBV in (4.24b), it is readily found that

$$\frac{\partial B}{\partial \omega} = \frac{4(W_m + W_e)}{VV^*} \quad (4.26)$$

and hence the susceptance is also an increasing function of frequency. The above relations also show that the frequency sensitivity of the reactance or susceptance is proportional to the total average energy stored. These relations are readily verified in the case of simple LC reactive networks, and a problem calling for this verification is given at the end of this chapter.

★4.4 Even and odd properties of Z_{in}

Before terminating the discussion dealing with one-port impedance functions, one further general property should be pointed out. This property is that the real part of $Z_{in} = R + jX$ is an even function of ω, whereas the imaginary part is an odd function. The physical necessity of this was pointed out in the previous chapter in the section dealing with group velocity. The property stems from the requirement that the response of a circuit to a real-time-dependent driving function must also be real. That is, if $\mathcal{U}(t)$ is the applied voltage at the terminal plane, the frequency spectrum is given by the Fourier transform:

$$V(\omega) = \int_{-\infty}^{\infty} e^{-j\omega t} \mathcal{V}(t)\, dt \tag{4.27}$$

The frequency spectrum of the current that flows is

$$I(\omega) = \frac{V(\omega)}{Z_{\text{in}}(\omega)} = \frac{V}{R + jX} \tag{4.28}$$

The current as a function of time is

$$\mathcal{I}(t) = \frac{1}{2\pi} \int_{-\infty}^{\infty} \frac{V(\omega)}{R(\omega) + jX(\omega)} e^{j\omega t}\, d\omega \tag{4.29}$$

and must be a real function. This will be the case if

$$\frac{V(-\omega)}{R(-\omega) + jX(-\omega)} = \frac{V^*(\omega)}{[R(\omega) + jX(\omega)]^*}$$

for then (4.29) becomes

$$\mathcal{I}(t) = \frac{1}{2\pi} \int_{-\infty}^{0} \frac{V(\omega)e^{j\omega t}}{R(\omega) + jX(\omega)}\, d\omega + \frac{1}{2\pi} \int_{0}^{\infty} \frac{V(\omega)e^{j\omega t}}{R(\omega) + jX(\omega)}\, d\omega$$

which we can show is a real function. In the first integral on the right replace ω by $-\omega$ to obtain

$$\mathcal{I}(t) = \frac{1}{2\pi} \int_{0}^{\infty} \left[\frac{V(-\omega)e^{-j\omega t}}{R(-\omega) + jX(-\omega)} + \frac{V(\omega)e^{j\omega t}}{R(\omega) + jX(\omega)} \right] d\omega \tag{4.30}$$

The two terms in the integrand are complex conjugates of each other, and hence the sum is real; i.e.,

$$\mathcal{I}(t) = \frac{1}{\pi} \operatorname{Re} \int_{0}^{\infty} \frac{V(\omega)e^{j\omega t}}{R(\omega) + jX(\omega)}\, d\omega \tag{4.31}$$

The condition specified on $V/(R + jX)$ is satisfied by V alone and by $R + jX$ alone. Clearly, from (4.27), $V(-\omega) = V^*(\omega)$. If

$$R(-\omega) + jX(-\omega) = [R(\omega) + jX(\omega)]^*$$

then

$$R(-\omega) = R(\omega) \qquad X(-\omega) = -X(\omega)$$

and R is an even function of ω, whereas X is an odd function of ω, as was to be proved. These even and odd properties are useful to know when approximate expressions for impedance functions are constructed from experimental data. For example, a series such as

$$a_1\omega + a_3\omega^3 + a_5\omega^5 + \cdots$$

could be used to represent X, but not to represent R, since the series is an odd function of ω.

4.5 N-port circuits

Figure 4.4 illustrates the junction of N waveguides or transmission lines (or a combination of the two) that terminate in a common region or junction. The region between the N chosen terminal planes may contain any arbitrary collection of passive elements. If each guide can support only one mode of propagation, this circuit constitutes an N-port microwave circuit. If one or more of the guides can support several independent modes of propagation, the number of electrical ports exceeds the number of mechanical ports. Each mode, since it carries power independently of all other modes, corresponds to an electrical port through which power may enter or leave the junction. To simplify the discussion it will be assumed that each guide supports only a single propagating mode. The extension of the theory to the case when some or all of the guides may support several propagating modes is more or less obvious.

Let the terminal planes be chosen sufficiently far from the junction so that the fields on the terminal planes are essentially just those of the incident and reflected dominant propagating modes. These fields may be uniquely defined at the terminal planes in terms of suitably defined equivalent voltages and currents. Clearly, the amplitudes of all the incident waves may be arbitrarily specified, i.e., chosen independently. The amplitudes of all the reflected waves are then determined by the physical properties of the junction; that is, all the V_n^- are linear functions of the V_n^+. When the V_n^+ and V_n^- are known, the corresponding currents I_n^+, I_n^- are known from the relations

$$I_n^+ = Y_n V_n^+ \qquad I_n^- = Y_n V_n^-$$

Since Maxwell's equations are linear and the junction is assumed to be linear in its behavior, any N linearly independent combinations of the $4N$ quantities V_n^+, V_n^-, I_n^+, and I_n^- may be chosen as the independent

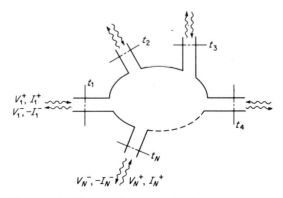

Fig. 4.4 An N-port microwave circuit.

variables to describe the electrical behavior of the junction. For an impedance description the total currents $I_n = I_n{}^+ - I_n{}^-$ at the terminal planes are chosen as independent variables. The N total terminal-plane voltages $V_n = V_n{}^+ + V_n{}^-$ are then the dependent variables, and are linearly related to the currents as follows:

$$\begin{bmatrix} V_1 \\ V_2 \\ \cdots \\ V_N \end{bmatrix} = \begin{bmatrix} Z_{11} & Z_{12} & \cdots & Z_{1N} \\ Z_{21} & Z_{22} & \cdots & Z_{2N} \\ \cdots & \cdots & \cdots & \cdots \\ Z_{N1} & Z_{N2} & \cdots & Z_{NN} \end{bmatrix} \begin{bmatrix} I_1 \\ I_2 \\ \cdots \\ I_N \end{bmatrix} \quad (4.32)$$

The matrix of elements Z_{ij} is the impedance matrix and provides a complete description of the electrical properties of the N-port circuit. Some of the properties of this impedance matrix are discussed below.

If the junction contains a nonreciprocal medium such as a plasma (ionized gas) or a ferrite with an applied d-c magnetic biasing field (ferrites are discussed in Chap. 6), then in general $Z_{ij} \neq Z_{ji}$; that is, the impedance matrix $[Z]$ is not symmetrical. The junction then requires $2N^2$ parameters to describe it completely since each Z_{ij} is complex and has two independent terms. If the junction does not contain any nonreciprocal media, $Z_{ij} = Z_{ji}$, and the impedance matrix is symmetrical. In this instance a total of only $2N^2 - (N^2 - N) = N(N + 1)$ independent parameters are required to describe the junction since $N^2 - N$ of the parameters are equal. Finally, if the junction is lossless—and many microwave junctions may be approximated as such with negligible error—then all the Z_{ij} must be pure imaginary since there can be no power loss within the junction. In this case there are only $\frac{1}{2}N(N + 1)$ independent parameters required for a complete description. Any network containing the required number of resistive and reactive elements may be used as a representation for the junction at a given frequency. However, it must be kept in mind that when the frequency is changed, the values of the network elements (resistance, capacitance, and inductance) must also be changed. Rarely would any one particular network representation provide a complete description of the junction over a band of frequencies unless the network parameters are changed in value when the frequency is changed.

The foregoing discussion applies also to the admittance matrix $[Y]$, which relates the currents to the total voltages at the terminal planes; i.e.,

$$\begin{bmatrix} I_1 \\ I_2 \\ \cdots \\ I_N \end{bmatrix} = \begin{bmatrix} Y_{11} & Y_{12} & \cdots & Y_{1N} \\ Y_{21} & Y_{22} & \cdots & Y_{2N} \\ \cdots & \cdots & \cdots & \cdots \\ Y_{N1} & Y_{N2} & \cdots & Y_{NN} \end{bmatrix} \begin{bmatrix} V_1 \\ V_2 \\ \cdots \\ V_N \end{bmatrix} \quad (4.33)$$

The impedance and admittance matrices are reciprocals of each other; so

$$[Y] = [Z]^{-1} \quad (4.34)$$

Proof of symmetry for the impedance matrix

The symmetry of the impedance matrix is readily proved when the junction contains media characterized by scalar parameters μ and ϵ. Let incident-wave amplitudes V_n^+ be so chosen that the total voltage V_n equals zero at all terminal planes except the ith plane. Let the corresponding field solution be $\mathbf{E}_i$, $\mathbf{H}_i$. Similarly, let a second solution $\mathbf{E}_j$, $\mathbf{H}_j$ correspond to the case when incident-wave amplitudes are chosen so that all V_n equal zero except V_j. The Lorentz reciprocity theorem [Eq. (2.135)] gives

$$\oint_S (\mathbf{E}_i \times \mathbf{H}_j - \mathbf{E}_j \times \mathbf{H}_i) \cdot \mathbf{n}\, dS = 0$$

when there are no sources within the closed surface S. Let S consist of the conducting walls bounding the junction and the N terminal planes. The integral over the walls vanishes if they are perfectly conducting or if they exhibit a surface impedance Z_m (Sec. 2.12). Therefore we obtain an integral over the terminal planes only, i.e.,

$$\sum_{n=1}^{N} \int_{t_n} (\mathbf{E}_i \times \mathbf{H}_j - \mathbf{E}_j \times \mathbf{H}_i) \cdot \mathbf{n}\, dS = 0 \qquad (4.35)$$

However, for the particular solutions considered here, $\mathbf{n} \times \mathbf{E}_i$ and $\mathbf{n} \times \mathbf{E}_j$, that is, $\mathbf{E}_{ti}$, $\mathbf{E}_{tj}$ are zero on all terminal planes except t_i and t_j, respectively, since all V_n except V_i and V_j have been chosen equal to zero. Thus (4.35) becomes

$$\int_{t_i} \mathbf{E}_i \times \mathbf{H}_j \cdot \mathbf{n}\, dS = \int_{t_j} \mathbf{E}_j \times \mathbf{H}_i \cdot \mathbf{n}\, dS$$

or

$$V_i(I_i)_j = V_j(I_j)_i \qquad (4.36)$$

where $(I_i)_j$ is the current at the terminal plane i arising from an applied voltage at plane j, and similarly for $(I_j)_i$. From the admittance description (4.33) of the junction we have

$$I_i = (I_i)_j = Y_{ij} V_j \qquad \text{for } V_n = 0,\ n \neq j$$
$$I_j = (I_j)_i = Y_{ji} V_i \qquad \text{for } V_n = 0,\ n \neq i$$

Hence (4.36) gives

$$V_i V_j Y_{ij} = V_j V_i Y_{ji}$$

or

$$Y_{ij} = Y_{ji} \qquad (4.37)$$

which proves the symmetry of the admittance matrix. Since the reciprocal of a symmetrical matrix is a symmetrical matrix also, it follows that the impedance matrix is also symmetrical. The symmetry of the impedance and admittance matrices is a consequence of reciprocity. For nonreciprocal media, μ or ϵ (or both) are nonsymmetrical matrices, and (2.135) no longer applies. In this case the impedance matrix is no longer symmetrical. Nonreciprocal microwave devices are discussed in Chap. 6; so no further comments on these are made in this section.

Proof of imaginary nature of [Z] for a lossless junction

For a lossless junction all the elements in the impedance and admittance matrices are pure imaginary. Let $[V]$ and $[I]$ be column matrices representing the terminal voltages and currents, respectively. The transposed matrices $[V]_t$, $[I]_t$ are row matrices of the form

$$[V]_t = [V_1 \quad V_2 \quad \cdots \quad V_N]$$
$$[I]_t = [I_1 \quad I_2 \quad \cdots \quad I_N]$$

The total complex power into the junction is

$$\tfrac{1}{2}[I^*]_t[V] = \tfrac{1}{2}[I^*]_t[Z][I]$$
$$= \tfrac{1}{2} \sum_{n=1}^{N} \sum_{m=1}^{N} I_n^* Z_{nm} I_m = P_l + 2j\omega(W_m - W_e) \qquad (4.38)$$

For a lossless junction $P_l = 0$ and the double sum must be pure imaginary. Since the I_n can be chosen as independent variables, they may all be chosen as zero except for the nth one. In this case

$$\text{Re } (I_n^* Z_{nn} I_n) = 0$$

or

$$\text{Re } Z_{nn} = 0$$

If all but I_n and I_m are chosen equal to zero, we obtain

$$\text{Re } [(I_n^* I_m + I_m^* I_n) Z_{nm} + I_n I_n^* Z_{nn} + I_m I_m^* Z_{mm}] = 0$$

But $I_n I_n^*$, $I_m I_m^*$, and $I_n^* I_m + I_n I_m^*$ are all real quantities and Z_{nn}, Z_{mm} are imaginary; so this equation can hold only if

$$\text{Re } Z_{nm} = 0$$

Therefore all Z_{nm} are pure imaginary for a lossless junction.

4.6 Two-port junctions

At this point it seems advisable to examine the special case of the two-port junction rather than continue with the general theory of N-port circuits. The derivation of some further properties of the N-port junction is called for in some of the problems at the end of this chapter. Three examples of two-port junctions are shown in Fig. 4.5. The first is the junction of two rectangular guides of unequal height (called an E-plane step since the E vector of the TE_{10} mode lies in the plane containing the step geometry). The second is a symmetrical junction consisting of two similar rectangular guides joined by an intermediate guide of greater width. The third two-port junction consists of a typical coaxial-line–waveguide junction, where the center conductor of the coaxial line extends into the rectangular guide to provide an antenna radiating energy into or coupling energy out of the rectangular guide. A discussion of these particular junctions will serve to develop the general impedance description of two-port junctions or circuits.

Since evanescent modes are excited at each discontinuity, the terminal planes are chosen far enough away so that these decaying waves have negligible amplitudes at the terminal planes. Equivalent voltages and currents are introduced proportional to the total transverse electric and magnetic fields, respectively, at each terminal plane. For example, for the junction in Fig. 4.5a, let the incident and reflected transverse fields of the TE_{10} mode at each terminal plane be (coordinates x, y, z refer to the left-hand-side guide, and the primed coordinates x', y', z' refer to the

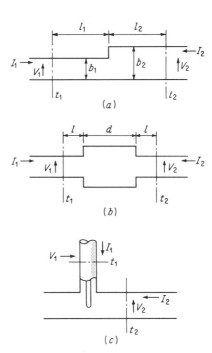

Fig. 4.5 Examples of two-port junctions.

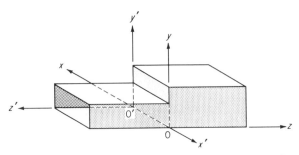

Fig. 4.6 Coordinates used to describe the junction in Fig. 4.5a.

right-hand-side guide, as in Fig. 4.6)

$$\mathbf{E}_t = (C_1{}^+ e^{j\beta l_1} + C_1{}^- e^{-j\beta l_1})\mathbf{a}_y \sin \frac{\pi x}{a} \qquad \text{at } t_1$$

$$\mathbf{H}_t = -Y_w(C_1{}^+ e^{j\beta l_1} - C_1{}^- e^{-j\beta l_1})\mathbf{a}_x \sin \frac{\pi x}{a} \qquad \text{at } t_1$$

$$\mathbf{E}_t = (C_2{}^+ e^{j\beta l_2} + C_2{}^- e^{-j\beta l_2})\mathbf{a}_y \sin \frac{\pi x'}{a} \qquad \text{at } t_2$$

$$\mathbf{H}_i = +Y_w(C_2{}^+ e^{j\beta l_2} - C_2{}^- e^{-j\beta l_2})\mathbf{a}_x \sin \frac{\pi x'}{a} \qquad \text{at } t_2$$

These expressions are obtained by choosing $h_z = (j\pi Y_w/\beta a) \cos(\pi x/a)$ in order to simplify the expressions for the transverse fields. Let the equivalent voltages and currents be chosen as

$$V_1{}^+ = K_1 C_1{}^+ e^{j\beta l_1} \qquad V_1{}^- = K_1 C_1{}^- e^{-j\beta l_1}$$
$$I_1{}^+ = Y_w K_1 C_1{}^+ e^{j\beta l_1} \qquad I_1{}^- = Y_w K_1 C_1{}^- e^{-j\beta l_1}$$
$$V_2{}^+ = K_2 C_2{}^+ e^{j\beta l_2} \qquad V_2{}^- = K_2 C_2{}^- e^{-j\beta l_2}$$
$$I_2{}^+ = Y_w K_2 C_2{}^+ e^{j\beta l_2} \qquad I_2{}^- = Y_w K_2 C_2{}^- e^{-j\beta l_2}$$

Thus the characteristic impedance of the equivalent transmission line is equal to the wave impedance $Z_w = Y_w^{-1}$ of the TE_{10} mode. To conserve power it is necessary to choose K_1 and K_2 so that

$$V_1{}^+(I_1{}^+)^* = Y_w K_1{}^2 |C_1{}^+|^2 = Y_w |C_1{}^+|^2 \int_0^a \int_0^{b_1} \sin^2 \frac{\pi x}{a} \, dx \, dy$$

or $K_1 = \sqrt{ab_1/2}$ and

$$V_2{}^+(I_2{}^+)^* = Y_w K_2{}^2 |C_2{}^+|^2 = Y_w |C_2{}^+|^2 \int_0^a \int_0^{b_2} \sin^2 \frac{\pi x'}{a} \, dx' \, dy'$$

or $K_2 = \sqrt{ab_2/2}$.

If we use the above equivalent voltages and currents, we have

$$\begin{bmatrix} V_1 \\ V_2 \end{bmatrix} = \begin{bmatrix} Z_{11} & Z_{12} \\ Z_{12} & Z_{22} \end{bmatrix} \begin{bmatrix} I_1 \\ I_2 \end{bmatrix} \tag{4.39}$$

as a suitable description of the E-plane step. An equivalent circuit consisting of a T network joining two transmission lines as in Fig. 4.7a provides a convenient equivalent circuit for the junction. Other equivalent circuits are also possible. In particular, if the junction is lossless, any circuit consisting of three independent parameters may be used. Figure 4.7b illustrates a circuit consisting of two lengths of transmission lines of length d_1 and d_2 connected by an ideal transformer of turns ratio $n:1$. Figure 4.7c is a variation of this circuit, where the transmission lines are replaced by reactive elements jX_1 and jX_2. The parameters of any one of these circuits may be expressed in terms of the parameters of any of the others. The required derivations may be carried out in the usual manner.

The foregoing discussion applies equally well to the other two-port junctions of Fig. 4.5 provided suitably defined voltages and currents are introduced. Likewise, the equivalent circuits of Fig. 4.7 may be used to describe the behavior of these other junctions. Although general forms for the equivalent networks can be readily specified, the values of the parameters are not so easily found. In some cases the network parameters can be evaluated analytically, whereas in many other cases they must be determined by experimental measurements.†

For the junction of Fig. 4.5b perfect symmetry exists about the midplane, and hence the equivalent impedance matrix has $Z_{11} = Z_{22}$. For a lossless junction of this type, the equivalent circuit can be any circuit

† For typical analytical solutions and the methods employed, see R. E. Collin, "Field Theory of Guided Waves," McGraw-Hill Book Company, New York, 1960.

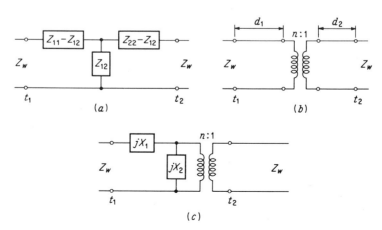

Fig. 4.7 Equivalent circuits for a lossless two-port circuit.

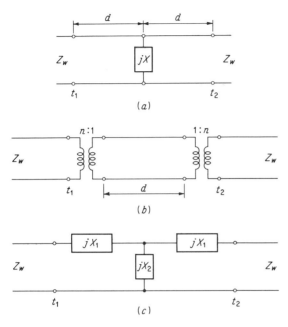

Fig. 4.8 Equivalent circuits for a symmetrical two-port junction.

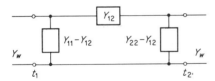

Fig. 4.9 Equivalent circuit for admittance matrix of a two-port junction.

containing two independent parameters in a symmetrical connection. Some typical circuits that may be used are illustrated in Fig. 4.8.

The foregoing discussion could be rephrased so as to apply to the admittance-matrix representation as well; i.e.,

$$\begin{bmatrix} I_1 \\ I_2 \end{bmatrix} = \begin{bmatrix} Y_{11} & Y_{12} \\ Y_{12} & Y_{22} \end{bmatrix} \begin{bmatrix} V_1 \\ V_2 \end{bmatrix} \tag{4.40}$$

The basic equivalent circuit described by (4.40) is the Π network illustrated in Fig. 4.9.

Example 4.1 To illustrate the use of equivalent circuits (assuming that their parameters are known at each frequency of interest) consider the coaxial-line–waveguide junction of Fig. 4.5c. Let a generator of internal impedance Z_g be connected to the coaxial line a distance l from the terminal plane t_1. Let the output guide be connected to a load that is matched to the guide, i.e., that presents an impedance Z_w at the terminal plane t_2. The overall circuit is that illustrated in Fig. 4.10a. We wish to evaluate the power transmitted to the load and the standing-wave ratio on the input line. The equivalent transformed load impedance at the plane t_1 is found by conventional circuit analysis to be

$$Z_L = Z_{11} - \frac{Z_{12}^2}{Z_{22} + Z_w} \tag{4.41}$$

This impedance is transformed by the length l of coaxial line into an impedance

$$Z'_L = Z_c \frac{Z_L + jZ_c \tan \beta l}{Z_c + jZ_L \tan \beta l} \tag{4.42}$$

at the generator terminals as in Fig. 4.10b. This reduced circuit is easily solved. The current supplied by the generator is

$$I_g = \frac{V_g}{Z_g + Z'_L} \tag{4.43}$$

and the power delivered to Z'_L is

$$P = \tfrac{1}{2}|I_g|^2 \operatorname{Re} Z'_L \tag{4.44}$$

If the coaxial line and junction have negligible loss, this is also the power that is delivered to the load.

To compute the standing-wave ratio note that the effective impedance terminating the coaxial line at t_1 is Z_L. Thus a reflection coefficient Γ given by

$$\Gamma = \frac{Z_L - Z_c}{Z_L + Z_c} \tag{4.45a}$$

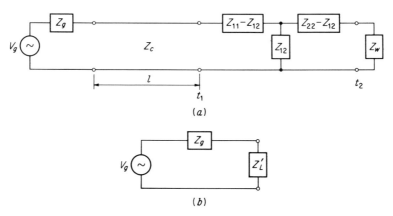

Fig. 4.10 Equivalent circuit for generator connected to a coaxial-line-fed waveguide.

is produced. This reflection results in a standing-wave ratio S given by

$$S = \frac{1 + |\Gamma|}{1 - |\Gamma|} \tag{4.45b}$$

The reflection coefficient and standing-wave ratio depend only on the effective terminating impedance Z_L.

Maximum power will be delivered to the load if Z_g is made equal to the complex conjugate of Z'_L, just as in the case of a low-frequency circuit. Other low-frequency network theorems may also be applied. For example, Thévenin's theorem may be applied at the terminal plane t_2 to reduce the circuit to an equivalent generator with a new voltage V'_g and a new internal impedance Z'_g. The new impedance is readily found by transforming Z_g to an equivalent impedance

$$Z_{ge} = Z_c \frac{Z_g + jZ_c \tan \beta l}{Z_c + jZ_g \tan \beta l} \tag{4.46}$$

at the plane t_1. When viewed through the junction, this impedance appears as an impedance

$$Z'_g = Z_{22} - \frac{Z_{12}{}^2}{Z_{11} + Z_{ge}} \tag{4.47}$$

at the terminal plane t_2.

The evaluation of the open-circuit voltage at t_2 is not quite as straightforward. However, by using Thévenin's theorem twice in succession, the desired result can be deduced with a minimum of labor. We first construct a Thévenin equivalent circuit at plane t_1. The equivalent internal generator impedance to use here is Z_{ge} given above. Let the voltage waves produced by the generator when the coaxial line is open-circuited at t_1 be

$$V^+ e^{-j\beta z} + V^- e^{j\beta z}$$

where z is measured from the generator. At $z = l$ the coaxial line is open-circuited; so the total current

$$I = Y_c(V^+ e^{-j\beta z} - V^- e^{j\beta z})$$

must vanish at $z = l$. Thus $V^- = V^+ e^{-2j\beta l}$. Hence, at the generator end where $z = 0$,

$$V(0) = V^+(1 + e^{-2j\beta l})$$
$$I(0) = Y_c V^+(1 - e^{-2j\beta l})$$

But $I(0) = I_g$ and $V_g = I_g Z_g + V(0)$; so

$$V(0) = V^+(1 + e^{-2j\beta l}) = V_g - I_g Z_g = V_g - I(0)Z_g$$
$$= V_g - Y_c V^+ Z_g (1 - e^{-2j\beta l})$$

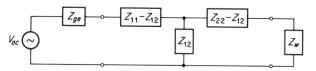

Fig. 4.11 Thévenin equivalent circuit at terminal plane t_1.

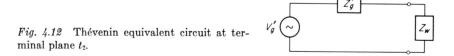

Fig. 4.12 Thévenin equivalent circuit at terminal plane t_2.

When we solve for V^+ we obtain

$$V^+ = \frac{V_g}{(1 + Y_c Z_g) + (1 - Y_c Z_g)e^{-2j\beta l}} \tag{4.48}$$

The open-circuit voltage at t_1 is now readily found to be

$$\begin{aligned} V_{oc} &= V^+ e^{-j\beta l} + V^- e^{j\beta l} = 2V^+ e^{-j\beta l} \\ &= \frac{2V_g e^{-j\beta l}}{(1 + Y_c Z_g) + (1 - Y_c Y_g)e^{-2j\beta l}} \end{aligned} \tag{4.49}$$

The Thévenin equivalent circuit at t_1 is that illustrated in Fig. 4.11.

Application of Thévenin's theorem once more in the usual manner leads readily to the circuit of Fig. 4.12, where Z_g' is given by (4.47) and V_g' is given by

$$V_g' = \frac{Z_{12} V_{oc}}{Z_{11} + Z_{ge}} \tag{4.50}$$

The reader may readily verify that the power delivered to Z_w as computed from the circuit of Fig. 4.12 is the same as that given by (4.44) (assuming no circuit losses), that is,

$$P = \frac{1}{2} \left| \frac{V_g'}{Z_g' + Z_w} \right|^2 Z_w \tag{4.51}$$

and is equal to (4.44). When there are circuit losses, not all the power delivered to the equivalent load impedance Z_L' of (4.44) is absorbed in the load Z_w. However, (4.51) does give the correct power delivered to the load even if other circuit losses are present. An alternative way of solving the problem is to replace the coaxial line by an equivalent T network also (Prob. 4.9), in which case the circuit is reduced to a conventional lumped-parameter network.

168 Foundations for microwave engineering

Some equivalent two-port circuits†

Figure 4.13 illustrates a number of useful equivalent circuits and some of their duals for representing lossless two-port junctions. The impedance parameters Z_{11}, Z_{22}, and Z_{12} are given below in terms of the network parameters, and vice versa. The same equations apply for the admittance parameters for the dual network (replace Z_{11} by Y_{11}, Z_0 by Y_0, etc.).

† The material in this section has been reproduced in modified form from C. G. Montgomery, R. H. Dicke, and E. M. Purcell, "Principles of Microwave Circuits," pp. 105-108, McGraw-Hill Book Company, New York, 1948.

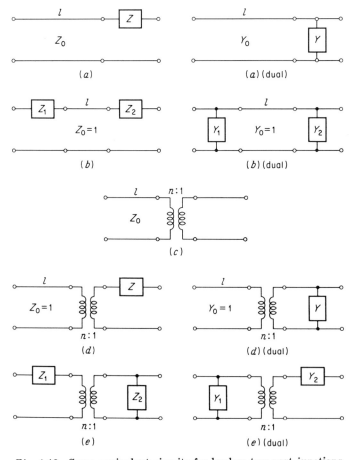

Fig. 4.13 Some equivalent circuits for lossless two-port junctions.

Note that Z_0 in these circuits is an independent parameter or characteristic impedance and does not equal $(\mu_0/\epsilon_0)^{1/2}$.

(a) $Z_{11} = -jZ_0 \cot \beta l$ $Z = Z_{22} - Z_{11}$

$Z_{22} = Z - jZ_0 \cot \beta l$ $\cos \beta l = \dfrac{Z_{11}}{Z_{12}}$

$Z_{12} = \pm jZ_0 \csc \beta l$ $Z_0 = jZ_{12}\left[1 - \left(\dfrac{Z_{11}}{Z_{12}}\right)^2\right]^{1/2}$

(b) $Z_{11} = Z_1 - j \cot \beta l$ $Z_1 = Z_{11} + \sqrt{1 + Z_{12}^2}$

$Z_{22} = Z_2 - j \cot \beta l$ $Z_2 = Z_{22} + \sqrt{1 + Z_{12}^2}$

$Z_{12} = \pm j \csc \beta l$ $\sin \beta l = \pm \dfrac{j}{Z_{12}}$

If $\beta l = \pi/2$, choose

$Z_{11} = Z_1$ $Z_1 = Z_{11}$
$Z_{22} = Z_2$ $Z_2 = Z_{22}$
$Z_{12} = jZ_0$ $Z_0 = -jZ_{12} \neq 1$

(c) $Z_{11} = -jZ_0 \cot \beta l$ $\cos \beta l = \dfrac{\sqrt{Z_{11}Z_{22}}}{Z_{12}}$

$Z_{22} = -j\dfrac{Z_0}{n^2} \cot \beta l$ $Z_0 = -jZ_{11}\sqrt{\dfrac{Z_{12}^2}{Z_{11}Z_{22}} - 1}$

$Z_{12} = j\dfrac{Z_0}{n} \csc \beta l$ $n = \sqrt{\dfrac{Z_{11}}{Z_{22}}}$

(d) $Z_{11} = -j \cot \beta l$ $\cot \beta l = jZ_{11}$

$Z_{22} = Z - \dfrac{j}{n^2} \cot \beta l$ $n = \dfrac{\sqrt{1 + Z_{11}^2}}{Z_{12}}$

$Z_{12} = \dfrac{j}{n} \csc \beta l$ $Z = Z_{22} - \dfrac{Z_{11}Z_{12}^2}{1 + Z_{11}^2}$

(e) $Z_{11} = Z_1 + n^2 Z_2$ $Z_1 = Z_{11} - \dfrac{Z_{12}^2}{Z_{22}}$

$Z_{22} = Z_2$ $Z_2 = Z_{22}$

$Z_{12} = \pm n Z_2$ $n = \dfrac{Z_{12}}{Z_{22}}$

An equivalent circuit incorporating a length of transmission line is particularly convenient to use since a shift in one (or both) of the terminal planes will reduce the circuit to a very simple form. For example, let the equivalent circuit of Fig. 4.13c be used to represent the junction in Fig. 4.14 between the terminal planes t_1 and t_2. If the terminal plane t_1 is shifted a distance l' to the left so that $\beta(l + l') = \pi$, that is, $l + l' = \lambda_g/2$, then the equivalent circuit has a section of transmission line one-half guide wavelength long. But since impedance is invariant to a transformation through a half-wavelength-long section of trans-

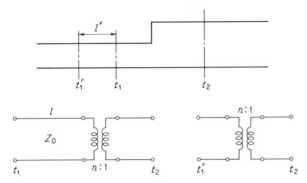

Fig. 4.14 A junction and its equivalent circuit.

mission line, this being equivalent to a 1:1 turns-ratio transformer, the section may be removed and the equivalent circuit reduces to a single ideal transformer. This new circuit represents the junction between the new terminal planes t_1' and t_2.

4.7 Scattering-matrix formulation

The preceding section dealing with the impedance description of microwave circuits is in many respects an abstraction since the voltages, currents, and impedances cannot be measured in a direct manner at microwave frequencies. The quantities must therefore be regarded as secondary, or derived, quantities. The quantities that are directly measurable, by means of a small probe used to sample the relative field strength, are the standing-wave ratio, location of a field minimum position, and power. The first two quantities lead directly to a knowledge of the reflection coefficient. The measurement of power is needed only if the absolute value of the field in the device needs to be known. Another parameter that is directly measurable is the transmission coefficient through a circuit or junction, this being a relative measurement of the amplitude and phase of the transmitted wave as compared with those of the incident wave. In other words, the directly measurable quantities are the amplitudes and phase angles of the waves reflected, or scattered, from a junction relative to the incident-wave amplitudes and phase angles. Again, in view of the linearity of the field equations and most microwave devices, the scattered-wave amplitudes are linearly related to the incident-wave amplitudes. The matrix describing this linear relationship is called the scattering matrix.

Consider the N-port junction of Fig. 4.15. If a wave with an associated equivalent voltage V_1^+ is incident on the junction at terminal plane t_1, a reflected wave $S_{11}V_1^+ = V_1^-$ will be produced in line 1, where S_{11} is the reflection coefficient, or scattering coefficient, for line 1, with a wave incident on line 1. Waves will also be transmitted, or scattered, out of the other junctions and will have amplitudes proportional to V_1^+. These amplitudes can be expressed as $V_n^- = S_{n1}V_1^+$, $n = 2, 3, \ldots, N$, where S_{n1} is a transmission coefficient to line n from line 1. When waves are incident in all lines, the scattered wave in each line has contributions arising from all the incident waves. Thus, in general, we can write

$$\begin{bmatrix} V_1^- \\ V_2^- \\ \ldots \\ V_N^- \end{bmatrix} = \begin{bmatrix} S_{11} & S_{12} & S_{13} & \cdots & S_{1N} \\ S_{21} & S_{22} & S_{23} & \cdots & S_{2N} \\ \ldots & \ldots & \ldots & \ldots & \ldots \\ S_{N1} & S_{N2} & S_{N3} & \cdots & S_{NN} \end{bmatrix} \begin{bmatrix} V_1^+ \\ V_2^+ \\ \ldots \\ V_N^+ \end{bmatrix} \quad (4.52a)$$

or

$$[V^-] = [S][V^+] \quad (4.52b)$$

where $[S]$ is called the scattering matrix.

When dealing with the scattering-matrix description of a junction, it is convenient to choose all the equivalent voltages (and currents, which, however, do not enter the picture explicitly) so that the power transmitted is given by $\frac{1}{2}|V_n^+|^2$ for all values of n. This corresponds to choosing the equivalent characteristic impedances equal to unity.† The main reason for doing this is to obtain a symmetrical scattering matrix for reciprocal structures. If this normalization is not used, then because of different impedance levels in different lines, the scattering matrix

† Any value different from unity would also be suitable, the only requirement being that all lines have the same characteristic impedance, so that power will always be equal to some constant times $|V_n^+|^2$.

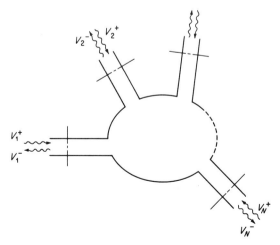

Fig. 4.15 An N-port junction illustrating scattered waves.

cannot be symmetrical. Note that, with the assumed normalization, $V = V^+ + V^-$ and $I = I^+ - I^- = V^+ - V^-$, and hence $V^+ = \frac{1}{2}(V + I)$ and $V^- = \frac{1}{2}(V - I)$. Thus the new variables V^+ and V^- are linear combinations of the variables V and I used in the impedance description. For this reason the currents do not enter into the scattering-matrix formulation. If desired, they may be calculated from the relation $I = V^+ - V^-$.

At any particular frequency and for a given location of the terminal planes, the scattering-matrix elements S_{nm} have definite values. If the frequency is changed, these elements change values also, in a manner not readily deduced analytically in general. However, at a fixed frequency the change in the scattering-matrix elements arising from a shift in the terminal-plane location is readily found. For example, let terminal plane t_n be shifted outward an amount l_n corresponding to an electrical phase shift of $\theta_n = \beta_n l_n$, where β_n is the propagation phase constant for the nth line. If the incident-wave voltage is still denoted by V_n^+ at this new terminal plane, all the transmission coefficients S_{mn}, $m \neq n$, for transmission into line m from line n must be multiplied by $e^{-j\theta_n}$ to account for the additional path length the waves must travel over. The reflected wave in line n has traveled a distance $2l_n$ more relative to the incident wave at the new terminal plane. Thus the new value of S_{nn} is $e^{-j2\theta_n}S_{nn}$. Likewise, waves traveling from line m to line n must travel a distance l_n farther, and thus S_{nm} is changed to $e^{-j\theta_n}S_{nm}$. These results are readily expressed in the general case by the following transformation of the $[S]$ matrix into the new $[S']$ matrix:

$$[S'] = \begin{bmatrix} e^{-j\theta_1} & & & \\ & e^{-j\theta_2} & & \\ & & \cdot & \\ & & & \cdot \\ & & & & e^{-j\theta_N} \end{bmatrix} \begin{bmatrix} S_{11} & S_{12} & \cdots & S_{1N} \\ S_{21} & S_{22} & \cdots & S_{2N} \\ \cdots & \cdots & \cdots & \cdots \\ S_{N1} & S_{N2} & \cdots & S_{NN} \end{bmatrix} \begin{bmatrix} e^{-j\theta_1} & & & \\ & e^{-j\theta_2} & & \\ & & \cdot & \\ & & & \cdot \\ & & & & e^{-j\theta_N} \end{bmatrix} \quad (4.53)$$

where $\theta_n = \beta_n l_n$ is the outward electrical phase shift of the nth terminal plane.

Symmetry of scattering matrix

For a reciprocal junction the scattering matrix is symmetrical, that is, $S_{nm} = S_{mn}$, provided the equivalent voltages have been chosen so that power is given by $\frac{1}{2}|V_n^+|^2$ for all modes. The latter condition is equivalent to choosing the characteristic impedance of all equivalent transmission lines used to represent the waveguides equal to unity. If the voltages are not chosen in this fashion, $[S]$ will, in general, not be symmetrical. Problem 4.15 gives an example of a nonsymmetrical scattering matrix.

The proof of the symmetry property of the scattering matrix is readily obtained by utilizing the known symmetry property of the impedance matrix. Thus the symmetry of the scattering matrix is basically a consequence of reciprocity. For the normalization used in the present section,

$$V_n = V_n^+ + V_n^- \qquad I_n = I_n^+ - I_n^- = V_n^+ - V_n^-$$

Thus, since $[V] = [V^+] + [V^-] = [Z][I] = [Z][V^+] - [Z][V^-]$, we have

$$([Z] + [U])[V^-] = ([Z] - [U])[V^+]$$

or

$$[V^-] = ([Z] + [U])^{-1}([Z] - [U])[V^+] \tag{4.54}$$

where $[U]$ is the unit matrix. Comparing this result with (4.52b) shows that the scattering matrix is related to the impedance matrix in the following manner:

$$[S] = ([Z] + [U])^{-1}([Z] - [U]) \tag{4.55a}$$

Alternatively, we have

$$[V^+] = \tfrac{1}{2}([V] + [I]) = \tfrac{1}{2}([Z] + [U])[I]$$

and

$$[V^-] = \tfrac{1}{2}([V] - [I]) = \tfrac{1}{2}([Z] - [U])[I]$$

and this gives

$$[V^-] = ([Z] - [U])([Z] + [U])^{-1}[V^+]$$

or

$$[S] = ([Z] - [U])([Z] + [U])^{-1} \tag{4.55b}$$

The transpose of (4.55a) is

$$[S]_t = ([Z] - [U])_t([Z] + [U])_t^{-1}$$

But since the matrices in parentheses are symmetrical, they are equal to their transpose; e.g.,

$$([Z] - [U])_t = [Z] - [U]$$

Hence

$$[S]_t = ([Z] - [U])([Z] + [U])^{-1}$$

and using (4.55b) now gives

$$[S]_t = [S] \tag{4.56}$$

a result that can hold only if $[S]$ is a symmetrical matrix.

Scattering matrix for a lossless junction

For a lossless junction the total power leaving the N ports must equal the total incident power. The mathematical statement of this power-conservation condition is

$$\sum_{n=1}^{N} |V_n^-|^2 = \sum_{n=1}^{N} |V_n^+|^2 \tag{4.57}$$

This condition will impose a number of restrictions on the scattering-matrix parameters such as to reduce the total number of independent parameters to $\frac{1}{2}N(N+1)$, the same number of independent parameters as in the impedance matrix for a lossless junction. Replacing V_n^- by

$$V_n^- = \sum_{i=1}^{N} S_{ni} V_i^+$$

the power-conservation condition may be expressed as

$$\sum_{n=1}^{N} \left| \sum_{i=1}^{N} S_{ni} V_i^+ \right|^2 = \sum_{n=1}^{N} |V_n^+|^2 \tag{4.58}$$

The V_n^+ are all independent incident voltages; so if we choose all $V_n^+ = 0$ except V_i^+, we obtain

$$\sum_{n=1}^{N} |S_{ni} V_i^+|^2 = |V_i^+|^2 \tag{4.59}$$

or

$$\sum_{n=1}^{N} |S_{ni}|^2 = \sum_{n=1}^{N} S_{ni} S_{ni}^* = 1 \tag{4.60}$$

The index i is arbitrary; so (4.60) must hold for all values of i. Equation (4.60) states that for a lossless junction the product of any column of the scattering matrix with the conjugate of this same column equals unity.

In addition to the above constraint on the S_{nm}, a number of additional constraints may be derived. If we choose all $V_n^+ = 0$ except V_s^+ and V_r^+, (4.58) gives

$$\sum_{n=1}^{N} |S_{ns} V_s^+ + S_{nr} V_r^+|^2 = \sum_{n=1}^{N} (S_{ns} V_s^+ + S_{nr} V_r^+)(S_{ns} V_s^+ + S_{nr} V_r^+)^*$$

$$= |V_s^+|^2 + |V_r^+|^2$$

Expanding the left-hand side gives

$$\sum_{n=1}^{N} |S_{ns}V_s^+|^2 + \sum_{n=1}^{N} |S_{nr}V_r^+|^2 + \sum_{n=1}^{N} S_{ns}S_{nr}^* V_s^+(V_r^+)^*$$
$$+ \sum_{n=1}^{N} S_{nr}S_{ns}^* V_r^+(V_s^+)^* = |V_s^+|^2 + |V_r^+|^2$$

Using (4.59) results in a number of terms canceling, and we are left with

$$\sum_{n=1}^{N} [S_{ns}S_{nr}^* V_s^+(V_r^+)^* + S_{ns}^* S_{nr} V_r^+(V_s^+)^*] = 0$$

In view of the independent nature of V_s^+ and V_r^+, choose, first of all, $V_s^+ = V_r^+$. We then obtain

$$|V_s^+|^2 \sum_{n=1}^{N} (S_{ns}S_{nr}^* + S_{ns}^* S_{nr}) = 0 \qquad (4.61a)$$

If, instead, we choose $V_s^+ = jV_r^+$, with V_r^+ real, we obtain

$$j|V_r^+|^2 \sum_{n=1}^{N} (S_{ns}S_{nr}^* - S_{ns}^* S_{nr}) = 0 \qquad (4.61b)$$

Since neither V_s^+ nor V_r^+ is zero, both (4.61a) and (4.61b) can hold only if

$$\sum_{n=1}^{N} S_{ns}S_{nr}^* = 0 \qquad s \neq r \qquad (4.62)$$

This equation states that the product of any column of the scattering matrix with the complex conjugate of any other different column is zero.

The conditions (4.60) and (4.62) are sufficient to restrict the number of independent parameters in the scattering matrix to $\frac{1}{2}N(N+1)$. A matrix with elements that satisfy these two conditions is called a unitary matrix. To illuminate this unitary property further it will be instructive to rederive the above results by means of matrix algebra. The power-conservation condition (4.57) can be expressed as

$$[V^-]_t[V^-]^* = [V^+]_t[V^+]^*$$
$$= ([S][V^+])_t([S][V^+])^*$$
$$= [V^+]_t[S]_t[S]^*[V^+]^*$$

Upon factoring this equation, we obtain

$$[V^+]_t([U] - [S]_t[S]^*)[V^+]^* = 0$$

This equation can hold only if

$$[S]_t[S]^* = [U] \tag{4.63a}$$

or

$$[S]^* = [S]_t^{-1} \tag{4.63b}$$

since $[V^+]$ is not zero. The result (4.63b) is the definition of a unitary matrix. The conditions (4.60) and (4.62) are obtained by carrying out the matrix multiplication called for in (4.63a).

4.8 Scattering matrix for a two-port junction

Since many common microwave circuits are two-port junctions, the scattering-matrix description of these is examined in greater detail. With reference to Fig. 4.16, let the scattering-matrix parameters of the junction be S_{11}, S_{21}, S_{12}, and S_{22}. The incident- and scattered-wave amplitudes are related by

$$[V^-] = [S][V^+] \tag{4.64a}$$

or

$$V_1^- = S_{11}V_1^+ + S_{12}V_2^+ \tag{4.64b}$$
$$V_2^- = S_{21}V_1^+ + S_{22}V_2^+ \tag{4.64c}$$

If the output guide is terminated in a matched load, $V_2^+ = 0$. From (4.64b) it is seen that S_{11} is the reflection coefficient in the input guide 1, with guide 2 terminated in a matched load. Also, S_{21} is the transmission coefficient into guide 2 from guide 1. Similar remarks, of course, apply to the parameters S_{22} and S_{12}.

If guide 2 is terminated in a normalized impedance $\bar{Z}_2$ at the terminal plane t_2, then V_2^- may be regarded as the incident wave on $\bar{Z}_2$, and V_2^+ is the wave reflected from $\bar{Z}_2$. The ratio must be equal to the reflection coefficient of the load; hence

$$\frac{V_2^+}{V_2^-} = \frac{\bar{Z}_2 - 1}{\bar{Z}_2 + 1} = S_L \tag{4.65}$$

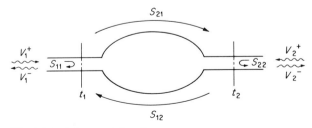

Fig. 4.16 A two-port junction.

Substituting into (4.64), we obtain

$$V_1^- - S_{11}V_1^+ = S_{12}V_2^+ = S_{12}S_L V_2^-$$
$$-S_{21}V_1^+ = S_{22}S_L V_2^- - V_2^-$$

Solving for V_1^-/V_1^+ gives

$$\frac{V_1^-}{V_1^+} = S_{11} - \frac{S_{12}S_{21}S_L}{S_{22}S_L - 1} \tag{4.66}$$

which shows how the input reflection coefficient in guide 1 is modified when the output guide is not terminated in a matched load.

For a reciprocal junction, $S_{12} = S_{21}$, and the scattering matrix contains, at most, six independent parameters, which are the magnitudes and phase angles of S_{11}, S_{12}, and S_{22}. If the junction is lossless, the scattering matrix contains only three parameters, since the S_{nm} are related by conditions (4.60) and (4.62), which in the present case become

$$S_{11}S_{11}^* + S_{12}S_{12}^* = 1 \tag{4.67a}$$
$$S_{22}S_{22}^* + S_{12}S_{12}^* = 1 \tag{4.67b}$$
$$S_{11}S_{12}^* + S_{12}S_{22}^* = 0 \tag{4.67c}$$

The first two equations show that

$$|S_{11}| = |S_{22}| \tag{4.68}$$

and hence the reflection coefficients in the input and output guides are equal in magnitude for a lossless junction. In addition, (4.67a) shows that

$$|S_{12}| = \sqrt{1 - |S_{11}|^2} \tag{4.69}$$

If we let $S_{11} = |S_{11}|e^{j\theta_1}$, $S_{22} = |S_{11}|e^{j\theta_2}$, and $S_{12} = (1 - |S_{11}|^2)^{\frac{1}{2}}e^{j\phi}$, then (4.67c) gives

$$|S_{11}|(1 - |S_{11}|^2)^{\frac{1}{2}}(e^{j\theta_1 - j\phi} + e^{j\phi - j\theta_2}) = 0$$

or, equivalently,

$$e^{j(\theta_1 + \theta_2)} = -e^{2j\phi}$$

Thus

$$\theta_1 + \theta_2 = 2\phi - \pi \pm 2n\pi$$

and

$$\phi = \frac{\theta_1 + \theta_2}{2} + \frac{\pi}{2} \mp n\pi \tag{4.70}$$

The two results (4.69) and (4.70) completely specify the transmission coefficient S_{12} in terms of the reflection coefficients S_{11} and S_{22}. Since S_{11} and S_{22} are readily measured and a knowledge of these suffices for the complete description of a lossless junction, the scattering matrix is a particularly convenient way of describing a lossless microwave two-port circuit.

The direct evaluation of the scattering-matrix parameters is illustrated by considering two simple examples. In Fig. 4.17a a shunt susceptance jB is connected across a transmission line with characteristic impedance Z_c. To find S_{11} we assume the output line to be matched, so that $V_2^+ = 0$. The reflection coefficient on the input side is

$$S_{11} = \frac{Y_c - Y_{\text{in}}}{Y_c + Y_{\text{in}}} = \frac{Y_c - Y_c - jB}{2Y_c + jB} = \frac{-jB}{2Y_c + jB}$$

From symmetry considerations it is clear that $S_{22} = S_{11}$. The third parameter S_{12} can be evaluated using (4.69) and (4.70) or by finding the transmitted voltage V_2^- with the output line matched. For a pure shunt element we must have $V_1^+ + V_1^- = V_2^- = V_1^+(1 + S_{11})$. Since $V_2^- = S_{21}V_1^+$ also, we obtain

$$S_{21} = 1 + S_{11} = S_{12} = \frac{2Y_c}{2Y_c + jB}$$

For the second example we consider a series reactance jX connecting lines with characteristic impedances Z_1 and Z_2, as in Fig. 4.17b. In this example the characteristic impedances of the two lines are different; so we must first choose normalized voltages. Let V_1^+, V_1^-, V_2^+, V_2^- be the actual transmission-line voltages for the waves that can exist on the input and output sides. Power flow for a single propagating wave is given by $\frac{1}{2}Y_1|V_1^+|^2$ and $\frac{1}{2}Y_2|V_2^+|^2$. If we choose normalized voltages $\bar{V}_1^+ = Y_1^{\frac{1}{2}}V_1^+$ and $\bar{V}_2^+ = Y_2^{\frac{1}{2}}V_2^+$, then power flow is given by $\frac{1}{2}|\bar{V}_1^+|^2$ and $\frac{1}{2}|\bar{V}_2^+|^2$ and is directly proportional to the voltage wave amplitude squared.

If the output line is matched, we have

$$\frac{V_1^-}{V_1^+} = \frac{\bar{V}_1^-}{\bar{V}_1^+} = S_{11} = \frac{Z_{\text{in}} - Z_1}{Z_{\text{in}} + Z_1} = \frac{Z_2 - Z_1 + jX}{Z_2 + Z_1 + jX}$$

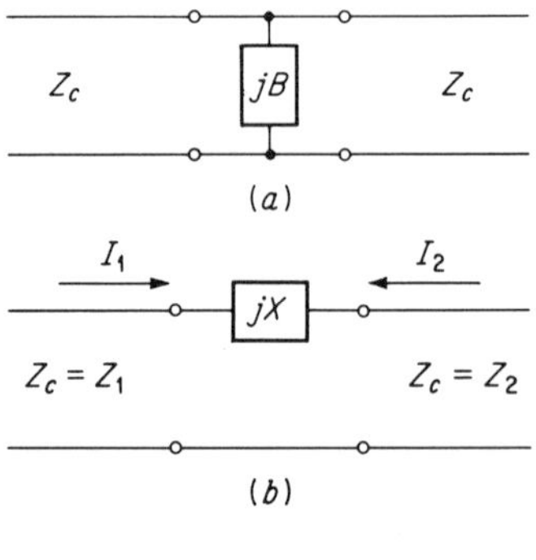

Fig. 4.17 Shunt and series elements on a transmission line.

With the input line matched, we find

$$\frac{V_2^-}{V_2^+} = \frac{\bar{V}_2^-}{\bar{V}_2^+} = S_{22} = \frac{Z_1 - Z_2 + jX}{Z_2 + Z_1 + jX}$$

To find S_{21}, again consider the output line matched. On the input line we have $V_1 = V_1^+ + V_1^- = V_1^+(1 + S_{11})$ and

$$I_1 = Y_1(V_1^+ - V_1^-) = Y_1 V_1^+(1 - S_{11})$$

The current is continuous through a series element, and hence

$$-I_2 = I_2^- = I_1 = Y_1 V_1^+(1 - S_{11})$$

But $I_2^- = Y_2 V_2^-$; so $Y_2 V_2^- = Y_1 V_1^+(1 - S_{11})$. We now obtain

$$S_{21} = S_{12} = \frac{\bar{V}_2^-}{\bar{V}_1^+} = \left(\frac{Y_2}{Y_1}\right)^{\frac{1}{2}} \frac{V_2^-}{V_1^+} = \left(\frac{Z_2}{Z_1}\right)^{\frac{1}{2}} (1 - S_{11})$$

$$= \left(\frac{Z_2}{Z_1}\right)^{\frac{1}{2}} \frac{2Z_1}{Z_1 + Z_2 + jX} = \frac{2\sqrt{Z_1 Z_2}}{Z_1 + Z_2 + jX}$$

The equality of S_{12} and S_{21} occurs because of the symmetrical manner in which Z_1 and Z_2 enter into this expression. If unnormalized voltages were used, the same expressions would be obtained for S_{11} and S_{22}, but for S_{21} and S_{12} we would obtain instead

$$S_{21} = \frac{2Z_2}{Z_1 + Z_2 + jX} \qquad S_{12} = \frac{2Z_1}{Z_1 + Z_2 + jX}$$

4.9 Transmission-matrix representation

When a number of microwave circuits are connected together in cascade, it is more convenient to represent each junction or circuit by a transmission matrix that gives the output quantities in terms of the input quantities. The reason for this is that, with such a representation, the matrix which describes the complete cascade connection may be obtained simply by multiplying the matrices describing each junction together. The independent variables may be chosen as the input voltages V_n and currents I_n, the incident- and reflected-wave amplitudes V_n^+, V_n^- on the input side, or any other convenient linearly independent quantities. When voltages and currents are chosen, we shall call the corresponding matrix the voltage-current transmission matrix. If incident- and reflected-wave amplitudes are chosen, we shall refer to the matrix as a wave-amplitude transmission matrix. To simplify the discussion we shall consider the cascade connection of two-port circuits only. However, the general formulation is readily extended to cover the cascade connection of N-port circuits.

The transmission-matrix formulation is of great value in analyzing

180 Foundations for microwave engineering

infinitely long periodic structures such as those used in slow-wave circuits for traveling-wave tubes and linear accelerators. Since examples of these are analyzed in Chap. 8, we shall consider only the basic formulation in this section.

Voltage-current transmission matrix

Figure 4.18a illustrates a two-port junction with input total voltage and current V_1, I_1 and output quantities V_2, I_2. Since V_2 and I_2 may be chosen as the independent variables and the junction is linear, the dependent variables V_1, I_1 are linearly related to V_2, I_2. Consequently, we may write

$$V_1 = \mathcal{A}V_2 + \mathcal{B}I_2 \tag{4.71a}$$
$$I_1 = \mathcal{C}V_2 + \mathcal{D}I_2 \tag{4.71b}$$

where $\mathcal{A}$, $\mathcal{B}$, $\mathcal{C}$, and $\mathcal{D}$ are suitable constants that characterize the junction. Note that we have chosen the positive direction of current to be to the right at all terminals. This is done so that the output current I_2 becomes the input current to the next junction, etc., in a cascade connection as illustrated in Fig. 4.18b.

In matrix form (4.71) becomes

$$\begin{bmatrix} V_1 \\ I_1 \end{bmatrix} = \begin{bmatrix} \mathcal{A} & \mathcal{B} \\ \mathcal{C} & \mathcal{D} \end{bmatrix} \begin{bmatrix} V_2 \\ I_2 \end{bmatrix} \tag{4.72}$$

The relationship of the voltage-current transmission matrix to the impedance matrix is readily found by rewriting the following equations in the form (4.71):

$$V_1 = I_1 Z_{11} - I_2 Z_{12}$$
$$V_2 = I_1 Z_{12} - I_2 Z_{22}$$

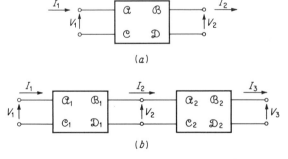

Fig. 4.18 (a) A two-port junction; (b) cascade connection of two-port junctions.

Sec. 4.9 Circuit theory for waveguiding systems 181

These equations may be solved to give

$$\begin{bmatrix} V_1 \\ I_1 \end{bmatrix} = \begin{bmatrix} Z_{11}/Z_{12} & (Z_{11}Z_{22} - Z_{12}{}^2)/Z_{12} \\ 1/Z_{12} & Z_{22}/Z_{12} \end{bmatrix} \begin{bmatrix} V_2 \\ I_2 \end{bmatrix} \qquad (4.73)$$

The $\mathcal{ABCD}$ parameters of the junction are readily identified in terms of the Z_{nm} from this relation. The determinant of the voltage-current transmission matrix is

$$\mathcal{AD} - \mathcal{BC} = 1 \qquad (4.74)$$

for a reciprocal junction, as is readily verified from (4.73).

For a cascade connection as illustrated in Fig. 4.18b we may write

$$\begin{bmatrix} V_1 \\ I_1 \end{bmatrix} = \begin{bmatrix} \mathcal{A}_1 & \mathcal{B}_1 \\ \mathcal{C}_1 & \mathcal{D}_1 \end{bmatrix} \begin{bmatrix} V_2 \\ I_2 \end{bmatrix}$$

$$\begin{bmatrix} V_2 \\ I_2 \end{bmatrix} = \begin{bmatrix} \mathcal{A}_2 & \mathcal{B}_2 \\ \mathcal{C}_2 & \mathcal{D}_2 \end{bmatrix} \begin{bmatrix} V_3 \\ I_3 \end{bmatrix}$$

and therefore

$$\begin{aligned} \begin{bmatrix} V_1 \\ I_1 \end{bmatrix} &= \begin{bmatrix} \mathcal{A}_1 & \mathcal{B}_1 \\ \mathcal{C}_1 & \mathcal{D}_1 \end{bmatrix} \begin{bmatrix} \mathcal{A}_2 & \mathcal{B}_2 \\ \mathcal{C}_2 & \mathcal{D}_2 \end{bmatrix} \begin{bmatrix} V_3 \\ I_3 \end{bmatrix} \\ &= \begin{bmatrix} \mathcal{A}_1\mathcal{A}_2 + \mathcal{B}_1\mathcal{C}_2 & \mathcal{A}_1\mathcal{B}_2 + \mathcal{B}_1\mathcal{D}_2 \\ \mathcal{C}_1\mathcal{A}_2 + \mathcal{D}_1\mathcal{C}_2 & \mathcal{C}_1\mathcal{B}_2 + \mathcal{D}_1\mathcal{D}_2 \end{bmatrix} \begin{bmatrix} V_3 \\ I_3 \end{bmatrix} \end{aligned} \qquad (4.75)$$

Thus the input quantities are readily found in terms of the output variables simply by multiplying the transmission matrices together. The ratio of the output voltage to current is determined by the load impedance.

Wave-amplitude transmission matrix

The wave-amplitude transmission matrix relates the incident- and reflected-wave amplitudes on the input side of the junction to those on the output side. It bears the same relationship to the scattering matrix as the voltage-current transmission matrix does to the impedance matrix. Just as in the case of the voltage-current transmission-matrix representation, it is convenient to choose the variables in such a fashion that the output variables from one junction become the input variables for the next junction. With reference to Fig. 4.19a, we thus choose

$$c_1{}^+ = V_1{}^+ \qquad (4.76a)$$
$$c_1{}^- = V_1{}^- \qquad (4.76b)$$
$$c_2{}^+ = V_2{}^- \qquad (4.76c)$$
$$c_2{}^- = V_2{}^+ \qquad (4.76d)$$
$$c_3{}^+ = V_3{}^- \qquad (4.76e)$$
$$c_3{}^- = V_3{}^+ \qquad (4.76f)$$

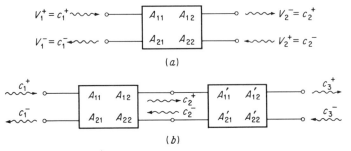

Fig. 4.19 Wave-amplitude transmission-matrix representation of a junction.

The superscript + refers to the amplitude of the wave propagating to the right, and the superscript − refers to the amplitude of the wave propagating to the left. The input and output quantities are linearly related; so we may write

$$\begin{bmatrix} c_1^+ \\ c_1^- \end{bmatrix} = \begin{bmatrix} A_{11} & A_{12} \\ A_{21} & A_{22} \end{bmatrix} \begin{bmatrix} c_2^+ \\ c_2^- \end{bmatrix} \qquad (4.77)$$

where the A_{nm} are suitable constants that describe the junction. For the cascade connection of Fig. 4.19b we have

$$\begin{bmatrix} c_1^+ \\ c_1^- \end{bmatrix} = \begin{bmatrix} A_{11} & A_{12} \\ A_{21} & A_{22} \end{bmatrix} \begin{bmatrix} A'_{11} & A'_{12} \\ A'_{21} & A'_{22} \end{bmatrix} \begin{bmatrix} c_3^+ \\ c_3^- \end{bmatrix} \qquad (4.78)$$

In terms of the scattering matrix for the single junction we have

$$\begin{bmatrix} V_1^- \\ V_2^- \end{bmatrix} = \begin{bmatrix} c_1^- \\ c_2^+ \end{bmatrix} = \begin{bmatrix} S_{11} & S_{12} \\ S_{12} & S_{22} \end{bmatrix} \begin{bmatrix} V_1^+ \\ V_2^+ \end{bmatrix}$$

$$= \begin{bmatrix} S_{11} & S_{12} \\ S_{12} & S_{22} \end{bmatrix} \begin{bmatrix} c_1^+ \\ c_2^- \end{bmatrix} \qquad (4.79)$$

These equations may be solved for c_1^+ and c_1^- to give

$$\begin{bmatrix} c_1^+ \\ c_1^- \end{bmatrix} = \begin{bmatrix} 1/S_{12} & -S_{22}/S_{12} \\ S_{11}/S_{12} & (S_{12}^2 - S_{11}S_{22})/S_{12} \end{bmatrix} \begin{bmatrix} c_2^+ \\ c_2^- \end{bmatrix} \qquad (4.80)$$

from which the A_{nm} are readily identified in terms of the S_{nm}. Note that the determinant of the [A] matrix equals unity, i.e.,

$$A_{11}A_{22} - A_{12}A_{21} = 1 \qquad (4.81)$$

as is readily verified from (4.80). However, if the wave amplitudes had not been normalized, so that power was given by $\frac{1}{2}|c_n^+|^2$, etc., the determinant of the [A] matrix would be different from unity in general (this follows from the nonsymmetry of the scattering matrix in this case). Some of the problems given at the end of this chapter illustrate these points.

★4.10 Excitation of waveguides

The preceding sections have dealt with the circuit aspects of passive microwave junctions. In order to complete the picture, it is necessary to consider also the equivalent-circuit representations for typical sources that are used to excite waves in a waveguide or transmission line. This particular aspect of waveguide theory is somewhat specialized, and it is not possible to give a complete analysis in this text without departing too far from the main theme. However, we shall present certain aspects of the excitation problem that provide a basis for choosing appropriate equivalent circuits and generators for representing typical sources and, in addition, make it possible to solve a number of coupling problems of engineering importance. The theory is, for the most part, developed by considering specific examples.

Probe coupling in a rectangular waveguide

Figure 4.20 illustrates a typical coaxial-line–waveguide probe coupling. The short-circuit position l and probe length d can be adjusted to achieve maximum power transfer from the coaxial line into the waveguide. The center conductor of the coaxial line extends into the waveguide to form an electric probe. Any waveguide mode that has a nonzero electric field along the probe will excite currents on the probe. By reciprocity, when the probe current is produced by a TEM wave incident from the coaxial line, the same waveguide modes will be excited.† It is thus easy to see that, for maximum coupling to the dominant TE_{10} mode in a rectangular guide, the probe should extend into the guide through the center

---

† This reciprocity principle is very useful for determining what modes a given probe can excite.

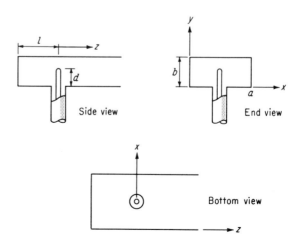

Fig. 4.20 Coaxial-line probe coupling to a waveguide.

184 Foundations for microwave engineering

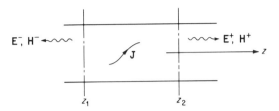

Fig. 4.21 A current source in a waveguide.

of the broad face so as to coincide with the position of maximum electric field for the TE_{10} mode. The evanescent modes that are also excited are localized fields that store reactive energy. These give the junction its reactive properties. The section of short-circuited waveguide provides an adjustable reactance that may be used to tune out the probe reactance. The probe reactance can be evaluated by determining the amplitudes of the evanescent modes that are excited and computing the net reactive energy stored in these nonpropagating modes.† Since the details are rather lengthy, we shall evaluate only the amplitude of the radiated TE_{10} mode.

The current on the probe must be zero at the end of the probe. For a thin probe a sinusoidal standing-wave current distribution is a reasonable approximation to make for the probe current. Thus let the probe current be considered as an infinitely thin filamentary current of the form

$$I = I_0 \sin k_0(d - y) \qquad 0 \leq y \leq d \qquad x = \frac{a}{2} \qquad z = 0 \qquad (4.82)$$

We wish to determine the amplitude of the TE_{10} mode excited by this current. A general technique for accomplishing this is a mathematical formulation of the reciprocity principle invoked earlier to determine which waveguide modes a given source will excite. The required results are derived below.

Figure 4.21 illustrates an infinitely long waveguide in which a current source **J** is located in the region between z_1 and z_2. The field radiated by this source may be expressed as an infinite sum of waveguide modes as follows:

$$\mathbf{E}^+ = \sum_n C_n^+(\mathbf{e}_n + \mathbf{e}_{zn})e^{-j\beta_n z} \qquad z > z_2 \qquad (4.83a)$$

$$\mathbf{H}^+ = \sum_n C_n^+(\mathbf{h}_n + \mathbf{h}_{zn})e^{-j\beta_n z} \qquad z > z_2 \qquad (4.83b)$$

$$\mathbf{E}^- = \sum_n C_n^-(\mathbf{e}_n - \mathbf{e}_{zn})e^{j\beta_n z} \qquad z < z_1 \qquad (4.83c)$$

$$\mathbf{H}^- = \sum_n C_n^-(-\mathbf{h}_n + \mathbf{h}_{zn})e^{j\beta_n z} \qquad z < z_1 \qquad (4.83d)$$

† See, for example, R. E. Collin, "Field Theory of Guided Waves," chap. 7, McGraw-Hill Book Company, New York, 1960.

In (4.83) n is a general summation index and implies a summation over all possible TE and TM modes. The unknown amplitudes C_n may be determined by an application of the Lorentz reciprocity formula (2.135). For the volume V, choose that bounded by the waveguide walls and cross-sectional planes located at z_1 and z_2 in Fig. 4.21. Let the field $\mathbf{E}_1$, $\mathbf{H}_1$, to be used in the Lorentz reciprocity formula, be the field radiated by the current source. This field is given by (4.83). For the field $\mathbf{E}_2$, $\mathbf{H}_2$, choose the nth waveguide mode $\mathbf{E}_n^-$, $\mathbf{H}_n^-$; that is,

$$\mathbf{E}_2 = \mathbf{E}_n^- = (\mathbf{e}_n - \mathbf{e}_{zn})e^{j\beta_n z}$$

$$\mathbf{H}_2 = \mathbf{H}_n^- = (-\mathbf{h}_n + \mathbf{h}_{zn})e^{j\beta_n z}$$

Equation (2.135) gives

$$\oint_S (\mathbf{E}_1 \times \mathbf{H}_n^- - \mathbf{E}_n^- \times \mathbf{H}_1) \cdot \mathbf{n}\, dS = \int_V \mathbf{E}_n^- \cdot \mathbf{J}\, dV$$

since the field $\mathbf{E}_2$, $\mathbf{H}_2$ is a source-free solution ($\mathbf{J}_2 = 0$) within V. The surface integral is zero over the waveguide walls by virtue of the boundary condition $\mathbf{n} \times \mathbf{E}_1 = \mathbf{n} \times \mathbf{E}_n^- = 0$. Since the modes are orthogonal, i.e.,

$$\int_{S_0} \mathbf{E}_m^\pm \times \mathbf{H}_n^\pm \cdot \mathbf{n}\, dS = 0 \qquad n \neq m$$

all the terms except the nth in the expansion of $\mathbf{E}_1$, $\mathbf{H}_1$ vanish when integrated over the waveguide cross section S_0. Thus we have

$$\int_{z_2} C_n^+[(\mathbf{e}_n + \mathbf{e}_{zn}) \times (-\mathbf{h}_n + \mathbf{h}_{zn}) - (\mathbf{e}_n - \mathbf{e}_{zn}) \times (\mathbf{h}_n + \mathbf{h}_{zn})] \cdot \mathbf{a}_z\, dS$$

$$- \int_{z_1} C_n^-[(\mathbf{e}_n - \mathbf{e}_{zn}) \times (-\mathbf{h}_n + \mathbf{h}_{zn}) - (\mathbf{e}_n - \mathbf{e}_{zn}) \times (-\mathbf{h}_n + \mathbf{h}_{zn})] \cdot \mathbf{a}_z\, dS$$

$$= -2C_n^+ \int_{z_2} \mathbf{e}_n \times \mathbf{h}_n \cdot \mathbf{a}_z\, dS - \int_V \mathbf{E}_n^- \cdot \mathbf{J}\, dV$$

since the integral over the cross section at z_1 vanishes identically. Hence C_n^+ is given by

$$C_n^+ = -\frac{1}{P_n} \int_V \mathbf{E}_n^- \cdot \mathbf{J}\, dV = -\frac{1}{P_n} \int_V (\mathbf{e}_n - \mathbf{e}_{zn}) \cdot \mathbf{J} e^{j\beta_n z}\, dV \qquad (4.84a)$$

If $\mathbf{E}_n^+$, $\mathbf{H}_n^+$ is chosen for the field $\mathbf{E}_2$, $\mathbf{H}_2$, we obtain

$$C_n^- = -\frac{1}{P_n} \int_V \mathbf{E}_n^+ \cdot \mathbf{J}\, dV = -\frac{1}{P_n} \int_V (\mathbf{e}_n + \mathbf{e}_{zn}) \cdot \mathbf{J} e^{-j\beta_n z}\, dV \qquad (4.84b)$$

where

$$P_n = 2 \int_{S_0} \mathbf{e}_n \times \mathbf{h}_n \cdot \mathbf{a}_z\, dS \qquad (4.84c)$$

and S_0 is a cross-sectional surface of the waveguide. The normalization constant P_n depends on the choice of expressions used for $\mathbf{e}_n$ and $\mathbf{h}_n$, the latter being arbitrary.

The above results are now applied to the probe problem introduced earlier. For the TE_{10} mode with fields given by

$$E_y = e_y e^{-j\beta z} = \sin \frac{\pi x}{a} e^{-j\beta z} \tag{4.85a}$$

$$H_x = h_x e^{-j\beta z} = -Y_w \sin \frac{\pi x}{a} e^{-j\beta z} \tag{4.85b}$$

we have

$$P_{10} = 2 \int_0^a \int_0^b Y_w \sin^2 \frac{\pi x}{a} \, dx \, dy = ab Y_w \tag{4.86}$$

where Y_w is the wave admittance for the TE_{10} mode and β is the propagation constant.

The probe in the short-circuited guide is equivalent to the original probe plus its image at $z = -2l$ placed in an infinite guide, as in Fig. 4.22. If we assume that the field radiated into the $z > 0$ region is

$$E_y^+ = C^+ \sin \frac{\pi x}{a} e^{-j\beta z} \tag{4.87}$$

then application of (4.84a) gives

$$C^+ = -\frac{1}{abY_w} \left[\int_0^d I_0 \sin k_0(d-y) \, dy - \int_0^d I_0 \sin k_0(d-y) e^{-j2\beta l} \, dy \right]$$

$$= \frac{I_0 Z_w}{abk_0} (e^{-2j\beta l} - 1)(1 - \cos k_0 d) \tag{4.88}$$

since

$$E_{10}^- = \begin{cases} \sin \frac{\pi x}{a} e^{j\beta z} = 1 & \text{at } z = 0, \, x = \frac{a}{2} \\ e^{-j2\beta l} & \text{at } x = \frac{a}{2}, \, z = -2l \end{cases}$$

Note also that the direction of the current in the image probe is reversed. This is necessary so that the fields radiated by the probe and its image will give a zero tangential electric field at the short-circuit position.

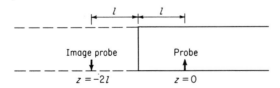

Fig. 4.22 Probe and its image.

The total transverse field of the TE_{10} mode radiated by the probe is thus, for $z > 0$,

$$E_y = \frac{I_0 Z_w}{abk_0}(e^{-2j\beta l} - 1)(1 - \cos k_0 d) \sin \frac{\pi x}{a} e^{-j\beta z} \tag{4.89a}$$

$$H_x = -Y_w E_y \tag{4.89b}$$

The total radiated power is given by

$$P = \frac{Y_w}{2} \int_0^a \int_0^b |E_y|^2 \, dx \, dy$$

$$= \frac{I_0^2 Z_w}{4abk_0^2} |e^{-2j\beta l} - 1|^2 (1 - \cos k_0 d)^2 \tag{4.90}$$

At the base of the probe antenna ($y = 0$), the total coaxial-line current is, from (4.82),

$$I = I_0 \sin k_0 d$$

Let the input impedance seen from the coaxial line, referred to the base, be $Z_{in} = R_0 + jX$. The complex Poynting vector theorem then gives [Eq. (4.14)]

$$Z_{in} = R_0 + jX = \frac{P + 2j\omega(W_m - W_e)}{\frac{1}{2}II^*}$$

where P is the power radiated into the guide and $W_m - W_e$ is the reactive energy stored in the vicinity of the probe owing to the excitation of nonpropagating (evanescent) modes. Since P has been evaluated and is given by (4.90), we can compute the input resistance. We obtain

$$R_0 = \frac{2P}{I_0^2 \sin^2 k_0 d} = \frac{Z_w}{2abk_0^2}|1 - e^{-2j\beta l}|^2 \tan^2 \frac{k_0 d}{2} \tag{4.91}$$

upon using the identities $1 - \cos 2\theta = 2 \sin^2 \theta$ and $\sin 2\theta = 2 \sin \theta \cos \theta$. This input resistance is called the radiation resistance of the probe. Note that its value can be varied by varying the parameters l and d, that is, the short-circuit position and probe length. Varying these parameters thus enables an optimum amount of power transfer to be achieved by adjusting R_0 to equal the characteristic impedance of the coaxial line and introducing a suitable reactance to tune out the reactance jX. Suitable techniques for reactance cancellation are discussed in the next chapter.

Radiation from linear current elements

Figure 4.23a and b illustrates linear current elements in a waveguide. For the case of the transverse current element, (4.84) shows that

$$C_n^+ = C_n^- = -\frac{1}{P_n} \int \mathbf{e} \cdot \mathbf{J} \, dl \tag{4.92}$$

This result may be interpreted to mean that a transverse-current element is equivalent to a shunt voltage source connected across an equivalent transmission line representing the waveguide when only a single mode, say the $n = 1$ mode, propagates. The reason for this is that the transverse current radiates a field with transverse electric field components that are equal on adjacent sides of the current source, and this is equivalent to continuity of the equivalent voltage across the source region. The transverse magnetic field is discontinuous across the source region, and thus the equivalent current is also discontinuous across the equivalent voltage generator. Figure 4.24 illustrates the equivalent circuit for this type of source for the dominant propagating mode. The ideal transformer provides a means of adjusting the coupling between the voltage generator and the transmission line so that the same amount of power is radiated as in the waveguide. The shunt susceptance jB represents the net reactive energy stored in the field of the evanescent modes that are excited.

For an axial current located at $z = 0$, (4.84) gives

$$C_n{}^+ = \frac{1}{P_n} \int \mathbf{J} \cdot \mathbf{e}_{zn} e^{j\beta_n z} \, dl$$

$$C_n{}^- = -\frac{1}{P_n} \int \mathbf{J} \cdot \mathbf{e}_{zn} e^{-j\beta_n z} \, dl$$

If the current is a symmetrical function of z between $-l < z < l$, then,

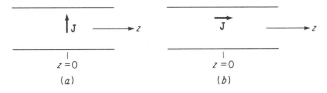

Fig. 4.23 Linear current elements in a waveguide.

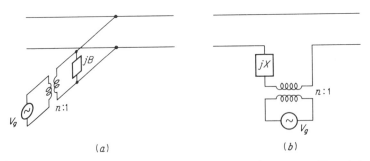

Fig. 4.24 Equivalent circuits for current sources in a waveguide. (a) Transverse current source; (b) axial current source.

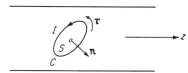

Fig. 4.25 A current loop in a waveguide.

since $\mathbf{e}_{zn}$ is not a function of z, we have

$$C_n^+ = -C_n^- = \frac{1}{P_n} \int_{-l}^{l} \mathbf{J} \cdot \mathbf{e}_{zn} \cos \beta_n z \, dz \tag{4.93}$$

For this case the radiated transverse magnetic field is continuous across the source but the transverse electric field is discontinuous. The source is thus equivalent to a voltage generator connected in series with an equivalent transmission line, as illustrated in Fig. 4.24b.

A linear current element may be viewed as an equivalent oscillating electric dipole. From Maxwell's equation we have

$$\nabla \times \mathbf{H} = j\omega\epsilon\mathbf{E} + \mathbf{J} = j\omega\epsilon_0\mathbf{E} + j\omega\mathbf{P} + \mathbf{J}$$

and hence $\mathbf{J}$ enters into the field equations in the same manner as the polarization current $j\omega\mathbf{P}$. Thus $\mathbf{J}$ may be considered equivalent to an electric dipole $\mathbf{P}$ given by

$$\mathbf{P} = \frac{\mathbf{J}}{j\omega}$$

Radiation from current loops

Figure 4.25 illustrates a linear current loop in a waveguide. The amplitude of the nth radiated mode is given by

$$C_n^+ = -\frac{1}{P_n} \oint_C \mathbf{E}_n^- \cdot \boldsymbol{\tau} I \, dl$$

where $\boldsymbol{\tau} I$ is the vector current flowing around the contour C. $\boldsymbol{\tau}$ is a unit vector along C. By Stokes' law we obtain

$$C_n^+ = -\frac{I}{P_n} \oint_C \mathbf{E}_n^- \cdot d\mathbf{l} = -\frac{I}{P_n} \int_S \nabla \times \mathbf{E}_n^- \cdot \mathbf{n} \, dS$$

But $\nabla \times \mathbf{E}_n^- = -j\omega\mathbf{B}_n^- = -j\omega\mu_0\mathbf{H}_n^-$, and hence

$$C_n^+ = \frac{j\omega I}{P_n} \int_S \mathbf{B}_n^- \cdot \mathbf{n} \, dS \tag{4.94a}$$

Similarly,

$$C_n^- = \frac{j\omega I}{P_n} \int_S \mathbf{B}_n^+ \cdot \mathbf{n} \, dS \tag{4.94b}$$

It is seen that the excitation amplitude of the nth mode is proportional to the total magnetic flux of this mode passing through the loop.

190 Foundations for microwave engineering

If the current loop is so small that the field $\mathbf{B}_n$ of the nth mode may be considered constant over the area of the loop, we obtain

$$C_n{}^+ = \frac{j\omega I}{P_n} \mathbf{B}_n{}^+ \cdot \int_S \mathbf{n}\, dS = \frac{j\omega I}{P_n} \mathbf{B}_n{}^+ \cdot \mathbf{S}_0$$

Now $I\mathbf{S}_0$ is the magnetic dipole moment $\mathbf{M}$ of the loop, where $\mathbf{S}_0$ is the vector area of the loop; so we obtain

$$C_n{}^+ = \frac{j\omega}{P_n} \mathbf{B}_n{}^- \cdot \mathbf{M} \tag{4.95a}$$

and similarly

$$C_n{}^- = \frac{j\omega}{P_n} \mathbf{B}_n{}^+ \cdot \mathbf{M} \tag{4.95b}$$

Radiation from a small current loop may be considered to be magnetic dipole radiation, as these equations show. For an axial magnetic dipole (transverse current loop) the equivalent source is a shunt-connected voltage source, whereas a transverse magnetic dipole is equivalent to a series-connected voltage source.

★4.11 Waveguide coupling by apertures †

The foregoing formulation of the radiation from currents in a waveguide in terms of radiation from equivalent electric and magnetic dipoles is directly applicable to the coupling of waveguides by small apertures, or holes, in a common wall. To a first approximation a small aperture in a conducting wall is equivalent to an electric dipole normal to the aperture and having a strength proportional to the normal component of the exciting electric field, plus a magnetic dipole in the plane of the aperture and having a strength proportional to the exciting tangential magnetic field. The constants of proportionality are parameters that depend on the aperture size and shape. These constants are called the electric and magnetic polarizabilities of the aperture and characterize the coupling or radiating properties of the aperture.‡ A qualitative argument to demonstrate the physical reasonableness of these properties of an aperture is given below.

Figure 4.26a illustrates the normal electric field of strength $\mathbf{E}$ at a conducting surface without an aperture. When an aperture is cut in the screen, the electric field lines fringe through the aperture in the manner indicated in Fig. 4.26b. But this field distribution is essentially that produced by an equivalent electric dipole as shown in Fig. 4.26c. Note that the dipole is oriented normal to the aperture.

† The theory was originally developed by H. A. Bethe, Theory of Diffraction by Small Holes, *Phys. Rev.*, vol. 66, pp. 163–182, 1944.
‡ For a derivation of these results, see Collin, *loc. cit.*

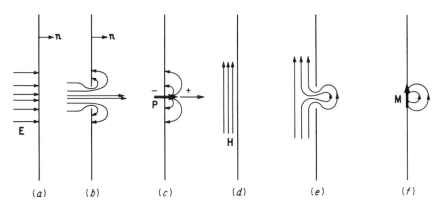

Fig. 4.26 Aperture in a conducting wall.

In a similar manner the tangential magnetic field lines shown in Fig. 4.26d will fringe through the aperture as in Fig. 4.26e. These fringing field lines are equivalent to those produced by a magnetic dipole located in the plane of the aperture.

For a small circular aperture of radius $r_0 \ll \lambda_0$, the dipole moments are related to the incident electromagnetic field, in the absence of the aperture, as follows:

$$\mathbf{P} = -\epsilon_0 \alpha_e (\mathbf{n} \cdot \mathbf{E}) \mathbf{n} \tag{4.96a}$$

$$\mathbf{M} = -\alpha_m \mathbf{H}_t \tag{4.96b}$$

where $\mathbf{n} \cdot \mathbf{E}$ is the normal electric field and $\mathbf{H}_t$ is the tangential magnetic field at the center of the aperture. The electric polarizability α_e is given by

$$\alpha_e = -\tfrac{2}{3} r_0^3 \tag{4.97a}$$

and the magnetic polarizability α_m is given by

$$\alpha_m = \tfrac{4}{3} r_0^3 \tag{4.97b}$$

The presence of an aperture also perturbs the field on the incident side of the screen. This perturbed field is that radiated by equivalent dipoles which are the negative of those given by (4.96) and located on the input side of the screen. It is important to note that when the aperture is replaced by equivalent electric and magnetic dipoles, the field radiated by these is computed by assuming that the aperture is now closed by a conducting wall. The equivalent dipoles correctly account for the field coupled through the aperture in the conducting screen. It should also be pointed out that the theory is an approximate one, valid for small apertures only. The examples that follow will clarify the application of the theory.

Aperture in a transverse wall

Figure 4.27a illustrates a small circular aperture in a transverse wall in a rectangular waveguide. To determine the exciting field, assume that the aperture is closed. A TE_{10} mode incident from $z < 0$ is reflected by the conducting wall at $z = 0$ to produce a standing-wave field in the region $z < 0$. This field is

$$E_y = C(e^{-j\beta z} - e^{j\beta z}) \sin \frac{\pi x}{a} \tag{4.98a}$$

$$H_x = -CY_w(e^{-j\beta z} + e^{j\beta z}) \sin \frac{\pi x}{a} \tag{4.98b}$$

plus a z component of magnetic field which is not required to be known for the present problem.

The normal electric field at the aperture is zero; so no induced electric dipole is produced. The tangential magnetic field at the center of the aperture is, from (4.98b),

$$H_x = -2CY_w$$

and hence an induced x-directed magnetic dipole M is produced and is

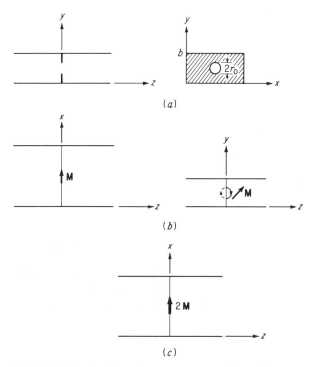

Fig. 4.27 Aperture in a transverse waveguide wall.

given by

$$M = -\alpha_m H_x = 2CY_w \alpha_m = \tfrac{8}{3} r_0^3 C Y_w \tag{4.99}$$

The field radiated into the region $z > 0$ is that radiated by the magnetic dipole M, as illustrated in Fig. 4.27b. This dipole is equivalent to a half circular current loop in the yz plane as illustrated. To find the field radiated by this dipole in the presence of the conducting transverse wall, image theory may be used. Since the image of the half circular current loop in the transverse wall is the other half of the current loop, the image of M is another magnetic dipole of moment M. The effect of the transverse wall is equivalent to removing the wall and doubling the strength of the dipole, as depicted in Fig. 4.27c. If the field radiated into the region $z > 0$ is

$$E_y^+ = A e^{-j\beta z} \sin \frac{\pi x}{a} = A e_y e^{-j\beta z}$$

$$H_x^+ = -A Y_w e^{-j\beta z} \sin \frac{\pi x}{a} = A h_x e^{-j\beta z}$$

application of formula (4.95a) gives

$$A = \frac{j\omega\mu_0}{P_{10}} H_x^-(2M)$$

$$= \frac{j\omega\mu_0}{P_{10}} Y_w (\tfrac{16}{3} r_0^3 C Y_w)$$

since the field B_n^- is $-\mu_0 h_x = \mu_0 Y_w \sin(\pi x/a)$ in the present case. The constant P_{10} is given by

$$P_{10} = -2 \int_0^a \int_0^b e_y h_x \, dx \, dy$$

$$= 2Y_w \int_0^a \int_0^b \sin^2 \frac{\pi x}{a} \, dx \, dy = ab Y_w$$

Hence we obtain

$$A = \tfrac{16}{3} r_0^3 \frac{j\omega\mu_0}{ab Z_w} C \tag{4.100}$$

Since the incident TE_{10} mode has an amplitude C, the aperture has a transmission coefficient

$$T = \frac{A}{C} = \tfrac{16}{3} j r_0^3 \frac{\omega\mu_0}{ab Z_w} = j \tfrac{16}{3} r_0^3 \frac{k_0 Z_0}{ab Z_w} \tag{4.101}$$

The presence of the aperture causes a field to be scattered into the region $z < 0$ also. For radiation into this region the effective magnetic dipole moment is the negative of that given by (4.99). Application of

(4.95b) now gives

$$E_y = A \sin \frac{\pi x}{a} e^{j\beta z}$$

with A given by (4.100) for the perturbed electric field in the region $z < 0$. The total electric field for $z < 0$ is now

$$E_y = [Ce^{-j\beta z} + (A - C)e^{j\beta z}] \sin \frac{\pi x}{a}$$

Therefore the aperture in a transverse wall produces a reflection coefficient

$$\Gamma = \frac{A - C}{C} = \frac{A}{C} - 1 = j\tfrac{1\,6}{3}r_0^3 \frac{\beta}{ab} - 1 \tag{4.102}$$

upon replacing Z_w by $k_0 Z_0/\beta$ in (4.101). A normalized shunt susceptance $j\bar{B}$ would produce a reflection coefficient

$$\Gamma = \frac{1 - Y_{\text{in}}}{1 + Y_{\text{in}}} = \frac{1 - (1 + j\bar{B})}{1 + (1 + j\bar{B})} = \frac{-j\bar{B}}{2 + j\bar{B}}$$

when connected across a transmission line. When $\bar{B}$ is very large, we have

$$\Gamma = \frac{-j\bar{B}}{j\bar{B}}\left(1 + \frac{2}{j\bar{B}}\right)^{-1} \approx -1 - j\frac{2}{\bar{B}}$$

Comparing with (4.102) shows that the aperture is equivalent to a normalized inductive susceptance

$$j\bar{B} = -j\frac{3ab}{8r_0^3 \beta} \tag{4.103}$$

Aperture in broad wall of a waveguide

Figure 4.28 illustrates a circular aperture of radius r_0 placed in the broad wall separating two rectangular waveguides. The incident field is a

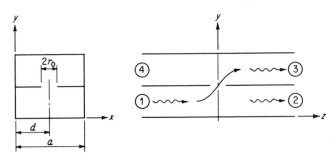

Fig. 4.28 Aperture in a broad wall separating two waveguides.

Sec. 4.11   Circuit theory for waveguiding systems   195

TE_{10} mode in the lower guide, and is given by

$$E_y = C \sin \frac{\pi x}{a} e^{-j\beta z} \tag{4.104a}$$

$$H_x = -CY_w \sin \frac{\pi x}{a} e^{-j\beta z} \tag{4.104b}$$

$$H_z = j\frac{\pi Y_w}{\beta a} C \cos \frac{\pi x}{a} e^{-j\beta z} \tag{4.104c}$$

At the center of the aperture located at $z = 0$, $x = d$, the exciting field is

$$E_y = C \sin \frac{\pi d}{a}$$

$$\mathbf{H} = CY_w \left(-\mathbf{a}_x \sin \frac{\pi d}{a} + j\frac{\pi}{\beta a} \mathbf{a}_z \cos \frac{\pi d}{a} \right)$$

Using (4.96), the equivalent dipoles for radiation into the upper guide are

$$\mathbf{P} = \epsilon_0 \tfrac{2}{3} r_0^3 C \sin \frac{\pi d}{a} \mathbf{a}_y \tag{4.105a}$$

$$\mathbf{M} = -\tfrac{4}{3} r_0^3 CY_w \left(-\mathbf{a}_x \sin \frac{\pi d}{a} + j\frac{\pi}{\beta a} \mathbf{a}_z \cos \frac{\pi d}{a} \right) \tag{4.105b}$$

Let the fields $\mathbf{E}_{10}^+$, $\mathbf{E}_{10}^-$, $\mathbf{B}_{10}^+$, and $\mathbf{B}_{10}^-$ be chosen as

$$\mathbf{E}_{10}^+ = \mathbf{a}_y \sin \frac{\pi x}{a} e^{-j\beta z}$$

$$\mathbf{E}_{10}^- = \mathbf{a}_y \sin \frac{\pi x}{a} e^{j\beta z}$$

$$\mu_0 \mathbf{H}_{10}^+ = \mathbf{B}_{10}^+ = -\mu_0 Y_w \left(\mathbf{a}_x \sin \frac{\pi x}{a} - j\frac{\pi}{\beta a} \mathbf{a}_z \cos \frac{\pi x}{a} \right) e^{-j\beta z}$$

$$\mu_0 \mathbf{H}_{10}^- = \mathbf{B}_{10}^- = \mu_0 Y_w \left(\mathbf{a}_x \sin \frac{\pi x}{a} + j\frac{\pi}{\beta a} \mathbf{a}_z \cos \frac{\pi x}{a} \right) e^{j\beta z}$$

Also, let the field radiated by the electric dipole be

$$\mathbf{E} = \begin{cases} A_1 \mathbf{E}_{10}^+ & z > 0 \\ A_2 \mathbf{E}_{10}^- & z < 0 \end{cases}$$

$$\mathbf{H} = \begin{cases} A_1 \mathbf{H}_{10}^+ & z > 0 \\ A_2 \mathbf{H}_{10}^- & z < 0 \end{cases}$$

whereas that radiated by the magnetic dipole is

$$\mathbf{E} = \begin{cases} A_3 \mathbf{E}_{10}^+ & z > 0 \\ A_4 \mathbf{E}_{10}^- & z < 0 \end{cases}$$

$$\mathbf{H} = \begin{cases} A_3 \mathbf{H}_{10}^+ & z > 0 \\ A_4 \mathbf{H}_{10}^- & z < 0 \end{cases}$$

The electric dipole **P** is equivalent to an electric current **J** given by $j\omega\mathbf{P}$. Since the dipole is oriented in the transverse plane, (4.84) gives

$$A_1 = A_2 = -\frac{1}{P_{10}}(j\omega\mathbf{P}) \cdot \mathbf{a}_y \sin\frac{\pi d}{a}$$

$$= -\frac{j\omega\epsilon_0}{abY_w}\frac{2}{3}r_0^3 C \sin^2\frac{\pi d}{a} \qquad (4.106)$$

Note that no integration is necessary since **P** is an infinitesimal dipole. The constant P_{10} is equal to abY_w.

The field radiated by the magnetic dipole may be found by using (4.95). Thus

$$A_3 = \frac{j\omega\mu_0 Y_w}{abY_w}\left(\mathbf{a}_x \sin\frac{\pi d}{a} + j\frac{\pi}{\beta a}\cos\frac{\pi d}{a}\mathbf{a}_z\right)$$

$$\cdot (-\tfrac{4}{3}r_0^3 CY_w)\left(-\mathbf{a}_x \sin\frac{\pi d}{a} + j\frac{\pi}{\beta a}\mathbf{a}_z \cos\frac{\pi d}{a}\right)$$

$$= \frac{j\omega\mu_0 Y_w}{ab}\frac{4}{3}r_0^3 C\left(\sin^2\frac{\pi d}{a} + \frac{\pi^2}{\beta^2 a^2}\cos^2\frac{\pi d}{a}\right) \qquad (4.107)$$

Similarly, it is found that

$$A_4 = \frac{j\omega\mu_0 Y_w}{ab}\frac{4}{3}r_0^3 C\left(-\sin^2\frac{\pi d}{a} + \frac{\pi^2}{\beta^2 a^2}\cos^2\frac{\pi d}{a}\right) \qquad (4.108)$$

With these expressions for the amplitudes, the total field radiated into the upper waveguide is readily evaluated. It is given by

$$\mathbf{E} = \begin{cases}(A_1 + A_3)\mathbf{E}_{10}^+ & z > 0 \quad (4.109a)\\ (A_2 + A_4)\mathbf{E}_{10}^- & z < 0 \quad (4.109b)\end{cases}$$

$$\mathbf{H} = \begin{cases}(A_1 + A_3)\mathbf{H}_{10}^+ & z > 0 \quad (4.109c)\\ (A_2 + A_4)\mathbf{H}_{10}^- & z < 0 \quad (4.109d)\end{cases}$$

Note that the electric dipole radiates the same in both directions but the magnetic dipole does not. By choosing the aperture position d correctly, it is possible to obtain zero radiation in one direction; that is, $A_2 + A_4$ can be made to vanish. Equating (4.106) with the negative of (4.108) gives

$$\sin\frac{\pi d}{a} = \frac{\lambda_0}{\sqrt{6}\,a} \qquad (4.110)$$

With this aperture position $A_2 + A_4 = 0$ but $A_1 + A_3 \neq 0$. This is an interesting situation since it means that power entering port 1 in Fig. 4.28 flows out through ports 2 and 3 but not port 4. If the wave was incident through port 2, power will leave through ports 1 and 4 but not port 3, as is apparent from symmetry considerations. A four-port junction with these properties is called a directional coupler, about which more is said in a later chapter.

To complete the analysis for the present example, it is necessary also to find the field scattered into the lower guide by the presence of the aperture. This field is that radiated by dipoles which are the negative of those given by (4.105). Since the analysis is similar to that used to evaluate A_1, A_2, A_3, and A_4, it is not carried out.

Problems

4.1 For TM modes in a waveguide, show that the line integral of the transverse electric field between any two points on the boundary is zero.

Hint: Note that $\nabla_t e_z \cdot d\mathbf{l} = (de_z/dl)\, dl$ = directional derivative of e_z along the path. Integrate this and use the boundary conditions for e_z. As an alternative, note that there is no axial magnetic flux, so that the line integral around a closed path in the transverse plane must vanish.

4.2 For TE modes show that the line integral of the transverse electric field between two points located on the guide boundary depends on the path of integration chosen.

Hint: Note that, because there is an axial magnetic field, the line integral around a closed path does not vanish.

4.3 An obstacle located at $z = 0$ excites evanescent H modes that decay exponentially away from the obstacle in the positive z direction. Integrate the complex Poynting vector over cross-sectional planes at $z = 0$ and $z = \infty$ and the guide walls for the nmth evanescent H mode, and show that there is no power transmitted into the region $z > 0$. Show also that the reactive energy stored in the nonpropagating H mode in the region $z > 0$ is predominantly magnetic.

Hint: Note from (2.59) that the total inward flux of the complex Poynting vector equals the power loss (which is to be taken equal to zero in this problem) plus $2j\omega(W_m - W_e)$.

4.4 Repeat Prob. 4.3 for the case of an E mode and show that nonpropagating E modes store predominantly electric energy.

★4.5 For the circuits illustrated, verify that the slope of the reactance function is given by (4.25).

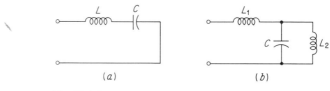

Fig. P 4.5

★4.6 For the N-port junction choose an excitation such that all $I_n = 0$ except I_j; thus $V_i = Z_{ij}I_j$ for all i. Show that all Z_{ij} must have real parts that are even functions of ω and imaginary parts that are odd functions of ω.

198 Foundations for microwave engineering

★4.7 Generalize the result (4.25) to show that for a lossless N-port junction

$$[I^*]_t \left[\frac{\delta Z}{\delta \omega}\right] [I] = \sum_{n=1}^{N} \sum_{m=1}^{N} I_n^* \frac{\delta Z_{nm}}{\delta \omega} I_m = 4j(W_e + W_m)$$

4.8 Verify that (4.51) and (4.44) are equal.

4.9 Show that a length l of transmission line of characteristic impedance Z_c is equivalent to a T network with parameters

$$Z_{11} = Z_{22} = -jZ_c \cot \beta l \qquad Z_{12} = \pm jZ_c \csc \beta l$$

4.10 Let $Z_{sc}{}^1$, $Z_{sc}{}^2$, $Z_{oc}{}^1$, $Z_{oc}{}^2$ be the input impedance of a T network when terminals 2 are short-circuited, when terminals 1 are short-circuited, when terminals 2 are open-circuited, and when terminals 1 are open-circuited, respectively. In terms of these impedances show that the parameters of the T network are given by

$$Z_{11} = Z_{oc}{}^1 \qquad Z_{22} = Z_{oc}{}^2 \qquad Z_{12}{}^2 = (Z_{oc}{}^1 - Z_{sc}{}^1)Z_{oc}{}^2 = (Z_{oc}{}^2 - Z_{sc}{}^2)Z_{oc}{}^1$$

Use these relations to verify the equations for the circuits of Fig. 4.13d and e.

4.11 For the following microwave circuit, evaluate the power transmitted to the load Z_L. Find the standing-wave ratio in the two transmission-line sections. Assume $Z_L = 2Z_1$, $X_1 = X_2 = Z_1$, $V_g = 5$ volts (peak).

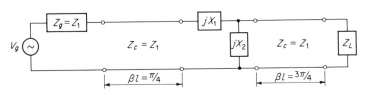

Fig. P 4.11

4.12 For a particular microwave junction the equivalent-T-network parameters are $Z_{11} = j2$, $Z_{12} = j/\sqrt{2}$, $Z_{22} = -j0.25$. Find the parameters for the alternative equivalent circuit illustrated.

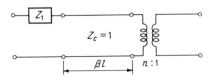

Fig. P 4.12

4.13 For the illustrated three-port junction, compute the power delivered to the loads $Z_1 = 40$ ohms and $Z_2 = 60$ ohms. Assume that $V_g = 10$ volts peak.

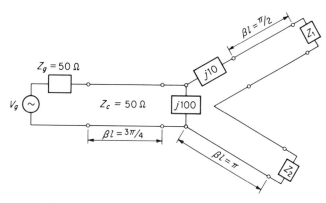

Fig. P 4.13

4.14 For the circuit illustrated, evaluate the fraction of the total power delivered by the generator that is delivered to Z_1 and Z_2. What value should the characteristic impedance of the input line have in order to make the standing-wave ratio on this line a minimum (assume that Z_c is to be real)?

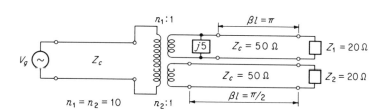

Fig. P 4.14

4.15 Consider the junction of two transmission lines with characteristic impedances Z_1 and Z_2 as illustrated. When the usual transmission-line voltages and currents are

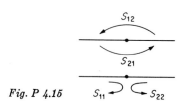

Fig. P 4.15

used, show that the scattering-matrix parameters are given by

$$S_{11} = \frac{Z_2 - Z_1}{Z_2 + Z_1} \qquad S_{22} = -S_{11}$$
$$S_{12} = \frac{2Z_1}{Z_2 + Z_1} \qquad S_{21} = \frac{2Z_2}{Z_2 + Z_1}$$

Show that the normalized voltages $(V_1^+)'$ and $(V_2^+)'$ are given by $(V_1^+)' = \sqrt{Y_1}\, V_1^+$, $(V_2^+)' = \sqrt{Y_2}\, V_2^+$, where the unprimed quantities are the usual transmission-line voltages. When normalized voltages are used, show that the scattering-matrix parameters are

$$S_{11} = -S_{22} = \frac{Z_2 - Z_1}{Z_2 + Z_1} \qquad S_{12} = S_{21} = \frac{2\sqrt{Z_1 Z_2}}{Z_2 + Z_1}$$

4.16 The field of a TE_{11} mode in a rectangular guide of width a and height b is derived from

$$h_z = C \cos\frac{\pi x}{a} \cos\frac{\pi y}{b}$$

Determine the expressions for the equivalent-transmission-line voltage V^+ and current I^+ for the two cases (1) when $Z_c = Z_w$ = wave impedance of the TE_{11} mode, (2) when $Z_c = 1$.

4.17 Apply the complex Poynting vector theorem to show that, for a one-port microwave termination, the reflection coefficient Γ satisfies the relation

$$(1 + \Gamma)(1 - \Gamma^*) = \frac{2j\omega(W_m - W_e) + P_l}{\frac{1}{2}|V^+|^2}$$

when the wave amplitudes are normalized, so that $VV^* = II^*$, that is, the equivalent characteristic impedance is unity.

4.18 Show that the $\mathcal{ABCD}$ parameters for a section of transmission line of length l and characteristic impedance Z_c are given by $\mathcal{A} = \mathcal{D} = \cos\beta l$, $\mathcal{B} = jZ_c \sin\beta l$, $\mathcal{C} = jY_c \sin\beta l$.

4.19 For a section of transmission line of length l, show that the wave-amplitude transmission matrix is a diagonal matrix with elements

$$A_{11} = e^{j\beta l} \qquad A_{22} = e^{-j\beta l} \qquad A_{12} = A_{21} = 0$$

4.20 Consider the junction of two transmission lines as in Prob. 4.15. Using conventional transmission-line voltages, show that the $[A]$-matrix parameters describing the junction are $A_{11} = A_{22} = (Z_1 + Z_2)/2Z_1$, $A_{12} = A_{21} = (Z_2 - Z_1)/2Z_1$. When normalized wave amplitudes (voltages) are used, show that $A_{11} = A_{22} = (Z_1 + Z_2)/(2\sqrt{Z_1 Z_2})$, $A_{12} = A_{21} = (Z_2 - Z_1)/(2\sqrt{Z_1 Z_2})$.

4.21 Find the $[A]$-matrix parameters for a shunt susceptance jB connected across a transmission line of unit characteristic impedance. Repeat for a reactance jX connected in series with the line.

4.22 Show that when normalized voltages are used the scattering-matrix parameters of a two-port junction are given in terms of the equivalent-T-network parameters by

$$S_{11} = \frac{\Delta - 1 + Z_{11} - Z_{22}}{\Delta + 1 + Z_{11} + Z_{22}} \qquad S_{12} = S_{21} = \frac{2Z_{12}}{\Delta + 1 + Z_{11} + Z_{22}}$$
$$S_{22} = \frac{\Delta - 1 + Z_{22} - Z_{11}}{\Delta + 1 + Z_{11} + Z_{22}}$$

Circuit theory for waveguiding systems   201

where

$$\Delta = Z_{11}Z_{22} - Z_{12}^2$$

4.23 Show that the T-network parameters are related to the scattering-matrix parameters as follows:

$$Z_{11} = \frac{(1 + S_{11})(1 - S_{22}) + S_{12}^2}{W} \qquad Z_{22} = \frac{(1 - S_{11})(1 + S_{22}) + S_{12}^2}{W}$$

$$Z_{12} = \frac{2S_{12}}{W}$$

where

$$W = (1 - S_{11})(1 - S_{22}) - S_{12}^2$$

4.24 For a discontinuity in a waveguide, the following scattering-matrix parameters were measured:

$$S_{11} = \tfrac{1}{3} + j\tfrac{2}{3} \qquad S_{12} = j\tfrac{2}{3} \qquad S_{22} = \tfrac{1}{3} - j\tfrac{2}{3}$$

Find the parameters of an equivalent T network that will represent the discontinuity (Prob. 4.23).

4.25 For an E-plane step (Fig. 4.6), the following were measured:

$$S_{11} = \frac{1-j}{3+j} \qquad S_{22} = \frac{-(1+j)}{3+j}$$

An equivalent circuit of the form illustrated is to be used to represent the junction. Determine the susceptance jB and the ideal transformer turns ratio $n:1$ from the above given data.

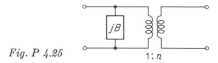

Fig. P 4.25

★4.26 Find the TE_{10} field radiated by the current loop illustrated. Consider the loop area to be so small that (4.95) is applicable. Area of loop equals S_0.

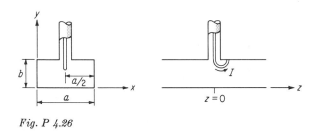

Fig. P 4.26

★4.27 Find the TE_{10} field radiated by the current loop of Prob. 4.26 if a short-circuit is placed at $z = -l$.

★4.28 A linear constant current I extends across the center of a rectangular waveguide at $x = a/2$, $z = 0$. Show that the total radiated electric field is

$$E_y = \frac{-j\omega\mu_0 I}{a} \sum_{n=1}^{\infty} \frac{1}{\gamma_n} \sin \frac{n\pi}{2} \sin \frac{n\pi x}{a} e^{-\gamma_n |z|}$$

where

$$\gamma_n = j\beta_n = \left(\frac{n^2\pi^2}{a^2} - k_0^2\right)^{\frac{1}{2}}$$

★4.29 Two rectangular waveguides have a common side wall in which a small circular aperture is located as illustrated. If the incident TE_{10} mode has an amplitude C, show that the amplitude of the TE_{10} mode radiated in both directions in the second guide is given by

$$A = \frac{-j\omega\mu_0}{abk_0 Z_0 \beta} \left(\frac{\pi}{a}\right)^2 \frac{4}{3} r_0^3 C$$

Show that the reflection coefficient in the input guide is A/C.

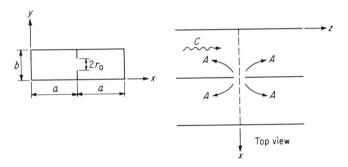

Fig. P 4.29

References

1. Ghose, R. N.: "Microwave Circuit Theory and Analysis," chaps. 4, 9, and 11, McGraw-Hill Book Company, New York, 1963.
2. Kerns, D. M.: Basis of Application of Network Equations to Waveguide Problems, *J. Res. Natl. Bur. Std.*, vol. 42, pp. 515–540, 1949.
3. Marcuvitz, N. (ed.): "Waveguide Handbook," McGraw-Hill Book Company, New York, 1951.
4. Montgomery, C. G., R. H. Dicke, and E. M. Purcell (eds.): "Principles of Microwave Circuits," McGraw-Hill Book Company, New York, 1948.
5. Pannenborg, A. E.: On the Scattering Matrix of Symmetrical Waveguide Junctions, *Philips Res. Rept.*, vol. 7, pp. 131–157, 1952.
6. Ramo, S., and J. R. Whinnery: "Fields and Waves in Modern Radio," 2d ed., chap. 11, John Wiley & Sons, Inc., New York, 1953.

5
Impedance transformation and matching

In this chapter we are concerned with the important problem of impedance matching, such as the matching of an arbitrary load impedance to a given transmission line or the matching of two lines with different characteristic impedances. Methods of impedance matching to obtain maximum power transfer are presented, along with broadband design methods for quarter-wave transformers and tapered transmission-line impedance transformers. To facilitate the development of the theory, the Smith chart, a graphical aid for the solution of many transmission-line and waveguide impedance problems, is described first.

5.1 Smith chart

In Sec. 3.5 it was shown that a load impedance Z_L was transformed into an impedance

$$Z_{\text{in}} = Z_c \frac{Z_L + jZ_c \tan \beta l}{Z_c + jZ_L \tan \beta l} \tag{5.1}$$

when viewed through a length l of transmission line with characteristic impedance Z_c. This formula is valid for any waveguiding system with phase constant β, provided the impedances are properly interpreted in terms of suitably defined equivalent voltages and currents. Alternatively, the reflection coefficient $\Gamma(l)$, a distance l from the termination, is uniquely given by

$$\Gamma(l) = \frac{Z_{\text{in}}(l) - Z_c}{Z_{\text{in}}(l) + Z_c} \tag{5.2}$$

with $Z_{\text{in}}(l)$ given by (5.1). The reflection coefficient is a physical quantity that can be measured, and the normalized impedances Z_{in}/Z_c and Z_L/Z_c may therefore be appropriately defined in terms of the reflection coefficient Γ at any point l on the line and the reflection coefficient Γ_L of the load, as follows:

$$\bar{Z}_{in} = \frac{Z_{in}}{Z_c} = \frac{1 + \Gamma(l)}{1 - \Gamma(l)} = \frac{1 + \Gamma_L e^{-2j\beta l}}{1 - \Gamma_L e^{-2j\beta l}} \tag{5.3a}$$

$$\bar{Z}_L = \frac{Z_L}{Z_c} = \frac{1 + \Gamma_L}{1 - \Gamma_L} \tag{5.3b}$$

The Smith chart is a graphical representation of the impedance-transformation property of a length of transmission line as given by (5.1). Clearly, it would be impractical to plot all values of Z_L and Z_{in} on a rectangular coordinate impedance plane, with one coordinate representing the real part, or resistance, and the other coordinate representing the reactance, since this would require a semi-infinite sheet of paper. On the other hand, all values of the reflection coefficient lie within a unit circle in the reflection-coefficient plane since $|\Gamma| \leq 1$. Furthermore, each value of Γ specifies a value of normalized input impedance by means of (5.3a), so that there is a one-to-one correspondence between reflection coefficient and input impedance. Instead of plotting contours of constant values of the reflection coefficient, contours of constant values of input resistance and input reactance are plotted on the reflection-coefficient plane. For a given value of the reflection coefficient, the corresponding input impedance can be read directly from the plot. In addition, a movement a distance d along the line corresponds to a change in the reflection coefficient by a factor $e^{-2j\beta d}$ only. This is represented by a simple rotation through an angle $2\beta d$; so the corresponding impedance point moves on a constant radius circle through this angle to its new value. The chart thus enables the transformation of impedance along a transmission line to be evaluated graphically in an efficient and straightforward manner. A more detailed description of the chart and its use is given below. In addition, a number of matching problems are solved with the aid of the Smith chart in later sections of this chapter.

Let the reflection coefficient Γ be expressed in polar form as

$$\Gamma = \rho e^{j\theta} \tag{5.4}$$

where $\rho = |\Gamma|$ and $\theta < \Gamma < \Gamma_L - 2\beta l$. Let the normalized input impedance be

$$\bar{Z}_{in} = \frac{Z_{in}}{Z_c} = \bar{R} + j\bar{X} = \frac{1 + \Gamma}{1 - \Gamma} = \frac{1 + \rho e^{j\theta}}{1 - \rho e^{j\theta}} \tag{5.5}$$

From (5.5) it is readily found that in the reflection-coefficient plane (ρ, θ plane), the contours of constant $\bar{R}$ and constant $\bar{X}$ are given by (Prob. 5.1)

$$\left(u - \frac{\bar{R}}{\bar{R} + 1}\right)^2 + v^2 = \frac{1}{(\bar{R} + 1)^2} \tag{5.6a}$$

$$(u - 1)^2 + \left(v - \frac{1}{\bar{X}}\right)^2 = \frac{1}{\bar{X}^2} \tag{5.6b}$$

Sec. 5.1 Impedance transformation and matching

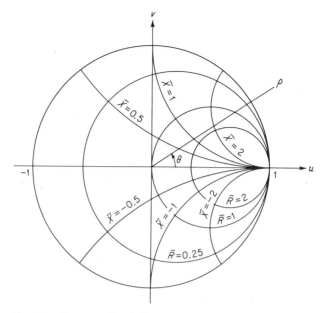

Fig. 5.1 Constant $\bar{R}$ and $\bar{X}$ circles in the reflection-coefficient plane.

where $u = \rho \cos \theta$ and $v = \rho \sin \theta$ and are rectangular coordinates in the ρ, θ plane. The above constant $\bar{R}$ and constant $\bar{X}$ contours are circles and plot as illustrated in Fig. 5.1.

For convenience in using the chart, a scale giving the angular rotation $2\beta l = 4\pi l/\lambda$ in terms of wavelength λ is attached along the circumference of the chart. Note that moving away from the load (toward the generator) corresponds to going around the chart in a clockwise direction, as illustrated in Fig. 5.2. A complete revolution around the chart is made in going a distance $l = \lambda/2$ along the transmission line. At these intervals the input impedance repeats itself. The origin for the angular scale is arbitrarily chosen at the left side of the circle.

To illustrate the use of the chart, let a line be terminated in a load impedance $\bar{R}_1 + j\bar{X}_1 = 0.5 + j0.5$. This point is located in Fig. 5.2 and labeled P_1. At a distance $l = 0.2\lambda$ away, the corresponding input impedance may be found as follows: A constant-radius circle through P_1 is constructed first. The new impedance point P_2 lies on this circle at an angle $2\beta l = 0.8\pi$ rad in a clockwise direction from P_1. This angular rotation is readily carried out by adding 0.2λ to the wavelength reading obtained from the intersection of the radius vector through P_1 and the angular scale at the circumference of the chart. From the chart it is found that the new value of normalized impedance is

$$\bar{R}_2 + j\bar{X}_2 = 2 - j1.04$$

If we begin at a point P_1, where the impedance is $\bar{R}_1 + j\bar{X}_1$, and move on a constant-radius circle an amount $\lambda/4$ to arrive at a point diametrically opposite, Γ_1 changes into $-\Gamma_1$ ($2\beta l$ changes by π), and we obtain an impedance

$$\bar{R}_2 + j\bar{X}_2 = \frac{1 - \Gamma_1}{1 + \Gamma_1} = \frac{1}{\bar{R}_1 + j\bar{X}_1} = \bar{G}_1 + j\bar{B}_1$$

Thus the input normalized admittance $\bar{G}_1 + j\bar{B}_1$ corresponding to a given input impedance $\bar{R}_1 + j\bar{X}_1$ may be found from the value of impedance at a point diametrically across from the first impedance point, provided $\bar{R}_2$ and $j\bar{X}_2$ are interpreted as the input conductance and susceptance. To clarify this, note that $\bar{R}_2 + j\bar{X}_2$ at P_2 is the normalized input impedance at point P_2 and equals the normalized input admittance at point P_1 at a distance $l = \lambda/4$ away.

The Smith chart may be used to find the transformation of admittances equally well. All that is required is to interpret the constant resistance and reactance contours (constant $\bar{R}$ and $j\bar{X}$ contours) as constant conductance $\bar{G}$ and susceptance $j\bar{B}$ contours. Note that a positive $\bar{X}$ corresponds to an inductive reactance but a positive $\bar{B}$ corresponds to a capacitive susceptance.

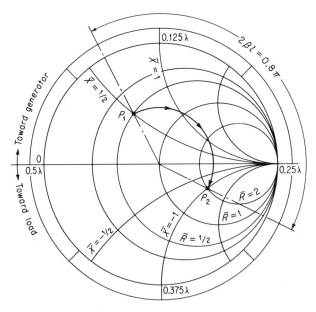

Fig. 5.2 The Smith chart.

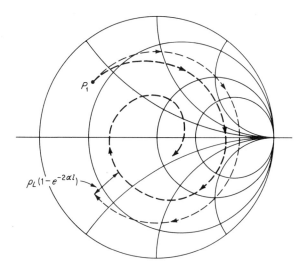

Fig. 5.3 Inward spiraling of the impedance point on a Smith chart for a lossy transmission line.

On a lossy line the reflection coefficient at any point is given by

$$\Gamma = \Gamma_L e^{-2\alpha l - 2j\beta l} = \rho_L e^{-2\alpha l - 2j\beta l + j\theta_L} \tag{5.7}$$

As we move from the load toward the generator, $\rho = \rho_L e^{-2\alpha l}$ continually decreases, and hence we move along a spiral that eventually terminates at the center, as in Fig. 5.3. In practice, we move on a constant ρ circle first through the angle $2\beta l$ and then move in radially until we are a distance $\rho_L e^{-2\alpha l}$ from the center. Many practical charts have convenient scales attached to them, so that the amount of inward spiraling is readily obtained. Note that the center of the chart represents a matched condition ($\rho = 0$).

5.2 Impedance matching with reactive elements

When a given load is to be connected to a generator by means of a transmission line or waveguide many wavelengths long, it is preferable to match the load and generator to the transmission line or waveguide at each end of the line. There are several reasons for doing this, perhaps the most important one being the great reduction in frequency sensitivity of the match. Although the transformed load impedance as seen from the generator end of the transmission line can be matched to the generator for maximum power transfer, a small change in the operating frequency will change the electrical length βl of a long line by an appreciable fraction of π rad, and hence greatly modify the effective load impedance seen at the generator end and thus modify the matching requirements as well.

To avoid this frequency sensitivity of the matching requirements, the load and generator should be individually matched to the transmission line or waveguide.

Another disadvantage of not matching the load to the transmission line is that when a matching network is used at the generator end only, there may be a large standing-wave field along the transmission line if the original load is badly mismatched. This reduces the power-handling capability of the system since, for a given power transfer, the maximum field strength before dielectric breakdown occurs is reached sooner. In addition, greater transmission losses are also incurred when there is a standing-wave current along the line.

The techniques that may be used to match a given load impedance to a transmission line or waveguide may equally well be used to match the generator to the line. Hence it suffices to limit the following discussion to that of matching an arbitrary load impedance to the transmission line. For convenience, normalized impedances are used. The first matching technique discussed employs short-circuited (or open-circuited) sections of transmission lines as reactive elements, and is referred to as *stub* matching. However, the principles involved are general in nature and may be applied to any waveguiding system by substituting suitable shunt or series reactive elements for the transmission-line stubs. A description of some typical reactive elements that may be used is given in a later section of this chapter.

Single-stub matching

Case 1 Shunt stub Consider a line terminated in a pure conductive load of normalized admittance $\bar{Y}_L = \bar{G}$, as in Fig. 5.4. At some point a distance d from the load, the normalized input admittance will be $\bar{Y}_{\text{in}} = 1 + j\bar{B}$. At this point we can connect a stub with normalized input susceptance $-j\bar{B}$ across the line to yield a resultant

$$\bar{Y}_{\text{in}} = 1 + j\bar{B} - j\bar{B} = 1$$

that is, to arrive at a matched condition. The stub should be connected

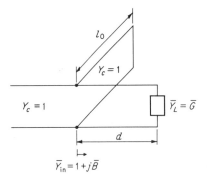

Fig. 5.4 Single-shunt-stub matching network.

at the smallest value of d that will give $\bar{Y}_{\text{in}} = 1 + j\bar{B}$ in order to keep the frequency sensitivity as small as possible. The stub may be either an open-circuited or a short-circuited section of line, the latter being the most commonly used version because of ease in adjustment and better mechanical rigidity.

To find the position d we must solve the equation

$$\bar{Y}_{\text{in}} = 1 + j\bar{B} = \frac{\bar{Y}_L + jt}{1 + j\bar{Y}_L t} \qquad t = \tan \beta d$$

If we assume that $\bar{Y}_L = \bar{G}$ is pure real, we require

$$(1 + j\bar{B})(1 + j\bar{G}t) = \bar{G} + jt$$

or by equating real and imaginary parts, we obtain

$$1 - \bar{B}\bar{G}t = \bar{G} \tag{5.8a}$$

$$j(\bar{B} + \bar{G}t) = jt \tag{5.8b}$$

Equation (5.8b) gives $\bar{B} = (1 - \bar{G})t$, and substitution of this into (5.8a) yields

$$t = \frac{1 - \bar{G}}{\bar{B}\bar{G}} = \frac{1 - \bar{G}}{(1 - \bar{G})\bar{G}t}$$

or

$$t^2 = \tan^2 \beta d = \frac{1}{\bar{G}}$$

Replacing $\tan^2 \beta d$ by $(1 - \cos^2 \beta d)/\cos^2 \beta d$ finally gives

$$d = \frac{\lambda}{2\pi} \cos^{-1} \sqrt{\frac{\bar{G}}{1 + \bar{G}}} \tag{5.9}$$

where $\beta = 2\pi/\lambda$. Note that two principal values of d are possible, depending on which sign is chosen for the square root. An alternative relation is obtained if we replace $2 \cos^2 \beta d$ by $1 + \cos 2\beta d$; thus

$$1 + \cos 2\beta d = \frac{2\bar{G}}{1 + \bar{G}}$$

and

$$\cos 2\beta d = \frac{\bar{G} - 1}{\bar{G} + 1}$$

which gives

$$d = \frac{\lambda}{4\pi} \cos^{-1} \frac{\bar{G} - 1}{\bar{G} + 1} \tag{5.10}$$

If d_1 is a solution of (5.10), then $\lambda/2 - d_1$ is another principal solution, since $\pm d_1 \pm n\lambda/2$ are all solutions of (5.10).

The value of the input susceptance $j\bar{B}$ is given by

$$\bar{B} = (1 - \bar{G})t = \frac{1 - \bar{G}}{\sqrt{\bar{G}}} \tag{5.11}$$

since $\tan^2 \beta d = 1/\bar{G}$. The required length l_0 of a short-circuited stub to give an input susceptance $-j\bar{B}$ is found from the relation

$$\bar{Y}_{\text{in}} = -j\bar{B} = -j \cot \beta l_0$$

and (5.11); thus

$$\cot \beta l_0 = \frac{1 - \bar{G}}{\sqrt{\bar{G}}}$$

or

$$l_0 = \frac{\lambda}{2\pi} \tan^{-1} \frac{\sqrt{\bar{G}}}{1 - \bar{G}} \tag{5.12}$$

where the sign of $\sqrt{\bar{G}}$ must be chosen to give the correct sign for $\bar{B}$ in (5.11). If $0 < d < \lambda/4$, the positive square root should be used, whereas if the other solution, $\lambda/4 < d < \lambda/2$, is chosen, the negative square root must be used.

A similar analysis may be carried out when $\bar{Y}_L$ is complex, but it gets more involved. The following procedure is usually followed instead. First locate a position of a voltage minimum from the load. At this point the reflection coefficient is a negative real quantity and the input admittance is pure real and given by

$$Y_{\text{in}} = \frac{1 - \Gamma}{1 + \Gamma} = \frac{1 + \rho}{1 - \rho} = S \tag{5.13}$$

where S is the standing-wave ratio on the line. Let d_0 be the distance from this voltage-minimum point to the point where $\bar{Y}_{\text{in}} = 1 + j\bar{B}$, as in Fig. 5.5. The equations to be solved for the stub position d_0 and stub length l_0 are the same as given earlier, but with S replacing $\bar{G}$. Hence

$$d_0 = \frac{\lambda}{4\pi} \cos^{-1} \frac{S - 1}{S + 1} \tag{5.14a}$$

$$l_0 = \frac{\lambda}{2\pi} \tan^{-1} \frac{\sqrt{S}}{S - 1} \tag{5.14b}$$

The position of the stub from the load is readily computed by finding the distance from the load to the $V_{\min}$ position and adding this to d_0. Note that a stub position $d_0 \pm \lambda/2$ is also a suitable one. Thus, if $d_0 - \lambda/2$ is still on the generator side of the load, the stub should be placed at this

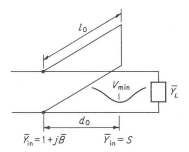

Fig. 5.5 Location of stub relative to a voltage minimum.

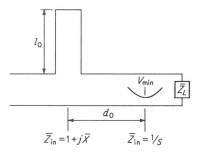

Fig. 5.6 The series stub.

point instead of at d_0 in order to reduce the frequency sensitivity of the match.

Case 2 Series stub At a position of a voltage minimum, $\bar{Z}_{in} = S^{-1}$. At some position d_0 from this point, $\bar{Z}_{in} = 1 + j\bar{X}$. By connecting a stub with a normalized input reactance of $-j\bar{X}$ in series with the line at this point, the resultant input impedance is reduced to unity and a matched condition is obtained. This series stub-matching network is illustrated in Fig. 5.6.

To find d_0 we must solve the equation

$$\bar{Z}_{in} = 1 + j\bar{X} = \frac{S^{-1} + j \tan \beta d_0}{1 + jS^{-1} \tan \beta d_0}$$

This is the same equation as considered earlier, with $\bar{X}$, S^{-1} replacing $\bar{B}$, S, and thus the solutions are

$$d_0 = \frac{\lambda}{4\pi} \cos^{-1} \frac{S^{-1} - 1}{S^{-1} + 1} = \frac{\lambda}{4\pi} \cos^{-1} \frac{1 - S}{1 + S}$$

$$= \pm \frac{\lambda}{4} + \frac{\lambda}{4\pi} \cos^{-1} \frac{S - 1}{S + 1} \quad (5.15a)$$

$$\bar{X} = \left(1 - \frac{1}{S}\right) \tan \beta d_0 = \frac{S - 1}{\sqrt{S}} \quad (5.15b)$$

where the sign of $\sqrt{S}$ must be chosen to yield the correct sign for $\tan \beta d_0$; that is, for $0 < d_0 < \lambda/4$, use $+\sqrt{S}$, and for $\lambda/4 < d_0 < \lambda/2$, use $-\sqrt{S}$.

The required stub length l_0 is determined from the relation

$$j \tan \beta l_0 = -j\bar{X}$$

and hence, from (5.15b), we obtain

$$l_0 = \frac{\lambda}{2\pi} \tan^{-1} \frac{1 - S}{\sqrt{S}} \quad (5.16)$$

The shunt stub is most commonly used for coaxial lines because it is easy to construct a shunt stub for a coaxial line whereas a series stub is difficult to build. A disadvantage with a single-stub-matching system is that every load requires a new stub position. The use of two stubs spaced by a fixed amount and located a fixed distance from the load may be used to overcome this disadvantage. However, a double-stub-matching system of this type will not match all possible values of load admittance. The theory of double-stub matching is presented in the next section.

5.3 Double-stub matching network

The double-stub tuner, or matching network, is illustrated schematically in Fig. 5.7. We may transform the normalized load admittance $\bar{Y}_L'$ into an equivalent load admittance $\bar{Y}_L = \bar{G}_L + j\bar{B}_L$ at the plane aa and treat the problem illustrated in Fig. 5.7b without loss in generality.

Let the point P_1 on the Smith chart in Fig. 5.8 represent $\bar{Y}_L$. The first stub adds a susceptance $j\bar{B}_1$ which moves P_1 along a *constant-conductance circle* to point P_2 in Fig. 5.8. At the plane bb just on the right-hand side of the second stub, the input admittance is $\bar{Y}_b = \bar{G}_b + j\bar{B}_b$, and is obtained by moving along a constant-radius circle from P_2 to P_3 through an angle $\phi = 2\beta d = 4\pi d/\lambda$ rad in a clockwise sense. The point P_3 must lie on the $\bar{G} = 1$ circle if the addition of a susceptance $j\bar{B}_2$ contributed by the second stub is to move point P_3 into the center of the chart (matched condition) along the $\bar{G} = 1$ circle.

From the description just given, it is clear that the first stub must add a susceptance of just the right amount, so that after the admittance at plane aa is transformed through a length of line d, we end up at a point on the $\bar{G} = 1$ circle. The required value of susceptance $j\bar{B}_1$ to be contributed by the first stub may be obtained by rotating the $\bar{G} = 1$ circle through an angle $-\phi$. The intersection of the rotated $\bar{G} = 1$ circle and the $\bar{G}_L$ circle determines the point P_2, and hence $j\bar{B}_1$, as illustrated in Fig. 5.9. A point P_2' would also be suitable; the location P_3' then corresponds to the admittance just to the right of the second stub.

From Fig. 5.9 it is clear that for all values of $\bar{Y}_L$ that lie within the

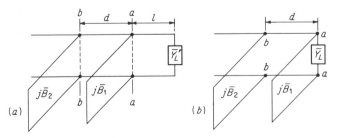

Fig. 5.7 The double-stub tuner.

Sec. 5.3   Impedance transformation and matching   213

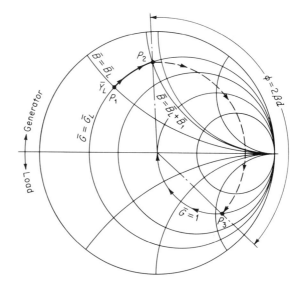

Fig. 5.8 Graphical representation of the operation of a double-stub tuner.

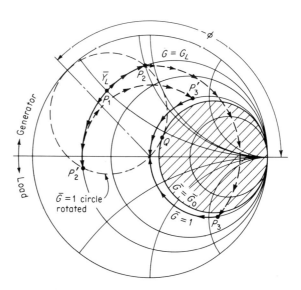

Fig. 5.9 Graphical determination of required susceptance for the first stub in a double-stub tuner.

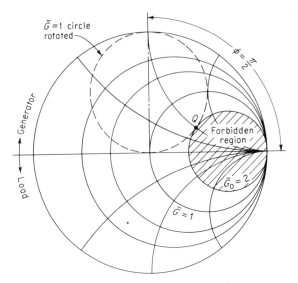

Fig. 5.10 Illustration of range of load impedance which cannot be matched when $d = \lambda/8$.

$\tilde{G} = \tilde{G}_0$ circle, a match cannot be obtained since all values $\tilde{G} > \tilde{G}_0$ will not intersect the rotated $\tilde{G} = 1$ circle. The conductance circle $\tilde{G} = \tilde{G}_0$ is tangent to the rotated $\tilde{G} = 1$ circle at the point Q. It is easy to see that the smaller the distance d, the larger the range of load admittances that may be matched (see Fig. 5.10 for the case of $d = \lambda/8$, $\phi = \pi/2$). Also note that $\tilde{G}_0$ will always be greater than unity; so all loads with $\tilde{G}_L < 1$ can be matched.

At plane aa in Fig. 5.7 we have $\tilde{Y}_L = \tilde{G}_L + j\bar{B}_L$. Just to the left of the first stub we have $\tilde{Y}_a = \tilde{G}_L + j\bar{B}_L + j\bar{B}_1$. Just to the right of the second stub we have

$$\tilde{Y}_b = \frac{\tilde{G}_L + j\bar{B}_L + j\bar{B}_1 + jt}{1 + jt(\tilde{G}_L + j\bar{B}_L + j\bar{B}_1)} \qquad t = \tan\beta d \qquad (5.17)$$

Since $\tilde{Y}_b$ must equal $1 + j\bar{B}$, (5.17) gives, upon equating the real part to unity,

$$\tilde{G}_L^2 - \tilde{G}_L \frac{1+t^2}{t^2} + \frac{(1 - \bar{B}_L t - \bar{B}_1 t)^2}{t^2} = 0 \qquad (5.18a)$$

or

$$\tilde{G}_L = \frac{1+t^2}{2t^2}\left[1 \pm \sqrt{1 - \frac{4t^2(1 - \bar{B}_L t - \bar{B}_1 t)^2}{(1+t^2)^2}}\right] \qquad (5.18b)$$

In (5.18b) we note that the term under the radical sign equals one minus a positive quantity. Since $\tilde{G}_L$ must be real, the term under the radical sign must be positive, or zero. Hence the value of the square-root term lies between zero and one. The corresponding limits on $\tilde{G}_L$ are

$$0 \leq \tilde{G}_L \leq \frac{1 + t^2}{t^2} = \frac{1}{\sin^2 \beta d} \tag{5.19}$$

For any given choice of d the whole range of load admittances outside the circle $\tilde{G}_0 = \csc^2 \beta d$ may be matched. As an example, for $d = \lambda/8$, $\beta d = \pi/4$, $\phi = 2\beta d = \pi/2$, and all values of load admittance outside the $\tilde{G}_0 = 2$ circle may be matched, as shown in Fig. 5.10. For $d = \lambda/4$, all values of $\tilde{Y}_L$ with $\tilde{G}_L \leq 1$ may be matched.

Although the theory predicts that virtually all load impedances may be matched by choosing d near zero, or $\lambda/2$, so that $\csc^2 \beta d$ becomes infinite, this is not true in practice. The maximum value of stub susceptance that can be obtained is limited by the finite attenuation of the transmission line used. If $j\beta$ were replaced by $j\beta + \alpha$, it would be found that, even with $\lambda/2$ spacing, all values of load admittance could not be matched. In addition, stub spacings near $\lambda/2$ lead to very frequency-sensitive matching networks, so that in practice spacings of $\lambda/8$ or $3\lambda/8$ are preferred. The larger spacing is used at the higher frequencies, where the wavelength is too small to permit use of $\lambda/8$ spacing.

A complete analytical solution to the double-stub matching network is readily obtained. Solution of (5.18a) for the susceptance $\bar{B}_1$ of the first stub gives

$$\bar{B}_1 = -\bar{B}_L + \frac{1 \pm \sqrt{(1 + t^2)\tilde{G}_L - \tilde{G}_L^2 t^2}}{t} \tag{5.20}$$

where $\bar{B}_L$, $\tilde{G}_L$, and $t = \tan \beta d$ are all known. Equating the imaginary part of (5.17) to $j\bar{B}$ gives

$$\bar{B} = \frac{(1 - \bar{B}_L t - \bar{B}_1 t)(\bar{B}_L + \bar{B}_1 + t) - \tilde{G}_L^2 t}{(1 - \bar{B}_L t - \bar{B}_1 t)^2 + \tilde{G}_L^2 t^2}$$

Substituting for $\bar{B}_1$ into this equation yields

$$\bar{B} = \frac{\mp \sqrt{\tilde{G}_L(1 + t^2) - \tilde{G}_L^2 t^2} - \tilde{G}_L}{\tilde{G}_L t} \tag{5.21}$$

The upper and lower signs in (5.20) and (5.21) go together. The susceptance of the second stub must be chosen as $-j\bar{B}$ in order to provide a matched condition.

5.4 Triple-stub tuner

The disadvantage of not being able to match all load admittances with a double-stub tuner may be overcome by using a triple-stub tuner, as illustrated in Fig. 5.11. Stub 1 provides a susceptance $j\bar{B}_1$ such that $\bar{Y}_L + j\bar{B}_1$ transforms to some new admittance $\bar{Y}'_L$ just to the right of stub 2. Stubs 2 and 3 provide a conventional double-stub tuner for matching $\bar{Y}'_L$ to the line. These two stubs will match all values of $\bar{Y}'_L$ for which $\bar{G}'_L < \csc^2 \beta d$. Thus the function of stub 1 is to ensure that a susceptance $j\bar{B}_1$ is added to $\bar{Y}_L$ such that the transformed admittance $\bar{Y}'_L$ has a $\bar{G}'_L$ less than $\csc^2 \beta d$. To find a suitable value of $j\bar{B}_1$ we note that, after moving a distance d from $\bar{Y}_L$, the admittance $\bar{Y}_L + j\bar{B}_1$ must transform into a point P_1 outside the circle $\bar{G}_0 = \csc^2 \beta d$, as in Fig. 5.12. If we rotate the $\bar{G}_0 = \csc^2 \beta d$ circle through an angle $-\phi = -2\beta d$, we can readily see at once the range of susceptances $j\bar{B}_1$ that may be added to $\bar{Y}_L$ to keep the resulting $\bar{Y}'_L$ outside the $\bar{G}_0 = \csc^2 \beta d$ circle. The procedure is illustrated in Fig. 5.13. In this example $\bar{Y}_L$ falls within the rotated $\bar{G}_0 = \csc^2 \beta d$ circle, and hence a susceptance $j\bar{B}_1$ must be added to move the resultant load $\bar{Y}_L + j\bar{B}_1$ to some point beyond P_1 or P_2, say

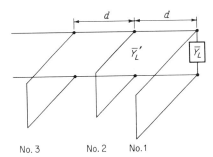

Fig. 5.11 Triple-stub tuner.

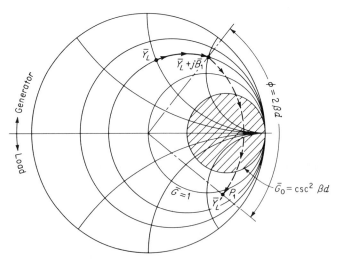

Fig. 5.12 Transformation of $\bar{Y}_L$ into $\bar{Y}'_L$.

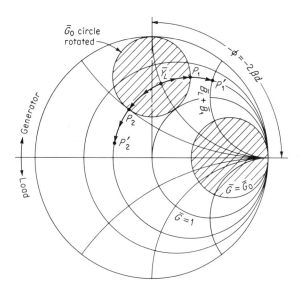

Fig. 5.13 Graphical solution for $j\bar{B}_1$ for a triple-stub tuner.

to P_1' or P_2', which provides for a margin of safety. The resultant load when transformed to the position of stub 2 will lie outside the circle $\bar{G}_0 = \csc^2 \beta d$, and hence can be matched by stubs 2 and 3.

A triple-stub tuner can match all values of load admittances. It may be considered to be two double-stub tuners in series, that is,

$$j\bar{B}_2 = j\bar{B}_2' + j\bar{B}_2''$$

so part of $j\bar{B}_2$ is associated with each end stub of the two double-stub tuners even though only one physical stub is present in the center.

5.5 Waveguide reactive elements

In the place of transmission-line stubs, any other element that acts as a shunt susceptance may be used for the purpose of matching an arbitrary load impedance to a waveguide or transmission line. A number of such reactive elements for use in rectangular waveguides supporting the dominant TE_{10} propagating mode are described in this section.† The formulas given for the normalized susceptance of these elements are approximate ones, with accuracies of the order of 10 percent or better.

† Detailed information on susceptance values and equivalent circuits are given in N. Marcuvitz (ed.), "Waveguide Handbook," McGraw-Hill Book Company, New York, 1951.

The derivation of these formulas requires the detailed solution of boundary-value problems and is outside the scope of this text.†

Shunt inductive elements

Figure 5.14 illustrates a number of rectangular waveguide elements that act as shunt inductive susceptances for the TE_{10} mode. These consist of thin metallic windows extending across the narrow dimension of the guide as in Fig. 5.14a and b, a very thin cylindrical post as in Fig. 5.14c, and a small circular aperture as in Fig. 5.14d. When a TE_{10} mode is incident on any of these discontinuities, evanescent TE_{n0} modes are excited in order to provide a total field that will satisfy the required boundary condition of a vanishing tangential electric field on the obstacle. These nonpropagating modes store predominantly magnetic energy and give the obstacle its inductive characteristics.

Approximate values for the normalized inductive susceptance of these obstacles are:

For Fig. 5.14a,

$$\bar{B} = \frac{2\pi}{\beta a} \cot^2 \frac{\pi d}{2a} \left(1 + \frac{a\gamma_3 - 3\pi}{4\pi} \sin^2 \frac{\pi d}{a}\right) \quad (5.22)$$

where $\beta = [k_0^2 - (\pi/a)^2]^{\frac{1}{2}}$ and $\gamma_3 = [(3\pi/a)^2 - k_0^2]^{\frac{1}{2}}$.

For Fig. 5.14b,

$$\bar{B} = \frac{2\pi}{\beta a} \cot^2 \frac{\pi d}{2a} \left(1 + \csc^2 \frac{\pi d}{2a}\right) \quad (5.23)$$

† For typical methods employed, see:
  R. E. Collin, "Field Theory of Guided Waves," chaps. 6–8, McGraw-Hill Book Company, New York, 1960.
  L. Lewin, "Advanced Theory of Waveguides," Iliffe Books, Ltd., London, 1951.
  H. Motz, "Electromagnetic Problems of Microwave Theory," Methuen & Co., Ltd., London, 1951.

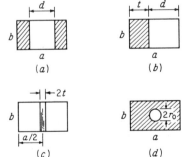

Fig. 5.14 Shunt inductive elements. (a) Symmetrical diaphragm; (b) asymmetrical diaphragm; (c) thin circular post; (d) small circular aperture.

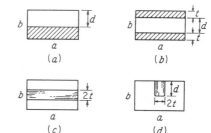

Fig. 5.15 Shunt capacitive elements. (a) Asymmetrical capacitive diaphragm; (b) symmetrical diaphragm; (c) capacitive rod; (d) capacitive post.

For the thin inductive post of Fig. 5.14c,

$$\bar{B} = \frac{4\pi}{\beta a}\left[\ln\frac{a}{\pi t} - 1 + 2\left(\frac{a}{\pi t}\right)^2 \sum_{n=3,5,\ldots}^{\infty}\left(\frac{\pi}{a\gamma_n} - \frac{1}{n}\right)\sin^2\frac{n\pi t}{a}\right]^{-1} \quad (5.24)$$

where $\gamma_n = [(n\pi/a)^2 - k_0^2]^{\frac{1}{2}}$ and t is the post radius. For the small centered circular aperture of Fig. 5.14d,

$$\bar{B} = \frac{3ab}{8\beta r_0^3} \quad (5.25)$$

Shunt capacitive elements

Typical shunt capacitive elements that may be used for matching purposes are illustrated in Fig. 5.15. These consist of thin metal septa extending across the broad dimension of the guide to form capacitive diaphragms as in Fig. 5.15a and b, a thin circular rod extending across the guide as in Fig. 5.15c, and a short thin circular post extending into the guide as in Fig. 5.15d. The post illustrated in Fig. 5.15d behaves more like an LC series network connected across a transmission line. When the depth of penetration is between $0.7b$ and $0.9b$, it becomes resonant and acts almost like an ideal short circuit. For lengths greater than this resonant length, the post is equivalent to a shunt inductive susceptance. Actually, for a post of finite thickness, the equivalent circuit is a T network, but for small-diameter posts, the series elements in this T network are negligible (for post diameters less than about $0.05a$).

Approximate expressions for the normalized susceptance of the obstacles illustrated in Fig. 5.15 are:

For the asymmetrical diaphragm of Fig. 5.15a,

$$\bar{B} = \frac{4\beta b}{\pi}\left[\ln\csc\frac{\pi d}{2b} + \left(\frac{\pi}{b\gamma_1} - 1\right)\cos^4\frac{\pi d}{2b}\right] \quad (5.26)$$

where $\beta = [k_0^2 - (\pi/a)^2]^{\frac{1}{2}}$ and $\gamma_1 = [(\pi/b)^2 - \beta^2]^{\frac{1}{2}}$.

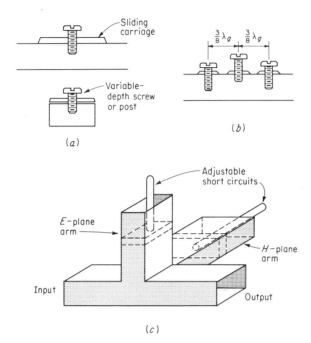

Fig. 5.16 Waveguide tuners. (a) Sliding-screw tuner; (b) triple-screw tuner; (c) E-H tuner.

For the symmetrical diaphragm of Fig. 5.15b,

$$\bar{B} = \frac{2\beta b}{\pi}\left[\ln\csc\frac{\pi d}{2b} + \left(\frac{2\pi}{b\gamma_2} - 1\right)\cos^4\frac{\pi d}{2b}\right] \tag{5.27}$$

where $\gamma_2 = [(2\pi/b)^2 - \beta^2]^{\frac{1}{2}}$.

For the capacitive rod of Fig. 5.15c and the post illustrated in Fig. 5.15d, no simple approximate formulas are available. Analytical expressions for the T-network parameters for the capacitive rod are given by Lewin, but are not reproduced here.†

Waveguide stub tuners

An approximate equivalent of a single-stub matching network is the *sliding-screw* tuner illustrated in Fig. 5.16a. This consists of a variable-depth screw mounted on a sliding carriage free to move longitudinally along the guide over a distance of at least a half guide wavelength. The screw penetrates into the guide through a centered narrow slot in the broad wall of the guide. This slot is cut along the current flow lines so that it has a negligible perturbing effect on the internal field. Since the position of the screw is adjustable over at least a half guide wavelength,

† *Op. cit.*, chap. 2. This text contains many excellent and instructive derivations of equivalent circuit parameters for a variety of waveguide structures.

its penetration does not need to be so great that it will behave as an inductive element; i.e., a match can be obtained with a shunt capacitive susceptance in all cases, as a review of single-stub-matching theory will verify.

Three variable-depth screws spaced a fixed distance of about $3\lambda_g/8$ apart as in Fig. 5.16b are essentially equivalent to a triple-stub tuner and can match a large variety of loads even though the range of susceptance values obtainable from a single screw is limited.

Circuits that are physically more like the actual short-circuited transmission-line stub are also possible. Figure 5.16c illustrates combined E-plane and H-plane stubs, referred to as an E-H tuner. The positions of the sliding short circuits in the E- and H-plane arms are variable, so that a wide range of load impedances may be matched. The equivalent circuit of either an E-plane or H-plane junction is, however, much more elaborate than a simple shunt- or series-connected transmission line because the junctions are of the order of a wavelength in size and hence produce a very complicated field structure in their vicinity. Nevertheless, since no power flow is possible through the arms terminated in the short circuits, these do still provide adjustable reactance elements that may be used for matching purposes.

5.6 Quarter-wave transformers

Quarter-wave transformers are primarily used as intermediate matching sections when it is desired to connect two waveguiding systems of different characteristic impedance. Examples are the connection of two transmission lines with different characteristic impedances, connection of an empty waveguide to a waveguide partially or completely filled with dielectric, connection of two guides of different width, height, or both, and the matching of a dielectric medium such as a microwave lens to free space. If a match over a narrow band of frequencies suffices, a single section transformer may be used. To obtain a good match over a broad band of frequencies, two, three, or even more intermediate quarter-wave sections are commonly used. The optimum design of such multisection quarter-wave transformers is presented in this section.

The essential principle involved in a quarter-wave transformer is readily explained by considering the problem of matching a transmission line of characteristic impedance Z_1 to a pure resistive load impedance Z_L, as illustrated in Fig. 5.17. If an intermediate section of transmission

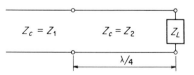

Fig. 5.17 A quarter-wave transformer.

line with a characteristic impedance Z_2 and a quarter wavelength long is connected between the main line and the load, the effective load impedance presented to the main line is

$$Z = Z_2 \frac{Z_L + jZ_2 \tan(\beta\lambda/4)}{Z_2 + jZ_L \tan(\beta\lambda/4)} = \frac{Z_2^2}{Z_L} \qquad (5.28)$$

If Z_2 is chosen equal to $\sqrt{Z_1 Z_L}$, then $Z = Z_1$ and the load is matched to the main line. In other words, the intermediate section of transmission line of length $\lambda/4$ transforms the load impedance Z_L into an impedance Z_1 and hence acts as an ideal transformer of turns ratio $\sqrt{Z_1/Z_L}$. A perfect match is obtained only at that frequency for which the transformer is a quarter wavelength (or $n\lambda/2 + \lambda/4$) long.

Let θ be the electrical length of the transformer at the frequency f, that is, $\beta(f)l = \theta$, where the phase constant β has been written as a function of frequency. For a TEM wave in an air-filled line, $\beta l = 2\pi f l/c$. At any frequency the input impedance presented to the main line is

$$Z_{\text{in}} = Z_2 \frac{Z_L + jZ_2 t}{Z_2 + jZ_L t} \qquad (5.29)$$

where $t = \tan\theta = \tan\beta l$. Consequently, the reflection coefficient is

$$\Gamma = \frac{Z_{\text{in}} - Z_1}{Z_{\text{in}} + Z_1} = \frac{Z_2(Z_L - Z_1) + jt(Z_2^2 - Z_1 Z_L)}{Z_2(Z_L + Z_1) + jt(Z_2^2 + Z_1 Z_L)}$$

$$= \frac{Z_L - Z_1}{Z_L + Z_1 + jt2\sqrt{Z_1 Z_L}} \qquad (5.30)$$

The latter form is obtained by using the relation $Z_2^2 = Z_1 Z_L$. The magnitude of Γ, denoted by ρ, is readily evaluated and is given by

$$\rho = \frac{|Z_L - Z_1|}{[(Z_L + Z_1)^2 + 4t^2 Z_1 Z_L]^{\frac{1}{2}}} = \frac{1}{\left[1 + \left(\frac{2\sqrt{Z_1 Z_L}}{Z_L - Z_1}\sec\theta\right)^2\right]^{\frac{1}{2}}} \qquad (5.31)$$

For θ near $\pi/2$, this equation is well approximated by

$$\rho = \frac{|Z_L - Z_1|}{2\sqrt{Z_1 Z_L}} |\cos\theta| \qquad (5.32)$$

A plot of ρ versus θ is given in Fig. 5.18, and this is essentially a plot of ρ versus frequency. The variation of ρ with frequency, or θ, is periodic because of the periodic variation of the input impedance with frequency; i.e., the impedance repeats its value every time the electrical length of the transformer changes by π. If ρ_m is the maximum value of reflection coefficient that can be tolerated, the useful bandwidth provided by the transformer is that corresponding to the range $\Delta\theta$ in Fig. 5.18. Because of the rapidly increasing values of ρ on either side of $\theta = \pi/2$, the useful bandwidth is small. The value of θ at the edge of the useful passband

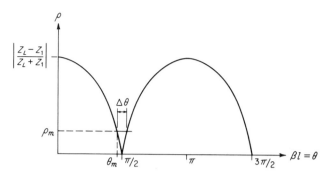

Fig. 5.18 Bandwidth characteristic for a single-section quarter-wave transformer.

may be found from (5.31) by equating ρ to ρ_m; thus

$$\theta_m = \cos^{-1}\left|\frac{2\rho_m \sqrt{Z_1 Z_L}}{(Z_L - Z_1)\sqrt{1 - \rho_m^2}}\right| \tag{5.33}$$

In the case of a TEM wave, $\theta = \beta l = \frac{f}{f_0}\frac{\pi}{2}$, where f_0 is the frequency for which $\theta = \pi/2$. In this case the bandwidth is given by

$$\Delta f = 2(f_0 - f_m) = 2\left(f_0 - \frac{2f_0}{\pi}\theta_m\right)$$

and the fractional bandwidth is given by

$$\frac{\Delta f}{f_0} = 2 - \frac{4}{\pi}\cos^{-1}\left|\frac{2\rho_m \sqrt{Z_1 Z_L}}{(Z_L - Z_1)\sqrt{1 - \rho_m^2}}\right| \tag{5.34}$$

where that solution of (5.33) that gives $\theta_m < \pi/2$ is to be chosen.

Although there are a number of instances when the bandwidth provided by a single-section transformer may be adequate, there are also a number of situations in which much greater bandwidths must be provided for. The required increase in bandwidth can be obtained by using multisection quarter-wave transformers. The approximate theory of these multisection transformers is discussed first, in order to develop a theory that also has application in the design of other microwave devices, such as directional couplers and antenna arrays. This is followed by a discussion and presentation of results obtainable from a more exact analysis.

It should be noted that in the previous discussion it was assumed that the characteristic impedances Z_1 and Z_2 were independent of frequency. For transmission lines this is a good approximation, but for waveguides the wave impedance varies with frequency, and this complicates the analysis considerably. In addition, for both transmission lines and waveguides, there are reactive fields excited at the junctions of the different sections, brought about because of the change in geometrical cross section necessary to achieve the required characteristic impedances. These junction effects can often be represented by a pure shunt susceptance at each junction.† The susceptive elements will also vary the performance of any practical transformer from the predicted performance based on an ideal model where junction effects are neglected. In spite of all these limitations, only the theory for ideal transformers is developed here; i.e., junction effects and the frequency dependency of the equivalent characteristic impedances are neglected. Thus the theory given will be indicative of the performance that can be obtained in the nonideal case only.‡

5.7 Theory of small reflections

As a preliminary to the approximate analysis of multisection quarter-wave transformers, some results pertaining to the overall reflection coefficient arising from several small reflecting obstacles are required. Consider the case of a load impedance Z_L connected to a transmission line of characteristic impedance Z_1 through an intermediate section of line of electrical length $\beta l = \theta$ and characteristic impedance Z_2, as illustrated in Fig. 5.19. For each junction the reflection and transmission coef-

---

† S. B. Cohn, Optimum Design of Stepped Transmission Line Transformers, *IRE Trans.*, vol. MTT-3, pp. 16–21, April, 1955. This paper presents an approximate theory of Chebyshev transformers, together with a method of accounting for the reactances introduced at each step.

‡ For typical application to waveguide transformers see:
  R. E. Collin and J. Brown, The Design of Quarter-wave Matching Layers for Dielectric Surfaces, *Proc. IEE*, vol. 103, pt. C, pp. 153–158, March, 1956.
  L. Young, Optimum Quarter-wave Transformers, *IRE Trans.*, vol. MTT-8, pp. 478–482, September, 1960; also Inhomogeneous Quarter-wave Transformers of Two Sections, *ibid.*, pp. 645–649, November, 1960.
  E. S. Hensperger, Broad-band Stepped Transformers from Rectangular to Double-ridged Waveguide, *IRE Trans.*, vol. MTT-6, pp. 311–314, July, 1958.

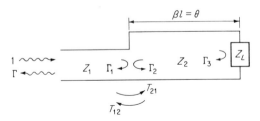

Fig. 5.19 A microwave circuit with two reflecting junctions.

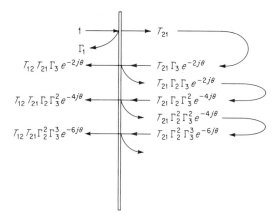

Fig. 5.20 Multiple reflection of waves for a circuit with two reflecting junctions.

ficients are

$$\Gamma_1 = \frac{Z_2 - Z_1}{Z_2 + Z_1} \qquad \Gamma_2 = -\Gamma_1$$

$$T_{21} = 1 + \Gamma_1 = \frac{2Z_2}{Z_1 + Z_2} \qquad T_{12} = 1 + \Gamma_2 = \frac{2Z_1}{Z_1 + Z_2}$$

$$\Gamma_3 = \frac{Z_L - Z_2}{Z_L + Z_2}$$

A wave of unit amplitude is incident, and the total reflected wave has a complex amplitude Γ equal to the total reflection coefficient. When the incident wave strikes the first junction, a partial reflected wave of amplitude Γ_1 is produced. A transmitted wave of amplitude T_{21} is then incident on the second junction. A portion of this is reflected to give a wave of amplitude $\Gamma_3 T_{21} e^{-2j\theta}$ incident from the right on the first junction. A portion $T_{12} T_{21} \Gamma_3 e^{-2j\theta}$ is transmitted, and a portion $\Gamma_2 \Gamma_3 T_{12} T_{21} e^{-2j\theta}$ is reflected back toward Z_L. Figure 5.20 illustrates the first few of the infinite number of multiply reflected waves that occur. The total reflected wave of amplitude Γ is the sum of all the partial waves transmitted past the first junction toward the left. This sum is given by

$$\Gamma = \Gamma_1 + T_{12}T_{21}\Gamma_3 e^{-2j\theta} + T_{12}T_{21}\Gamma_3{}^2\Gamma_2 e^{-4j\theta} + \cdots$$

$$= \Gamma_1 + T_{12}T_{21}\Gamma_3 e^{-2j\theta} \sum_{n=0}^{\infty} \Gamma_2{}^n \Gamma_3{}^n e^{-2jn\theta}$$

This geometric series is readily summed to give [note that $\sum_{n=0}^{\infty} r^n =$

$(1 - r)^{-1}]$

$$\Gamma = \Gamma_1 + \frac{T_{12}T_{21}\Gamma_3 e^{-2j\theta}}{1 - \Gamma_2\Gamma_3 e^{-2j\theta}}$$

Replacing T_{12} by $1 + \Gamma_2 = 1 - \Gamma_1$ and T_{21} by $1 + \Gamma_1$ gives

$$\Gamma = \frac{\Gamma_1 + \Gamma_3 e^{-2j\theta}}{1 + \Gamma_1\Gamma_3 e^{-2j\theta}} \tag{5.35}$$

If $|\Gamma_1|$ and $|\Gamma_3|$ are both small compared with unity, an excellent approximation to Γ is

$$\Gamma = \Gamma_1 + \Gamma_3 e^{-2j\theta} \tag{5.36}$$

This result states that, for small reflections, the resultant reflection coefficient is just that obtained by taking only first-order reflections into account. This is the result that will be used to obtain a first-order theory for multisection quarter-wave transformers. As an indication of the accuracy of the approximate formula, note that if $|\Gamma_1| = |\Gamma_3| = 0.2$, the error in Γ does not exceed 4 percent.

5.8 Approximate theory for multisection quarter-wave transformers

Figure 5.21 illustrates an N-section quarter-wave transformer. At the first junction the reflection coefficient is

$$\Gamma_0 = \frac{Z_1 - Z_0}{Z_1 + Z_0} = \rho_0 \tag{5.37a}$$

Similarly, at the nth junction, the reflection coefficient is

$$\Gamma_n = \frac{Z_{n+1} - Z_n}{Z_{n+1} + Z_n} = \rho_n \tag{5.37b}$$

The last reflection coefficient is

$$\Gamma_N = \frac{Z_L - Z_N}{Z_L + Z_N} = \rho_N \tag{5.37c}$$

Note that Z_0 is a characteristic impedance, and not necessarily equal to $(\mu_0/\epsilon_0)^{\frac{1}{2}}$ here. Each section has the same electrical length $\beta l = \theta$, and

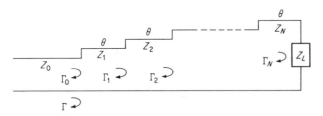

Fig. 5.21 A multisection quarter-wave transformer.

will be a quarter wave long at the matching frequency f_0. The load Z_L is assumed to be a pure resistance, and may be greater or smaller than Z_0. In this analysis it is chosen greater, so that all $\Gamma_n = \rho_n$, where ρ_n is the magnitude of Γ_n. If Z_L is smaller than Z_0, all Γ_n are negative real numbers and the only modification required in the theory is replacing all ρ_n by $-\rho_n$.

For a first approximation the total reflection coefficient is the sum of the first-order reflected waves only. This is given by

$$\Gamma = \rho_0 + \rho_1 e^{-2j\theta} + \rho_2 e^{-4j\theta} + \cdots + \rho_N e^{-2jN\theta} \tag{5.38}$$

where $e^{-2jn\theta}$ accounts for the phase retardation introduced because of the different distances the various partial waves must travel.

At this point it is expedient to assume that the transformer is symmetrical, so that $\rho_0 = \rho_N$, $\rho_1 = \rho_{N-1}$, $\rho_2 = \rho_{N-2}$, etc. In this case (5.38) becomes

$$\Gamma = e^{-jN\theta}[\rho_0(e^{jN\theta} + e^{-jN\theta}) + \rho_1(e^{j(N-2)\theta} + e^{-j(N-2)\theta}) + \cdots] \tag{5.39}$$

where the last term is $\rho_{(N-1)/2}(e^{j\theta} + e^{-j\theta})$ for N odd and $\rho_{N/2}$ for N even. It is thus seen that for a symmetrical transformer the reflection coefficient Γ is given by a Fourier cosine series:

$$\Gamma = 2e^{-jN\theta}[\rho_0 \cos N\theta + \rho_1 \cos(N-2)\theta \\ + \cdots \rho_n \cos(N-2n)\theta + \cdots] \tag{5.40}$$

In (5.40) the last term is $\rho_{(N-1)/2} \cos \theta$ for N odd and $\tfrac{1}{2}\rho_{N/2}$ for N even. It should now be apparent that by a proper choice of the reflection coefficients ρ_n, and hence the Z_n, a variety of passband characteristics can be obtained. Since the series is a cosine series, the periodic function that it defines is periodic over the interval π corresponding to the frequency range over which the length of each transformer section changes by a half wavelength. The specification of ρ_n to obtain a maximally flat and an equal-ripple passband characteristic is given in the following two sections.

5.9 Binomial transformer

A maximally flat passband characteristic is obtained if $\rho = |\Gamma|$ and the first $N-1$ derivatives with respect to frequency (or θ) vanish at the matching frequency f_0, where $\theta = \pi/2$. Such a characteristic is obtained if we choose

$$\Gamma = A(1 + e^{-2j\theta})^N \tag{5.41a}$$

for which

$$\rho = |\Gamma| = |A2^N(\cos\theta)^N| \tag{5.41b}$$

When $\theta = 0$ or π, we have $\Gamma = (Z_L - Z_0)/(Z_L + Z_0)$, and from (5.41a) we obtain $\Gamma = A2^N$. Thus the constant A is given by

$$A = 2^{-N} \frac{Z_L - Z_0}{Z_L + Z_0} \tag{5.42}$$

Expanding (5.41a) by the binomial expansion gives

$$\Gamma = 2^{-N} \frac{Z_L - Z_0}{Z_L + Z_0}(1 + e^{-2j\theta})^N = 2^{-N} \frac{Z_L - Z_0}{Z_L + Z_0} \sum_{n=0}^{N} C_n^N e^{-j2n\theta} \tag{5.43}$$

where the binomial coefficients are given by

$$C_n^N = \frac{N(N-1)(N-2)\cdots(N-n+1)}{n!} = \frac{N!}{(N-n)!n!} \tag{5.44}$$

Note that $C_n^N = C_{N-n}^N$, $C_0^N = 1$, $C_1^N = N = C_{N-1}^N$, etc. Comparing (5.43) with (5.38) shows that we must choose

$$\rho_n = 2^{-N} \frac{Z_L - Z_0}{Z_L + Z_0} C_n^N = \rho_{N-n} \tag{5.45}$$

since $C_n^N = C_{N-n}^N$.

To obtain a simple solution for the characteristic impedances Z_n, it is convenient to make a further approximation. Since we have already specified that all ρ_n are to be small, we can use the result

$$\ln \frac{Z_{n+1}}{Z_n} \approx 2 \frac{Z_{n+1} - Z_n}{Z_{n+1} + Z_n} = 2\rho_n$$

Thus we have

$$\ln \frac{Z_{n+1}}{Z_n} = 2\rho_n = 2^{-N} C_n^N \ln \frac{Z_L}{Z_0} \tag{5.46}$$

where we have also used the approximation

$$\ln \frac{Z_L}{Z_0} = 2\frac{Z_L - Z_0}{Z_L + Z_0} + \frac{2}{3}\left(\frac{Z_L - Z_0}{Z_L + Z_0}\right)^3 + \cdots \approx 2\frac{Z_L - Z_0}{Z_L + Z_0}$$

Equation (5.46) gives the solution for the logarithm of the impedances, and since these are proportional to the binomial coefficients, the transformer is called a binomial transformer. Since the theory is approximate, the range of Z_L is restricted to about

$$0.5Z_0 < Z_L < 2Z_0$$

for accurate results.

As an example, consider a two-section transformer. From (5.46) we obtain

$$\ln \frac{Z_1}{Z_0} = \frac{1}{4} \ln \frac{Z_L}{Z_0} \quad \text{or} \quad Z_1 = Z_L^{\frac{1}{4}} Z_0^{\frac{3}{4}}$$

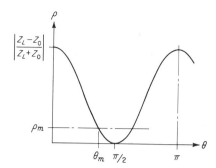

Fig. 5.22 Passband characteristic for a maximally flat transformer.

and

$$\ln \frac{Z_2}{Z_1} = \frac{1}{2} \ln \frac{Z_L}{Z_0} \quad \text{or} \quad Z_2 = Z_L^{\frac{3}{4}} Z_0^{\frac{1}{4}}$$

since $C_0^2 = 1$ and $C_1^2 = 2$. Although the approximate theory was used, it turns out that the above values for Z_1 and Z_2 for the special case of the two-section transformer are the correct nonapproximate solutions, a result that gives an indication of the accuracy of the approximate theory.

The type of passband characteristic obtained with a maximally flat transformer is illustrated in Fig. 5.22. Let ρ_m be the maximum value of ρ that can be tolerated. The angle θ_m that gives $\rho = \rho_m$ is given by

$$\theta_m = \cos^{-1} \left| \frac{2\rho_m}{\ln (Z_L/Z_0)} \right|^{1/N} \tag{5.47}$$

as obtained from (5.41b). In the case of transmission-line sections, $\theta = \pi f/2f_0$, and hence the fractional bandwidth is given by

$$\frac{\Delta f}{f_0} = \frac{2(f_0 - f_m)}{f_0} = 2 - \frac{4}{\pi} \cos^{-1} \left| \frac{2\rho_m}{\ln (Z_L/Z_0)} \right|^{1/N} \tag{5.48}$$

since $\theta_m = \pi f_m/2f_0$. Note that in (5.48) the solution to the inverse cosine function is chosen so that $\theta_m < \pi/2$. By comparing Figs. 5.18 and 5.22, it is clear that a multisection maximally flat transformer can provide a much greater useful bandwidth than a single-section transformer.

5.10 Chebyshev transformer

Instead of a maximally flat passband characteristic, an equally useful characteristic is one that may permit ρ to vary between 0 and ρ_m in an oscillatory manner over the passband. A transformer designed to yield an equal-ripple characteristic as illustrated in Fig. 5.23 is of this type and provides a considerable increase in bandwidth over the binomial transformer design. The equal-ripple characteristic is obtained by making ρ behave according to a Chebyshev polynomial, and hence the name Chebyshev transformer. It is possible to have ρ vanish at as

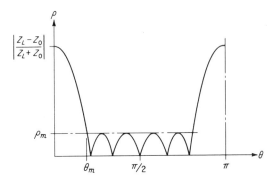

Fig. 5.23 Equal-ripple characteristic obtained from a Chebyshev transformer.

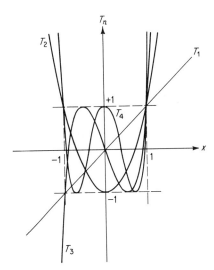

Fig. 5.24 Chebyshev polynomials.

many different frequencies in the passband as there are transformer sections. To see how Chebyshev polynomials may be used in the design, it is necessary to consider the basic properties of these polynomials first.

The Chebyshev polynomial of degree n, denoted by $T_n(x)$, is an nth-degree polynomial in x. The first four polynomials and the recurrence relation are

$T_1(x) = x$

$T_2(x) = 2x^2 - 1$

$T_3(x) = 4x^3 - 3x$

$T_4(x) = 8x^4 - 8x^2 + 1$

$T_n(x) = 2xT_{n-1} - T_{n-2}$

The polynomials T_n oscillate between ± 1 for x in the range $|x| \leq 1$ and increase in magnitude indefinitely for x outside this range. Figure 5.24 gives a sketch of the first four polynomials. If x is replaced by $\cos\theta$, we have

$$T_n(\cos\theta) = \cos n\theta \tag{5.49}$$

which clearly shows that $|T_n| \leq 1$ for $-1 \leq x \leq 1$. As θ varies from 0 to π, the corresponding range of x is from $+1$ to -1. Since we wish to make ρ have the equal-ripple characteristic only over the range θ_m to $\pi - \theta_m$, we cannot use $T_n(\cos\theta)$ directly. If we consider instead

$$T_n\left(\frac{\cos\theta}{\cos\theta_m}\right) = \cos n\left(\cos^{-1}\frac{\cos\theta}{\cos\theta_m}\right) \tag{5.50}$$

we see that the argument will become equal to unity when $\theta = \theta_m$ and will be less than unity for $\theta_m < \theta < \pi - \theta_m$. This function will therefore confine the equal-ripple oscillations of T_n to the desired passband.

The function given in (5.50) is an nth-degree polynomial in the variable $\cos\theta/\cos\theta_m$. Since $(\cos\theta)^n$ can be expanded into a series of cosine terms such as $\cos\theta$, $\cos 2\theta$, ..., $\cos n\theta$, it follows that (5.50) is a series of the form (5.40). Hence we may choose

$$\Gamma = 2e^{-jN\theta}[\rho_0 \cos N\theta + \rho_1 \cos(N-2)\theta + \cdots$$
$$\qquad\qquad + \rho_n \cos(N-2n)\theta + \cdots]$$
$$= Ae^{-jN\theta}T_N(\sec\theta_m \cos\theta) \tag{5.51}$$

where A is a constant to be determined. When $\theta = 0$, we have

$$\Gamma = \frac{Z_L - Z_0}{Z_L + Z_0} = AT_N(\sec\theta_m)$$

and so

$$A = \frac{Z_L - Z_0}{(Z_L + Z_0)T_N(\sec\theta_m)}$$

Consequently, we have

$$\Gamma = e^{-jN\theta}\frac{Z_L - Z_0}{Z_L + Z_0}\frac{T_N(\sec\theta_m \cos\theta)}{T_N(\sec\theta_m)} \tag{5.52}$$

In the passband the maximum value of $T_N(\sec\theta_m \cos\theta)$ is unity, and hence

$$\rho_m = \frac{Z_L - Z_0}{(Z_L + Z_0)T_N(\sec\theta_m)} \tag{5.53a}$$

If the passband, and hence θ_m, is specified, the passband tolerance ρ_m is fixed, and vice versa. From (5.53a) we obtain

$$T_N(\sec\theta_m) = \frac{Z_L - Z_0}{Z_L + Z_0}\rho_m^{-1}$$

or by using (5.50) for $\cos\theta = 1$,

$$\sec\theta_m = \cos\left(\frac{1}{N}\cos^{-1}\frac{Z_L - Z_0}{Z_L + Z_0}\rho_m^{-1}\right) \quad (5.53b)$$

which gives θ_m in terms of the passband tolerance on ρ, that is, ρ_m.

In order to solve (5.51) for the unknown ρ_n, we need the following results:

$$(\cos\theta)^n = 2^{-n}e^{-jn\theta}(1 + e^{2j\theta})^n = 2^{-n}e^{-jn\theta}\sum_{m=0}^{n} C_m^n e^{j2m\theta}$$

$$= 2^{-n+1}[C_0^n \cos n\theta + C_1^n \cos(n-2)\theta + \cdots$$
$$+ C_m^n \cos(n-2m)\theta + \cdots] \quad (5.54)$$

The last term in (5.54) is $\frac{1}{2}C_{n/2}^n$ for n even and $C_{(n-1)/2}^n \cos\theta$ for n odd. Using (5.54) and the earlier expression for $T_n(x)$, we can obtain the following:

$$T_1(\sec\theta_m \cos\theta) = \sec\theta_m \cos\theta \quad (5.55a)$$

$$T_2(\sec\theta_m \cos\theta) = 2(\sec\theta_m \cos\theta)^2 - 1 = \sec^2\theta_m(1 + \cos 2\theta) - 1 \quad (5.55b)$$

$$T_3(\sec\theta_m \cos\theta) = \sec^3\theta_m(\cos 3\theta + 3\cos\theta) - 3\sec\theta_m \cos\theta \quad (5.55c)$$

$$T_4(\sec\theta_m \cos\theta) = \sec^4\theta_m(\cos 4\theta + 4\cos 2\theta + 3) - 4\sec^2\theta_m(\cos 2\theta + 1) \quad (5.55d)$$

These results are sufficient for designing transformers up to four sections in length. A greater number of sections would rarely be required in practice.

As an example, consider the design of a two-section transformer to match a line with $Z_0 = 1$ to a line or load with $Z_L = 2$. Let the maximum tolerable value of ρ be $\rho_m = 0.05$. Using (5.53a), we obtain

$$T_2(\sec\theta_m) = 2\sec^2\theta_m - 1 = \frac{1}{3(0.05)} = 6.67$$

and hence $\sec\theta_m = 1.96$, and $\theta_m = 1.04$. Thus the fractional bandwidth that is obtained is

$$\frac{\Delta\theta}{\pi/2} = \frac{\Delta f}{f_0} = \frac{4}{\pi}\left(\frac{\pi}{2} - 1.04\right) = 0.675$$

From (5.51), (5.52), and (5.55b) we obtain [refer to remarks following (5.40) as regards the last term in the cosine series for ρ]

$$2\rho_0 \cos 2\theta + \rho_1 = \rho_m T_2(\sec\theta_m \cos\theta)$$
$$= \rho_m \sec^2\theta_m \cos 2\theta + \rho_m(\sec^2\theta_m - 1)$$

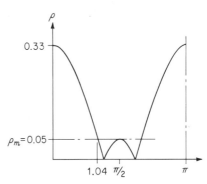

Fig. 5.25 Passband characteristic for a two-section Chebyshev transformer with $\rho_m = 0.05$, $Z_L/Z_0 = 2$.

and hence

$$\rho_0 = \tfrac{1}{2}\rho_m \sec^2 \theta_m = \rho_2 = 0.096$$
$$\rho_1 = \rho_m(\sec^2 \theta_m - 1) = 0.142$$

The impedances Z_1 and Z_2 are given by

$$Z_1 = \frac{1 + \rho_0}{1 - \rho_0} Z_0 = 1.21 \qquad Z_2 = \frac{1 + \rho_1}{1 - \rho_1} Z_1 = 1.62$$

A plot of the passband characteristic is given in Fig. 5.25.

★5.11 Chebyshev transformer (exact results)

An exact theory for a multisection transformer having an equal-ripple passband characteristic has also been developed (see references at the end of this chapter). Since the analysis is rather long, only the final results for the two- and three-section transformers are given here.

In the exact theory of multisection ideal transformers it is convenient to introduce the power loss ratio P_{LR}, which is defined as the available power (incident power) divided by the actual power delivered to the load. If the incident power is P_i, the reflected power is $\rho^2 P_i$ and the power delivered to the load is $(1 - \rho^2)P_i$. Hence

$$P_{\mathrm{LR}} = \frac{P_i}{(1 - \rho^2)P_i} = \frac{1}{1 - \rho^2} \tag{5.56a}$$

and

$$\rho = \sqrt{\frac{P_{\mathrm{LR}} - 1}{P_{\mathrm{LR}}}} \tag{5.56b}$$

If T is the overall transmission coefficient, then $|T|^2 = 1 - \rho^2$.

For any transformer an expression for Z_{in} is readily obtained, and from this ρ, and hence P_{LR}, can be computed. When this is done it is found

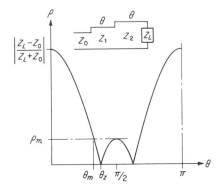

Fig. 5.26 Passband characteristic for a two-section Chebyshev transformer.

that P_{LR} can be expressed in the form

$$P_{LR} = 1 + Q_{2N}(\cos \theta) \tag{5.57}$$

where $Q_{2N}(\cos \theta)$ is an even polynomial of degree $2N$ in $\cos \theta$, with coefficients that are functions of the various impedances Z_n. To obtain an equal-ripple characteristic, P_{LR} is now specified to be

$$P_{LR} = 1 + k^2 T_N^2 (\sec \theta_m \cos \theta) \tag{5.58}$$

where k^2 is the passband tolerance on P_{LR}; that is, the maximum value of P_{LR} in the passband is $1 + k^2$, since T_N^2 has a maximum value of unity. By equating (5.57) and (5.58), algebraic equations that can be solved for the various characteristic impedances are obtained.

Figure 5.26 is a plot of ρ versus θ for a two-section transformer. For this transformer

$$P_{LR} = 1 + \frac{(Z_L - Z_0)^2}{4 Z_L Z_0} \frac{(\sec^2 \theta_z \cos^2 \theta - 1)^2}{\tan^4 \theta_z} \tag{5.59}$$

where θ_z is the value of θ at the lower zero where ρ vanishes. The maximum value of P_{LR} in the passband is

$$1 + \frac{(Z_L - Z_0)^2}{4 Z_L Z_0} \cot^4 \theta_z$$

and hence

$$\rho_m = \left(\frac{k^2}{1 + k^2} \right)^{\frac{1}{2}} \tag{5.60}$$

where $k^2 = \cot^4 \theta_z (Z_L - Z_0)^2 / 4 Z_L Z_0$. The required values of Z_1 and Z_2 are given by

$$Z_1^2 = Z_0^2 \left[\frac{(Z_L - Z_0)^2}{4 Z_0^2 \tan^4 \theta_z} + \frac{Z_L}{Z_0} \right]^{\frac{1}{2}} + \frac{(Z_L - Z_0) Z_0}{2 \tan^2 \theta_z} \tag{5.61a}$$

$$Z_2 = \frac{Z_L}{Z_1} Z_0 \tag{5.61b}$$

The value of θ_m is given by

$$\theta_m = \cos^{-1}\sqrt{2}\cos\theta_z \tag{5.62a}$$

and

$$\frac{\Delta f}{f_0} = 2 - \frac{4}{\pi}\cos^{-1}\sqrt{2}\cos\theta_z \tag{5.62b}$$

provided $2\,\Delta\theta/\pi = \Delta f/f_0$ [if not, (5.62b) gives $2\,\Delta\theta/\pi$]. If the bandwidth is specified, then θ_z, and from (5.62a) θ_m, are fixed. Equation (5.60) then specifies ρ_m. On the other hand, if ρ_m is given, the bandwidth is determined.

In the limit as θ_z approaches $\pi/2$, the two zeros of ρ coalesce to give a maximally flat transformer. From (5.61) it is found that, for this case (compare with the approximate theory),

$$Z_1 = Z_L^{\frac{1}{4}}Z_0^{\frac{3}{4}} \tag{5.63a}$$
$$Z_2 = Z_L^{\frac{3}{4}}Z_0^{\frac{1}{4}} \tag{5.63b}$$

For the maximally flat transformer, the value of θ_m at the point where $\rho = \rho_m$ is given by

$$\theta_m = \cos^{-1}\cot\theta_z \tag{5.64}$$

where θ_z is the previously defined quantity for the Chebyshev transformer. Equations (5.62) and (5.64) provide a comparison of the relative bandwidths obtainable from the Chebyshev transformer and the maximally flat transformer. This comparison is illustrated in Fig. 5.27 for $N = 2$ and $N = 3$ and shows that the Chebyshev transformer can give bandwidths that are considerably greater for the same maximum tolerable value ρ_m.

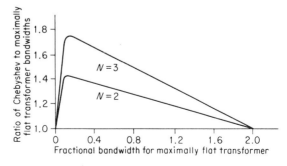

Fig. 5.27 Comparison of bandwidths for Chebyshev and maximally flat transformers.

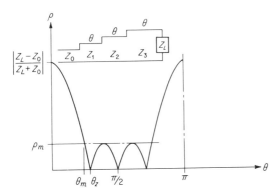

Fig. 5.28 Passband characteristic for three-section Chebyshev transformer.

Figure 5.28 illustrates the passband characteristic for a three-section Chebyshev transformer. The power loss ratio is given by

$$P_{LR} = 1 + \frac{(Z_L - Z_0)^2}{4 Z_L Z_0} \frac{(\sec^2 \theta_z \cos^2 \theta - 1)^2 \cos^2 \theta}{\tan^4 \theta_z} \tag{5.65}$$

The passband tolerance k^2 is given by

$$k^2 = \frac{(Z_L - Z_0)^2}{4 Z_L Z_0} \left(\frac{2 \cos \theta_z}{3 \sqrt{3} \tan^2 \theta_z} \right)^2 \tag{5.66}$$

from which ρ_m may be found by using (5.60). Again the general result that specifying k^2 determines the bandwidth, and vice versa, holds. The value of θ_m is given by

$$\theta_m = \cos^{-1} \frac{2}{\sqrt{3}} \cos \theta_z \tag{5.67a}$$

and for transmission lines for which $\Delta f/f_0 = 2 \Delta \theta/\pi$,

$$\frac{\Delta f}{f_0} = \frac{2(\pi/2 - \theta_m)}{\pi/2} = 2 - \frac{4}{\pi} \cos^{-1} \frac{2}{\sqrt{3}} \cos \theta_z \tag{5.67b}$$

The characteristic impedance Z_1 is determined by solving

$$\frac{Z_L - Z_0}{\tan^2 \theta_z} = \frac{Z_1^2}{Z_0} + 2 \left(\frac{Z_L}{Z_0} \right)^{\frac{1}{2}} Z_1 - \frac{Z_L Z_0^2}{Z_1^2} - 2 \left(\frac{Z_L}{Z_0} \right)^{\frac{1}{2}} Z_1^{-1} Z_0^2 \tag{5.68a}$$

and Z_2 and Z_3 are given by

$$Z_2 = (Z_L Z_0)^{\frac{1}{2}} \tag{5.68b}$$

$$Z_3 = \frac{Z_L Z_0}{Z_1} \tag{5.68c}$$

When θ_z approaches $\pi/2$, all three zeros coalesce at $\pi/2$, and a maximally flat transformer is obtained. The required value of Z_1 may be obtained from (5.68a) by equating the left-hand side to zero. It will be found that $Z_1 = Z_L{}^\alpha Z_0{}^{1-\alpha}$, where $\tfrac{1}{8} \le \alpha < \tfrac{1}{4}$. With Z_L/Z_0 near unity, α will be close to $\tfrac{1}{8}$, and for large values of Z_L/Z_0, α will approach $\tfrac{1}{4}$. By picking various values of α, a solution for Z_1 can be found quite readily by a trial-and-error process (note that the equation for Z_1 is a fourth-degree equation). For the maximally flat transformer the value of θ_m and the bandwidth are given by

$$\theta_m = \cos^{-1}\left[\left(\frac{2}{3\sqrt{3}}\right)^{\frac{1}{2}} \frac{\cos\theta_z}{(\sin\theta_z)^{\frac{3}{2}}}\right] \quad (5.69a)$$

$$\frac{\Delta f}{f_0} = 2 - \frac{4}{\pi}\theta_m \quad (5.69b)$$

The Chebyshev transformer represents an optimum design in that no other design can give a greater bandwidth with a smaller passband tolerance. If it is assumed that some choice, other than (5.58), for the polynomial Q_{2N} in (5.57) can give a smaller passband tolerance for the same bandwidth, it will be found that a plot of the polynomial Q_{2N} will intersect the polynomial $T_N{}^2$ in at least $N+1$ points. Since the polynomials are even in $\cos\theta$, they have at most $N+1$ coefficients. Thus Q_{2N} must be equal to $T_N{}^2$ since they have $N+1$ points in common. But this equality contradicts the original assumption that Q_{2N} could yield a better result, and hence proves that the Chebyshev transformer is an optimum one.

5.12 Tapered transmission lines

In a multisection quarter-wave transformer used to match two transmission lines with different characteristic impedances, the change in impedance level is obtained in a number of discrete steps. An alternative is to use a tapered transition which has a characteristic impedance that varies continuously in a smooth fashion from the impedance of one line to that of the other line. A transition, or matching section, of this type is referred to as a tapered transmission line. An approximate theory of tapered transmission lines, analogous to the approximate theory presented earlier for multisection transformers, is readily developed. This approximate theory is presented below. A following section gives a derivation of the exact differential equation for the reflection coefficient on a tapered transmission line and also gives a brief evaluation of the validity of the approximate theory.

Figure 5.29a illustrates schematically a tapered transmission line used to match a line with normalized impedance unity to a load with normalized impedance $\bar{Z}_L$ (assumed to be a pure resistive load). The taper has a normalized impedance $\bar{Z}$ which is a function of the distance z along

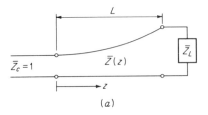

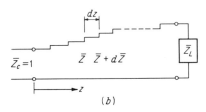

Fig. 5.29 Tapered-transmission-line matching section.

the taper. Figure 5.29b illustrates an approximation to the continuous taper by considering it to be made up of a number of sections of line of differential length dz and for which the impedance changes by differential amounts $d\bar{Z}$ from section to section.

The step change $d\bar{Z}$ in impedance at z produces a differential reflection coefficient

$$d\Gamma_0 = \frac{\bar{Z} + d\bar{Z} - \bar{Z}}{\bar{Z} + d\bar{Z} + \bar{Z}} \approx \frac{d\bar{Z}}{2\bar{Z}} = \tfrac{1}{2} d(\ln \bar{Z})$$
$$= \frac{1}{2}\frac{d}{dz}(\ln \bar{Z})\, dz \tag{5.70}$$

At the input to the taper the contribution to the input reflection coefficient from this step is

$$d\Gamma_i = e^{-j2\beta z}\frac{1}{2}\frac{d}{dz}(\ln \bar{Z})\, dz$$

If it is assumed that the total reflection coefficient can be computed by summing up all the individual contributions, as was done in the approximate theory of the multisection quarter-wave transformer, the input reflection coefficient is given by

$$\Gamma_i = \tfrac{1}{2}\int_0^L e^{-j2\beta z}\frac{d}{dz}(\ln \bar{Z})\, dz \tag{5.71}$$

where L is the total taper length. If the variation in $\bar{Z}$ with z is known, Γ_i may be readily evaluated from the above. A problem of much greater practical importance is the synthesis problem, where $\bar{Z}(z)$ is to be determined to give Γ_i the desired characteristics as a function of frequency. Before taking up the synthesis problem, two examples of practical taper designs are presented.

Exponential taper

The exponential taper is one for which $\ln \bar{Z}$ varies linearly, and hence $\bar{Z}$ varies exponentially, from unity to $\ln \bar{Z}_L$; that is,

$$\ln \bar{Z} = \frac{z}{L} \ln \bar{Z}_L \tag{5.72a}$$

$$\bar{Z} = e^{(z/L)\ln \bar{Z}_L} \tag{5.72b}$$

Substituting (5.72) into (5.71) gives

$$\Gamma_i = \tfrac{1}{2} \int_0^L \frac{\ln \bar{Z}_L}{L} e^{-2j\beta z}\, dz = \tfrac{1}{2} e^{-j\beta L} \ln \bar{Z}_L \frac{\sin \beta L}{\beta L} \tag{5.73}$$

where it has been assumed that we are dealing with a transmission line for which $\beta = k = 2\pi/\lambda$ and is not a function of z. A plot of $\rho_i = |\Gamma_i|$ versus βL is given in Fig. 5.30. For a fixed length of taper this is a plot of ρ_i as a function of frequency since $k = 2\pi f/(\mu\epsilon)^{\frac{1}{2}}$. Note that when L is greater than $\lambda/2$, the reflection coefficient is quite small, the first minor lobe being about 22 percent of the major-lobe maximum.

Taper with triangular distribution

If $d(\ln \bar{Z})/dz$ is chosen as a triangular function of the form

$$\frac{d(\ln \bar{Z})}{dz} = \begin{cases} \dfrac{4z}{L^2} \ln \bar{Z}_L & 0 \leq z \leq \dfrac{L}{2} \\ \dfrac{4}{L^2}(L-z) \ln \bar{Z}_L & \dfrac{L}{2} \leq z \leq L \end{cases} \tag{5.74}$$

a matching section with more desirable properties is obtained. Inte-

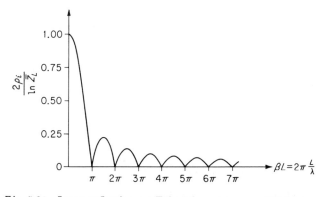

Fig. 5.30 Input reflection coefficient for an exponential taper.

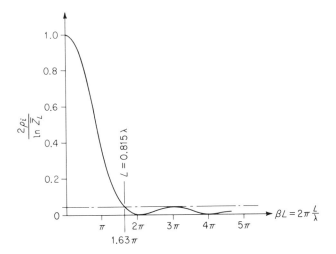

Fig. 5.31 Input reflection coefficient for a taper with a triangular distribution of reflections.

grating (5.74) gives

$$\bar{Z} = \begin{cases} e^{[2(z/L)^2]\ln \bar{Z}_L} & 0 \leq z \leq \dfrac{L}{2} \\ e^{(4z/L - 2z^2/L^2 - 1)\ln \bar{Z}_L} & \dfrac{L}{2} \leq z \leq L \end{cases} \quad (5.75)$$

Substituting (5.74) into (5.71) and performing the straightforward integration give

$$\Gamma_i = \tfrac{1}{2} e^{-j\beta L} \ln \bar{Z}_L \left[\frac{\sin(\beta L/2)}{\beta L/2} \right]^2 \quad (5.76)$$

A plot of ρ_i versus βL is given in Fig. 5.31. Note that, by comparison with the exponential taper, this taper has a first minor-lobe maximum which is less than 5 percent of the major-lobe peak. However, this small value of reflection coefficient occurs for a taper length of about $3\lambda/2$, or for a length twice that for the exponential taper. If $\bar{Z}_L$ is considerably greater than unity, this latter taper will be preferable because of the much smaller values of ρ_i obtained for all frequencies, such that the taper length is greater than 0.815λ, which corresponds to the lower edge of the passband in Fig. 5.31.

★5.13 Synthesis of transmission-line tapers

Equation (5.71) is repeated here for convenience:

$$\Gamma_i(2\beta) = \tfrac{1}{2} \int_0^L e^{-j2\beta z} \frac{d(\ln \bar{Z})}{dz} dz \quad (5.77)$$

This equation may be interpreted as the Fourier transform of a function $d(\ln \bar{Z})/dz$, which is zero outside the range $0 \leq z \leq L$.† As such the Fourier inversion formula gives

$$\frac{1}{2}\frac{d(\ln \bar{Z})}{dz} = \frac{1}{2\pi}\int_{-\infty}^{\infty} e^{j2\beta z}\, \Gamma_i(2\beta) 2\, d\beta \tag{5.78}$$

This formula, in principle, solves the synthesis problem since it gives the required value of $d(\ln \bar{Z})/dz$ to yield the specified $\Gamma_i(2\beta)$. To simplify the discussion to follow, it will be convenient to introduce the following normalized variables:

$$p = 2\pi\frac{z - L/2}{L} \tag{5.79a}$$

$$u = \frac{\beta L}{\pi} = \frac{2L}{\lambda} \tag{5.79b}$$

In this case (5.77) becomes

$$\Gamma_i = \tfrac{1}{2}e^{-j\beta L}\int_{-\pi}^{\pi} e^{-jpu}\,\frac{d(\ln \bar{Z})}{dp}\, dp \tag{5.80}$$

Now define $g(p)$ to be

$$g(p) = \frac{d(\ln \bar{Z})}{dp} \tag{5.81a}$$

and $F(u)$ by

$$F(u) = \int_{-\pi}^{\pi} e^{-jpu} g(p)\, dp \tag{5.81b}$$

Thus

$$\Gamma_i = \tfrac{1}{2}e^{-j\beta L} F(u) \tag{5.82}$$

The Fourier transform pair (5.77) and (5.78) now may be expressed as

$$F(u) = \int_{-\pi}^{\pi} e^{-jpu} g(p)\, dp \tag{5.83a}$$

$$g(p) = \frac{1}{2\pi}\int_{-\infty}^{\infty} e^{jpu} F(u)\, du \tag{5.83b}$$

The synthesis problem may now be stated as follows: Specify a reflection-coefficient characteristic $F(u)$ that will give the desired taper performance and yet be such that the $g(p)$ computed by (5.83b) will be a function identically zero outside the range $|p| > \pi$. This latter restriction corresponds to the physical requirement that $d(\ln \bar{Z})/dz$ be different from zero only in the range $0 \leq z \leq L$. Obviously, any arbitrary $F(u)$ cannot be specified, for in general this would lead to a $g(p)$ that exists

† The Fourier transform relation was first pointed out by F. Bolinder, Fourier Transforms in the Theory of Inhomogeneous Transmission Lines, *Proc. IRE*, vol. 38, p. 1354, November, 1950.

over the whole infinite range $-\infty < p < \infty$. For example, if $F(u)$ were chosen to be equal to unity for $-1 \leq u \leq 1$ and zero otherwise, then (5.83b) would give

$$g(p) = \frac{\sin p}{\pi p} \qquad -\infty < p < \infty$$

To realize such a $g(p)$ would require an infinitely long taper, clearly an impractical solution. Before further progress with the synthesis problem can be made, restrictions to be imposed on $F(u)$, to obtain a physically realizable solution, must be deduced.

In order to derive suitable restrictions on $F(u)$, let $g(p)$ be expanded in a complex Fourier series as follows:

$$g(p) = \begin{cases} \sum_{n=-\infty}^{\infty} a_n e^{jnp} & -\pi \leq p \leq \pi \\ 0 & |p| > \pi \end{cases} \qquad (5.84)$$

where the a_n are as yet unspecified coefficients subject to the restriction $a_n = a_{-n}^*$ so that g will be a real function. Substitution into (5.83a) now gives

$$F(u) = 2\pi \sum_{n=-\infty}^{\infty} a_n \frac{\sin \pi(u-n)}{\pi(u-n)}$$

$$= 2\pi \sum_{n=-\infty}^{\infty} a_n (-1)^n \frac{\sin \pi u}{\pi(u-n)} = 2\pi \frac{\sin \pi u}{\pi u} \sum_{n=-\infty}^{\infty} a_n (-1)^n \frac{u}{u-n} \qquad (5.85)$$

The coefficients a_n can be related to $F(u = n)$, for when u equals an integer n,

$$\lim_{u \to n} \frac{\sin \pi(u-n)}{\pi(u-n)} = 1$$

and

$$\lim_{u \to n} \frac{\sin \pi(u-m)}{\pi(u-m)} = 0$$

Thus $F(n) = 2\pi a_n$, or $a_n = F(n)/2\pi$; so

$$F(u) = \sum_{n=-\infty}^{\infty} F(n) \frac{\sin \pi(u-n)}{\pi(u-n)} \qquad (5.86)$$

This result is a statement of the well-known sampling theorem used in communication theory and states that $F(u)$ is uniquely reconstructed from a knowledge of the sample values of $F(u)$ at $u = n$,

$n = 0, \pm 1, \pm 2, \ldots$

by means of the interpolation formula (5.86). One possible way to restrict $F(u)$ is now seen to be a relaxation on the specification of $F(u)$; that is, specify $F(u)$ at all integer values of u only. This, however, is not an entirely satisfactory solution, because we have no a priori knowledge that if we specify $F(u)$ at the integer values of u only, the resultant $F(u)$ given by (5.86) will be an acceptable reflection-coefficient characteristic for all values of u, even though it can be realized by a $g(p)$ given by (5.84) with $a_n = F(n)/2\pi$.

We should like to obtain greater flexibility in the choice of $F(u)$, and to see how this may be accomplished, let it be assumed that all a_n for $|n| > N$ are zero. In this case

$$g(p) = \sum_{n=-N}^{N} a_n e^{jnp} \tag{5.87a}$$

$$F(u) = 2\pi \frac{\sin \pi u}{\pi u} \sum_{n=-N}^{N} (-1)^n a_n \frac{u}{u-n} \tag{5.87b}$$

The series in (5.87b) can be recognized as the partial-fraction expansion of a function of the form

$$\frac{Q(u)}{\prod_{n=1}^{N} (u^2 - n^2)}$$

where $Q(u)$ is, apart from the restriction $Q(-u) = Q^*(u)$ so that $a_n = a_{-n}^*$, an arbitrary polynomial of degree $2N$ in u, and the denominator is the product of the N terms $(u^2 - 1)$, $(u^2 - 4)$, etc. Using the partial-fraction-expansion formula, we have

$$\frac{Q(u)}{\prod_{n=1}^{N} (u^2 - n^2)} = \sum_{m=-N}^{N}{}' \frac{Q(m)}{(u-m)2m \prod_{\substack{n=1 \\ n \neq m}}^{N} (m^2 - n^2)} + \lim_{u \to \infty} \frac{Q(u)}{u^{2N}}$$

$$= \frac{uQ(u)}{u \prod_{n=1}^{N} (u^2 - n^2)} = \sum_{m=-N}^{N}{}' \frac{uQ(m)}{(u-m)2m^2 \prod_{\substack{n=1 \\ n \neq m}}^{N} (m^2 - n^2)}$$

$$+ \frac{Q(0)}{\prod_{n=1}^{N} (-n^2)} \tag{5.88}$$

where the prime means omission of the term $m = 0$. This is clearly of the same form as the series in (5.87b), with

$$(-1)^m a_m = \frac{Q(m)}{2m^2 \prod_{\substack{n=1 \\ n \neq m}}^{N} (m^2 - n^2)} \tag{5.89a}$$

$$a_0 = \frac{Q(0)}{\prod_{n=1}^{N} (-n^2)} \tag{5.89b}$$

The expression for $F(u)$ now can be written as

$$F(u) = 2\pi \frac{\sin \pi u}{\pi u} \frac{Q(u)}{\prod_{n=1}^{N} (u^2 - n^2)} \tag{5.90}$$

where $Q(u)$ is an arbitrary polynomial of degree $2N$ in u, subject to the restriction $Q(-u) = Q^*(u)$. This result states that the first $2N$ zeros of $\sin \pi u$, which are canceled by the denominator in (5.90), can be replaced by $2N$ new arbitrarily located zeros by proper choice of $Q(u)$. If $g(p)$ were a constant (exponential taper), we should have $F(u)$ proportional to $(\sin \pi u)/\pi u$. But with $2N + 1$ coefficients available in the expansion of $g(p)$, we are at liberty to rearrange $2N$ of the zeros of $(\sin \pi u)/\pi u$ to obtain a more desirable $F(u)$. We have now reduced the synthesis problem to one of specifying an arbitrary polynomial $Q(u)$. To illustrate the theory, two examples are discussed below.

A qualitative insight into how $Q(u)$ should be specified may be obtained by imagining that $F(u)$ is a rubber band stretched horizontally at some height above the βL or u axis. The zeros of $F(u)$ may then be thought of as points at which this rubber band is pinned down to the u axis. If the band is pinned down at a number of closely spaced points, it will not rise much above the u axis in the regions between. The corresponding reflection coefficient will then also be small in this region. At a double zero the band is pinned down in such a fashion that its slope is zero at the point as well. This results in a less rapid increase in the height with distance away from the point. The number of zeros available (the points at which the band may be pinned down) is fixed and equal to those in the $\sin \pi u$ function. The polynomial $Q(u)$ permits only a relocation of these zeros.

With reference to Fig. 5.30, which illustrates the characteristic for an exponential taper, let the zero at $\beta L = \pi$ be moved to 2π to form a double zero at this point. Likewise, let the zero at 3π be moved to form a double zero at 4π, and so on. The function $Q(u)$ that will provide this shift in every other zero so as to produce double zeros at $u = \pm 2, \pm 4, \pm 6, \pm 8$,

Sec. 5.13   Impedance transformation and matching

etc., can be chosen as

$$Q(u) = \prod_{n=1}^{N} (u^2 - 4n^2)^2$$

From (5.90) we obtain

$$F(u) = 2\pi \frac{\sin \pi u}{\pi u} \frac{\prod_{n=1}^{N} (u^2 - 4n^2)^2}{\prod_{n=1}^{N} (u^2 - n^2)}$$

We now wish to let N go to infinity. However, the products do not converge in this case; so we must modify the expression for $F(u)$ to the following:

$$F(u) = C \frac{\sin \pi u}{\pi u} \frac{\prod_{n=1}^{N} (1 - u^2/4n^2)^2}{\prod_{n=1}^{N} (1 - u^2/n^2)}$$

This modification is permissible since Q contains an arbitrary constant multiplier. All constants can be incorporated in the one constant C. If the following infinite-product representations for the sine functions are noted,

$$\frac{\sin \pi u}{\pi u} = \prod_{n=1}^{\infty} \left(1 - \frac{u^2}{n^2}\right)$$

$$\left[\frac{\sin (\pi u/2)}{\pi u/2}\right]^2 = \prod_{n=1}^{\infty} \left(1 - \frac{u^2}{4n^2}\right)^2$$

it is readily seen that as $N \to \infty$ we obtain

$$F(u) = C \left[\frac{\sin (\pi u/2)}{\pi u/2}\right]^2 \qquad (5.91)$$

This is the reflection-coefficient characteristic for the taper with a triangular $d(\ln \bar{Z})/dz$ function discussed earlier. In the present case we have arrived at this solution for a taper with equally spaced double zeros by a direct-synthesis procedure. As Fig. 5.31 shows, the specification of double zeros holds the values of $F(u)$ (that is, ρ_i) to much smaller values in the region between zeros.

As shown earlier, the coefficients a_n in the Fourier series expansion of $g(p)$ are given by

$$a_n = \frac{1}{2\pi} F(n) = \frac{C}{2\pi} \left[\frac{\sin (n\pi/2)}{n\pi/2}\right]^2 \qquad (5.92)$$

from (5.91). The reader may readily verify that the expansion (5.87a) for $g(p)$, with the above coefficients, is a triangular wave. To fix the constant C, we integrate (5.81a) to obtain

$$\int_{-\pi}^{\pi} g(p)\,dp = \int_{-\pi}^{\pi} \frac{d(\ln \bar{Z})}{dp}\,dp = \ln \bar{Z}\,|_{-\pi}^{\pi} = \ln \bar{Z}_L$$

But from (5.84) we have

$$\int_{-\pi}^{\pi} g(p)\,dp = 2\pi a_0$$

and hence

$$a_0 = \frac{1}{2\pi} \ln \bar{Z}_L \tag{5.93}$$

From (5.92) we now find that

$$a_0 = \frac{1}{2\pi} \ln \bar{Z}_L = \frac{C}{2\pi}$$

so $C = \ln \bar{Z}_L$. With this value of C, the reflection coefficient corresponding to the $F(u)$ in (5.91) is easily verified to be the same as that given by (5.76).

As a second example, consider the synthesis of a taper with an $F(u)$ having a triple zero at $u = \pm 2$. This can be accomplished by moving the zeros at ± 1 and ± 3 into the points $u = \pm 2$. The resultant reflection coefficient should remain very small for a considerable region around $|u| = 2$. In the present case $N = 3$, and we choose

$$Q(u) = C(u^2 - 4)^3$$

Thus

$$F(u) = 2\pi C \frac{\sin \pi u}{\pi u} \frac{(u^2 - 4)^3}{(u^2 - 1)(u^2 - 4)(u^2 - 9)}$$

A plot of $|F(u)/F(0)|$ is given in Fig. 5.32. As anticipated, $F(u)$ remains small in a considerable region around the point $|u| = 2$. Since the zeros at $|u| = 3$ have been removed, $F(u)$ reaches a relatively large value at this point. However, for a range of frequencies around which $L \approx \lambda$, this taper represents a very good design.

The coefficients in the expansion for $g(p)$ are given by

$$a_0 = \frac{1}{2\pi} F(0) = \frac{1}{2\pi} \ln \bar{Z}_L$$

$$a_1 = a_{-1} = \frac{1}{2\pi} F(1) = \frac{0.316}{2\pi} \ln \bar{Z}_L$$

$$a_2 = a_{-2} = 0$$

$$a_3 = a_{-3} = \frac{1}{2\pi} F(3) = -\frac{0.098}{2\pi} \ln \bar{Z}_L$$

$$a_n = a_{-n} = 0 \qquad n > 3$$

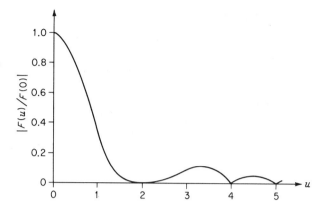

Fig. 5.32 Reflection-coefficient characteristic for a taper with a triple zero at $|u| = 2$.

Hence

$$g(p) = \frac{d(\ln \bar{Z})}{dp} = \frac{\ln \bar{Z}_L}{2\pi}(a_0 + 2a_1 \cos p + 2a_3 \cos 3p)$$

$$= \frac{\ln \bar{Z}_L}{2\pi}(1 + 0.632 \cos p - 0.196 \cos 3p)$$

Integrating gives

$$\ln \bar{Z} = \frac{\ln \bar{Z}_L}{2\pi}(p + 0.632 \sin p - 0.0653 \sin 3p) + C$$

The constant of integration C is determined from the requirement that $\ln \bar{Z} = 0$ at $p = -\pi$, or $\ln \bar{Z}_L$ at $p = \pi$; thus $C = \tfrac{1}{2} \ln \bar{Z}_L$ and

$$\ln \bar{Z} = \frac{\ln \bar{Z}_L}{2\pi}(p + \pi + 0.632 \sin p - 0.0653 \sin 3p)$$

Replacing p by $2\pi(z - L/2)/L$ from (5.79a) now specifies $\ln \bar{Z}$ as a function of z and completes the design of the taper.

The foregoing synthesis procedure must be used with some caution, stemming from the approximations involved. The theory is valid as long as $d(\ln \bar{Z})/dz$ is small; that is, $\ln \bar{Z}$ must be a slowly varying function of z in order that the reflection coefficient everywhere along the taper be small; that is, $|\Gamma(z)|^2 \ll 1$. This means that $|g(p)|$, and hence all $|a_n|$, must not be permitted to assume excessively large values. Consequently, $|F(n)|$ must not be permitted to become too large. If too many zeros are closely grouped together around a particular value of u, then outside this range $F(u)$ may become excessively large and the accuracy of the theory will suffer. Such "supermatched" designs must be avoided.

★5.14 Chebyshev taper

If the number of sections in a Chebyshev transformer is increased indefinitely, with the overall length L kept fixed, we obtain a Chebyshev taper. This taper has equal-amplitude minor lobes and is an optimum design in the sense that it gives the smallest minor-lobe amplitudes for a fixed taper length, and conversely, for a specified minor-lobe amplitude it has the shortest length. As such it is a good taper by which to judge how far other tapers depart from an optimum design. It has been shown that, in the limit as the number of sections in a Chebyshev transformer goes to infinity, the reflection coefficient becomes†

$$\Gamma_i = \tfrac{1}{2} e^{-j\beta L} \ln \bar{Z}_L \frac{\cos L \sqrt{\beta^2 - \beta_0^2}}{\cosh \beta_0 L} \qquad (5.94)$$

where β_0 is the value of β at the lower edge of the passband, as illustrated in Fig. 5.33. As β increases from zero to β_0, the magnitude ρ_i of Γ_i decreases to a final value of $(\ln \bar{Z}_L)/(2 \cosh \beta_0 L)$, since in this region $\cos L \sqrt{\beta^2 - \beta_0^2} = \cosh L \sqrt{\beta_0^2 - \beta^2}$. Beyond this point the function in the numerator is the cosine function that oscillates between ± 1 and produces the equal-amplitude minor lobes. The major-lobe to minor-lobe amplitude ratio equals $\cosh \beta_0 L$. Hence, if this is specified so as to keep ρ_i less than or equal to some maximum value ρ_m in the passband, the taper length L is fixed for a given choice of the frequency of the lower edge of the passband which determines β_0. We have

$$\cosh \beta_0 L = \frac{\ln \bar{Z}_L}{2\rho_m} \qquad (5.95)$$

Conversely, if β_0 and the taper length L are given, the passband tolerance ρ_m is fixed.

† R. E. Collin, The Optimum Tapered Transmission Line Matching Section, *Proc. IRE*, vol. 44, pp. 539–548, April, 1956.

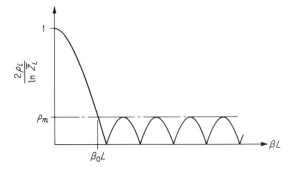

Fig. 5.33 Reflection-coefficient characteristic for a Chebyshev taper.

The theory given earlier may be used to determine the function $g(p)$ that will produce the reflection coefficient given by (5.94). Introducing the u variable again, we find that the function $F(u)$ is

$$F(u) = (\ln \bar{Z}_L) \frac{\cos \pi \sqrt{u^2 - u_0^2}}{\cosh \pi u_0} \tag{5.96}$$

where $\pi u = \beta L$, $\pi u_0 = \beta_0 L$. The function $\cos \pi \sqrt{u^2 - u_0^2}$ can be expressed in infinite-product form as

$$\cos \pi \sqrt{u^2 - u_0^2} = \cosh \pi u_0 \prod_{n=1}^{\infty} \left[1 - \frac{u^2}{u_0^2 + (n - \tfrac{1}{2})^2} \right]$$

and this is the limiting value of the polynomial $Q(u)$ in (5.90) as $N \to \infty$. The $\sin \pi u$ term has been canceled by the infinite product $\prod_{n=1}^{\infty}(1 - u^2/n^2)$ as in the first example presented on taper synthesis. However, we do not need this product expansion since $2\pi a_n = F(n)$ in any case. From (5.96) we have

$$a_n = a_{-n} = \frac{1}{2\pi} F(n) = \frac{\ln \bar{Z}_L}{2\pi} \frac{\cos \pi \sqrt{n^2 - u_0^2}}{\cosh \pi u_0} \tag{5.97}$$

Thus

$$g(p) = \frac{\ln \bar{Z}_L}{2\pi \cosh \pi u_0} (\cosh \pi u_0 + 2 \cos \pi \sqrt{1 - u_0^2} \cos p$$
$$+ 2 \cos \pi \sqrt{4 - u_0^2} \cos 2p + \cdots)$$
$$= \frac{\ln \bar{Z}_L}{2\pi \cosh \pi u_0} \left(\cosh \pi u_0 + 2 \sum_{n=1}^{\infty} \cos \pi \sqrt{n^2 - u_0^2} \cos np \right) \tag{5.98}$$

Integrating with respect to p gives

$$\ln \bar{Z} = \frac{\ln \bar{Z}_L}{2\pi \cosh \pi u_0} \left(p \cosh \pi u_0 + 2 \sum_{n=1}^{\infty} \frac{\cos \pi \sqrt{n^2 - u_0^2}}{n} \sin np \right) + C \tag{5.99}$$

where C is a constant of integration. To render this result more suitable for computation, it is expedient to add and subtract a similar series; i.e.,

$$\ln \bar{Z} = \frac{p \ln \bar{Z}_L}{2\pi} + \frac{\ln \bar{Z}_L}{\pi \cosh u_0 \pi} \sum_{n=1}^{\infty} \frac{\cos n\pi}{n} \sin np$$
$$+ \frac{\ln \bar{Z}_L}{\pi \cosh u_0 \pi} \sum_{n=1}^{\infty} \frac{\cos \pi \sqrt{n^2 - u_0^2} - \cos n\pi}{n} \sin np + C$$

The second series converges rapidly because $\cos \pi \sqrt{n^2 - u_0^2}$ approaches $\cos n\pi$ as n becomes large. The first series may be recognized as the

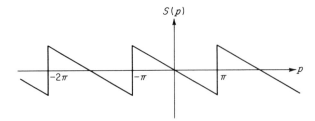

Fig. 5.34 Sawtooth function.

Fourier sine series for the sawtooth function $S(p)$,

$$S(p) = \begin{cases} -\dfrac{p}{2\pi}\dfrac{\ln \bar{Z}_L}{\cosh \pi u_0} & -\pi < p < \pi \\ 0 & p = \pm\pi \end{cases} \qquad (5.100)$$

and the periodic continuation of this function as illustrated in Fig. 5.34. Hence we obtain

$$\ln \bar{Z} = \frac{p \ln \bar{Z}_L}{2\pi} + S(p)$$

$$+ \frac{\ln \bar{Z}_L}{\pi \cosh \pi u_0} \sum_{n=1}^{\infty} \frac{\cos \pi \sqrt{n^2 - u_0^2} - \cos n\pi}{n} \sin np + C$$

At $p = \pi$, we have $\ln \bar{Z} = \ln \bar{Z}_L$, and since $S(p)$ and $\sin np$ are zero at this point, $C + \tfrac{1}{2}\ln \bar{Z}_L = \ln \bar{Z}_L$, or $C = \tfrac{1}{2}\ln \bar{Z}_L$. Our final result is

$$\ln \bar{Z} = \left(\frac{p}{2\pi} + \frac{1}{2} - \frac{p}{2\pi \cosh \pi u_0}\right) \ln \bar{Z}_L$$

$$+ \frac{\ln \bar{Z}_L}{\pi \cosh \pi u_0} \sum_{n=1}^{\infty} \frac{\cos \pi \sqrt{n^2 - u_0^2} - \cos n\pi}{n} \sin np \qquad (5.101)$$

for $-\pi < p < \pi$. An interesting feature of the above result is that $\ln \bar{Z}$ changes in a stepwise fashion from 0 to $(\ln \bar{Z}_L)/(2 \cosh \pi u_0)$ as p changes from $-\pi - \epsilon$ to $-\pi + \epsilon$, where $\epsilon \ll 1$. Likewise, at the other end of the taper, $\ln \bar{Z}$ changes abruptly from a value $\ln \bar{Z}_L - (\ln \bar{Z}_L)/(2 \cosh \pi u_0)$ to $\ln \bar{Z}_L$ as the point $p = \pi$ is passed. This means that the optimum taper has a step change in impedance at each end. The physical basis for this is readily understood by noting that when the frequency is very high, so that the taper is many wavelengths long, the reflection from the smooth part of the taper vanishes. Thus, in order still to maintain equal-amplitude minor lobes, the two-step changes in impedance must be provided to give a reflection coefficient

$$\rho_i = \frac{\ln \bar{Z}_L}{2} \frac{\cos \beta L}{\cosh \beta_0 L} \qquad \text{for } \beta \gg \beta_0$$

As an indication of the superiority of the Chebyshev taper, computations show that it is 27 percent shorter than the taper with $d(\ln \bar{Z})/dz$ a triangular function, for the same passband tolerance and lower cutoff frequency. If the tapers are made the same length, the Chebyshev taper provides a major- to minor-lobe ratio of 84 as compared with 21 for the taper with a triangular distribution.

★5.15 Exact equation for the reflection coefficient

The basic equation (5.71) for the input reflection coefficient Γ_i was derived by neglecting all multiple reflections between individual differential sections. The exact equation, derived below, enables an estimate of the range of validity of the approximate theory to be made. First of all, the differential equation describing the total reflection coefficient Γ at any point z on the line, according to the approximate theory, is derived for later comparison.

With reference to Fig. 5.35, let $d\Gamma_0$ be the reflection coefficient arising from the change $d\bar{Z}$ in characteristic impedance in the interval dz at z. This differential reflection was shown earlier to be given by [see (5.70)]

$$d\Gamma_0 = \frac{1}{2} \frac{d(\ln \bar{Z})}{dz} dz$$

The total reflection coefficient at z is the sum of all differential contributions $d\Gamma_0$ from z to L and is

$$\Gamma(z) = \frac{1}{2} \int_z^L e^{-j2\beta(u-z)} \frac{d(\ln \bar{Z})}{du} du$$

where u is a dummy variable that measures the distance from the point $z = 0$ toward the load end. The phase angle of the reflected wave arising at u is $2\beta(z - u)$ relative to the forward propagating wave at z.

Differentiating $\Gamma(z)$ with respect to z gives

$$\begin{aligned} \frac{d\Gamma}{dz} &= \frac{2j\beta}{2} \int_z^L e^{j2\beta(z-u)} \frac{d(\ln \bar{Z})}{du} du - \tfrac{1}{2} e^{-j2\beta(u-z)} \frac{d(\ln \bar{Z})}{du} \bigg|_{u=z} \\ &= 2j\beta\Gamma - \frac{1}{2} \frac{d(\ln \bar{Z})}{dz} \end{aligned} \qquad (5.102)$$

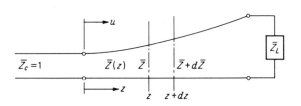

Fig. 5.35 Tapered transmission line.

This is the approximate differential equation for the total reflection coefficient at any point z along the taper.

To find the exact differential equation for Γ, let $\bar{Z}_{in}$ be the input impedance at z and $\bar{Z}_{in} + d\bar{Z}_{in}$ be the input impedance at $z + dz$. We then have

$$\bar{Z}_{in} = \bar{Z} \frac{\bar{Z}_{in} + d\bar{Z}_{in} + j\bar{Z} \tan(\beta\, dz)}{\bar{Z} + j(\bar{Z}_{in} + d\bar{Z}_{in}) \tan(\beta\, dz)}$$

$$\approx \bar{Z} \frac{\bar{Z}_{in} + d\bar{Z}_{in} + j\bar{Z}\beta\, dz}{\bar{Z} + j(\bar{Z}_{in} + d\bar{Z}_{in})\beta\, dz}$$

$$\approx (\bar{Z}_{in} + d\bar{Z}_{in} + j\bar{Z}\beta\, dz)\left(1 - j\frac{\bar{Z}_{in}}{\bar{Z}} \beta\, dz\right)$$

$$\approx \bar{Z}_{in} + d\bar{Z}_{in} + j\beta\bar{Z}\, dz - j\frac{\bar{Z}_{in}^2}{\bar{Z}} \beta\, dz$$

upon replacing the tangent by its argument $\beta\, dz$, neglecting products of differential terms, and replacing the denominator by

$$[\bar{Z} + j(\bar{Z}_{in}\beta\, dz)]^{-1} \approx \bar{Z}^{-1}\left(1 - j\frac{\bar{Z}_{in}}{\bar{Z}} \beta\, dz\right)$$

The above gives

$$\frac{d\bar{Z}_{in}}{dz} = j\beta\left(\frac{\bar{Z}_{in}^2}{\bar{Z}} - \bar{Z}\right)$$

Now we also have

$$\bar{Z}_{in} = \frac{1+\Gamma}{1-\Gamma} \bar{Z}$$

and hence

$$\frac{d\bar{Z}_{in}}{dz} = \frac{1+\Gamma}{1-\Gamma} \frac{d\bar{Z}}{dz} + \frac{2\bar{Z}}{(1-\Gamma)^2} \frac{d\Gamma}{dz}$$

Combining these two equations for $d\bar{Z}_{in}/dz$ and replacing $\bar{Z}_{in}$ by $\bar{Z}(1+\Gamma)/(1-\Gamma)$ finally give

$$\frac{d\Gamma}{dz} = 2j\beta\Gamma - \tfrac{1}{2}(1-\Gamma^2)\frac{d(\ln \bar{Z})}{dz} \tag{5.103}$$

upon using the relation $\bar{Z}^{-1}\, d\bar{Z}/dz = d(\ln \bar{Z})/dz$. If we compare (5.103) with (5.102), we see that the approximate equation differs only by the factor $1 - \Gamma^2$ that multiplies the $d(\ln \bar{Z})/dz$ term in the exact equation. If $|\Gamma^2| \ll 1$ everywhere along the line, the approximate equation would be expected to yield good results.

The exact equation (5.103) is called a Riccati equation. It is a nonlinear equation because of the term in Γ^2 and does not have a known general solution. This equation can be integrated only in certain special cases (one such case is for the exponential taper, as in Prob. 5.28). However, the practical difficulty in applying the exact theory does not stem so much from the lack of a general solution of (5.103), since numerical integration or iteration schemes can always be employed, as from being unable to specify what the characteristic impedance $\bar{Z}$ is along a general taper. If the taper is very gradual, then $\bar{Z}(z)$ can be taken as the characteristic impedance of a uniform line having the same cross-sectional dimensions as the taper has at the plane z. But for such gradual tapers the approximate theory is valid and gives good results, so that the more complicated Riccati equation is not required. For more rapidly varying tapers, the field structure along the taper is perturbed to such an extent that no simple method of specifying $\bar{Z}(z)$ exists. In actual fact, the new boundary-value problem will have to be solved to determine $\bar{Z}(z)$, and this solution will also provide the values of the reflection coefficient along the taper. Thus one concludes that the inability to specify $\bar{Z}(z)$, except for the case of gradual tapers, makes the exact equation of minor importance in practice.

In the case of waveguide tapers, β is a function of the cross-sectional geometry and hence a function of z along the taper. In order to integrate the approximate equation (5.102), in this case, an auxiliary variable θ is introduced as follows:

$$\theta = \int_0^z 2\beta(z)\,dz \qquad d\theta = 2\beta\,dz$$

and hence

$$\frac{d}{dz} = \frac{d}{d\theta}\frac{d\theta}{dz} = 2\beta\frac{d}{d\theta}$$

We now have

$$\frac{d\Gamma}{dz} = 2\beta\frac{d\Gamma}{d\theta} = 2j\beta\Gamma - \beta\frac{d(\ln \bar{Z})}{d\theta}$$

or

$$\frac{d\Gamma}{d\theta} = j\Gamma - \frac{1}{2}\frac{d(\ln \bar{Z})}{d\theta} \tag{5.104}$$

This is readily integrated to give (multiply both sides by $e^{-j\theta}$ and note that $d\Gamma e^{-j\theta}/d\theta = -j\Gamma e^{-j\theta} + e^{-j\theta}d\Gamma/d\theta$)

$$\Gamma(\theta)e^{-j\theta}\Big|_0^{\theta_0} = -\tfrac{1}{2}\int_0^{\theta_0} e^{-j\theta}\frac{d(\ln \bar{Z})}{d\theta}\,d\theta$$

where

$$\theta_0 = \int_0^L 2\beta \, dz$$

If $\bar{Z} = \bar{Z}_L$ at $z = L$, then $\Gamma(\theta_0) = 0$, and since $\Gamma(0) = \Gamma_i$, we have

$$\Gamma_i = \tfrac{1}{2} \int_0^{\theta_0} e^{-j\theta} \frac{d(\ln \bar{Z})}{d\theta} \, d\theta \qquad (5.105)$$

In terms of the new variable θ, the problem is formally the same as before. However, the specification of $\bar{Z}(\theta)$ as a function of θ may be rather difficult to translate back into a function of z. Consequently, a general synthesis procedure applied to (5.105) may be difficult to carry out, although in principle it is formally the same problem as considered earlier, except for the last step, i.e., expressing $\bar{Z}(\theta)$ as a function of z in order to determine the shape of the taper.

Problems

5.1 Show that the R = constant and X = constant contours in the reflection-coefficient plane are circles given by (5.6a) and (5.6b).

5.2 In the following circuit, what is the smallest value of d that will make the resistive part of Z_{in} equal to 50 ohms at the plane t_1? Find the required value of jX to make Z_{in} equal to 50 ohms at the plane t_2.

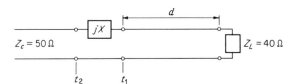

Fig. P 5.2

5.3 On a certain line terminated in a normalized load impedance $\bar{Z}_L$ it was found that the standing-wave ratio S was equal to 2 and that a voltage minimum occurred at $\lambda/4$ from the load. What is $\bar{Z}_L$? Find the position and length of a single shunt stub that will match the load to the line.

5.4 What are the length and position required for a series stub to match the load specified in Prob. 5.3?

5.5 A normalized load $\bar{Z}_L = 2$ terminates a transmission line. Does a voltage maximum or minimum occur at the load? Find the position and length of a shunt stub that will match the load to the line at a frequency f_1, where the wavelength is λ_1. With the stub parameters and the load fixed, let the wavelength be increased to $1.1\lambda_1$. What is the new value of stub susceptance and the standing-wave ratio on the line? If at the wavelength λ_1 the stub is placed $\lambda_1/2$ farther toward the generator from its original position, what is the standing-wave ratio on the line when the wavelength is increased to $1.1\lambda_1$? This illustrates the greater frequency sensitivity of the match when the stub is placed farther from the load.

5.6 A double stub with spacing 0.25λ is used to match a normalized load admittance $\bar{Y}_L = 0.5 + j1$. Find the required stub susceptances.

5.7 Can a normalized load admittance $\bar{Y}_L = 2.5 + j1$ be matched with a double-stub tuner with stub spacing $\lambda/10$?

5.8 What is the minimum stub spacing in a double-stub tuner that will permit a normalized load admittance $\bar{Y}_L = 1.5 + j2$ to be matched if the spacing is restricted to be greater than $\lambda/4$? If the spacing must be less than $\lambda/4$, what is the maximum stub spacing that can be used?

5.9 Show that, for a double-stub tuner with $d = \lambda/4$, the required values of stub susceptances are given by

$$B_1 = -B_L \pm [G_L(1 - G_L)]^{\frac{1}{2}}$$

$$B_2 = \pm \left(\frac{1 - G_L}{G_L}\right)^{\frac{1}{2}}$$

Hint: Take the limit of (5.20) and (5.21) as $\tan \beta d$ becomes infinite.

5.10 A normalized load $\bar{Y}_L = 4 + j4$ is to be matched with a double-stub tuner with a spacing $d = 0.49\lambda$. Use (5.20) and (5.21) to find the required stub susceptances. Find the standing-wave ratio S on the line if the frequency is now increased so that λ becomes equal to λ_1, for which $d = 0.5\lambda_1$. This problem illustrates the high-frequency sensitivity associated with stub spacings near $\lambda/2$.

5.11 A line with an attenuation constant $\alpha = 0.01$ neper/m is used as a short-circuited stub. Using the formula $Y_{in} = Y_c \coth(\alpha + j\beta)l$, find the maximum value of normalized susceptance this stub can give. If $\alpha = 0.02$, what is the maximum value of stub susceptance that can be obtained?

5.12 Consider the series-shunt matching circuit illustrated. What values of series stub reactance and shunt stub susceptance will match an arbitrary load $\bar{Y}_L = \bar{G}_L + j\bar{B}_L$ a distance d away?

Hint: Note that $\bar{Y}_{in}$ at the position of the shunt stub equals $(j\bar{X} + \bar{Z}'_{in})^{-1}$, where $\bar{Z}'_{in}$ is the input impedance just to the right of the series stub. Next impose the condition $\bar{Y}_{in} = 1 + j\bar{B}$ to find the required value of $j\bar{X}$.

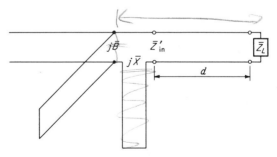

Fig. P 5.12

5.13 Develop a graphical method, using the Smith chart, to solve Prob. 5.12.

Hint: $\bar{Y}_{in}$ must lie on the $\bar{G} = 1$ circle. But $\bar{Y}_{in}$ is obtained by moving 180° around the chart from the $\bar{Z}_{in} = j\bar{X} + \bar{Z}'_{in}$ point. Hence, to make $\bar{Y}_{in}$ lie on the $\bar{G} = 1$ circle, choose $j\bar{X}$ to make $\bar{Z}_{in}$ lie on the $\bar{R} = 1$ circle rotated through 180°.

5.14 A horn antenna is fed from a rectangular waveguide in which the TE_{10} mode propagates. At the junction a reflection coefficient $\Gamma = 0.3e^{j\pi/4}$ is produced. What is the normalized input admittance to the horn? Find the required normalized susceptance and the spacing in guide wavelengths from the junction of an inductive diaphragm that will match the waveguide to the horn.

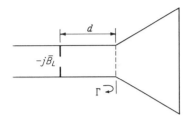

Fig. P 5.14

5.15 What are the required length and impedance of a quarter-wave transformer that will match a 100-ohm load to a 50-ohm line at $f = 10{,}000$ Mc (air-filled line)? What is the frequency band of operation over which the reflection coefficient remains less than 0.1?

5.16 Design a two-section binomial transformer to match the load given in Prob. 5.15. What bandwidth is obtained for $\rho_m = 0.1$?

5.17 Design a three-section binomial transformer to match a 100-ohm load to a 50-ohm line. The maximum VSWR that can be tolerated is 1.1. What bandwidth can be obtained? Plot ρ versus f.

5.18 Design a two-section Chebyshev transformer to match a 100-ohm load to a 50-ohm line. The maximum voltage standing-wave ratio (VSWR) that can be tolerated is 1.2. What bandwidth is obtained? Plot ρ versus f. Use the approximate theory.

★5.19 Use the exact theory to design a two-section Chebyshev transformer to match a normalized load $\bar{Z}_L = 5$ to a transmission line. The required fractional bandwidth is 0.6. What is the resultant value of ρ_m?

★5.20 Design a three-section Chebyshev transformer (exact theory) to match the load specified in Prob. 5.19. The same bandwidth is required. Compute ρ_m and note the improvement obtained.

★5.21 Let x_s be a zero of $T_N(x)$; that is, $T_N(x_s) = 0$. Let the corresponding value of $\cos\theta$ be $\cos\theta_z$. Note that $T_N(\sec\theta_m \cos\theta) = 0$ when $\sec\theta_m \cos\theta = \sec\theta_m \cos\theta_z$. Use the result

$$T_N(\cos\phi) = \cos N\phi$$

to compute the zeros of $T_2(x)$. Note that the zeros occur when $\cos 2\phi = 0$, or $\phi = \pi/4$, $3\pi/4$, and that $x_s = \cos\phi_z$. Using these results together with the relation $\sec\theta_m = x_s \sec\theta_z$ show that, for a two-section transformer,

$$P_{LR} = 1 + k^2 T_2{}^2(\sec\theta_m \cos\theta)$$

reduces to (5.59). Note that when $\theta = 0$, $\rho^2 = (Z_L - Z_0)^2/(Z_L + Z_0)^2$, and hence

$$k^2 = \frac{(Z_L - Z_0)^2}{4 Z_L Z_0 T_2{}^2(\sec\theta_m)}$$

5.22 For a particular application it is desired to obtain a reflection-coefficient characteristic

$$\rho_i = C\left(\theta - \frac{\pi}{2}\right)^2 \qquad 0 \leq \theta \leq \pi$$

The normalized load impedance equals 2. What must the value of C be? Use the approximate theory for an N-section transformer to design a four-section transformer that will approximate the above specified ρ_i. Expand ρ_i in a Fourier cosine series to determine the coefficients ρ_n. Plot the resultant ρ_i versus θ and compare with the specified characteristic. Show that the approximation to the specified ρ_i is a least-mean-square-error approximation. If the number of sections in the transformer were increased indefinitely, could the specified ρ_i be realized?

5.23 An empty rectangular waveguide is to be matched to a dielectric-filled rectangular guide by means of an intermediate quarter-wave transformer. Find the length and dielectric constant of the matching section. Use $a = 2.5$ cm, $f = 10{,}000$ Mc, $\kappa_0 = 2.56$. Plot ρ_i versus frequency f. Note that the appropriate impedances to use here are the wave impedances for the TE_{10} mode. κ_0 is the dielectric constant in the output guide.

Fig. P 5.23

5.24 Design a two-section binomial transformer to match the empty guide to the dielectric-filled guide in Prob. 5.23. Use intermediate sections with dielectric constants κ_1 and κ_2. Plot the input reflection coefficient as a function of frequency.

5.25 Obtain an expression for ρ_i for a taper which has $d(\ln Z)/dz = C \sin \pi(z/L)$. Determine C so that $\ln Z = \ln Z_L$ at $z = L$ and equals zero at $z = 0$. Plot ρ_i versus βL.

★5.26 Design a taper that has double zeros at $\beta L = \pm 2\pi$ and $\pm 3\pi$. Plot $F(u)$ versus u for this taper. Determine $\ln Z$ as a function of z. Achieve the design by moving the zeros at $\pm \pi$ and $\pm 4\pi$ into the points at $\pm 2\pi$, $\pm 3\pi$.

★5.27 Design a taper with single zeros at $\beta L = \pm \pi$, $\pm 1.25\pi$, $\pm 1.5\pi$, $\pm 1.75\pi$ by moving the zeros at $\pm 2\pi$, $\pm 3\pi$, $\pm 4\pi$ into the specified points. Plot the resultant $F(u)$. Determine the expansion for $\ln Z$ as a function of z for this taper. Note that the close spacing of zeros keeps ρ_i small in the range $\pi < \beta l < 2\pi$, but that removal of the zeros at $\pm 3\pi$ and $\pm 4\pi$ lets ρ_i rise to a large value in this region.

★5.28 Show that, for the exponential taper, the exact solution of (5.103) for the input reflection coefficient is

$$\Gamma_i = \frac{A \sin (BL/2)}{B \cos (BL/2) + 2j\beta \sin (BL/2)}$$

where $A = (\ln Z_L)/L$, $B = \sqrt{4\beta^2 - A^2}$.

★5.29 Show that the approximate differential equation for the input impedance along a slowly varying tapered line is ($Z_\text{in} \approx Z$ at all points)

$$\frac{dZ_\text{in}}{dz} = 2j\beta(Z - Z_\text{in})$$

Integrate this to obtain the input impedance to an exponential line terminated in a load impedance Z_L.

References

Quarter-wave transformers

1. Collin, R. E.: Theory and Design of Wide Band Multisection Quarter-wave Transformers, *Proc. IRE*, vol. 43, pp. 179–185, February, 1955.
2. Cohn, S. B.: Optimum Design of Stepped Transmission Line Transformers, *IRE Trans.*, vol. MTT-3, pp. 16–21, April, 1955.
3. Riblet, H. J.: General Synthesis of Quarter-wave Impedance Transformers, *IRE Trans.*, vol. MTT-5, pp. 36–43, January, 1957.
4. Young, L.: Tables for Cascaded Homogeneous Quarter-wave Transformers, *IRE Trans.*, vol. MTT-7, pp. 233–237, April, 1959. See also *IRE Trans.*, vol. MTT-8, pp. 243–244, for corrections.
5. Solymar, L.: Some Notes on the Optimum Design of Stepped Transmission Line Transformers, *IRE Trans.*, vol. MTT-6, pp. 374–378, October, 1958.

Tapered lines

6. Bolinder, F.: Fourier Transforms in the Theory of Inhomogeneous Transmission Lines, *Proc. IRE*, vol. 38, p. 1354, November, 1950.
7. Klopfenstein, R. W.: A Transmission Line Taper of Improved Design, *Proc. IRE*, vol. 44, pp. 31–35, January, 1956.
8. Collin, R. E.: The Optimum Tapered Transmission Line Matching Section, *Proc. IRE*, vol. 44, pp. 539–548, April, 1956.
9. Bolinder, F.: Fourier Transforms in the Theory of Inhomogeneous Transmission Lines, *Trans. Roy. Inst. Technol., Stockholm*, no. 48, 1951.
10. Matsumaru, K.: Reflection Coefficient of E-plane Tapered Waveguides, *IRE Trans.*, vol. MTT-6, pp. 143–149, April, 1958.
11. Johnson, R. C.: Design of Linear Double Tapers in Rectangular Waveguides, *IRE Trans.*, vol. MTT-7, pp. 374–378, July, 1959.

6
Passive microwave devices

This chapter examines a number of microwave devices that are in common use both in the laboratory and in microwave communication systems. For some of the devices examined the treatment is only qualitative in nature. Of particular interest are the nonreciprocal ferrite devices discussed in the latter part of the chapter.

6.1 Terminations

Two types of waveguide and transmission-line terminations are in common use. One is the matched load, and the other is a variable short circuit that produces an adjustable reactive load. These terminations are extensively used in the laboratory when measuring the impedance or scattering parameters of a microwave circuit element. The matched load provides a termination that absorbs all the incident power, and hence is equivalent to terminating the line in its characteristic impedance. The variable short circuit is a termination that reflects all the incident power. The phase of the reflected wave is varied by changing the position of the short circuit, and this is equivalent to changing the reactance of the termination.

The usual matched load for a waveguide is a tapered wedge or slab of lossy material inserted into the guide, as illustrated in Fig. 6.1. Since the material is lossy, the incident power is absorbed. Reflections are avoided by tapering the lossy material into a wedge. Thus the termination may be viewed as a lossy tapered transmission line. An overall length of one or more wavelengths is usually sufficient to provide a matched load with an input standing-wave ratio of 1.01 or smaller.

Variable short circuit

The simplest form of adjustable short circuit for use in a waveguide is a sliding block of copper or some other good conductor that makes a snug fit in the guide, as illustrated in Fig. 6.2. The position of the block is varied by means of a micrometer drive. This simple form of adjustable short circuit, however, is not very satisfactory in its electrical performance. The erratic contact between the sliding block and the waveguide walls causes the equivalent electrical short-circuit position to deviate in a random fashion from the physical short-circuit position which is the front face of the sliding block. In addition, some power leakage past the block may occur, and this results in a reflection coefficient less than unity. These problems may be overcome by using a choke-type plunger, as discussed below.

The choke-type plunger is an example of the use of the impedance transformation properties of a quarter-wave transformer. Consider, for example, a load impedance Z_s that is approximately zero. If this impedance is viewed through a two-section quarter-wave transformer, as in Fig. 6.3, the impedance seen at the input is

$$Z'_s = \left(\frac{Z_1}{Z_2}\right)^2 Z_s \tag{6.1}$$

If Z_2 is chosen much greater than Z_1, the new impedance Z'_s will approximate a short circuit by a factor $(Z_1/Z_2)^2$ better than Z_s does. This is essentially the principle used in choke-type plungers. The improvement factor of course deteriorates when the frequency is changed, so that the transformer sections are no longer a quarter wave long. However, by proper design, a bandwidth of 10 percent or more can be achieved. In very critical applications more than two sections may be used. The

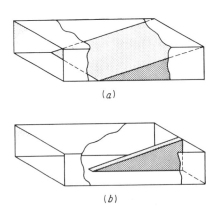

Fig. 6.1 Matched loads for a waveguide. (a) Lossy wedge; (b) tapered resistive card.

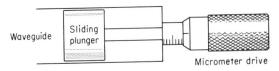

Fig. 6.2 A simple short circuit for a waveguide.

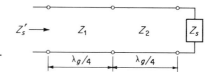

Fig. 6.3 Two-section quarter-wave transformer.

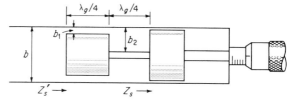

Fig. 6.4 Choke-type adjustable short circuit (side view).

foregoing theory is now applied to a choke-type plunger for use in a rectangular guide.

For the TE_{10} dominant mode in a rectangular guide, the surface currents on the interior wall flow up and down along the side walls and both across and in the axial direction on the broad walls. In the simple type of short circuit illustrated in Fig. 6.2, the axial current must flow across the gap between the upper and lower waveguide walls and vertically across the front face of the sliding block. The currents that flow along the side walls flow in the vertical direction and do not need to cross the gap between the waveguide walls and the front face of the plunger. Consequently, the erratic performance arises only from the axial currents flowing on the upper and lower walls. To avoid this erratic behavior, the plunger may be made in the form illustrated in Fig. 6.4. The width of the plunger is uniform and slightly less than the interior guide width. However, the height of the plunger is made nonuniform. The front section is a quarter guide wavelength long and less than the guide height b by an amount $2b_1$. The gap b_1 is made as small as possible consistent with the requirement that the front section must not touch the upper and lower waveguide walls. The second section is also a quarter guide wavelength long, but with the gap b_2 made as large as possible consistent with maintaining the mechanical strength of the plunger. The final back section makes a sliding fit in the guide. The quarter-wave sections have equivalent characteristic impedances proportional to $2b_1/b$ and $2b_2/b$ relative to that of the input guide. Thus application of (6.1) gives

$$Z'_s = \left(\frac{b_1}{b_2}\right)^2 Z_s$$

for the normalized input impedance. By good mechanical design b_2 can be made about ten times greater than b_1, and hence an improvement in performance by a factor of 100 over the non-choke-type short circuit may be achieved.

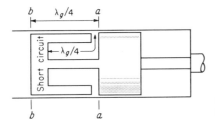

Fig. 6.5 Alternative choke-type plunger design.

A somewhat different design, as illustrated in Fig. 6.5, is also frequently used. In this plunger a two-section folded quarter-wave transformer is used. The inner line transforms the short-circuit impedance to an ideal open circuit at the plane *aa*. At this point, i.e., at an open-circuit or infinite impedance point, the axial current is zero. Hence there is no current present to flow across the gap between the waveguide wall and the plunger at the contact point *aa*. The next, or outer, quarter-wave transformer transforms the open-circuit impedance at *aa* into a short-circuit impedance at the front end of the plunger, i.e., at plane *bb* in Fig. 6.5. Short-circuit plungers of this type give very satisfactory performance.

The above application of quarter-wave transformers is also used in the construction of choke joints for joining two waveguide sections together, in rotary joints, for plungers used to tune cavity resonators, etc.†

6.2 Attenuators

Attenuators may be of the fixed or the variable type. The first is used only if a fixed amount of attenuation is to be provided. For bridge set-ups used to measure transmission coefficients, the variable attenuator is used. There are many ways of constructing a variable attenuator; only one type, the rotary attenuator, is considered in detail. A simple form of attenuator consists of a thin tapered resistive card, of the type used for matched loads, whose depth of penetration into the waveguide is adjustable. The card is inserted into the guide through a longitudinal slot cut in the center of the broad wall of a rectangular guide. An attenuator of this form has a rather complicated attenuation variation with depth of insertion and frequency.

A better precision type of attenuator utilizes an adjustable length of waveguide operated below its cutoff frequency. The disadvantage of this type of attenuator is that the output is attenuated by reducing the coupling between the input and output guides, and not by absorption of the incident power. As a result, a high degree of attenuation corresponds to a reflection coefficient near unity in the input guide, and this is often undesirable.

† See, for example, G. L. Ragan (ed.), "Microwave Transmission Circuits," McGraw-Hill Book Company, New York, 1948.

Perhaps the most satisfactory precision attenuator developed to date is the rotary attenuator, which we now examine in some detail. The basic components of this instrument consist of two rectangular-to-circular waveguide tapered transitions, together with an intermediate section of circular waveguide that is free to rotate, as in Fig. 6.6. A very thin tapered resistive card is placed at the output end of each transition section and oriented parallel to the broad walls of the rectangular guide. A similar resistive card is located in the intermediate circular-guide section. The incoming TE_{10} mode in the rectangular guide is transformed into the TE_{11} mode in the circular guide with negligible reflection by means of the tapered transition. The polarization of the TE_{11} mode is such that the electric field is perpendicular to the thin resistive card in the transition section. As such, this resistive card has a negligible effect on the TE_{11} mode. Since the resistive card in the center section can be rotated, its orientation relative to the electric field of the incoming TE_{11} mode can be varied so that the amount by which this mode is attenuated is adjustable.

With reference to Fig. 6.7, let the center resistive card be oriented at an angle θ relative to the direction of the electric field polarization of the TE_{11} mode. The TE_{11} mode polarized in the x direction may be decomposed into the sum of two TE_{11} modes polarized along the u and v directions, as illustrated in Fig. 6.7. That portion which is parallel to the resistive slab will be absorbed, whereas the portion which is polarized

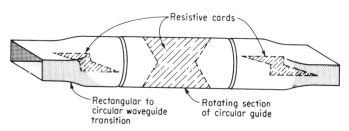

Fig. 6.6 Basic construction of a rotary attenuator.

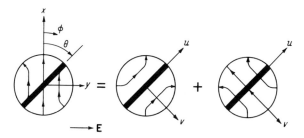

Fig. 6.7 Decomposition of TE_{11} mode into two orthogonally polarized modes.

perpendicular to the slab will be transmitted. However, upon entering the transition section, the transmitted mode is again not perpendicular to the resistive card in this section; so some additional attenuation will occur.

To derive the expression for the dependence of the attenuation on the rotation angle θ, consider the analytic expression for the TE_{11}-mode electric field. For polarization in the x direction the electric field is given by (Table 3.4)

$$\mathbf{E} = \frac{J_1(p'_{11}r/a)}{r} \mathbf{a}_r \cos \phi - \frac{p'_{11}}{a} J'_1\left(\frac{p'_{11}r}{a}\right) \mathbf{a}_\phi \sin \phi \tag{6.2}$$

apart from the propagation factor $e^{-j\beta z}$. Since

$$\sin \phi = \sin(\phi - \theta + \theta) = \cos \theta \sin(\phi - \theta) + \sin \theta \cos(\phi - \theta)$$

and similarly

$$\cos \phi = \cos(\phi - \theta + \theta) = \cos \theta \cos(\phi - \theta) - \sin \theta \sin(\phi - \theta)$$

the above expression may be written as

$$\mathbf{E} = \cos \theta \left[\frac{J_1}{r} \mathbf{a}_r \cos(\phi - \theta) - \frac{p'_{11}}{a} J'_1 \mathbf{a}_\phi \sin(\phi - \theta) \right]$$

$$- \sin \theta \left[\frac{J_1}{r} \mathbf{a}_r \sin(\phi - \theta) + \frac{p'_{11}}{a} J'_1 \mathbf{a}_\phi \cos(\phi - \theta) \right] \tag{6.3}$$

which is equivalent to referring the angle variable ϕ to a new origin at θ. The first term in brackets in (6.3) is a TE_{11} mode polarized with the electric field along the u axis as in Fig. 6.7, and the second term in brackets is a TE_{11} mode polarized along the v axis. Since the first part is completely absorbed, only the portion multiplied by $\sin \theta$ is transmitted into the output transition section. If we assume that the resistive card in this section is parallel to the y axis, only the component of the transmitted field which is polarized along the x axis is transmitted. We have, at the input to the transition section,

$$\mathbf{E} = -\sin \theta \left[\frac{J_1}{r} \mathbf{a}_r \sin(\phi - \theta) + \frac{p'_{11}}{a} J'_1 \mathbf{a}_\phi \cos(\phi - \theta) \right]$$

$$= \sin^2 \theta \left(\frac{J_1}{r} \mathbf{a}_r \cos \phi - \frac{p'_{11}}{a} J'_1 \mathbf{a}_\phi \sin \phi \right)$$

$$- \sin \theta \cos \theta \left(\frac{J_1}{r} \mathbf{a}_r \sin \phi + \frac{p'_{11}}{a} J'_1 \mathbf{a}_\phi \cos \phi \right) \tag{6.4}$$

of which the first part is a TE_{11} mode polarized along the x direction and is transmitted. Comparing (6.4) with (6.2) shows that the transmitted field is reduced by a factor $\sin^2 \theta$ from the amplitude of the incident field. Hence the attenuation produced is given in decibels by

$$\alpha = -20 \log(\sin^2 \theta) = -40 \log(\sin \theta) \tag{6.5}$$

A notable feature is that the attenuation depends only on the angle of rotation θ, a result that is verified in practice to a high degree of approximation.

6.3 Phase changers

A phase changer is an instrument that produces an adjustable change in the phase angle of the wave transmitted through it. Ideally, it should be perfectly matched to the input and output lines and should produce zero attenuation. These requirements can be met to within a reasonable degree of approximation. There are a variety of designs for phase changers.† Two types, a linear phase changer and the rotary phase changer, are discussed in this section.

Linear phase changer

The linear phase changer consists of three dielectric slabs placed in a rectangular guide, as illustrated in Fig. 6.8. The center slab is free to move longitudinally, and is moved by a suitable drive mechanism to which it is keyed by means of a dielectric key that protrudes through a long centered slot cut in one broad face of the guide. Each end of the dielectric slabs is cut in a stepwise fashion to provide a broadband multisection quarter-wave transformer to match the partially filled guide to the empty and completely filled guides. When the center slab is about $0.3a$ wide, it is found that the sum of the two propagation constants β_1 and β_3 is greater than $\beta_2 + \beta_4$. If the center slab is displaced a distance x to the right, the effect is to lengthen lines 1 and 3 by an amount x and to shorten lines 2 and 4 by this same amount x. As a result, the phase change undergone by a wave propagating through the structure will be

$$\Delta\phi = [(\beta_1 + \beta_3) - (\beta_2 + \beta_4)]x \tag{6.6}$$

A phase change linearly proportional to displacement is thus produced. The amount of phase change obtained increases when the dielectric

† R. E. Collin, Waveguide Phase Changer, *Wireless Eng.*, vol. 32, pp. 82–88, March, 1955.

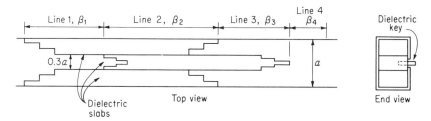

Fig. 6.8 A linear phase changer.

constant of the slabs is increased. For material with a dielectric constant of 2.56 and a 3-cm guide with $a = 0.9$ in., the phase change obtained is about 0.4 rad/cm of displacement. About 16 cm of displacement will provide a phase shift in excess of 360°. Because of the differential action, the scale length and displacement are sufficiently great to provide for very accurate setting and reading. Detailed design data for an instrument for use at a wavelength of 3 cm are given in the literature.†

Rotary phase changer

The rotary phase changer is a precision instrument that is widely used in microwave measurements. Its basic construction is similar to that of the rotary attenuator, except that the center resistive card is replaced by a half-wave plate and the two outer resistive cards are replaced by quarter-wave plates. The quarter-wave plates convert a linearly polarized TE_{11} mode into a circularly polarized mode, and vice versa. The half-wave plate produces a phase shift equal to twice the angle θ through which it is rotated. The analysis of the principles of operation is given below.

A circularly polarized field is a field with x and y components of electric field that are equal in magnitude but 90° apart in time phase.‡ A quarter-wave plate is a device that will produce a circularly polarized wave when a linearly polarized wave is incident upon it. Figure 6.9 illustrates the basic components of the rotary phase changer. The quarter-wave plate may be constructed from a slab of dielectric material, as illustrated in Fig. 6.9b. When the TE_{11} mode is polarized parallel to

† *Ibid.*
‡ Circularly polarized fields are discussed in greater detail in Sec. 6.6.

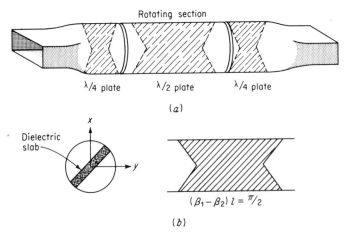

Fig. 6.9 (a) Rotary phase changer; (b) quarter-wave plate.

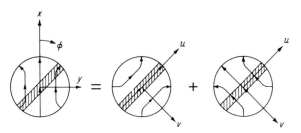

Fig. 6.10 Decomposition of incident TE$_{11}$ mode.

the slab, the propagation constant β_1 is greater than for the case when the mode is polarized perpendicular to the slab; that is, $\beta_1 > \beta_2$, where β_2 is the propagation constant for perpendicular polarization. The length l of a quarter-wave plate is chosen to obtain a differential phase change $(\beta_1 - \beta_2)l$ equal to 90°. The ends of the dielectric slab are tapered to reduce reflections to a negligible value. The half-wave plate is similar in construction, except that its length is increased to produce a differential phase change of 180°.

In the rotary phase changer the quarter-wave plates are oriented at an angle of 45° relative to the broad wall of the rectangular guide. The incoming TE$_{11}$ mode may be decomposed into two modes polarized parallel and perpendicular to the quarter-wave plate, as illustrated in Fig. 6.10. The incident mode is assumed given by (6.2) as

$$\mathbf{E} = \frac{J_1}{r} \mathbf{a}_r \cos \phi - \frac{p'_{11}}{a} J'_1 \mathbf{a}_\phi \sin \phi$$

If we replace $\cos \phi$ by

$$\cos\left(\phi - \frac{\pi}{4} + \frac{\pi}{4}\right) = \frac{\sqrt{2}}{2}\left[\cos\left(\phi - \frac{\pi}{4}\right) - \sin\left(\phi - \frac{\pi}{4}\right)\right]$$

and $\sin \phi$ by

$$\frac{\sqrt{2}}{2}\left[\cos\left(\phi - \frac{\pi}{4}\right) + \sin\left(\phi - \frac{\pi}{4}\right)\right]$$

the above expression for the incident field may be written as

$$\mathbf{E} = \mathbf{E}_1 + \mathbf{E}_2 \tag{6.7a}$$

where

$$\mathbf{E}_1 = \frac{\sqrt{2}}{2}\left[\frac{J_1}{r} \mathbf{a}_r \cos\left(\phi - \frac{\pi}{4}\right) - \frac{p'_{11}}{a} J'_1 \mathbf{a}_\phi \sin\left(\phi - \frac{\pi}{4}\right)\right] \tag{6.7b}$$

$$\mathbf{E}_2 = -\frac{\sqrt{2}}{2}\left[\frac{J_1}{r} \mathbf{a}_r \sin\left(\phi - \frac{\pi}{4}\right) + \frac{p'_{11}}{a} J'_1 \mathbf{a}_\phi \cos\left(\phi - \frac{\pi}{4}\right)\right] \tag{6.7c}$$

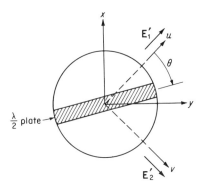

Fig. 6.11 Orientation of half-wave plate.

The field $\mathbf{E}_1$ is polarized parallel to the slab, and $\mathbf{E}_2$ is polarized perpendicular to the slab. After propagation through the quarter-wave plate, these fields become

$$\mathbf{E}_1' = \mathbf{E}_1 e^{-j\beta_1 l} \tag{6.8a}$$

$$\mathbf{E}_2' = \mathbf{E}_2 e^{-j\beta_2 l} = \mathbf{E}_2 e^{-j\beta_1 l} e^{-j(\beta_2-\beta_1)l} = j\mathbf{E}_2 e^{-j\beta_1 l} \tag{6.8b}$$

since $(\beta_2 - \beta_1)l = -\pi/2$. The resultant field consists of two orthogonally polarized TE_{11} modes of equal amplitude and 90° apart in time phase, and hence constitutes a circularly polarized field.

Consider next the action of the half-wave plate on the above circularly polarized field. Let the half-wave plate be rotated by an angle θ past the quarter-wave plate, as in Fig. 6.11. The field $\mathbf{E}_1' + \mathbf{E}_2'$ may be expressed in terms of TE_{11} modes polarized parallel and perpendicular to the half-wave plate by changing the origin of the angle variable ϕ to $\pi/4 + \theta$; that is, we use the relations

$$\cos\left(\phi - \frac{\pi}{4}\right) = \cos\left(\phi - \frac{\pi}{4} - \theta + \theta\right) = \cos\theta\cos\left(\phi - \theta - \frac{\pi}{4}\right)$$
$$- \sin\theta\sin\left(\phi - \theta - \frac{\pi}{4}\right)$$

and

$$\sin\left(\phi - \frac{\pi}{4}\right) = \sin\theta\cos\left(\phi - \theta - \frac{\pi}{4}\right) + \cos\theta\sin\left(\phi - \theta - \frac{\pi}{4}\right)$$

Thus we obtain

$$\mathbf{E}_1' = \frac{\sqrt{2}}{2} e^{-j\beta_1 l} \left\{ \cos\theta \left[\frac{J_1}{r} \mathbf{a}_r \cos\left(\phi - \theta - \frac{\pi}{4}\right) \right. \right.$$
$$\left. - \frac{p_{11}'}{a} J_1' \mathbf{a}_\phi \sin\left(\phi - \theta - \frac{\pi}{4}\right) \right] - \sin\theta \left[\frac{J_1}{r} \mathbf{a}_r \sin\left(\phi - \theta - \frac{\pi}{4}\right) \right.$$
$$\left.\left. + \frac{p_{11}'}{a} J_1' \mathbf{a}_\phi \cos\left(\phi - \theta - \frac{\pi}{4}\right) \right] \right\} \tag{6.9a}$$

$$E_2' = \frac{-j\sqrt{2}}{2} e^{-j\beta_1 l} \left\{ \cos\theta \left[\frac{J_1}{r} \mathbf{a}_r \sin\left(\phi - \theta - \frac{\pi}{4}\right) \right.\right.$$
$$\left. + \frac{p_{11}'}{a} J_1' \mathbf{a}_\phi \cos\left(\phi - \theta - \frac{\pi}{4}\right) \right] + \sin\theta \left[\frac{J_1}{r} \mathbf{a}_r \cos\left(\phi - \theta - \frac{\pi}{4}\right) \right.$$
$$\left.\left. - \frac{p_{11}'}{a} J_1' \mathbf{a}_\phi \sin\left(\phi - \theta - \frac{\pi}{4}\right) \right] \right\} \quad (6.9b)$$

The field polarized parallel to the half-wave plate has an r component varying as $\cos(\phi - \theta - \pi/4)$, whereas the perpendicularly polarized mode has an r component of electric field varying as $\sin(\phi - \theta - \pi/4)$. Hence, from (6.9) we obtain

$$E_1' + E_2' = E_1'' + E_2'' \quad (6.10a)$$

where

$$E_1'' = \frac{\sqrt{2}}{2} e^{-j\beta_1 l}(\cos\theta - j\sin\theta) \left[\frac{J_1}{r} \mathbf{a}_r \cos\left(\phi - \theta - \frac{\pi}{4}\right) \right.$$
$$\left. - \frac{p_{11}'}{a} J_1' \mathbf{a}_\phi \sin\left(\phi - \theta - \frac{\pi}{4}\right) \right]$$
$$= \frac{\sqrt{2}}{2} e^{-j\beta_1 l - j\theta} \left[\frac{J_1}{r} \mathbf{a}_r \cos\left(\phi - \theta - \frac{\pi}{4}\right) - \frac{p_{11}'}{a} J_1' \mathbf{a}_\phi \sin\left(\phi - \theta - \frac{\pi}{4}\right) \right]$$
$$(6.10b)$$

$$E_2'' = \frac{-j\sqrt{2}}{2} e^{-j\beta_1 l - j\theta} \left[\frac{J_1}{r} \mathbf{a}_r \sin\left(\phi - \theta - \frac{\pi}{4}\right) \right.$$
$$\left. + \frac{p_{11}'}{a} J_1' \mathbf{a}_\phi \cos\left(\phi - \theta - \frac{\pi}{4}\right) \right] \quad (6.10c)$$

After propagating through the half-wave plate of length $2l$, this field becomes

$$E_3 = E_1'' e^{-2j\beta_1 l} \quad (6.11a)$$
$$E_4 = E_2'' e^{-2j\beta_2 l} = E_2'' e^{-2j(\beta_2 - \beta_1)l - 2j\beta_1 l} = -E_2'' e^{-2j\beta_1 l} \quad (6.11b)$$

since $2(\beta_2 - \beta_1)l = -\pi$.

This new field may now be decomposed once again into two TE_{11} modes polarized parallel and perpendicular to the quarter-wave plate in the output guide. If we assume that this plate is parallel to the input quarter-wave plate, we can obtain the required decomposition by referring the angle variable ϕ to $\pi/4$ as the origin. If we follow the procedure used earlier, we obtain

$$E_3 + E_4 = E_3' + E_4' \quad (6.12a)$$

where

$$E_3' = \frac{\sqrt{2}}{2} e^{-3j\beta_1 l - j2\theta} \left[\frac{J_1}{r} \mathbf{a}_r \cos\left(\phi - \frac{\pi}{4}\right) - \frac{p_{11}'}{a} J_1' \mathbf{a}_\phi \sin\left(\phi - \frac{\pi}{4}\right) \right] \quad (6.12b)$$

$$E_4' = \frac{j\sqrt{2}}{2} e^{-3j\beta_1 l - j2\theta} \left[\frac{J_1}{r} \mathbf{a}_r \sin\left(\phi - \frac{\pi}{4}\right) + \frac{p_{11}'}{a} J_1' \mathbf{a}_\phi \cos\left(\phi - \frac{\pi}{4}\right) \right] \tag{6.12c}$$

Finally, after propagating through the second quarter-wave plate, the output field becomes

$$\mathbf{E}_0 = \mathbf{E}_3'' + \mathbf{E}_4'' \tag{6.13a}$$

where

$$\mathbf{E}_3'' = \mathbf{E}_3' e^{-j\beta_1 l} \tag{6.13b}$$

$$\mathbf{E}_4'' = \mathbf{E}_4' e^{-j\beta_2 l} = j\mathbf{E}_4' e^{-j\beta_1 l} \tag{6.13c}$$

When the fields $\mathbf{E}_3''$ and $\mathbf{E}_4''$ are combined, we obtain

$$\mathbf{E}_0 = e^{-4j\beta_1 l - 2j\theta} \left(\frac{J_1}{r} \mathbf{a}_r \cos\phi - \frac{p_{11}'}{a} J_1' \mathbf{a}_\phi \sin\phi \right) \tag{6.14}$$

which is again a linearly polarized TE_{11} mode having the same direction of polarization as the incident field given by (6.2) and (6.7). Note, however, that the phase has been changed by an amount $4\beta_1 l + 2\theta$. Thus rotation of the half-wave plate through an angle θ changes the phase of the transmitted wave by an amount 2θ. This simple dependence of the phase change on a mechanical rotation is the chief advantage of the rotary phase changer.

Besides dielectric slabs, the circular guide may be loaded with metallic fins or rods to produce 90 and 180° differential phase-shift sections. These methods are discussed in a paper by Fox.†

6.4 Directional couplers

A directional coupler is a four-port microwave junction with the properties discussed below. With reference to Fig. 6.12, which is a schematic illustration of a directional coupler, the ideal directional coupler has the property that a wave incident in port 1 couples power into ports 2 and 3 but not into port 4. Similarly, power incident in port 4 couples into ports 2 and 3 but not into port 1. Thus ports 1 and 4 are uncoupled. For waves incident in port 2 or 3, the power is coupled into ports 1 and

† A. G. Fox, An Adjustable Waveguide Phase Changer, *Proc. IRE*, vol. 35, pp. 1489–1498, December, 1947.

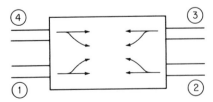

Fig. 6.12 A directional coupler. Arrows indicate the direction of power flow.

Fig. 6.13 Aperture-coupled-waveguide directional coupler.

4 only, so that ports 2 and 3 are also uncoupled. In addition, all four ports are matched. That is, if three ports are terminated in matched loads, the fourth port appears terminated in a matched load, and an incident wave in this port suffers no reflection.

Directional couplers are widely used in impedance bridges for microwave measurements and for power monitoring. For example, if a radar transmitter is connected to port 1, the antenna to port 2, a microwave crystal detector to port 3, and a matched load to port 4, the power received in port 3 is proportional to the power flowing from the transmitter to the antenna in the forward direction only. Since the reflected wave from the antenna, if it exists, is not coupled into port 3, the detector monitors the power output of the transmitter.

The usual directional coupler consists of two waveguides with suitable coupling apertures located in a common wall, as illustrated in Fig. 6.13. Since these devices are required to operate over a band of frequencies, it is not possible to obtain ideal performance over the whole frequency band. The performance of a directional coupler is measured by two parameters, the *coupling* and the *directivity*. Let P_i be the incident power in port 1, and let P_f be the coupled power in the forward direction in arm 3. The coupling in decibels is then given by

$$C = 10 \log \frac{P_i}{P_f} \qquad (6.15)$$

Ideally, the power P_b coupled in the backward direction in arm 4 should be zero. The extent to which this is achieved is measured by the directivity D, which is defined as

$$D = 10 \log \frac{P_f}{P_b} \qquad (6.16)$$

The directivity is a measure of how well the power can be coupled in the desired direction in the second waveguide.

A number of properties of the ideal directional coupler may be deduced from the symmetry and unitary properties of its scattering matrix (Sec. 4.7). The least stringent definition of a directional coupler as illustrated in Fig. 6.12 is that it is a four-port junction with

$S_{14} = S_{23} = 0$

$S_{11} = S_{22} = 0$, that is, ports 1 and 2 matched and the coupling elements S_{12}, S_{13}, S_{24}, and S_{34} not equal to zero. The scattering matrix then has the form

$$[S] = \begin{bmatrix} 0 & S_{12} & S_{13} & 0 \\ S_{12} & 0 & 0 & S_{24} \\ S_{13} & 0 & S_{33} & S_{34} \\ 0 & S_{24} & S_{34} & S_{44} \end{bmatrix} \quad (6.17)$$

If we form the product of row 1 with the complex conjugate of row 3, and also the product of row 2 with the conjugate of row 4, we obtain

$$S_{13}S_{33}^* = 0 \qquad S_{24}S_{44}^* = 0$$

because of the unitary nature of the scattering matrix. Since S_{13} and S_{24} are assumed to be nonzero, these equations show that $S_{33} = S_{44} = 0$; that is, all four ports are matched. Thus the scattering matrix becomes

$$[S] = \begin{bmatrix} 0 & S_{12} & S_{13} & 0 \\ S_{12} & 0 & 0 & S_{24} \\ S_{13} & 0 & 0 & S_{34} \\ 0 & S_{24} & S_{34} & 0 \end{bmatrix} \quad (6.18)$$

If we take the product of row 1 with the conjugate of row 4, and similarly row 2 with the conjugate of row 3, we now find that

$$S_{12}S_{24}^* + S_{13}S_{34}^* = 0 \qquad S_{12}S_{13}^* + S_{24}S_{34}^* = 0$$

If we note that $|S_{12}S_{24}^*| = |S_{12}||S_{24}|$, these equations are seen to give

$$|S_{12}||S_{24}| = |S_{13}||S_{34}| \quad (6.19a)$$
$$|S_{12}||S_{13}| = |S_{24}||S_{34}| \quad (6.19b)$$

When we divide the first equation by the second equation, we obtain

$$\frac{|S_{24}|}{|S_{13}|} = \frac{|S_{13}|}{|S_{24}|}$$

or

$$|S_{13}| = |S_{24}| \quad (6.20a)$$

Thus the coupling between ports 1 and 3 is the same as that between ports 2 and 4. Use of (6.20a) in (6.19a) also gives

$$|S_{12}| = |S_{34}| \quad (6.20b)$$

so that the coupling between ports 1 and 2 equals that between ports 3 and 4 also.

The product of the first row with its conjugate equals unity, so that

$$|S_{12}|^2 + |S_{13}|^2 = 1 \quad (6.21a)$$

Similarly,

$$|S_{12}|^2 + |S_{24}|^2 = 1 \quad (6.21b)$$

By choosing the terminal plane in arm 1 properly, we can adjust the phase angle of S_{12} so that S_{12} is real [see (4.53)]. Thus let S_{12} be a real positive number C_1. Similarly, by choosing the terminal plane in arm 3 properly, we can make S_{13} a positive imaginary quantity jC_2, where C_2 is real and positive. We now have

$$C_1^2 + C_2^2 = 1 \tag{6.22}$$

We can also choose the reference plane in arm 4 so as to make S_{34} real and thus equal to C_1 by virtue of (6.20b). It is now necessary for S_{24} to be equal to jC_2 since $S_{12}S_{24}^* + S_{13}S_{34}^* = 0$, as given earlier. Thus the simplest form for the scattering matrix of an ideal directional coupler is

$$[S] = \begin{bmatrix} 0 & C_1 & jC_2 & 0 \\ C_1 & 0 & 0 & jC_2 \\ jC_2 & 0 & 0 & C_1 \\ 0 & jC_2 & C_1 & 0 \end{bmatrix} \tag{6.23}$$

where $C_2 = (1 - C_1^2)^{\frac{1}{2}}$ from (6.22).

It may also be shown from the unitary properties of the scattering matrix of a lossless reciprocal four-port junction that if all four ports are matched, the device must be a directional coupler.†

Directional-coupler designs

There are a great variety of ways of constructing directional couplers. Some of the more common aperture-coupled types are described below. Their design is based on the small-aperture-coupling theory presented in Sec. 4.11. This theory was originally developed by Bethe.‡

Bethe-hole coupler

The Bethe-hole directional coupler consists of two rectangular waveguides coupled by means of a small circular aperture located in the center of the common broad wall. To achieve directional coupling, the axis of the two guides must be at an angle θ, as illustrated in Fig. 6.14a. A variation of this design consists of a similar arrangement, with $\theta = 0$, but an offset aperture as in Fig. 6.14b.

The theory for the coupler in Fig. 6.14b was given in Sec. 4.11. An incident TE$_{10}$ mode in guide 1, with an amplitude A, produces a normal electric dipole in the aperture plus a tangential magnetic dipole proportional and in the same direction as the magnetic field of the incident wave. In the upper guide the normal electric dipole and the axial component of the magnetic dipole radiate symmetrically in both directions. The transverse component of the magnetic dipole radiates antisym-

---

† C. G. Montgomery, R. H. Dicke, and E. M. Purcell, "Principles of Microwave Circuits," sec. 9.10, McGraw-Hill Book Company, New York, 1948.

‡ H. A. Bethe, Theory of Diffraction by Small Holes, *Phys. Rev.*, vol. 66, pp. 163–182, 1944.

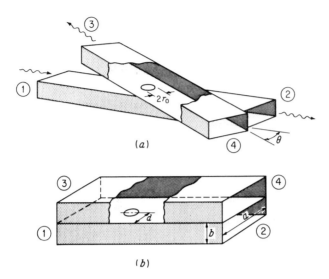

Fig. 6.14 Bethe-hole directional coupler.

metrically, so that by proper adjustment of the angle θ or the aperture position d, the radiation in the direction of port 4 can be canceled and that in the direction of port 3 enhanced. From (4.106), (4.107), and (4.108), the field radiated by the electric dipole is found to have an amplitude

$$B_1 = -\frac{j\omega\epsilon_0}{abY_w}\frac{2}{3}r_0^3 A \sin^2\frac{\pi d}{a} \quad (6.24a)$$

in both directions, whereas the magnetic dipole radiates a field with amplitude

$$B_2 = \frac{j\omega\mu_0 Y_w}{ab}\frac{4}{3}r_0^3 A \left(\sin^2\frac{\pi d}{a} + \frac{\pi^2}{\beta^2 a^2}\cos^2\frac{\pi d}{a}\right) \quad (6.24b)$$

in the direction of port 4, and

$$B_3 = \frac{j\omega\mu_0 Y_w}{ab}\frac{4}{3}r_0^3 A \left(-\sin^2\frac{\pi d}{a} + \frac{\pi^2}{\beta^2 a^2}\cos^2\frac{\pi d}{a}\right) \quad (6.24c)$$

in the direction of port 3. To cancel the radiation in the latter direction we must have $B_1 + B_3 = 0$. Solution of this equation for the aperture position d gives

$$\sin\frac{\pi d}{a} = \frac{\lambda_0}{\sqrt{6}\,a} \quad (6.25a)$$

The radiation can be canceled at port 4 instead, in which case we equate

$B_1 + B_2$ to zero and obtain

$$\sin \frac{\pi d}{a} = \frac{\lambda_0}{\sqrt{2(\lambda_0^2 - a^2)}} \qquad (6.25b)$$

If the aperture is centered, $d = a/2$ and $\cos \pi d/a = 0$. A directional coupler with $\theta = 0$ is obtained in this case only for $\lambda_0 = \sqrt{2}\, a$. For other values of λ_0 the upper guide may be rotated by an angle θ to reduce the coupling of the transverse magnetic field to the magnetic dipole by a factor $\cos \theta$. The required angle is found by equating $B_2 \cos \theta$ to $-B_1$ to obtain

$$\cos \theta = \frac{k_0^2}{2\beta^2} = \frac{1}{2}\left(\frac{\lambda_g}{\lambda_0}\right)^2 \qquad (6.26)$$

With a centered aperture, there is only one possibility for obtaining a directional coupler.

For the coupler of Fig. 6.14a the coupled wave in port 3 has an amplitude

$$B_1 + B_3 \cos \theta$$

and hence the coupling is given by

$$C = -20 \log \left|\frac{B_1 + B_3 \cos \theta}{A}\right| = -20 \log \frac{4}{3}\frac{\beta r_0^3}{ab}\left(\cos \theta + \frac{k_0^2}{2\beta^2}\right) \qquad (6.27)$$

The directivity is given by

$$D = 20 \log \left|\frac{B_1 + B_3 \cos \theta}{B_1 + B_2 \cos \theta}\right| = 20 \log \frac{2\beta^2 \cos \theta + k_0^2}{2\beta^2 \cos \theta - k_0^2} \qquad (6.28)$$

If the waveguide wall has an appreciable thickness, the amount of coupling a given aperture will produce is reduced. The attenuation due to wall thickness may be accounted for in an approximate way by considering the aperture in a thick wall as a circular guide operated below cutoff.[†]

Two-hole couplers

Two-hole couplers consist of two rectangular waveguides coupled by two identical apertures spaced a quarter guide wavelength $\lambda_g/4$ apart as in Fig. 6.15. The aperture may, in general, have directive properties, i.e., radiate a field with different amplitudes in the forward and reverse directions. With a wave of unit amplitude incident at port 1, let the field coupled into the second guide have an amplitude B_f in the forward direction and B_b in the backward direction. Since B_f and B_b are the amplitudes of the coupled fields for an incident wave of unit amplitude, they are called the aperture-coupling coefficients. If only a small amount

† C. G. Montgomery, "Technique of Microwave Measurements," sec. 14.3, McGraw-Hill Book Company, New York, 1947.

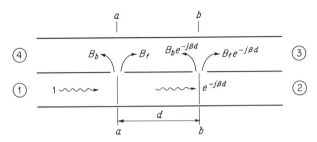

Fig. 6.15 Two-hole directional coupler.

of the incident power is coupled by the first aperture, the amplitude of the incident wave is essentially unity at the second aperture also. Thus this aperture couples the same amount of power into the second guide. Note, however, that because of the difference in path length, the phase of the field coupled by the second aperture is $-\beta d$ relative to that coupled by the first aperture. The total forward wave in the upper guide at the plane bb is $2B_f e^{-j\beta d}$, and the total backward wave at the plane aa is $B_b(1 + e^{-2j\beta d})$. Hence, since the forward-path lengths in the two guides are always the same, the forward waves always add in phase. The backward waves will add out of phase whenever $2\beta d = n\pi$, $n = 1, 3, 5, \ldots$. In particular, a value of $d = \lambda_g/4$ will result in cancellation of the backward wave. The coupling is given by

$$C = -20 \log 2|B_f| \tag{6.29}$$

and the directivity is given by

$$D = 20 \log \frac{2|B_f|}{|B_b| \, |1 + e^{-2j\beta d}|} = 20 \log \frac{|B_f|}{|B_b| \, |\cos \beta d|}$$

$$= 20 \log \left|\frac{B_f}{B_b}\right| + 20 \log |\sec \beta d| \tag{6.30}$$

The directivity is the sum of the inherent directivity of the single aperture plus a directivity associated with the array (in this case a two-element array only). Since B_f and B_b are the aperture-coupling parameters and are generally slowly varying functions of frequency, the coupling C is not particularly frequency-sensitive. However, the directivity is a sensitive function of frequency because of the sensitivity of the array factor $\sec \beta d$.

Schwinger reversed-phase coupler

The Schwinger reversed-phase coupler is designed to interchange the frequency sensitivity of the coupling C and directivity D. This is accomplished by making one aperture radiate a field which is the negative of that radiated by the other. With reference to Fig. 6.15, let the first aperture radiate fields B_f, B_b and the second aperture $-B_f$, $-B_b$. At plane bb in the upper guide, the total field is now $B_f - B_f = 0$ under all conditions. Hence port 3 is not coupled to port 1. At the plane aa the total field is, after accounting for the phase change due to propagation,

$$B_b - B_b e^{-2j\beta d} = e^{-j\beta d} B_b 2j \sin \beta d$$

Thus the coupling between ports 1 and 4 is

$$C = -20 \log 2|B_b \sin \beta d| \tag{6.31}$$

and is a maximum for $d = \lambda_g/4$. For this coupler the directivity D is theoretically infinite and independent of frequency, whereas the coupling C is quite frequency-sensitive, although not as frequency-sensitive as the directivity D given by (6.30), since $\sin \beta d$ varies more slowly around $\pi/2$ than does $\cos \beta d$. Actually, in practice, the directivity D is not infinite since, in the foregoing discussion, it was assumed that the same incident field was present at each aperture, and each aperture radiated the same field into the upper guide. Because of interaction effects between the two apertures, the assumption of equal-amplitude fields coupled by both apertures is an approximation valid for a small amount of coupling only.

Figure 6.16a illustrates a typical reversed-phase coupler. The TE_{10} mode has a zero normal electric field and transverse magnetic field at the narrow wall, and hence the coupling to this mode in the upper guide

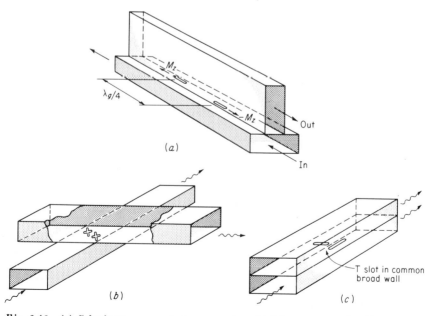

Fig. 6.16 (a) Schwinger reversed-phase coupler; (b) Moreno crossed-guide coupler; (c) Riblet T-slot coupler.

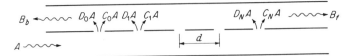

Fig. 6.17 A multielement directional coupler.

is through the induced axial magnetic dipole moment of the aperture only. In the lower guide the axial magnetic field of the TE$_{10}$ mode is of opposite sign on the two sides of the center, so that induced dipoles M_z and $-M_z$ are produced. These dipoles radiate symmetrically in both directions, but are phase-reversed to obtain the desired reversed-phase directional coupler.

Two other double-aperture-coupled directional couplers in common use are the Moreno crossed-guide coupler and the Riblet T-slot coupler, illustrated in Fig. 6.16 also. Design nomograms for these couplers, as well as for the Schwinger reversed-phase coupler, are given in a paper by Anderson.†

Multielement couplers

To achieve good directivity over a band of frequencies, couplers with many apertures may be used. The theory and design of such couplers parallel those given for multisection quarter-wave transformers in Chap. 5. Figure 6.17 illustrates an $N + 1$ element coupler with all aperture spacings equal to d. If we assume that the total power coupled is relatively small compared with the incident power, the incident wave can be considered to have essentially the same amplitude A at each coupling aperture apart from the additional phase change. Let the apertures have coupling coefficients C_n, $n = 0, 1, 2, \ldots, N$, in the forward direction, and D_n, $n = 0, 1, 2, 3, \ldots, N$, in the reverse direction. At the position of the Nth aperture the total forward wave in the upper guide is

$$B_f = A e^{-j\beta Nd} \sum_{n=0}^{N} C_n \tag{6.32}$$

At the plane of the first aperture the total backward wave has an amplitude

$$B_b = A \sum_{n=0}^{N} D_n e^{-j\beta 2nd} \tag{6.33}$$

† T. N. Anderson, Directional Coupler Design Nomograms, *Microwave J.*, vol. 2, pp. 34–38, May, 1959.

The coupling and directivity are given by

$$C = -20 \log \left| \sum_{n=0}^{N} C_n \right| \tag{6.34}$$

$$D = -20 \log \frac{\left| \sum_{n=0}^{N} D_n e^{-j\beta 2nd} \right|}{\left| \sum_{n=0}^{N} C_n \right|} = -C - 20 \log \left| \sum_{n=0}^{N} D_n e^{-j\beta 2nd} \right| \tag{6.35}$$

We now assume that all the apertures are similar, so that the coupling coefficients D_n can be expressed as a product of a frequency-independent amplitude constant d_n and a frequency-dependent factor T_b. Thus

$$D = -C - 20 \log |T_b| - 20 \log \left| \sum_{n=0}^{N} d_n e^{-j\beta 2nd} \right| \tag{6.36}$$

The first two terms give the directivity associated with the individual apertures, and the last term is the directivity arising from the array. In the following discussion we shall be concerned only with designing the array factor

$$F = \left| \sum_{n=0}^{N} d_n e^{-j\beta 2nd} \right| \tag{6.37}$$

to keep it as small as possible over as broad a band as possible.

The array factor of (6.37) is essentially identical with the expression for the reflection coefficient from an N-section quarter-wave transformer. Thus a maximally flat passband characteristic can be obtained by choosing the d_n proportional to the binomial coefficients $C_n{}^N$; that is,

$$d_n = K C_n{}^N \tag{6.38}$$

where K is a constant of proportionality.

When the D_n can be expressed in the form $T_b d_n$, then usually the C_n can be expressed in the form $c_n T_f$, where c_n is a frequency-independent amplitude constant. The c_n and d_n can be adjusted by varying the size of the aperture while maintaining its shape and position fixed. For a given type of aperture, T_f, T_b, and the relationship between c_n and d_n would be known. Usually T_f and T_b can be chosen, so that $c_n = d_n$. Thus we have

$$C = -20 \log |T_f| \left| \sum_{n=0}^{N} d_n \right| = -20 \log K|T_f| \left| \sum_{n=0}^{N} C_n{}^N \right| \tag{6.39}$$

which fixes the constant of proportionality in (6.38) when the coupling C is given at the center of the band.

To obtain an equal-ripple characteristic in the passband, the array factor F is made proportional to a Chebyshev polynomial. If we choose a symmetrical array, with $d_0 = d_N$, $d_1 = d_{N-1}$, etc., we obtain [Eq. (5.40)]

$$F = \left| \sum_{n=0}^{M} 2d_n \cos(N - 2n)\beta d \right| \qquad (6.40)$$

where $M = (N - 1)/2$ for N odd and $N/2$ for N even. Note that for N odd there are an even number of apertures, since the first aperture has been labeled the zeroth aperture. In (6.40) the Mth term is d_M for N even and $2d_M \cos(N - 2M)\beta d$ for N odd. To obtain a Chebyshev-type response we now choose

$$F = \left| \sum_{n=0}^{M} 2d_n \cos(N - 2n)\theta \right| = K|T_N(\sec \theta_m \cos \theta)| \qquad (6.41)$$

as in (5.51). In this equation $\theta = \beta d$ and $\sec \theta_m$ is the value of $\sec \beta d$ at the upper and lower edges of the passband. At the center of the passband, $\theta = \pi/2$, corresponding to a spacing $d = \lambda_g/4$. The constant K is chosen to give the desired value of coupling C in the center of the band, where $\theta = \pi/2$. With $c_n = d_n$, we have

$$C = -20 \log |T_f| \left| \sum_{n=0}^{N} c_n \right| = -20 \log |T_f| \left| \sum_{n=0}^{N} d_n \right|$$

$$= -20 \log K|T_f| |T_N(\sec \theta_m)| \qquad (6.42)$$

since $\left| \sum_{n=0}^{N} d_n \right| = K|T_N(\sec \theta_m)|$ from (6.41).

If we use (6.36), (6.41), and (6.42), the expression for directivity may be written as

$$D = 20[\log K|T_f T_N(\sec \theta_m)| - \log |T_b| - \log K|T_N(\sec \theta_m \cos \theta)|]$$

$$= 20 \left[\log \left| \frac{T_f}{T_b} \right| + \log \left| \frac{T_N(\sec \theta_m)}{T_N(\sec \theta_m \cos \theta)} \right| \right] \qquad (6.43)$$

Since T_f/T_b is a function of frequency, D will not have a Chebyshev-type behavior. However, the departure from a true Chebyshev behavior will usually be small since T_f/T_b gives very little directivity, except perhaps near the center of the band. For a conservative design we choose the minimum value of D on the basis that T_f/T_b contributes negligible directivity. Certainly, for a broadband design, this will be the case at the edges of the passband. The minimum value D_m of directivity in the passband as contributed by the array factor F occurs when

$$T_N(\sec \theta_m \cos \theta) = 1$$

Hence, let D_m be defined as

$$D_m = 20 \log |T_N(\sec \theta_m)|$$

This equation shows that if we specify D_m, then sec θ_m is fixed, which in turn fixes the bandwidth, and vice versa. Thus we must specify either D_m or sec θ_m, and the other is fixed. We may then solve (6.42) for the constant K in terms of the given value of the coupling C at the center of the band. From (6.41) the coefficients d_n are found.

As an example consider the design of a three-hole Chebyshev coupler utilizing offset circular apertures in the common broad wall between two rectangular guides, as in Fig. 6.14b. For the nth aperture with radius r_n, the field coupled in the forward direction is the sum of (6.24a) and (6.24b). Thus, with $A = 1$, the coupled field amplitude is equal to the coupling parameter C_n; that is,

$$C_n = r_n^3 \left[-\frac{j\omega\epsilon_0}{abY_w} \frac{2}{3} \sin^2 \frac{\pi x_0}{a} + \frac{j\omega\mu_0 Y_w}{ab} \frac{4}{3} \left(+\sin^2 \frac{\pi x_0}{a} + \frac{\pi^2}{\beta^2 a^2} \cos^2 \frac{\pi x_0}{a} \right) \right]$$

$$= r_n^3 T_f \quad (6.44a)$$

where we have called the aperture position x_0 so as not to confuse it with the aperture axial spacing d. The field coupled in the backward direction is the sum of (6.24a) and (6.24b) and gives the coupling parameter D_n when $A = 1$. Thus

$$D_n = r_n^3 \left[\frac{-j\omega\epsilon_0}{abY_w} \frac{2}{3} \sin^2 \frac{\pi x_0}{a} - \frac{j\omega\mu_0 Y_w}{ab} \frac{4}{3} \left(\sin^2 \frac{\pi x_0}{a} - \frac{\pi^2}{\beta^2 a^2} \cos^2 \frac{\pi x_0}{a} \right) \right]$$

$$= r_n^3 T_b \quad (6.44b)$$

The factors T_f and T_b have been chosen so that $c_n = d_n = r_n^3$ and are functions of the aperture radius only.

If we specify D_m to equal 40 db, (6.43) gives

$40 = 20 \log |T_N(\sec \theta_m)|$

or

$|T_N(\sec \theta_m)| = 100$

when the contribution of the factor T_f/T_b is neglected.

Since $N = 2$ for a three-hole coupler and $T_2(x) = 2x^2 - 1$, we have

$\sec \theta_m = \sqrt{50.5} = 7.1$

and $\theta_m = 81.9$ and $98.1°$. The corresponding frequency at the edge of the passband is found by solving $\beta d = \theta_m$, with d chosen so that $\beta d = \pi/2$ at the center frequency.

If we specify that the coupling C is to be 20 db (1 percent of the incident power), then from (6.42) we have

$K|T_f T_N(\sec \theta_m)| = 0.1$

and hence

$$K = 10^{-3}|T_f|^{-1}$$

The next step is to evaluate T_f from (6.44a) in terms of the waveguide parameters at the center of the passband. Having carried out this step, we use (6.41) to obtain

$$2d_0 \cos 2\theta + d_1 = K[2(\sec \theta_m \cos \theta)^2 - 1]$$
$$= K[\sec^2 \theta_m (1 + \cos 2\theta) - 1]$$

As the final step we choose

$$r_3{}^3 = r_0{}^3 = d_0 = \tfrac{1}{2} K \sec^2 \theta_m$$
$$r_1{}^3 = d_1 = K(\sec^2 \theta_m - 1)$$

for the aperture radii. Our design is a conservative one since, with three apertures only, the available bandwidth is not particularly large and the factor T_f/T_b will contribute quite significantly to the directivity through the band if the aperture position x_0 is chosen according to (6.25) to give $T_b = 0$ at the center of the band.

In practice, we specify the frequencies f_1 and f_2 at the lower and upper edges of the desired passband. Let the corresponding values of β be β_1 and β_2. Since β is a nonlinear function of frequency for a waveguide, the frequency at the center of the passband is that corresponding to $\tfrac{1}{2}(\beta_1 + \beta_2)$, and not $\tfrac{1}{2}(f_1 + f_2)$. This result arises because β is linearly related to θ, and the electrical phase shift θ is the parameter entering into the design formulas.

6.5 Hybrid junctions

Figure 6.18 illustrates a hybrid coil of the type commonly used in telephone repeater circuits. It has the properties that, when properly terminated in external impedances, a signal input in port 1 is coupled equally into ports 2 and 3 but not into port 4. Similarly, a signal input at port 3 is coupled equally into ports 1 and 4 but not into port 2, etc.

A four-port microwave junction with these properties is called a hybrid junction. Figure 6.19a illustrates a hybrid T, which is a common form of hybrid junction. When a TE_{10} mode is incident in port 1, the electric field within the junction is like that sketched in Fig. 6.19b. This electric field has even symmetry about the midplane and hence cannot excite the

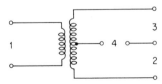

Fig. 6.18 Hybrid coil.

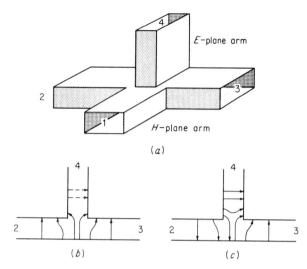

Fig. 6.19 (a) Hybrid-T junction; (b) electric field pattern for wave incident in port 1; (c) electric field pattern for wave incident in port 4.

TE_{10} mode in arm 4 since this mode must have an electric field with odd symmetry (shown dashed in Fig. 6.19b). Thus there is no coupling between ports 1 and 4. The coupling between ports 1 and 2, and 1 and 3, is clearly the same, as may be seen from the symmetry involved.

For a TE_{10} mode incident in arm 4, the electric field within the junction is sketched in Fig. 6.19c. Symmetry again shows that there is no coupling into port 1 (this is required by reciprocity as well). The coupling from port 4 into ports 2 and 3 is equal in magnitude but 180° out of phase. The scattering matrix of this hybrid T thus has the form

$$[S] = \begin{bmatrix} S_{11} & S_{12} & S_{12} & 0 \\ S_{12} & S_{22} & S_{23} & S_{24} \\ S_{12} & S_{23} & S_{33} & -S_{24} \\ 0 & S_{24} & -S_{24} & S_{44} \end{bmatrix} \qquad (6.45)$$

since $S_{12} = S_{13}$, $S_{24} = -S_{34}$, from symmetry.

Matching elements that do not destroy the symmetry of the junction may be placed in the E-plane and H-plane arms so as to make $S_{11} = S_{44} = 0$. For a lossless structure we may then show that the unitary properties of the scattering matrix require that $S_{22} = S_{33} = 0$, so that all ports are matched. In addition, $S_{23} = 0$; so ports 2 and 3 as well as ports 1 and 4 are uncoupled. The hybrid T now becomes a directional coupler with 3-db coupling, and is often called a *magic T*, even though there is nothing magic about its operation.

With $S_{11} = S_{44} = 0$, the scattering matrix becomes

$$[S] = \begin{bmatrix} 0 & S_{12} & S_{12} & 0 \\ S_{12} & S_{22} & S_{23} & S_{24} \\ S_{12} & S_{23} & S_{33} & -S_{24} \\ 0 & S_{24} & -S_{24} & 0 \end{bmatrix}$$

The product of the second row with its conjugate gives

$$|S_{12}|^2 + |S_{22}|^2 + |S_{23}|^2 + |S_{24}|^2 = 1 \tag{6.46a}$$

and the similar expression for row 3 is

$$|S_{12}|^2 + |S_{23}|^2 + |S_{33}|^2 + |S_{24}|^2 = 1 \tag{6.46b}$$

If we subtract these two equations, we obtain

$$|S_{22}|^2 - |S_{33}|^2 = 0 \tag{6.46c}$$

so $|S_{22}| = |S_{33}|$. From rows 1 and 4 we have

$$2|S_{12}|^2 = 1 \quad \text{or} \quad |S_{12}| = \frac{\sqrt{2}}{2}$$

$$2|S_{24}|^2 = 1 \quad \text{or} \quad |S_{24}| = \frac{\sqrt{2}}{2}$$

and thus

$$|S_{12}| = |S_{24}| = \frac{\sqrt{2}}{2} \tag{6.47}$$

Use of this relation in (6.46a) gives

$$1 + |S_{22}|^2 + |S_{23}|^2 = 1$$

or $|S_{22}|^2 + |S_{23}|^2 = 0$. This sum can equal zero only if both S_{22} and S_{23} vanish. From the relation (6.46c) it follows that S_{33} equals zero also. The reduced form of the scattering matrix becomes

$$[S] = \begin{bmatrix} 0 & S_{12} & S_{12} & 0 \\ S_{12} & 0 & 0 & S_{24} \\ S_{12} & 0 & 0 & -S_{24} \\ 0 & S_{24} & -S_{24} & 0 \end{bmatrix}$$

By proper choice of terminal planes in arms 1 and 4, we can make both S_{12} and S_{24} real. Thus the scattering matrix of a magic T can be exhibited in the form

$$[S] = \frac{\sqrt{2}}{2} \begin{bmatrix} 0 & 1 & 1 & 0 \\ 1 & 0 & 0 & 1 \\ 1 & 0 & 0 & -1 \\ 0 & 1 & -1 & 0 \end{bmatrix} \tag{6.48}$$

upon using the relations (6.47).

Another form of hybrid junction is the ring circuit (sometimes called a rat race) illustrated in Fig. 6.20. To understand its operation, consider a wave incident in port 1. This wave splits equally into two waves traveling around the ring circuit in opposite directions. The two waves will arrive in phase at ports 2 and 4 and out of phase at port 3. Thus ports 1 and 3 are uncoupled. Similarly, ports 2 and 4 are uncoupled since the two paths coupling these ports differ in length by $\lambda_g/2$.

The magic T is used in microwave impedance bridges (similar to the well-known Wheatstone bridge), in frequency discriminator circuits, and in balanced mixers, as well as in other applications. The balanced mixer is described below.

Balanced microwave mixer

The balanced microwave mixer is used in superheterodyne receivers to balance out the local-oscillator noise at the input to the intermediate-frequency, or i-f, amplifier. A typical arrangement is shown in Fig. 6.21. The microwave crystal diodes are located at the two ends of the through arm of the magic T. The local-oscillator signal is fed into the H-plane arm and will arrive at the diodes in phase. The signal is fed into the

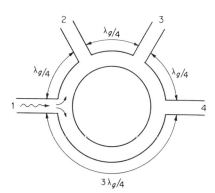

Fig. 6.20 Ring-type hybrid junction.

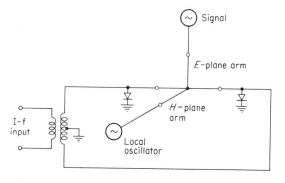

Fig. 6.21 Balanced microwave mixer.

E-plane arm and will arrive at the two diodes out of phase. The i-f signal, or difference frequency signal, produced by the mixing of the local-oscillator and signal frequencies in the nonlinear diodes, will be 180° out of phase. However, the local-oscillator noise will be in phase at the diodes. By feeding the diode outputs into a balanced i-f input (push-pull arrangement), the oscillator noise is seen to cancel but the i-f signals add up in phase. Another advantage obtained with the use of the magic T is the prevention of radiation by the local oscillator. The E- and H-plane arms are uncoupled, so that the local-oscillator signal cannot be coupled to the antenna, which is connected to the E-plane arm.

6.6 Microwave propagation in ferrites

The development of ferrite materials suitable for use at microwave frequencies has resulted in a large number of new microwave devices. A number of these have nonreciprocal electrical properties; i.e., the transmission coefficient through the device is not the same for different directions of propagation. An understanding of the operation of ferrite devices may be obtained once the basic nature of microwave propagation in an infinite unbounded ferrite medium is understood. In this section we consider plane-wave propagation in an infinite ferrite medium with a static biasing magnetic field B_0 present. It will be found that the natural modes of propagation in the direction of B_0 are left and right circularly polarized waves and that these modes have different propagation constants. In addition, we shall find that the permeability of the ferrite is not a single scalar quantity, but instead is a tensor, which can be represented as a matrix.

Ferrites are ceramiclike materials with specific resistivities that may be as much as 10^{14} greater than that of metals and with dielectric constants around 10 to 15 or greater. Ferrites are made by sintering a mixture of metallic oxides and have the general chemical composition $MO \cdot Fe_2O_3$, where M is a divalent metal such as manganese, magnesium, iron, zinc, nickel, cadmium, etc., or a mixture of these. Relative permeabilities of several thousand are common. The magnetic properties of ferrites arise mainly from the magnetic dipole moment associated with the electron spin. By treating the spinning electron as a gyroscopic top, a classical picture of the magnetization process and, in particular, the anisotropic magnetic properties may be obtained.

The electron has a number of intrinsic properties such as a charge of $-e = -1.602 \times 10^{-19}$ coul, a mass $w = 9.107 \times 10^{-31}$ kg, an angular momentum P equal in magnitude to $\frac{1}{2}\hbar$, or 0.527×10^{-34} joule-s ($\hbar$ is Planck's constant divided by 2π), and a magnetic dipole moment m equal to one Bohr magneton, that is, $m = e\hbar/2w = 9.27 \times 10^{-24}$ amp-m². For the electron, the angular momentum **P** and magnetic dipole moment **m** are antiparallel. The ratio of the magnetic moment to the angular

Sec. 6.6 — Passive microwave devices

momentum is called the gyromagnetic ratio γ; that is,

$$\gamma = \frac{m}{P} \tag{6.49}$$

If an electron is located in a uniform static magnetic field $\mathbf{B}_0$, a torque $\mathbf{T}$ given by

$$\mathbf{T} = \mathbf{m} \times \mathbf{B}_0 = -\gamma \mathbf{P} \times \mathbf{B}_0 \tag{6.50}$$

will be exerted on the dipole moment. This torque will cause the dipole axis to precess about an axis parallel to $\mathbf{B}_0$, as illustrated in Fig. 6.22. The equation of motion is obtained from the condition that the rate of change of angular momentum is equal to the torque and hence is

$$\frac{d\mathbf{P}}{dt} = \mathbf{T} = -\gamma \mathbf{P} \times \mathbf{B}_0 = \boldsymbol{\omega}_0 \times \mathbf{P} \tag{6.51a}$$

or

$$\gamma P B_0 \sin \phi = \omega_0 P \sin \phi = m B_0 \sin \phi \tag{6.51b}$$

where $\boldsymbol{\omega}_0$ is the vector-precession angular velocity directed along $\mathbf{B}_0$, and ϕ is the angle between $\mathbf{m}$ and $\mathbf{B}_0$. For free precession the angular velocity ω_0 is given by

$$\omega_0 = \gamma B_0$$

and is independent of the angle ϕ. The angular velocity ω_0 is often called the Larmor frequency.

If a small a-c magnetic field is superimposed on the static field $\mathbf{B}_0$, the magnetic dipole moment will undergo a forced precession. Of particular interest is the case where the a-c magnetic field is circularly polarized in the plane perpendicular to $\mathbf{B}_0$. A circularly polarized field results when the x and y components of the a-c field are equal in magnitude and 90°

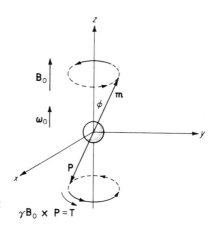

Fig. 6.22 Free precession of spinning electron.

apart in time phase. Thus let the a-c magnetic field be given by the phasor

$$\mathbf{B}_1^- = B_1(\mathbf{a}_x + j\mathbf{a}_y) \tag{6.52a}$$

If we assume B_1 to be real, the physical field is given by

$$\mathbf{B}_1^- = B_1 \operatorname{Re}(\mathbf{a}_x + j\mathbf{a}_y)e^{j\omega t} = B_1(\mathbf{a}_x \cos \omega t - \mathbf{a}_y \sin \omega t) \tag{6.52b}$$

The resultant field has a constant magnitude B_1, but the orientation of the field in space changes or rotates with time. At time t the resultant field vector makes an angle

$$\tan^{-1}\frac{B_y}{B_x} = -\tan^{-1}\tan \omega t = -\omega t$$

with the x axis and hence rotates at the rate $-\omega$, as in Fig. 6.23a. It is this rotation of the field vector in space that results in the field being called circularly polarized. If the above a-c magnetic field is that of a wave propagating in the z direction, it is said to be left circularly polarized. If the direction of rotation is clockwise, looking in the direction of propagation, the wave is called right circularly polarized. The latter type of wave would have an a-c magnetic field given by

$$\mathbf{B}_1^+ = B_1(\mathbf{a}_x - j\mathbf{a}_y) \tag{6.53}$$

With a left circularly polarized a-c magnetic field superimposed on the static field $\mathbf{B}_0 = B_0\mathbf{a}_z$, the resultant total field $\mathbf{B}_t$ is inclined at an angle $\theta = \tan^{-1}(B_1/B_0)$ with the z axis and rotates at a rate $-\omega$ about the z axis, as illustrated in Fig. 6.24a. Under steady-state conditions the magnetic dipole axis will be forced to precess about the z axis at the same rate. Thus the precession angle ϕ will have to be less than θ, as in Fig. 6.24a, in order to obtain a torque to cause precession in a counterclock-

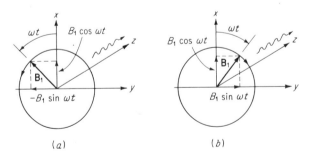

Fig. 6.23 Magnetic field for circularly polarized waves. (a) Left, or negative, circular polarization; (b) right, or positive, circular polarization.

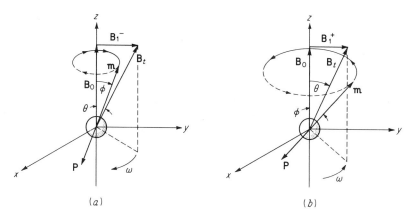

Fig. 6.24 Forced precession of spinning electron.

wise direction. The equation of motion (6.51) gives

$$\mathbf{T} = \mathbf{m} \times \mathbf{B}_t = -\gamma \mathbf{P} \times \mathbf{B}_t = \frac{d\mathbf{P}}{dt} = -\omega \mathbf{a}_z \times \mathbf{P}$$

or

$$-\gamma P B_t \sin(\theta - \phi) = -\omega P \sin \phi$$

Expanding $\sin(\theta - \phi)$, replacing $B_t \sin \theta$ by B_1, $B_t \cos \theta$ by B_0, and solving for $\tan \phi$ give

$$\tan \phi = \frac{\gamma B_1}{\gamma B_0 + \omega} = \frac{\gamma B_1}{\omega_0 + \omega} \tag{6.54}$$

The component of **m** which rotates in synchronism with $\mathbf{B}_1^-$ in the xy plane for the left (also called negative) circularly polarized a-c field is $m^- = m \sin \phi = m_0 \tan \phi$, where $m_0 = m \cos \phi$ is the z-directed component of **m**. For B_1 very small compared with B_0, the angle ϕ is small, so that m_0 is approximately equal to m. Using (6.54) gives

$$m^- = m_0 \tan \phi = \frac{\gamma m_0 B_1}{\omega_0 + \omega} \tag{6.55}$$

If we have a right (or positive) circularly polarized a-c field superimposed on the static field $\mathbf{B}_0$, the forced precession is in a clockwise sense about the z axis. A torque giving precession in this direction is obtained only if the angle ϕ is greater than the angle θ, as in Fig. 6.24b. In this case the equation of motion (6.51) gives

$$\gamma B_t \sin(\phi - \theta) = \omega \sin \phi$$

from which we obtain

$$\tan \phi = \frac{\gamma B_1}{\omega_0 - \omega} \tag{6.56}$$

The component of magnetization in the xy plane rotating in synchronism with the positive circularly polarized a-c field is

$$m^+ = m_0 \tan \phi = \frac{\gamma m_0 B_1}{\omega_0 - \omega} \tag{6.57}$$

The foregoing discussion has pointed out the essential features of the motion of a single spinning electron in a magnetic field consisting of a static field along the z axis and a small circularly polarized a-c field in the xy plane. A ferrite material may be regarded as a collection of N effective spinning electrons per unit volume. Since the spacing between electrons is of atomic dimensions, we may regard the density of magnetic dipoles per unit volume as a smeared-out continuous distribution from a macroscopic viewpoint. The total magnetic dipole moment per unit volume is $\mathbf{M} = N\mathbf{m}$. When the static field $\mathbf{B}_0$ is large enough to cause saturation of the magnetization in the ferrite, $\mathbf{M} = \mathbf{M}_s$. In a saturated ferrite all the spins are very tightly coupled, so that the whole sample acts essentially as a large single magnetic dipole. The magnetization $\mathbf{M}_s$ produces a contribution to the total internal $\mathbf{B}$ field according to the relation $\mathbf{B} = \mu_0(\mathbf{H}_0 + \mathbf{M}_s)$. The torque acting on $\mathbf{M}_s$ is due only to the field $\mathbf{B}_0 = \mu_0 \mathbf{H}_0$, since the cross product of $\mu_0 \mathbf{M}_s$ with $\mathbf{M}_s$ is zero and hence does not contribute to the torque. Thus, in the equation of motion for the magnetization, the field producing the torque is $\mu_0 \mathbf{H}_t$, where $\mathbf{H}_t$ is the total static plus a-c magnetic field intensity in the ferrite medium. That is,

$$\frac{d\mathbf{M}}{dt} = -\gamma(\mathbf{M} \times \mathbf{B}) = -\gamma\mu_0 \mathbf{M} \times (\mathbf{H} + \mathbf{M}) = -\gamma\mu_0 \mathbf{M} \times \mathbf{H}$$

If the magnetic field intensity in the ferrite is $\mathbf{H}_0 + \mathbf{H}_1^\pm$, where $\mathbf{H}_1^\pm$ is a circularly polarized a-c field, the resultant a-c magnetization will be given by expressions analogous to (6.55) and (6.57), but with m_0 replaced by $M_0 = Nm_0$. The total a-c magnetic field in the xy plane is the field $\mu_0 \mathbf{H}_1^\pm = \mathbf{B}_1^\pm$ plus the contribution from the a-c magnetization. Thus the total a-c fields for positive and negative circular polarization are [we are using the formula $\mathbf{B} = \mu_0(\mathbf{H} + \mathbf{M})$]

$$\mathbf{B}^+ = \mu_0 \mathbf{M}^+ + \mathbf{B}_1^+ = \mu_0(N\mathbf{m}^+ + \mathbf{H}_1^+) = \mu_0\left(1 + \frac{\mu_0 \gamma M_0}{\omega_0 - \omega}\right)\mathbf{H}_1^+ \tag{6.58a}$$

$$\mathbf{B}^- = \mu_0 \mathbf{M}^- + \mathbf{B}_1^- = \mu_0\left(1 + \frac{\mu_0 \gamma M_0}{\omega_0 + \omega}\right)\mathbf{H}_1^- \tag{6.58b}$$

where $M_0 = M \cos \phi$, $B_1 = \mu_0 H_1$, and $\mathbf{H}_1^+ = H_1(\mathbf{a}_x - j\mathbf{a}_y)$ in (6.58a), and $\mathbf{H}_1^-$ in (6.58b) equals $H_1(\mathbf{a}_x + j\mathbf{a}_y)$, as seen from (6.53) and (6.52a). The quantity M may be replaced by the saturation magnetization M_s in the ferrite since the static field B_0 is usually large enough to cause saturation.

If we assume that $B_1 \ll B_0$, so that $M_0 \approx M = M_s$, the effective permeabilities for positive and negative circularly polarized a-c fields are seen to be given by

$$\mu_+ = \mu_0 \left(1 + \frac{\gamma \mu_0 M_s}{\omega_0 - \omega}\right) \qquad (6.59a)$$

$$\mu_- = \mu_0 \left(1 + \frac{\gamma \mu_0 M_s}{\omega_0 + \omega}\right) \qquad (6.59b)$$

Plane circularly polarized TEM waves propagating in the direction of the static field $\mathbf{B}_0$ will have propagation constants

$$\beta_+ = \omega \sqrt{\epsilon \mu_+} \qquad (6.60a)$$

$$\beta_- = \omega \sqrt{\epsilon \mu_-} \qquad (6.60b)$$

where ϵ is the dielectric constant of the ferrite. The significance of the inequality of β_+ and β_- is discussed later. The results expressed by (6.60) are also derived in an alternative way later.

If small-signal conditions $B_1 \ll B_0$ are not assumed, we cannot replace M_0 by M_s. In place of (6.54) and (6.56), which give solutions for $\tan \phi$, we can solve for $\sin \phi$ to obtain, respectively,

$$\sin \phi = \frac{\tan \phi}{\sqrt{1 + \tan^2 \phi}} = \frac{\gamma B_1}{\sqrt{(\gamma B_1)^2 + (\omega_0 + \omega)^2}} \qquad (6.61a)$$

$$\sin \phi = \frac{\gamma B_1}{\sqrt{(\gamma B_1)^2 + (\omega_0 - \omega)^2}} \qquad (6.61b)$$

The magnetizations M^+ and M^- are given by

$$M^+ = M_s \sin \phi = \frac{\gamma \mu_0 M_s H_1}{\sqrt{(\gamma \mu_0 H_1)^2 + (\omega_0 - \omega)^2}} \qquad (6.62a)$$

$$M^- = \frac{\gamma \mu_0 M_s H_1}{\sqrt{(\gamma \mu_0 H_1)^2 + (\omega_0 + \omega)^2}} \qquad (6.62b)$$

It is seen that the a-c magnetization depends nonlinearly on the a-c field strength H_1, and hence, under large-signal conditions, μ_+ and μ_- will be functions of the applied a-c field strength. The nonlinear behavior of ferrites under large-signal conditions results in the generation of harmonics of the fundamental frequency ω. For this reason ferrites may be used as harmonic generators.†

For M^- it is clear that, if $B_1 \ll B_0$, that is, $\gamma B_1 \ll \gamma B_0 = \omega_0$, then (6.62b) is well approximated by

$$M^- = \frac{\gamma \mu_0 M_s H_1}{\omega_0 + \omega}$$

Similarly, if ω is not too close to the resonant frequency ω_0, we see that

† W. P. Ayres, P. H. Vartanian, and J. L. Melchor, Frequency Doubling in Ferrites, *J. Appl. Phys.*, vol. 27, p. 188, 1956; Microwave Frequency Doubling from 9 kmc to 18 kmc in Ferrites, *Proc. IRE*, vol. 45, pp. 643–646, May, 1957.

(6.62a) becomes

$$M^+ = \frac{\gamma\mu_0 M_s H_1}{\omega_0 - \omega}$$

These latter values of M^+ and M^- lead directly to the expressions (6.59) for μ_+ and μ_-. In any practical ferrite medium, damping effects are always present, so that M^+ will remain finite and small compared with M_s even when $\omega = \omega_0$. Thus, for small-signal conditions, we can assume that $M_0 \approx M_s$ in an actual ferrite medium. Damping effects are discussed in more detail later.

It will be instructive to study the propagation of a plane wave in an unbounded ferrite medium by solving Maxwell's equations directly, together with the equation of motion for the magnetization. This analysis will illustrate the general technique of linearization to be applied in the small-signal analysis of propagation through a medium such as a ferrite. However, it will not give as clear an insight into the physical reason why μ_+ and μ_- are different, as the analysis above did. That is, basically, μ_+ and μ_- differ because the precession angle ϕ must be greater than the angle θ in one case and less than θ in the other case, and hence the projection of the magnetic dipole moment onto the xy plane is different in the two cases.

Consider an infinite unbounded ferrite medium with an applied static magnetic field $\mathbf{B}_0 = \mu_0 \mathbf{H}_0 = B_0 \mathbf{a}_z$. Let the magnetization in the ferrite be $\mathbf{M}_s$ per unit volume when no time-varying magnetic field is applied. When a time-varying magnetic field $\mu_0 \mathbf{H}$ is also applied, a time-varying component $\mathbf{M}$ of magnetization will be produced. The equation of motion for the total magnetization per unit volume is similar to that for a single electron, and hence we have

$$\frac{d(\mathbf{M}_s + \mathfrak{M})}{dt} = \frac{d\mathfrak{M}}{dt} = -\gamma[(\mathbf{M}_s + \mathfrak{M}) \times (\mathbf{B}_0 + \mu_0 \mathfrak{IC})]$$

$$= -\gamma\mu_0(\mathbf{M}_s \times \mathbf{H}_0 + \mathbf{M}_s \times \mathfrak{IC} + \mathfrak{M} \times \mathbf{H}_0 + \mathfrak{M} \times \mathfrak{IC}) \quad (6.63)$$

If small-signal conditions are assumed, i.e.,

$$|\mathfrak{M}| \ll |\mathbf{M}_s| \quad \text{and} \quad |\mathfrak{IC}| \ll |\mathbf{H}_0|$$

the nonlinear term $\mathfrak{M} \times \mathfrak{IC}$ in (6.63) may be dropped. We then obtain for the equation of motion the linearized equation

$$\frac{d\mathfrak{M}}{dt} = -\gamma(\mu_0 \mathbf{M}_s \times \mathfrak{IC} + \mathfrak{M} \times \mathbf{B}_0) \quad (6.64)$$

since $\mathbf{M}_s \times \mathbf{B}_0 = 0$, because the saturation magnetization is in the same direction as the applied static field.

Let the time dependence be $e^{j\omega t}$, and let $\mathfrak{M}$ and $\mathfrak{IC}$ be represented by the phasors $\mathbf{M}$ and $\mathbf{H}$. From (6.64) we obtain

$$j\omega \mathbf{M} + \gamma \mathbf{M} \times \mathbf{B}_0 = j\omega \mathbf{M} + \omega_0 \mathbf{M} \times \mathbf{a}_z = -\gamma\mu_0 \mathbf{M}_s \times \mathbf{H}$$

where $\omega_0 = \gamma\mu_0 H_0 = \gamma B_0$. In component form we have

$$j\omega M_x + \omega_0 M_y = \gamma M_s \mu_0 H_y$$
$$j\omega M_y - \omega_0 M_x = -\gamma M_s \mu_0 H_x$$
$$j\omega M_z = 0$$

The solution of these equations for M_x, M_y, and M_z gives

$$M_x = \frac{\omega_0 \gamma \mu_0 M_s H_x + j\omega\gamma\mu_0 M_s H_y}{\omega_0^2 - \omega^2} \tag{6.65a}$$

$$M_y = \frac{\omega_0 \gamma \mu_0 M_s H_y - j\omega\gamma\mu_0 M_s H_x}{\omega_0^2 - \omega^2} \tag{6.65b}$$

$$M_z = 0 \tag{6.65c}$$

In the solution of Maxwell's equations it is convenient not to have to deal explicitly with the magnetization. The magnetization may be eliminated by introducing the magnetic susceptibility and permeability. In the scalar case this is done by means of the relations $M = \chi_m H$, $B = \mu_0(M + H) = \mu_0(1 + \chi_m)H = \mu H$. For a ferrite similar relations may be used, but χ_m and μ will not be scalar quantities. In matrix form (6.65) gives

$$\begin{bmatrix} M_x \\ M_y \\ M_z \end{bmatrix} = \begin{bmatrix} \chi_{xx} & \chi_{xy} & 0 \\ \chi_{yx} & \chi_{yy} & 0 \\ 0 & 0 & 0 \end{bmatrix} \begin{bmatrix} H_x \\ H_y \\ H_z \end{bmatrix} \tag{6.66}$$

where

$$\chi_{xx} = \chi_{yy} = \frac{\omega_0 \gamma \mu_0 M_s}{\omega_0^2 - \omega^2}$$

$$\chi_{xy} = -\chi_{yx} = \frac{j\omega\gamma\mu_0 M_s}{\omega_0^2 - \omega^2}$$

The matrix with the parameters χ_{xx}, χ_{xy}, χ_{yx}, and χ_{yy} in (6.66) represents the susceptibility tensor of the ferrite. The relation between the a-c **B** and **H** fields is

$$\mathbf{B} = \mu_0(\mathbf{H} + \mathbf{M})$$

or

$$\begin{bmatrix} B_x \\ B_y \\ B_z \end{bmatrix} = \mu_0 \begin{bmatrix} 1 + \chi_{xx} & \chi_{xy} & 0 \\ \chi_{yx} & 1 + \chi_{yy} & 0 \\ 0 & 0 & 1 \end{bmatrix} \begin{bmatrix} H_x \\ H_y \\ H_z \end{bmatrix} \tag{6.67}$$

The matrix relating the components of **H** to **B** in (6.67) is the permeability tensor for the ferrite. It will be denoted by a boldface μ with an overbar, i.e.,

$$\bar{\mu} = \mu_0 \begin{bmatrix} 1 + \chi_{xx} & \chi_{xy} & 0 \\ \chi_{yx} & 1 + \chi_{yy} & 0 \\ 0 & 0 & 1 \end{bmatrix}$$

In shorthand notation the matrix equation (6.67) will be written as

$$\mathbf{B} = \bar{\mu} \cdot \mathbf{H} \tag{6.68}$$

In the literature the minus sign in the equation of motion is often deleted, and this amounts to replacing γ by $-\gamma$ in the equations used in this text.

Losses present in a ferrite may be accounted for in a phenomenological way by introducing into the equation of motion a damping term that will produce a torque tending to reduce the precession angle ϕ. The following modified form of the equation has often been used in practice:

$$\frac{d\mathfrak{M}}{dt} = -\gamma\mu_0(\mathbf{M}_s + \mathfrak{M}) \times (\mathbf{H}_0 + \mathfrak{K}) + \frac{\alpha}{\mathfrak{M}} \mathfrak{M} \times \frac{d\mathfrak{M}}{dt} \tag{6.69}$$

where α is a dimensionless damping constant. With a small-signal analysis, the elements of the susceptibility matrix are now found to be†

$$\chi_{xx} = \chi_{yy} = \chi' - j\chi'' = \chi \tag{6.70a}$$

$$\chi_{xy} = -\chi_{yx} = j(K' - jK'') = jK \tag{6.70b}$$

where

$$\chi' = \frac{\omega_0\omega_m(\omega_0^2 - \omega^2) + \omega_m\omega_0\omega^2\alpha^2}{[\omega_0^2 - \omega^2(1 + \alpha^2)]^2 + 4\omega_0^2\omega^2\alpha^2}$$

$$\chi'' = \frac{\omega\omega_m\alpha[\omega_0^2 + \omega^2(1 + \alpha^2)]}{[\omega_0^2 - \omega^2(1 + \alpha^2)]^2 + 4\omega_0^2\omega^2\alpha^2}$$

$$K' = \frac{\omega\omega_m[\omega_0^2 - \omega^2(1 + \alpha^2)]}{[\omega_0^2 - \omega^2(1 + \alpha^2)]^2 + 4\omega_0^2\omega^2\alpha^2}$$

$$K'' = \frac{2\omega^2\omega_0\omega_m\alpha}{[\omega_0^2 - \omega^2(1 + \alpha^2)]^2 + 4\omega_0^2\omega^2\alpha^2}$$

and

$$\omega_m = \mu_0\gamma M_s$$

As we derived the permeability tensor to use in the constitutive equation relating $\mathbf{B}$ and $\mathbf{H}$, the only remaining task is to find solutions for Maxwell's equations in the form

$$\nabla \times \mathbf{E} = -j\omega\mathbf{B} = -j\omega\bar{\mu} \cdot \mathbf{H} \tag{6.71a}$$

$$\nabla \times \mathbf{H} = j\omega\epsilon\mathbf{E} \tag{6.71b}$$

$$\nabla \cdot \mathbf{B} = \nabla \cdot \mathbf{E} = 0 \tag{6.71c}$$

† R. F. Soohoo, "Theory and Application of Ferrites," chap. 5, Prentice-Hall, Inc., Englewood Cliffs, N.J., 1960.

Sec. 6.6 Passive microwave devices

For a TEM wave propagating in the z direction, i.e., along the direction of $\mathbf{B}_0$, solutions are readily found. Let the electric field be given by

$$\mathbf{E} = \mathbf{E}_0 e^{-j\beta z}$$

where $\mathbf{E}_0$ is a constant vector in the xy plane. Equation (6.71a) gives

$$\nabla \times \mathbf{E} = -\mathbf{E}_0 \times \nabla e^{-j\beta z} = j\beta \mathbf{E}_0 \times \mathbf{a}_z e^{-j\beta z} = -j\omega \bar{\mathbf{\mu}} \cdot \mathbf{H}$$

Let the solution for $\mathbf{H}$ be $\mathbf{H}_0 e^{-j\beta z}$, where $\mathbf{H}_0$ is also a constant vector in the xy plane. From (6.71b) we obtain

$$j\beta \mathbf{H}_0 \times \mathbf{a}_z = j\omega \epsilon \mathbf{E}_0$$

If this equation is substituted into the equation

$$j\beta \mathbf{E}_0 \times \mathbf{a}_z = -j\omega \bar{\mathbf{\mu}} \cdot \mathbf{H}_0$$

so as to eliminate $\mathbf{E}_0$, we obtain

$$j\beta \frac{j\beta}{j\omega\epsilon} (\mathbf{H}_0 \times \mathbf{a}_z) \times \mathbf{a}_z = -j\omega \bar{\mathbf{\mu}} \cdot \mathbf{H}_0$$

Expanding the left-hand side gives (note that $\mathbf{a}_z \cdot \mathbf{H}_0 = 0$)

$$\beta^2 \mathbf{H}_0 = \omega^2 \epsilon \bar{\mathbf{\mu}} \cdot \mathbf{H}_0 \tag{6.72a}$$

This equation may be written in the following matrix form:

$$\begin{bmatrix} \beta^2 - \omega^2 \epsilon \mu_0 (1+\chi) & -j\omega^2 \epsilon \mu_0 K \\ j\omega^2 \epsilon \mu_0 K & \beta^2 - \omega^2 \epsilon \mu_0 (1+\chi) \end{bmatrix} \begin{bmatrix} H_{0x} \\ H_{0y} \end{bmatrix} = 0 \tag{6.72b}$$

For a nontrivial solution for $\mathbf{H}_0$, the determinant must vanish. This condition yields the following eigenvalue equation for the propagation constant β:

$$[\beta^2 - \omega^2 \epsilon \mu_0 (1+\chi)]^2 - \omega^4 \epsilon^2 \mu_0^2 K^2 = 0$$

or

$$\beta^2 = \omega^2 \epsilon \mu_0 (1+\chi) \pm \omega^2 \epsilon \mu_0 K \tag{6.73}$$

If we substitute for χ and K and assume a lossless ferrite, so that $\chi'' = K'' = 0$, we readily find that the two solutions for β^2 are

$$\beta^2 = \beta_+^2 = \omega^2 \epsilon \mu_+ \tag{6.74a}$$

$$\beta^2 = \beta_-^2 = \omega^2 \epsilon \mu_- \tag{6.74b}$$

where μ_+ and μ_- are given by Eq. (6.59).

For each eigenvalue or solution for β^2, the ratio of H_{0x} to H_{0y} is determined. For the solution β_+^2 the first equation in the pair of equations (6.72b) gives

$$[\beta_+^2 - \omega^2 \epsilon \mu_0 (1+\chi)] H_{0x} - j\omega^2 \epsilon \mu_0 K H_{0y} = 0$$

or

$$\frac{H_{0x}}{H_{0y}} = j \tag{6.75a}$$

But this condition means that $\mathbf{H}_0 = H_0(\mathbf{a}_x - j\mathbf{a}_y)$, or is a positive circularly polarized wave. Similarly, the solution β_-^2 gives

$$\frac{H_{0x}}{H_{0y}} = -j \tag{6.75b}$$

which specifies a negative circularly polarized wave. Therefore it is seen that the natural modes of propagation along the direction of the static field in a ferrite are circularly polarized TEM waves. If directions of propagation other than along $\mathbf{B}_0$ were considered, it would be found that there are, again, two modes of propagation, but these are no longer circularly polarized TEM waves. For a linearly polarized wave propagating along $\mathbf{B}_0$, the plane of polarization rotates, since β_+ and β_- are not equal. This phenomenon is a nonreciprocal one, and is called Faraday rotation. It is discussed in the following section.

6.7 Faraday rotation

Consider an infinite lossless ferrite medium with a static field $\mathbf{B}_0$ applied along the z direction. Let a plane TEM wave that is linearly polarized along the x axis at $z = 0$ be propagating in the z direction, as in Fig. 6.25. We shall show that the plane of polarization of this wave rotates as it propagates (Faraday rotation). The linearly polarized wave may be decomposed into the sum of a left and right circularly polarized wave as follows:

$$\mathbf{E} = \mathbf{a}_x E_0 = (\mathbf{a}_x + j\mathbf{a}_y)\frac{E_0}{2} + (\mathbf{a}_x - j\mathbf{a}_y)\frac{E_0}{2} \qquad z = 0 \tag{6.76}$$

The component waves propagate with different phase constants β_+ and

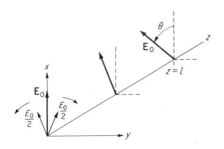

Fig. 6.25 Faraday rotation.

β_-, and hence the wave at $z = l$ becomes

$$\begin{aligned}\mathbf{E} &= (\mathbf{a}_x + j\mathbf{a}_y)\frac{E_0}{2}e^{-j\beta_-l} + (\mathbf{a}_x - j\mathbf{a}_y)\frac{E_0}{2}e^{-j\beta_+l}\\
&= \mathbf{a}_x\frac{E_0}{2}(e^{-j\beta_-l} + e^{-j\beta_+l}) + j\mathbf{a}_y\frac{E_0}{2}(e^{-j\beta_-l} - e^{-j\beta_+l})\\
&= \frac{E_0}{2}e^{-j(\beta_-+\beta_+)l/2}[\mathbf{a}_x(e^{-j(\beta_--\beta_+)l/2} + e^{j(\beta_--\beta_+)l/2})\\
&\qquad\qquad\qquad\qquad + j\mathbf{a}_y(e^{-j(\beta_--\beta_+)l/2} - e^{j(\beta_--\beta_+)l/2})]\\
&= E_0 e^{-j(\beta_-+\beta_+)l/2}\left[\mathbf{a}_x\cos(\beta_+ - \beta_-)\frac{l}{2} - \mathbf{a}_y\sin(\beta_+ - \beta_-)\frac{l}{2}\right]\quad(6.77)\end{aligned}$$

This resultant wave is a linearly polarized wave that has undergone a phase delay of $(\beta_- + \beta_+)l/2$. The new plane of polarization makes an angle

$$\theta = \tan^{-1}\frac{E_y}{E_x} = \tan^{-1}\left[-\tan(\beta_+ - \beta_-)\frac{l}{2}\right] = -(\beta_+ - \beta_-)\frac{l}{2} \quad (6.78)$$

with respect to the x axis. When $\omega < \omega_0$, that is, below the ferrite resonant frequency, β_+ is greater than β_- and the plane of polarization rotates counterclockwise, looking in the direction of propagation (Fig. 6.25). The rate of rotation is $(\beta_+ - \beta_-)/2$ rad/m. Rotation of 100° or more per centimeter is typical in ferrites at a frequency of 10,000 Mc/sec.

If the direction of propagation is reversed, the plane of polarization continues to rotate in the same direction. Thus, if we consider the propagation of the wave described by (6.77) back to the plane $z = 0$, the original polarization direction is not restored; instead the wave will arrive back at $z = 0$ polarized at an angle 2θ relative to the x axis. This result is easily derived by noting that if the component circularly polarized waves making up the linearly polarized wave in (6.77) are propagated from $z = l$ back to $z = 0$, they undergo additional phase delays of amount $\beta_+ l$ and $\beta_- l$ and become, at $z = 0$,

$$\mathbf{E} = (\mathbf{a}_x + j\mathbf{a}_y)\frac{E_0}{2}e^{-2j\beta_-l} + (\mathbf{a}_x - j\mathbf{a}_y)\frac{E_0}{2}e^{-2j\beta_+l} \quad (6.79)$$

By analogy with (6.77) it is now clear that the new direction of polarization at $z = 0$ makes an angle

$$2\theta = -(\beta_+ - \beta_-)l$$

with respect to the x axis. Thus Faraday rotation is a nonreciprocal effect.

A practical ferrite medium has finite losses, and this will have a significant influence on the propagation. The propagation constants $\gamma_+ = j\beta_+ + \alpha_+$ and $\gamma_- = j\beta_- + \alpha_-$ for circularly polarized waves will have unequal attenuation constants as well as unequal phase constants. When losses are present, the propagation constants are given by (6.73) if β^2 is replaced by $-\gamma^2$; thus

$$\gamma_+ = j\omega \sqrt{\mu_0 \epsilon}\, (1 + \chi' - j\chi'' + K' - jK'')^{\frac{1}{2}} \tag{6.80a}$$

$$\gamma_- = j\omega \sqrt{\mu_0 \epsilon}\, (1 + \chi' - j\chi'' - K' + jK'')^{\frac{1}{2}} \tag{6.80b}$$

where χ', χ'', K', K'' are given after (6.70). The solutions of (6.80) are

$$\beta_\pm = \frac{\omega \sqrt{\mu_0 \epsilon}}{\sqrt{2}} [1 + \chi' \pm K' + \sqrt{(1 + \chi' \pm K')^2 + (\chi'' \pm K'')^2}]^{\frac{1}{2}} \tag{6.81a}$$

$$\alpha_\pm = \omega^2 \mu_0 \epsilon \frac{\chi'' \pm K''}{2\beta_\pm} = \frac{\omega^2 \epsilon \mu''_\pm}{2\beta_\pm} \tag{6.81b}$$

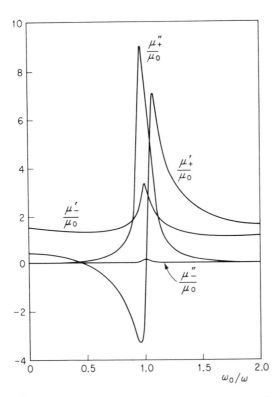

Fig. 6.26 Real and imaginary components of permeability for circularly polarized waves in a ferrite as a function of ω/ω_0 for $\omega/2\pi = 10$ kMc, $\omega_m/2\pi = 5.6$ kMc, $\alpha = 0.05$.

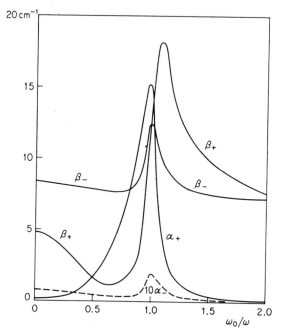

Fig. 6.27 Propagation and attenuation constants for circularly polarized waves in a ferrite, with parameters given in Fig. 6.26 ($\epsilon = 10\epsilon_0$). Note that $10\alpha_-$ is plotted since α_- is very small.

The permeabilities for circularly polarized waves are

$$\mu_\pm = \mu'_\pm - j\mu''_\pm = \mu_0(1 + x' - jx'' \pm K' \mp jK'') \tag{6.82}$$

The values of $\mu'_\pm$, $\mu''_\pm$ as given by (6.70) and (6.82) and the propagation factors $\beta_\pm$ and $\alpha_\pm$ are plotted in Figs. 6.26 and 6.27 for a typical ferrite with parameters $\omega_m = 2\pi \times 5.6 \times 10^9$, $\alpha = 0.05$, as a function of ω_0 at a frequency of 10,000 Mc. Note that ω_0 equals $2\pi \times 2.8$ Mc/oersted of applied field H_0 and that $4\pi \times 10^{-3}$ oersted is a field strength of 1 amp/m. The value of ω_m chosen corresponds to a saturation magnetization of 2,000 gauss, or $\mu_0 M_s$ equal to 0.2 weber/m². The curves in Fig. 6.27 show that α_- is always very small but that α_+ is large in the vicinity of the resonant frequency $\omega_0 \approx \omega$. For ω_0 considerably above ω, the attenuation α_+ becomes small, but in this region β_+ and β_- do not differ greatly, so that the rate of Faraday rotation would be small. At low values of ω_0, that is, small H_0, the rate of rotation would be much greater, particularly in the region where β_+ goes through a minimum, which occurs in the range where μ'_+ is negative. Note also that $\beta_+ > \beta_-$ when $\omega_0 > \omega$ but that $\beta_+ < \beta_-$ for $\omega_0 < \omega$. The direction of Faraday rotation is thus different in the two regions above and below the resonant frequency.

6.8 Microwave devices employing Faraday rotation

Gyrator

A gyrator is defined as a two-port device that has a relative difference in phase shift of 180° for transmission from port 1 to port 2 as compared with the phase shift for transmission from port 2 to port 1. A gyrator may be obtained by employing the nonreciprocal property of Faraday rotation. Figure 6.28 illustrates a typical microwave gyrator. It consists of a rectangular guide with a 90° twist connected to a circular guide, which in turn is connected to another rectangular guide at the other end. The two rectangular guides have the same orientation at the input ports. The circular guide contains a thin cylindrical rod of ferrite with the ends tapered to reduce reflections. A static axial magnetic field is applied so as to produce 90° Faraday rotation of the TE_{11} dominant mode in the circular guide. Consider a wave propagating from left to right. In passing through the twist the plane of polarization is rotated by 90° in a counterclockwise direction. If the ferrite produces an additional 90° of rotation, the total angle of rotation will be 180°, as indicated in Fig. 6.28. For a wave propagating from right to left, the Faraday rotation is still 90° in the same sense. However, in passing through the twist, the next 90° of rotation is in a direction to cancel the Faraday rotation. Thus, for transmission from port 2 to port 1, there is no net rotation of the plane of polarization. The 180° rotation for transmission from port 1 to port 2 is equivalent to an additional 180° of phase shift since it reverses the polarization of the field. It is apparent, then, that the device just described satisfies the definition of a gyrator.

If the inconvenience of having the input and output rectangular guides oriented at 90° can be tolerated, a gyrator without a 90° twist section can be built. With reference to Fig. 6.29, it is seen that if the ferrite produces

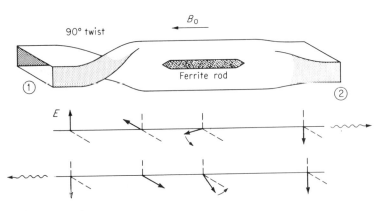

Fig. 6.28 A microwave gyrator.

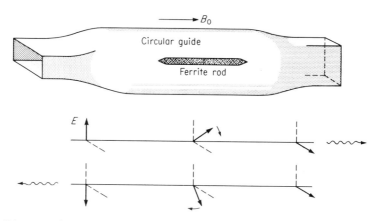

Fig. 6.29 A gyrator without a twist section.

90° of rotation and the output guide is rotated by 90° relative to the input guide, the emerging wave will have the right polarization to propagate in the output guide. When propagation is from port 2 to port 1, the wave arriving in guide 1 will have its polarization changed by 180°, as shown in Fig. 6.29. Hence a differential phase shift of 180° is again produced.

The solution for wave propagation in a circular guide with a longitudinal magnetized cylinder placed in the center can be carried out exactly.[†] However, the solution requires a great deal of algebraic manipulation, and it is very laborious to compute numerical values from the resultant transcendental equations for the propagation constants. The solution does verify that Faraday rotation takes place as would be expected, by analogy with propagation in an infinite ferrite medium.

Isolator

The isolator, or uniline, is a device that permits unattenuated transmission from port 1 to port 2 but provides very high attenuation for transmission in the reverse direction. The isolator is often used to couple a microwave signal generator to a load network. It has the great advantage that all the available power can be delivered to the load and yet reflections from the load do not get transmitted back to the generator output terminals. Consequently, the generator sees a matched load, and effects such as power output variation and frequency pulling (change in frequency), with variations in the load impedance, are avoided.

The isolator is similar to the gyrator in construction except that it employs a 45° twist section and 45° of Faraday rotation. In addition,

---

[†] A. A. Th. M. van Trier, Guided Electromagnetic Waves in Anisotropic Media, *Appl. Sci. Res.*, vol. B3, p. 305, 1953.

M. L. Kales, Modes in Waveguides That Contain Ferrites, *J. Appl. Phys.*, vol. 24, p. 604, 1953.

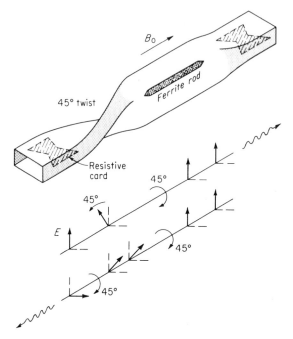

Fig. 6.30 A Faraday-rotation isolator.

thin resistive cards are inserted in the input and output guides to absorb the field that is polarized, with the electric vector parallel to the wide side of the guide, as shown in Fig. 6.30. The operation is as follows: A wave propagating from port 1 to port 2 has its polarization rotated 45° counterclockwise by the twist section and 45° clockwise by the Faraday rotator. It will emerge at port 2 with the correct polarization to propagate in the output guide. A wave propagating from port 2 to port 1 will have its plane of polarization rotated by 90° and will enter the guide at port 1 with the electric field parallel to the resistance card, and hence be absorbed. Without the resistance card, the wave would be reflected from port 1 because of the incorrect polarization, which cannot propagate in the guide constituting port 1. However, multiple reflections within the isolator will lead to transmission in both directions, and this makes it necessary to use resistance cards in both the input and output guides for satisfactory performance. Typical performance figures for an isolator are forward transmission loss of less than 1 db, reverse attenuation of 20 to 30 db, and bandwidth of operation approaching 10 percent.

Resonance isolator

If the curves in Fig. 6.27 for the propagation constants of circularly polarized waves in an infinite ferrite medium are examined, it will be seen that the attenuation constant for negative circular polarization is always very small whereas that for positive circular polarization is very large in the vicinity of the resonance point $\omega_0 \approx \omega$. This property may be used as the basis for a resonance isolator by using a negative circularly polarized wave for transmission in the low-loss direction and a positive circularly polarized wave for transmission in the reverse direction. In the latter case the wave is rapidly absorbed or attenuated.

The condition for circular polarization is an inherent property of the dominant TE_{10} mode in a rectangular guide at two positions within the guide. The TE_{10}-mode fields are

$$E_y = \sin \frac{\pi x}{a} e^{\pm j\beta z}$$

$$j\omega\mu_0 H_x = \pm j\beta \sin \frac{\pi x}{a} e^{\pm j\beta z}$$

$$j\omega\mu_0 H_z = -\frac{\pi}{a} \cos \frac{\pi x}{a} e^{\pm j\beta z}$$

Since H_x and H_z differ in phase by 90°, circular polarization occurs when $|H_x| = |H_z|$, or when $x = x_1$, where

$$\tan \frac{\pi x_1}{a} = \pm \frac{\pi}{\beta a} = \pm \frac{\lambda_g}{2a} \tag{6.83}$$

For the solution in the range $0 < x_1 < a/2$, the ratio of H_x to H_z is

$$\frac{H_x}{H_z} = \mp j \tag{6.84a}$$

and the solution occurring for $a/2 < x_1 < a$ gives

$$\frac{H_x}{H_z} = \pm j \tag{6.84b}$$

With respect to the y axis, the solution given by (6.84a) corresponds to a negative circularly polarized field for propagation in the $+z$ direction and to positive circular polarization for propagation in the $-z$ direction. If the solution given by (6.84b) is considered, the direction of polarization is reversed.

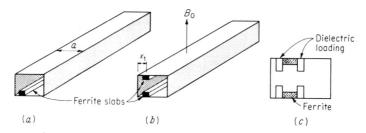

Fig. 6.31 Rectangular-waveguide resonance isolators.

The above property of the TE_{10} mode is utilized in the resonance isolator by locating a thin ferrite slab (or two slabs, as in Fig. 6.31b) in a rectangular guide at a position where the r-f magnetic field is circularly polarized. The ferrite is magnetized by a static field applied in the y direction, as in Fig. 6.31. Since the sense of the circular polarization depends on the direction of propagation, as (6.84) shows, it follows that, for propagation in one direction, the magnetic field is negative circularly polarized and suffers little attenuation, whereas in the reverse direction the field is positive circularly polarized and rapidly attenuated. By proper design the forward loss can be kept to under 0.5 db/in. at $\lambda_0 = 3$ cm, and the reverse loss can be as high as 6 to 10 db/in., or even more. Dielectric loading, as illustrated in Fig. 6.31c, gives an improved reverse-to-forward attenuation ratio.

6.9 Circulators

A circulator is a multiport device that has the property (Fig. 6.32) that a wave incident in port 1 is coupled into port 2 only, a wave incident in port 2 is coupled into port 3 only, and so on. The ideal circulator is also a matched device; i.e., with all ports except one terminated in matched loads, the input impedance of the remaining port is equal to the characteristic impedance of its input line, and hence presents a matched load.

A four-port circulator may be constructed from two magic T's or hybrid junctions and a gyrator as shown in Fig. 6.33. The gyrator produces an additional phase shift of 180° for propagation in the direction from a to b in Fig. 6.33. For propagation from b to a, and also from c to d or d to c, the electrical path lengths are equal.

Consider a wave incident in port 1. This wave is split into two equal-amplitude in-phase waves propagating in the side arms of the hybrid junction. The waves will arrive at points a and c in phase, and hence will

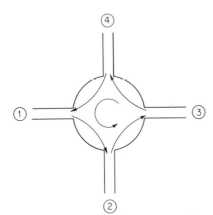

Fig. 6.32 Schematic diagram for a four-port circulator.

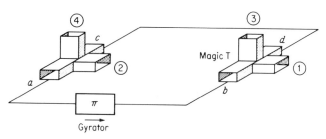

Fig. 6.33 A four-port circulator.

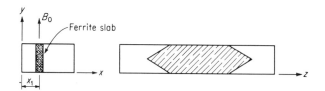

Fig. 6.34 A nonreciprocal phase shifter.

emerge from port 2. A wave incident in port 2 will be split into two waves, one arriving at d with a phase ϕ and the other arriving at b with a phase $\phi + \pi$ because of the presence of the gyrator. These partial waves have the right phase relationship to combine and emerge from port 3 in the hybrid junction. A wave incident in port 3 is split into two equal-amplitude waves, differing in phase by 180°, and hence will arrive at the other hybrid junction with the correct phase to combine and emerge from port 4. In a similar manner a wave incident in port 4 will split into two equal waves 180° apart in phase. But now the gyrator will restore phase equality, so that the waves will combine and emerge from port 1. Consequently, the microwave device illustrated in Fig. 6.33 has the required circulating transmission property.

A more compact form of four-port circulator may be constructed by employing 3-db side-hole directional couplers and rectangular-waveguide nonreciprocal phase shifters. The nonreciprocal phase shifter will be described first. It consists of a thin slab of ferrite placed in a rectangular guide at a point where the a-c magnetic field of the TE_{10} mode is circularly polarized, as in Fig. 6.34. A biasing field B_0 is applied in the y direction. Since the a-c magnetic field is right circularly polarized at x_1 for one direction of propagation and left circularly polarized for the opposite direction of propagation, the perturbing effect of the ferrite slab is different for the two directions of propagation. Consequently, the propagation phase constant β_+ for forward propagation is different from the propagation constant β_- for reverse propagation. By choosing the

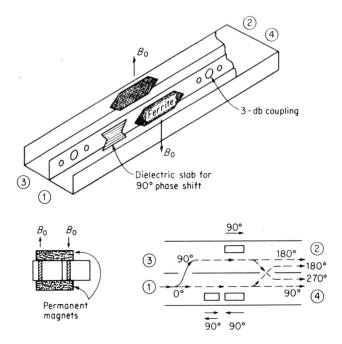

Fig. 6.35 A compact form of four-port circulator.

length of slab so that $(\beta_+ - \beta_-)l = \pi/2$, a differential phase shift of 90° for the two directions of propagation can be achieved.

A four-port circulator utilizing two 90° nonreciprocal phase shifter is illustrated in Fig. 6.35. The phase shifters are biased, with oppositely directed static fields—an arrangement easily achieved in practice with permanent magnets, as shown in Fig. 6.35. One guide is loaded with a dielectric insert to provide an additional 90° of reciprocal phase shift. The coupling holes are arranged to provide 3 db of coupling. The wave coupled through the apertures suffers a 90° change in phase, and this phase change is important in the operation of the circulator.

Consider a wave in port 1. This wave is split into two waves by the first 3-db coupler, the wave in the upper guide undergoing a 90° phase change because of the transmission properties of an aperture. The wave in the upper guide will arrive at the second coupler with a relative phase of 180°, and the wave in the lower guide with a relative phase of 90°. The second coupler splits these waves in the manner illustrated in Fig. 6.35. It is seen that the resultant waves are out of phase in port 4 but in phase at port 2. Thus transmission is from port 1 to port 2. A similar analysis will verify that a wave incident in port 2 emerges at port 3, or, in general, that the sequence $1 \to 2 \to 3 \to 4 \to 1$ is followed.

Three-port circulator

Carlin has shown that any lossless, matched, nonreciprocal three-port microwave junction is a perfect three-port circulator.† This theorem is readily proved from the properties of the scattering matrix. A perfectly matched three-port junction has a scattering matrix of the form

$$[S] = \begin{bmatrix} 0 & S_{12} & S_{13} \\ S_{21} & 0 & S_{23} \\ S_{31} & S_{32} & 0 \end{bmatrix} \quad (6.85)$$

For a nonreciprocal junction the scattering matrix is no longer symmetrical; that is, $S_{ij} \neq S_{ji}$. However, if the junction is lossless, conservation of power still requires that the $[S]$ matrix be unitary. Thus (4.63a) will hold for any lossless microwave junction independently of whether or not the junction is reciprocal. Applying the unitary condition to (6.85) gives

$$S_{12}S_{12}^* + S_{13}S_{13}^* = 1$$

$$S_{21}S_{21}^* + S_{23}S_{23}^* = 1$$

$$S_{31}S_{31}^* + S_{32}S_{32}^* = 1$$

$$S_{13}S_{23}^* = S_{12}S_{32}^* = S_{21}S_{31}^* = 0$$

Let us assume that $S_{21} \neq 0$. The fourth of the above equations then gives $S_{31} = 0$. The third equation now requires $|S_{32}| = 1$, and thus $S_{12} = 0$ from the fourth equation, $|S_{13}| = 1$ from the first equation, and $S_{23} = 0$ from the fourth equation again. Thus we see that $|S_{21}| = 1$ also from the second equation, so that

$$|S_{21}| = |S_{32}| = |S_{13}| = 1$$

$$S_{12} = S_{23} = S_{31} = 0$$

Consequently, there is perfect transmission from port 1 into port 2, from port 2 into port 3, and from port 3 into port 1. There is zero transmission in any other direction. The resultant scattering matrix of any *matched, lossless, nonreciprocal* three-port junction must then have the form

$$[S] = \begin{bmatrix} 0 & 0 & S_{13} \\ S_{21} & 0 & 0 \\ 0 & S_{32} & 0 \end{bmatrix} \quad (6.86)$$

If the locations of the terminal planes in the three input lines are properly

† H. J. Carlin, Principles of Gyrator Networks, *Polytech. Inst. Brooklyn, Microwave Res. Inst. Symp. Ser.*, vol. 4, p. 175, 1955.

chosen, the phase angles of S_{13}, S_{21}, and S_{32} can be made zero, and then $S_{13} = S_{21} = S_{32} = 1$.

If the above analysis were applied to a lossless, perfectly matched, reciprocal three-port junction, reciprocity would require that

$$S_{13} = S_{21} = S_{32} = 0$$

since $S_{31} = S_{12} = S_{32} = 0$. The resultant scattering matrix would then vanish. The conclusion to be drawn from this result is that it is impossible to construct a perfectly matched, lossless, *reciprocal* three-port junction. At least one of the reflection coefficients S_{11}, S_{22}, or S_{33} must be different from zero in the reciprocal case.

Practical realizations for three-port circulators usually involve the symmetrical junction (Y junction) of three identical waveguides or "strip-line" type of transmission lines, together with an axially magnetized ferrite rod or disks placed at the center. Figure 6.36 illustrates

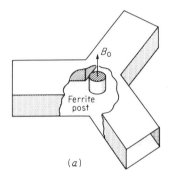

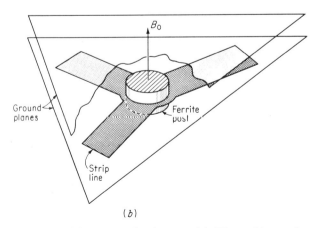

Fig. 6.36 Three-port circulators. (a) Waveguide version; (b) strip-line circulator.

both a waveguide version and a balanced strip-line version of the three-port circulator. The ferrite rod or disks are magnetized by a static B_0 field applied along the axis and give the junction the required nonreciprocal property. By placing suitable tuning elements in each arm (these can be identical in each arm because of the threefold symmetry involved) the junction can be matched; that is, S_{11}, S_{22}, and S_{33} can be made zero. The analysis given above then shows that the junction must, of necessity, be a perfect circulator if all losses are negligible. Losses are, of course, always present, and this limits the performance that can be achieved. Typical characteristics that can be obtained are insertion loss of less than 1 db, that is, $|S_{13}|$, $|S_{21}|$, $|S_{32}|$ greater than 0.89, isolation from 30 to 40 db, and input reflection coefficients less than 0.2. The isolation that can be obtained corresponds to values of $|S_{31}|$, $|S_{12}|$, and $|S_{23}|$ in the range 0.01 to 0.03.

6.10 Other ferrite devices

The devices utilizing ferrites for their operation described in the preceding sections represent only a small number of the large variety of devices that have been developed. In addition to the above, there are other forms of isolators, both reciprocal and nonreciprocal phase shifters, electronically controlled (by varying the current in the electromagnet that supplies the static biasing field) phase shifters and modulators, electronic switches and power limiters, etc. The nonlinear property of ferrites for high signal levels has also been used in harmonic generators, frequency mixers, and parametric amplifiers. A discussion of these devices, together with design considerations, performance data, and references to the original literature, is contained in the book by Lax and Button, listed in the references at the end of this chapter.

Problems

6.1 In the rotary phase changer, show that if the output quarter-wave plate, transition, and rectangular guide are rotated by an angle θ_1, an additional phase change of θ_1 is produced in the transmitted wave.

6.2 For a Bethe-hole directional coupler with the two guides aligned ($\theta = 0$) and a centered aperture, why does not (6.25a) give a useful solution for λ_0 as a function of a whereas (6.25b) does?

6.3 For a directional coupler of the type illustrated in Fig. 6.14b, derive an expression for the directivity for the two cases corresponding to the aperture positions given by (6.25). Which aperture position gives the least frequency-sensitive directivity?

6.4 Design a Bethe-hole directional coupler with a centered circular aperture. The waveguide size is $a = 0.9$ in., $b = 0.4$ in. The center frequency is 9.8 kMc. The required coupling is 20 db. Find the aperture radius and the frequency band over which the directivity remains greater than 40 db.

6.5 The accompanying figure illustrates two rectangular guides coupled by circular apertures in a common side wall. A TE_{10} mode of unit amplitude radiates a field of amplitude $-j\frac{4}{3}r_0^3(\pi/a)^2(ab\beta)^{-1}$ in both directions in the other guide. Design a five-hole directional coupler of the binomial type. The coupling required is 30 db at a frequency of 10,000 Mc. The guide width a is 2.5 cm, and the height $b = 1.2$ cm. Find the required aperture radii and the frequency band over which the directivity D remains greater than 50 db.

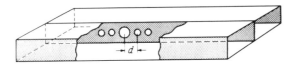

Fig. P 6.5

6.6 For the coupler described in Prob. 6.6, find the aperture radii to give a Chebyshev coupler. The minimum value of directivity required is $D_m = 50$ db. Find the corresponding frequency band. How much greater bandwidth is obtained as compared with that of the binomial coupler of Prob. 6.5?

6.7 Complete the numerical analysis of the three-hole Chebyshev coupler described in the text. Find the required aperture radii and the bandwidth of operation. Choose $f = 10,000$ Mc, $a = 2b = 2.4$ cm, $x_0 = a/2$. By how much does the directivity of the aperture increase the overall directivity at the center of the band and at the edges of the passband?

6.8 Consider an arbitrary lossless and reciprocal three-port Y junction as illustrated. Use the unitary and symmetry properties of the scattering matrix to show that all three ports cannot be simultaneously matched; that is, $S_{11} = S_{22} = S_{33} = 0$.

Hint: Assume that all S_{nn}, $n = 1, 2, 3$, are zero and show that this leads to inconsistent equations.

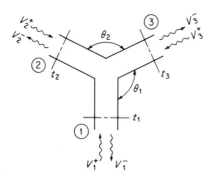

Fig. P 6.8

6.9 Let a short circuit be located a distance d from the terminal plane t_3 in arm 3 of the Y junction of Prob. 6.8. This constrains V_3^+ to equal $-V_3^- e^{-j2\beta d}$. Also $V_3^- = S_{13}V_1^+ + S_{23}V_2^+ + S_{33}V_3^+$. The resultant structure is a two-port junction. By using the given relations for V_3^+ and V_3^-, show that the scattering matrix of the new two-port junction has the elements

$$S'_{11} = S_{11} - \frac{S_{13}^2 e^{-j2\beta d}}{1 + S_{33} e^{-j2\beta d}}$$

$$S'_{12} = S'_{21} = S_{12} - \frac{S_{13} S_{23} e^{-j2\beta d}}{1 + S_{33} e^{-j2\beta d}}$$

$$S'_{22} = S_{22} - \frac{S_{23}^2 e^{-j2\beta d}}{1 + S_{33} e^{-j2\beta d}}$$

Show that a short-circuit position d can be found such that ports 1 and 2 become uncoupled, that is, $S'_{12} = 0$.

Hint: The terminal planes t_1, t_2, t_3 can be chosen so that S_{12} is pure imaginary, S_{13} is pure real, and $S_{13}S_{23}$ is pure imaginary. From $S'_{12} = 0$, obtain an equation for $\tan \beta d$ which always has a solution.

6.10 Consider a symmetrical Y junction, that is, $\theta_1 = \theta_2$ in Fig. P 6.8. Show that a short circuit can be placed in the symmetrical arm (arm 3) so that ports 1 and 2 are matched; that is, $S'_{11} = S'_{22} = 0$, where S'_{11} and S'_{22} are as given in Prob. 6.9. Note that in the present case $S_{13} = S_{23}$ and $S_{11} = S_{22}$. Thus perfect transmission between ports 1 and 2 is possible.

Hint: From $S'_{11} = 0$, solve for $e^{-j2\beta d}$ to obtain $S_{11} e^{j2\beta d} = S_{13}^2 - S_{11}S_{33}$. A solution is possible if $|S_{11}| = |S_{13}^2 - S_{11}S_{33}|$. Now choose terminal planes so that $S_{11} = S_{22} = a_{11}$, $S_{13} = a_{13}$ are real. From the product of row 1 and the conjugate of row 2, show that if $S_{12} = a_{12} + jb_{12}$, then $a_{12} = -a_{13}^2/2a_{11}$. From the product of the third row and the conjugate of the second row, show that $S_{33} = jb_{12} - a_{11} - a_{12}$. The product of the first row and its conjugate gives $a_{12}^2 + b_{12}^2 = 1 - a_{13}^2 - a_{11}^2$. With these relations for a_{12}, S_{33}, and b_{12}, show that $|S_{11}| = |S_{13}^2 - S_{11}S_{33}|$.

6.11 Use the equation of motion (6.63) to study second-harmonic generation in a ferrite. Assume that $\mathbf{M} = \mathbf{M}_1 e^{j\omega t} + \mathbf{M}_2 e^{2j\omega t}$ for the a-c magnetization and that $\mathbf{H}$ has only an x component with time dependence $e^{j\omega t}$; that is, the a-c magnetic field is $H_x e^{j\omega t} \mathbf{a}_x$. Neglect the third-harmonic term that occurs and show that

$$2j\omega \mathbf{M}_2 = \gamma\mu_0 H_x \mathbf{a}_x \times \mathbf{M}_1 + \gamma\mu_0 H_0 \mathbf{a}_z \times \mathbf{M}_2$$

Thus $2j\omega M_{2z} = \gamma\mu_0 H_x M_{1y}$. For M_{1y}, take the small-signal solution $\chi_{yx}H_x$ and use the value of χ_{yx} at resonance for a lossy ferrite to show that

$$M_{2z} = \frac{j\gamma\mu_0\omega_m H_x^2}{4\omega\omega_0\alpha}$$

Note that, for good efficiency, α must be small (small damping), so that the precession angle will be large at resonance.

6.12 From the unitary properties of the scattering matrix for a lossless nonreciprocal two-port microwave junction, show that it is not possible to have S_{21} zero while S_{12} is finite. Thus a lossless one-way transmission device cannot be built.

6.13 Continue the argument in the text to verify that the transmission sequence $1 \to 2 \to 3 \to 4 \to 1$ is followed in the circulator illustrated in Fig. 6.35.

6.14 Show that the scattering matrix for an ideal lossless N-port circulator can be put into the form

$$[S] = \begin{bmatrix} 0 & 0 & 0 & \cdots & 0 & 1 \\ 1 & 0 & 0 & \cdots & 0 & 0 \\ 0 & 1 & 0 & \cdots & 0 & 0 \\ \multicolumn{6}{c}{\dotfill} \\ 0 & 0 & 0 & \cdots & 1 & 0 \end{bmatrix}$$

by choosing proper terminal-plane locations in each port.

6.15 Show that, for TEM wave propagation in a direction perpendicular to B_0 in an infinite ferrite medium, the two solutions are linearly polarized waves with propagation constants

$$\gamma_+ = j\omega \sqrt{\mu_0 \epsilon} \left[\frac{(1+\chi)^2 - K^2}{1+\chi} \right]^{\frac{1}{2}}$$

$$\gamma_- = j\omega \sqrt{\mu_0 \epsilon}$$

Hint: Consider propagation along x, and in one case assume **H** to have only a y component, and for the other case assume **H** to have a z component.

References

1. Southworth, G. C.: "Principles and Applications of Waveguide Transmission," D. Van Nostrand Company, Inc., Princeton, N.J., 1950.
2. Reich, H. J., et al.: "Microwave Theory and Techniques," D. Van Nostrand Company, Inc., Princeton, N.J., 1953.
3. Montgomery, C. G., R. H. Dicke, and E. M. Purcell: "Principles of Microwave Circuits," McGraw-Hill Book Company, New York, 1948.
4. Ragan, G. L. (ed.): "Microwave Transmission Circuits," McGraw-Hill Book Company, New York, 1948.
5. Marcuvitz, N. (ed.): "Waveguide Handbook," McGraw-Hill Book Company, New York, 1951.
6. Harvey, A. F.: "Microwave Engineering," Academic Press Inc., New York, 1963. An outstanding handbook, offering a comprehensive survey of the whole microwave field, together with an extensive bibliography covering the international literature.

Microwave ferrites

7. Lax, B., and K. J. Button: "Microwave Ferrites and Ferrimagnetics," McGraw-Hill Book Company, New York, 1962. A very complete treatment of the physical properties of ferrites, electromagnetic propagation in ferrites, measurement techniques, and a discussion of many ferrite devices. Includes a 24-page bibliography.
8. Soohoo, R. F.: "Theory and Application of Ferrites," Prentice-Hall, Inc., Englewood Cliffs, N.J., 1960.
9. Clarricoats, P.: "Microwave Ferrites," John Wiley & Sons, Inc., New York, 1961.
10. Gurevich, A. G.: "Ferrites at Microwave Frequencies," transl. from the Russian by A. Tybulewicz, Consultants Bureau, New York, 1963.
11. Roberts, J.: "High Frequency Application of Ferrites," D. Van Nostrand Company, Inc., Princeton, N.J., 1960.

7

Electromagnetic resonators

7.1 Resonant circuits

Resonant circuits are of great importance for oscillator circuits, tuned amplifiers, frequency filter networks, wavemeters for measuring frequency, etc., at all frequencies from a few cycles per second up to and including light frequencies. Electric resonant circuits have many features in common, and it will be worthwhile to review some of these by using a conventional lumped-parameter RLC parallel network as an example. Figure 7.1 illustrates a typical low-frequency resonant circuit. The resistance R is usually only an equivalent resistance that accounts for the power loss in the inductor L and capacitor C and possibly the power extracted from the resonant system by some external load coupled to the resonant circuit.

At resonance the input impedance is pure real and equal to R. This implies that the average energies stored in the electric and magnetic fields are equal, since from (2.60)

$$Z_{\text{in}} = \frac{P_l + 2j\omega(W_m - W_e)}{\frac{1}{2}II^*} \tag{7.1}$$

This equation is valid for any one-port circuit provided a suitably defined equivalent terminal current I is used. Thus resonance always occurs when $W_m = W_e$, if we define resonance to be that condition which corresponds to a pure resistive input impedance. In the present case the time-average energy stored in the electric field in the capacitor is

$$W_e = \tfrac{1}{4}VV^*C$$

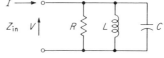

Fig. 7.1 Lumped-parameter resonant circuit.

and that stored in the magnetic field around the inductor is

$$W_m = \tfrac{1}{4} L I_L I_L^* = \tfrac{1}{4} L \left| \frac{V}{\omega L} \right|^2 = \frac{1}{4\omega^2 L} V V^*$$

The resonant frequency ω_0 is now found by equating W_m and W_e:

$$\omega_0 = (LC)^{-\frac{1}{2}} \tag{7.2}$$

An important parameter specifying the frequency selectivity, and performance in general, of a resonant circuit is the quality factor, or Q. A very general definition of Q that is applicable to all resonant systems is

$$Q = \frac{\omega(\text{time-average energy stored in system})}{\text{energy loss per second in system}} \tag{7.3}$$

At resonance $W_m = W_e$ and since the peak value of electric energy stored in the capacitor is $2W_e$ and occurs when the energy stored in the inductor is zero, and vice versa, the average energy W stored in the circuit is

$$W = W_m + W_e = 2W_m = 2W_e = \tfrac{1}{2} C V V^* \tag{7.4}$$

The power loss is $\tfrac{1}{2} G V V^*$ and is the energy loss per second. Hence, for the circuit of Fig. 7.1,

$$Q = \frac{\omega C}{G} = \omega R C = \frac{R}{\omega L} \tag{7.5}$$

since $\omega^2 L C = 1$ at resonance and $G = R^{-1}$.

In the vicinity of resonance, say $\omega = \omega_0 + \Delta\omega$, the input impedance can be expressed in a relatively simple form. We have

$$Z_{\text{in}} = \left(\frac{1}{R} + \frac{1}{j\omega L} + j\omega C \right)^{-1} = \left(\frac{1}{R} + j\omega_0 C + j\,\Delta\omega\, C + \frac{1 - \Delta\omega/\omega_0}{j\omega_0 L} \right)^{-1}$$

where the approximation $1/(\omega_0 + \Delta\omega) \approx (1 - \Delta\omega/\omega_0)/\omega_0$ has been used. Since $j\omega_0 C + 1/j\omega_0 L = 0$, we obtain

$$Z_{\text{in}} = \frac{\omega_0^2 R L}{\omega_0^2 L + j 2 R \,\Delta\omega} = \frac{R}{1 + j 2 Q (\Delta\omega/\omega_0)} \tag{7.6}$$

A plot of Z_{in} as a function of $\Delta\omega/\omega_0$ is given in Fig. 7.2, and is a typical resonance curve. When $|Z_{\text{in}}|$ has fallen to 0.707 of its maximum value, its phase angle is 45° if $\omega < \omega_0$ and $-45°$ if $\omega > \omega_0$. From (7.6) the corresponding value of $\Delta\omega$ is found to be given by

$$2Q \frac{\Delta\omega}{\omega_0} = 1$$

or

$$\Delta\omega = \frac{\omega_0}{2Q}$$

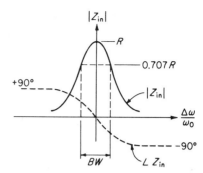

Fig. 7.2 Z_{in} for a parallel resonant circuit.

The fractional bandwidth BW between the 0.707R points is twice this; hence

$$Q = \frac{\omega_0}{2\,\Delta\omega} = \frac{1}{BW} \tag{7.7}$$

This relation provides an alternative definition of the Q; that is, the Q is equal to the fractional bandwidth between the points where $|Z_{in}|$ is equal to 0.707 of its maximum value (for a series resonant circuit this definition applies to $|Y_{in}|$).

If the resistor R in Fig. 7.1 represents the losses in the resonant circuit only, the Q given by (7.5) is called the unloaded Q. If the resonant circuit is coupled to an external load that absorbs a certain amount of power, this loading effect can be represented by an additional resistor R_L in parallel with R. The total resistance is now less, and consequently the new Q is also smaller. The Q, called the loaded Q and denoted Q_L, is

$$Q_L = \frac{RR_L/(R + R_L)}{\omega L}$$

The external Q, denoted Q_e, is defined to be the Q that would result if the resonant circuit were loss-free and only the loading by the external load were present. Thus

$$Q_e = \frac{R_L}{\omega L}$$

Use of these definitions shows that

$$\frac{1}{Q_L} = \frac{1}{Q_e} + \frac{1}{Q} \tag{7.8}$$

Another parameter of importance in connection with a resonant circuit is the damping factor δ. This parameter measures the rate at which the oscillations would decay if the driving source were removed. For a high-Q circuit, δ may be evaluated in terms of the Q, using a perturbation technique. With losses present, the energy stored in the resonant circuit will decay at a rate proportional to the average energy present at any time (since $P_l \propto VV^*$ and $W \propto VV^*$, we have $P_l \propto W$), so that

$$\frac{dW}{dt} = -2\delta W \tag{7.9}$$

or

$$W = W_0 e^{-2\delta t} \tag{7.10}$$

where W_0 is the average energy present at $t = 0$. But the rate of decrease of W must equal the power loss, so that

$$-\frac{dW}{dt} = 2\delta W = P_l$$

Consequently,

$$\delta = \frac{P_l}{2W} = \frac{\omega}{2}\frac{P_l}{\omega W} = \frac{\omega}{2Q} \tag{7.11}$$

upon using (7.3). The damping factor is seen to be inversely proportional to the Q. In place of (7.10) we now have

$$W = W_0 e^{-\omega t/Q} \tag{7.12}$$

In (7.12) Q must be replaced by Q_L if an external load is coupled to the circuit. The damping factor δ is also a measure of how fast the amplitude of oscillations in the resonant circuit can build up upon application of a driving source.

In microwave systems sections of transmission lines or metallic enclosures called cavities are used as resonators in place of the lumped-parameter circuit. The reason for this is that lumped-parameter circuits have too high losses, from both conductor loss and radiation loss, to be effective at microwave frequencies. In calculating the impedance of a microwave cavity, it is sometimes convenient to assume there are no losses present. The Q can be evaluated separately, and in terms of this parameter the impedance can be modified to account for losses by replacing the resonant frequency ω_0 by an equivalent complex resonant frequency $\omega_0(1 + j/2Q) = \omega_0 + j\delta$. Note that (7.6) can be written as

$$Z_{\text{in}} = \frac{\omega_0 R/2Q}{j[\omega - \omega_0(1 + j/2Q)]} \tag{7.13}$$

which shows that when losses are present this is equivalent to having a complex resonant frequency $\omega_0(1 + j/2Q)$. Equation (7.13) neglects the small change in resonant frequency that occurs when small losses are present.

7.2 Transmission-line resonant circuits

Series resonance; short-circuited line

At high frequencies, usually in the range 100 to 1,000 Mc, short-circuited or open-circuited sections of transmission lines are commonly used to replace the usual lumped LC resonant circuit. It is therefore of interest to consider the order of magnitude of Q and impedance that can be obtained. It will be assumed that air-filled lines are used, so that the only losses are those due to the series resistance R of the line. This is usually the case in practice since a dielectric-filled line has some shunt conductance loss and hence would result in a lower Q.

Consider a short-circuited line of length l, parameters R, L, C per unit length, as in Fig. 7.3. Let $l = \lambda_0/2$ at $f = f_0$, that is, at $\omega = \omega_0$. For f near f_0, say $f = f_0 + \Delta f$, $\beta l = 2\pi f l/c = \pi \omega/\omega_0 = \pi + \pi \Delta\omega/\omega_0$, since at ω_0, $\beta l = \pi$. The input impedance is given by

$$Z_{\text{in}} = Z_c \tanh(j\beta l + \alpha l) = Z_c \frac{\tanh \alpha l + j \tan \beta l}{1 + j \tan \beta l \tanh \alpha l}$$

But $\tanh \alpha l \approx \alpha l$ since we are assuming small losses, so that $\alpha l \ll 1$. Also $\tan \beta l = \tan(\pi + \pi \Delta\omega/\omega_0) = \tan \pi \Delta\omega/\omega_0 \approx \pi \Delta\omega/\omega_0$ since $\Delta\omega/\omega_0$ is small. Hence

$$Z_{\text{in}} = Z_c \frac{\alpha l + j\pi \Delta\omega/\omega_0}{1 + j\alpha l \pi \Delta\omega/\omega_0} \approx Z_c \left(\alpha l + j\pi \frac{\Delta\omega}{\omega_0} \right) \quad (7.14)$$

since the second term in the denominator is very small. Now

$$Z_c = \sqrt{\frac{L}{C}}$$

$\alpha = \frac{1}{2} R Y_c = (R/2)\sqrt{C/L}$, and $\beta l = \omega_0 \sqrt{LC}\, l = \pi$; so $\pi/\omega_0 = l\sqrt{LC}$, and the expression for Z_{in} becomes

$$Z_{\text{in}} = \sqrt{\frac{L}{C}} \left(\frac{l}{2} R \sqrt{\frac{C}{L}} + j\, \Delta\omega\, l\, \sqrt{LC} \right) = \tfrac{1}{2} R l + j l L\, \Delta\omega \quad (7.15)$$

It is of interest to compare (7.15) with a series $R_0 L_0 C_0$ circuit illustrated in Fig. 7.4. For this circuit

$$Z_{\text{in}} = R_0 + j\omega L_0 \left(1 - \frac{1}{\omega^2 L_0 C_0} \right)$$

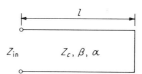

Fig. 7.3 Short-circuited transmission-line resonator.

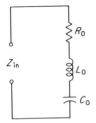

Fig. 7.4 A series resonant circuit.

If we let $\omega_0^2 = 1/L_0 C_0$, then

$$Z_{\text{in}} = R_0 + j\omega L_0 \frac{\omega^2 - \omega_0^2}{\omega^2}$$

Now $\omega^2 - \omega_0^2 = (\omega - \omega_0)(\omega + \omega_0) \approx \Delta\omega \, 2\omega$ if $\omega - \omega_0 = \Delta\omega$ is small. Thus

$$Z_{\text{in}} = R_0 + j\omega L_0 \frac{2\omega \, \Delta\omega}{\omega^2} = R_0 + j2L_0 \, \Delta\omega \tag{7.16}$$

By comparison with (7.15), we see that in the vicinity of the frequency for which $l = \lambda_0/2$, the short-circuited line behaves as a series resonant circuit with resistance $R_0 = \frac{1}{2}Rl$ and inductance $L_0 = \frac{1}{2}Ll$. We note that Rl, Ll are the total resistance and inductance of the line; so we might wonder why the factors $\frac{1}{2}$ arise. These enter because the current on the short-circuited line is a half sinusoid, and hence the effective circuit parameters R_0, L_0 are only one-half of the total line quantities.

The Q of the short-circuited line may be defined as for the circuit of Fig. 7.4:

$$Q = \frac{\omega_0 L_0}{R_0} = \frac{\omega_0 L}{R} = \frac{\beta}{2\alpha} \tag{7.17}$$

As an alternative, the general definition (7.3) may be used. We shall evaluate the Q of the short-circuited line from this definition by means of an approximate method valid for high-Q (i.e., low-loss) systems in order to illustrate a method of great utility in connection with many microwave devices (see Sec. 3.2, where this perturbation method is discussed). For small losses the energy stored in the system is, to first order, the same as if no losses were present. For a loss-free short-circuited line, the current on the line is a pure standing wave

$$I = I_0 \cos \beta z \, e^{j\omega t}$$

where z measures distance along the line from the short toward the input end. In a length dz, the energy stored in the magnetic field is

$$dW_m = \tfrac{1}{4} II^* L \, dz = \tfrac{1}{4} I_0^2 L \cos^2 \beta z \, dz$$

The total time-average stored energy in the magnetic field is

$$W_m = \tfrac{1}{4} I_0^2 L \int_0^{\lambda_0/2} \cos^2 \beta z \, dz = \frac{\lambda_0}{16} I_0^2 L$$

The energy stored in the electric field, i.e., in the line capacitance, is equal to W_m at resonance; so the time-average energy stored in the system is

$$W = W_m + W_e = \frac{\lambda_0}{8} L I_0^2$$

To a first approximation the losses do not modify the current distribution along the line. Hence the power loss is given by

$$P = \tfrac{1}{2} \int_0^{\lambda_0/2} R I I^* \, dz = \frac{R}{2} I_0^2 \int_0^{\lambda_0/2} \cos^2 \beta z \, dz = \frac{\lambda_0}{8} R I_0^2$$

Thus, at $\omega = \omega_0$,

$$Q = \frac{\omega_0 W}{P} = \frac{\omega_0 \lambda_0 L I_0^2/8}{R I_0^2 \lambda_0/8} = \frac{\omega_0 L}{R} \tag{7.18}$$

which checks with the earlier result. Typical values of Q range from several hundred up to about 10,000. As contrasted with low-frequency lumped-parameter circuits, the practical values of Q are very much higher for microwave resonators. It should be noted that in the above analysis the losses in the short circuit have been neglected. This does not introduce appreciable error if the length l is considerably greater than the conductor spacing.

Open-circuited line

By means of an analysis similar to that used earlier it is readily verified that an open-circuited transmission line is equivalent to a series resonant circuit in the vicinity of the frequency for which it is an odd multiple of a quarter wavelength long. The equivalent relations are (Fig. 7.5)

$$l = \frac{\lambda_0}{4} \quad \text{at } \omega_0 \tag{7.19a}$$

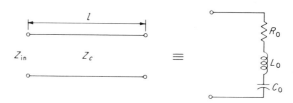

Fig. 7.5 Open-circuited transmission-line resonator.

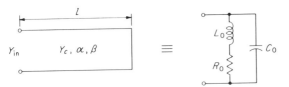

Fig. 7.6 Short-circuited antiresonant transmission line.

$$Z_{in} \approx \left(\alpha l + j\frac{\Delta\omega}{\omega_0}\frac{\pi}{2}\right) Z_c = \tfrac{1}{2}Rl + j\,\Delta\omega\, Ll \tag{7.19b}$$

$$R_0 = \tfrac{1}{2}Rl \tag{7.19c}$$

$$L_0 = \tfrac{1}{2}Ll \tag{7.19d}$$

$$\omega_0^2 = (LC)^{-1} \tag{7.19e}$$

Antiresonance

Short-circuited transmission lines behave as parallel resonant circuits in the frequency range where they are close to an odd multiple of a quarter wavelength long. The same property is true of open-circuited lines that are a multiple of a half wavelength long. When they behave as parallel resonant circuits, they are said to be antiresonant.

The case of parallel resonance is best analyzed on an admittance basis. With reference to Fig. 7.6, let l equal $\lambda_0/4$ at $\omega = \omega_0$. Then

$$\beta l = \omega\sqrt{LC}\, l = \omega_0 \sqrt{LC}\, l + \Delta\omega \sqrt{LC}\, l$$

at ω, and the input admittance is given by

$$Y_{in} = Y_c \coth(\alpha + j\beta)l = Y_c \frac{1 + j\tan\beta l \tanh\alpha l}{\tanh\alpha l + j\tan\beta l}$$

$$\approx Y_c \frac{1 - j\alpha l/(\Delta\omega\, l\sqrt{LC})}{\alpha l - j/(\Delta\omega\, l\sqrt{LC})}$$

since

$$\tan(\omega_0\sqrt{LC}\, l + \Delta\omega\sqrt{LC}\, l) = \tan\left(\frac{\pi}{2} + \Delta\omega\sqrt{LC}\, l\right)$$

$$\approx -(\Delta\omega\sqrt{LC}\, l)^{-1}$$

and $\tanh\alpha l \approx \alpha l$. A further approximation yields

$$Y_{in} = Y_c \frac{j\,\Delta\omega\sqrt{LC}\, l + \alpha l}{1 + j\,\Delta\omega\,\alpha l^2\sqrt{LC}} \approx Y_c(j\,\Delta\omega\sqrt{LC}\, l + \alpha l)$$

$$= \frac{RC}{2L}l + j\,\Delta\omega\, Cl \tag{7.20}$$

after replacing Y_c by $\sqrt{C/L}$ and α by $Y_c R/2$.

For the parallel resonant circuit of Fig. 7.6 we have

$$Y_{in} = \frac{1}{R_0 + j\omega L_0} + j\omega C_0 = \frac{j\omega C_0(R_0 + j\omega L_0) + 1}{R_0 + j\omega L_0}$$

$$\approx \frac{j\omega C_0 R_0 - \omega^2 L_0 C_0 + 1}{j\omega L_0}$$

since we assume $R_0 \ll \omega L_0$. If we define ω_0 by $\omega_0^2 L_0 C_0 = 1$, then

$$Y_{in} = \frac{C_0 R_0}{L_0} + \frac{\omega_0^2 L_0 C_0 - \omega^2 L_0 C_0}{j\omega L_0}$$

$$= R_0 \frac{C_0}{L_0} - jL_0 C_0 \frac{(\omega_0 - \omega)(\omega_0 + \omega)}{\omega L_0} \approx R_0 \frac{C_0}{L_0} + jC_0 2\,\Delta\omega \quad (7.21)$$

Comparison with (7.20) shows that the short-circuited line in the vicinity of a quarter wavelength long is equivalent to a parallel resonant circuit with

$$\frac{C_0 R_0}{L_0} = \frac{RC}{2L} l \qquad Cl = 2C_0 \qquad \frac{R_0}{L_0} = \frac{R}{L}$$

The Q of the circuit is given by

$$Q = \frac{\omega_0 L_0}{R_0} = \frac{\omega L}{R} = \frac{\beta}{2\alpha} \quad (7.22)$$

Although sections of transmission lines behave as simple lumped-parameter resonant circuits in the vicinity of a particular resonant frequency, they are in reality a much more complicated network having an infinite number of resonance and antiresonance frequencies. The resonance frequencies occur approximately when the short-circuited line is a multiple of a half wavelength long, that is, $f_n = nc/2l$, and the antiresonance frequencies (parallel resonance) when the line is an odd multiple of a quarter wavelength long, that is, $f_n = (2n + 1)c/4l$, where n is an integer. Thus the exact equivalent circuit would consist of an infinite number of resonant circuits coupled together. However, in practice, the frequency range of interest is normally such that a simple single-resonant-frequency circuit represents the transmission-line resonator with adequate accuracy.

7.3 Microwave cavities

At frequencies above 1,000 Mc transmission-line resonators have relatively low values of Q, and so it becomes preferable to use metallic enclosures, or cavities, instead. A cavity can be considered as a volume enclosed by a conducting surface and within which an electromagnetic field can be excited. The electric and magnetic energies are stored in the volume of the cavity. The finite conducting walls give rise to power loss and thus are equivalent to some effective resistance. The fields in the cavity may be excited, or coupled to an external circuit, by means of small coaxial-line probes or loops. Alternatively, the cavity may be coupled to a waveguide by means of a small aperture in a common wall. These coupling methods are illustrated in Fig. 7.7. Before considering the effects of coupling on cavity performance, the field solutions in rectangular and cylindrical cavities are presented.

Rectangular cavity

Figure 7.8 illustrates a rectangular cavity of height b, width a, and length d. It may be considered to be a section of rectangular waveguide terminated in a short circuit at $z = d$. If d equals a multiple of a half guide wavelength at the frequency f, the resultant standing-wave pattern is such that the x and y components of electric field are zero at $z = 0$. Consequently, a short circuit can be placed at $z = 0$ as well as in Fig. 7.9. The resultant structure is a rectangular cavity. This description of a cavity also shows that the field solution may be obtained directly from the corresponding waveguide solutions. For the nmth TE or TM mode, the propagation constant is given by

$$\beta_{nm}^2 = k_0^2 - \left(\frac{n\pi}{a}\right)^2 - \left(\frac{m\pi}{b}\right)^2 \tag{7.23a}$$

where $k_0 = 2\pi f_0/c$. We require $\beta_{nm}d = l\pi$, where l is an integer in order for the cavity to be a multiple of a half guide wavelength long. Thus, when d is specified, β_{nm} is given by

$$\beta_{nm} = \frac{l\pi}{d} \qquad l = 1, 2, \ldots \tag{7.23b}$$

However, this relation is consistent with the earlier one only for certain

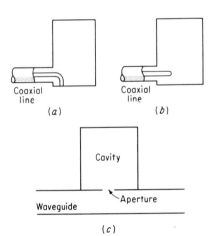

Fig. 7.7 Cavity-coupling methods. (a) Loop coupling; (b) probe coupling; (c) aperture coupling.

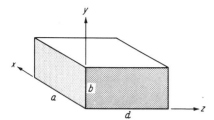

Fig. 7.8 A rectangular cavity.

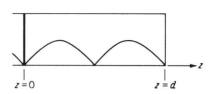

Fig. 7.9 Standing-wave pattern in a short-circuited waveguide.

discrete values of k_0. Only if $k_0 = k_{nml}$, where k_{nml} is given by

$$k_{nml} = \left[\left(\frac{l\pi}{d}\right)^2 + \left(\frac{m\pi}{b}\right)^2 + \left(\frac{n\pi}{a}\right)^2\right]^{\frac{1}{2}} \tag{7.24}$$

will (7.23a) and (7.23b) be satisfied. These particular values of k_0 give the resonant frequencies of the cavity; i.e.,

$$f_{nml} = \frac{ck_{nml}}{2\pi} = c\left[\left(\frac{l}{2d}\right)^2 + \left(\frac{m}{2b}\right)^2 + \left(\frac{n}{2a}\right)^2\right]^{\frac{1}{2}} \tag{7.25}$$

where c is the velocity of light. Note that there is a triply infinite number of resonant frequencies corresponding to different field distributions. Also note that there is more than one field solution for a given resonant frequency since (7.25) holds for both TE and TM modes. In addition, because of a lack of a preferential coordinate, in the case of a rectangular cavity, field solutions corresponding to TE and TM modes with respect to the x and y axes could also be constructed, and these would have the same resonant frequency. However, these latter modes are just a linear combination of a TE and a TM mode with respect to the z axis and therefore do not represent a new solution.

To illustrate the method of solution for the fields in a rectangular cavity and the evaluation of the unloaded Q, the TE_{101} mode is treated in detail. If $b < a < d$, this will be the mode with the lowest resonant frequency and corresponds to the TE_{10} mode in a rectangular waveguide. The mode subscripts nml pertain to the number of half-sinusoid variations in the standing-wave pattern along the x, y, and z axes, respectively. Using (3.107), the field solution for a TE_{10} mode is

$$H_z = (A^+ e^{-j\beta_{10}z} + A^- e^{j\beta_{10}z}) \cos\frac{\pi x}{a}$$

$$H_x = \frac{j\beta_{10}a}{\pi}(A^+ e^{-j\beta_{10}z} - A^- e^{j\beta_{10}z}) \sin\frac{\pi x}{a}$$

$$E_y = -j\frac{k_0 Z_0 a}{\pi}(A^+ e^{-j\beta_{10}z} + A^- e^{j\beta_{10}z}) \sin\frac{\pi x}{a}$$

$$H_y = E_x = E_z = 0$$

where A^+ and A^- are amplitude constants for the modes propagating in the $+z$ and $-z$ directions, respectively. To make E_y vanish at $z = 0, d$, we must choose $A^- = -A^+$ so that

$$A^+ e^{-j\beta_{10}z} + A^- e^{j\beta_{10}z} = -2jA^+ \sin \beta_{10} z$$

and also choose $\beta_{10} = \pi/d$. The corresponding value of k_0 is thus

$$k_0 = k_{101} = \left[\left(\frac{\pi}{a}\right)^2 + \left(\frac{\pi}{d}\right)^2\right]^{\frac{1}{2}} = \left[\left(\frac{\pi}{a}\right)^2 + \beta_{10}^2\right]^{\frac{1}{2}} \tag{7.26}$$

and this determines the resonant frequency. The solution for the fields may now be expressed as follows:

$$E_y = \frac{-2A^+ k_{101} Z_0 a}{\pi} \sin \frac{\pi x}{a} \sin \frac{\pi z}{d} \tag{7.27a}$$

$$H_x = \frac{2jA^+ a}{d} \sin \frac{\pi x}{a} \cos \frac{\pi z}{d} \tag{7.27b}$$

$$H_z = -2jA^+ \cos \frac{\pi x}{a} \sin \frac{\pi z}{d} \tag{7.27c}$$

Note that the magnetic field is $\pm 90°$ out of phase relative to the electric field. This is always the case in a lossless cavity and corresponds to voltage and current being $\pm 90°$ out of phase with each other in a lossless LC circuit.

At resonance the time-average electric and magnetic energy stored in the cavity are equal. The average stored electric energy is given by

$$W_e = \frac{\epsilon_0}{4} \int_0^a \int_0^b \int_0^d E_y E_y^* \, dx \, dy \, dz$$

$$= \frac{\epsilon_0}{4\pi^2} a^3 b d k_{101}^2 Z_0^2 |A^+|^2 \tag{7.28}$$

The reader may readily verify that

$$W_m = \frac{\mu_0}{4} \int_0^a \int_0^b \int_0^d (H_x H_x^* + H_z H_z^*) \, dx \, dy \, dz = W_e \tag{7.29}$$

In order to determine the cavity Q, the losses caused by the finite conductivity of the cavity walls must be evaluated. For small losses the surface currents are essentially those associated with the loss-free field solutions (7.27). Thus the surface current is given by

$$\mathbf{J}_s = \mathbf{n} \times \mathbf{H}$$

where $\mathbf{n}$ is a unit normal to the surface and directed into the cavity. Hence the power loss in the walls is given by

$$P_l = \frac{R_m}{2} \int_{\text{walls}} \mathbf{J}_s \cdot \mathbf{J}_s^* \, dS = \frac{R_m}{2} \int_{\text{walls}} |\mathbf{H}_{\tan}|^2 \, dS \tag{7.30}$$

where $R_m = 1/\sigma \delta_s$ is the resistive part of the surface impedance exhibited by the conducting wall having a conductivity σ and for which the skin depth is $\delta_s = (2/\omega\mu\sigma)^{\frac{1}{2}}$. In (7.30) $\mathbf{H}_{\text{tan}}$ is the tangential magnetic field at the surface of the cavity walls. Substituting from (7.27) into (7.30), a straightforward calculation gives

$$P_l = |A^+|^2 R_m \frac{2a^3b + 2d^3b + ad^3 + da^3}{d^2} \qquad (7.31)$$

With the use of (7.3), we find that the Q is given by

$$Q = \frac{\omega W}{P_l} = \frac{2\omega W_e}{P_l} = \frac{\omega k_{101}^2 Z_0^2 a^3 d^3 b \epsilon_0}{2\pi^2 R_m (2a^3b + 2d^3b + a^3d + d^3a)}$$

$$= \frac{(k_{101}ad)^3 b Z_0}{2\pi^2 R_m (2a^3b + 2d^3b + a^3d + d^3a)} \qquad (7.32)$$

upon replacing $\omega Z_0 \epsilon_0 = \omega \sqrt{\mu_0 \epsilon_0}$ by k_{101}.

As a typical example, consider a copper cavity ($\sigma = 5.8 \times 10^7$ mhos/m) with $a = b = d = 3$ cm. The resonant frequency is found to be 7,070 Mc/sec. The surface resistance $R_m = 0.022$ ohm, and the Q comes out equal to 12,700. The damping factor $\delta = \omega/2Q$ equals 1.74×10^6 nepers/sec, or about 2.5×10^{-4} neper/cycle of oscillation. Because of the high value of Q, 4,000 cycles of free oscillation can occur before the amplitude has decreased by a factor e^{-1}.

If the cavity is filled with a lossy dielectric material with permittivity $\epsilon = \epsilon' - j\epsilon''$, the time-average electric energy stored in the cavity volume is given by

$$W_e = \frac{\epsilon'}{4} \int_V |\mathbf{E}|^2 \, dV \qquad (7.33a)$$

The lossy dielectric has an effective conductivity $\omega \epsilon''$, and hence the power loss in the dielectric is

$$P_{ld} = \frac{1}{2} \int_V \mathbf{J} \cdot \mathbf{E}^* \, dV = \frac{\omega \epsilon''}{2} \int_V |\mathbf{E}|^2 \, dV \qquad (7.33b)$$

If Q_d is the Q when a lossy dielectric is present but the walls are perfectly conducting, then

$$Q_d = \frac{2\omega W_e}{P_{ld}} = \frac{\epsilon'}{\epsilon''} \qquad (7.34)$$

When wall losses are also present, the net Q is Q' and given by

$$Q' = \left(\frac{1}{Q_d} + \frac{1}{Q}\right)^{-1} \qquad (7.35)$$

where Q is the quality factor when lossy walls are present and $\epsilon'' = 0$. Q is given by (7.32), with ϵ_0 replaced by ϵ'. Also note that, for a cavity filled with dielectric, the resonant frequency is given by

$$f_{nml} = \sqrt{\frac{\epsilon_0}{\epsilon'}} \frac{c}{2\pi} k_{nml} \qquad (7.36)$$

Cylindrical cavity

The cylindrical cavity is a section of circular waveguide of length d and radius a, with short circuiting plates at each end, as in Fig. 7.10. This type of cavity is very commonly used in practice for wavemeters, to measure frequency, because of the high Q and wide frequency range of operation it provides. A high Q is necessary in a frequency meter in order to obtain a high degree of resolution or accuracy in the measurement of an unknown frequency. When the cavity is tuned to the frequency of the unknown source, it absorbs a maximum of power from the input line. A crystal detector coupled to the input line can be used to indicate this dip in power at resonance.

The fields in the cylindrical cavity may be determined from the corresponding waveguide solutions. The lowest resonant mode is the TE_{111} mode corresponding to the dominant TE_{11} mode in the circular guide. This mode is examined in detail below. Using the field solutions tabulated in Table 3.4 and combining a forward and backward propagating TE_{11} mode, we have

$$H_z = J_1\left(\frac{p'_{11}r}{a}\right) \cos \phi \, (A^+ e^{-j\beta_{11}z} + A^- e^{j\beta_{11}z}) \tag{7.37a}$$

$$H_r = \frac{-j\beta_{11}a}{p'_{11}} J'_1\left(\frac{p'_{11}r}{a}\right) \cos \phi \, (A^+ e^{-j\beta_{11}z} - A^- e^{j\beta_{11}z}) \tag{7.37b}$$

$$H_\phi = \frac{j\beta_{11}a^2}{(p'_{11})^2 r} J_1\left(\frac{p'_{11}r}{a}\right) \sin \phi \, (A^+ e^{-j\beta_{11}z} - A^- e^{j\beta_{11}z}) \tag{7.37c}$$

$$E_r = \frac{jk_0 Z_0 a^2}{(p'_{11})^2 r} J_1\left(\frac{p'_{11}r}{a}\right) \sin \phi \, (A^+ e^{-j\beta_{11}z} + A^- e^{j\beta_{11}z}) \tag{7.37d}$$

$$E_\phi = \frac{jk_0 Z_0 a}{p'_{11}} J'_1\left(\frac{p'_{11}r}{a}\right) \cos \phi \, (A^+ e^{-j\beta_{11}z} + A^- e^{j\beta_{11}z}) \tag{7.37e}$$

$$E_z = 0 \tag{7.37f}$$

where $p'_{11} = 1.841$. To make E_r and E_ϕ vanish at $z = 0, d$, we must

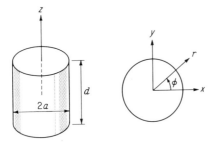

Fig. 7.10 Cylindrical cavity.

choose $A^- = -A^+$. The factor $A^+e^{-j\beta_{11}z} + A^-e^{j\beta_{11}z}$ becomes $-2jA^+ \sin \beta_{11}z$. β_{11} must be chosen equal to π/d to make $\sin \beta_{11}d$ vanish. The resonant frequency is determined from the relation

$$k_0 = \left[\beta_{11}^2 + \left(\frac{p'_{11}}{a}\right)^2\right]^{\frac{1}{2}} = \left[\left(\frac{\pi}{d}\right)^2 + \left(\frac{p'_{11}}{a}\right)^2\right]^{\frac{1}{2}} = \frac{2\pi f_{111}}{c} \tag{7.38}$$

To find the Q of the TE_{111} mode, the time-average energy W stored in the cavity and the power loss in the walls must be calculated. The general expressions are

$$W = 2W_e = \frac{\epsilon_0}{2}\int_0^a \int_0^{2\pi} \int_0^d (|E_r|^2 + |E_\phi|^2) r\, d\phi\, dr\, dz$$

$$P_l = \frac{R_m}{2}\int_{\text{walls}} |\mathbf{H}_{\tan}|^2 dS$$

These integrals may be evaluated by substituting the expressions given earlier for the fields. The final result obtained for the Q is

$$Q\frac{\delta_s}{\lambda_0} = \frac{\left[1 - \left(\frac{n}{p'_{nm}}\right)^2\right]\left[(p'_{nm})^2 + \left(\frac{l\pi a}{d}\right)^2\right]^{\frac{3}{2}}}{2\pi\left[(p'_{nm})^2 + \frac{2a}{d}\left(\frac{l\pi a}{d}\right)^2 + \left(1 - \frac{2a}{d}\right)\left(\frac{nl\pi a}{p'_{nm}d}\right)^2\right]} \tag{7.39}$$

where $l = n = m = 1$ for the TE_{111} mode. For the TE_{nml} mode, the Q is also given by (7.39), with the appropriate values of n, m, l, and p'_{nm} inserted. Note that all terms on the right are independent of frequency, and hence the Q varies as λ_0/δ_s for any given cavity and thus decreases as $f^{-\frac{1}{2}}$.

An analysis similar to the above may be carried out for the other TE_{nml} and TM_{nml} modes to obtain expressions for the fields. For the TE_{nml} modes it is necessary only to replace $\cos \phi$ and $\sin \phi$ by $\cos n\phi$ and $\sin n\phi$, J_1 by J_n, p'_{11} by p'_{nm}, and β_{11} by $l\pi/d$ in (7.37). Of particular interest is the TE_{011} mode for wavemeters because its Q is two to three times that of the TE_{111} mode. Another advantage of the TE_{011} mode is that $H_\phi = 0$, and hence there are no axial currents. This means that the end plate of the cavity can be free to move to adjust the cavity length d for tuning purposes without introducing any significant loss since no currents flow across the gap; i.e., the gap between the circular end plate and the cylinder wall is parallel to the current flow lines. However, the TE_{011} mode is not the dominant mode; so care must be exercised to choose a coupling scheme that does not excite the other possible modes that could resonate within the frequency tuning range of the cavity.

To determine what modes can resonate for a given value of $2a/d$ and frequency, it is convenient to construct a mode chart. For any given mode we have

$$f_{nml} = \frac{k_{nml}}{2\pi} c = \left[\left(\frac{x_{nm}}{a}\right)^2 + \left(\frac{l\pi}{d}\right)^2\right]^{\frac{1}{2}} \frac{c}{2\pi}$$

or

$$(2af_{nml})^2 = \left(\frac{cx_{nm}}{\pi}\right)^2 + \left(\frac{cl}{2}\right)^2 \left(\frac{2a}{d}\right)^2 \tag{7.40}$$

where $x_{nm} = p'_{nm}$ for TE modes and p_{nm} for TM modes. Figure 7.11 gives a plot of $(2af_{nml})^2$ against $(2a/d)^2$ for several modes and constitutes a mode chart. Examination of this chart shows over what range of frequency and $2a/d$ only a single mode can resonate (in the case of degenerate modes, two modes resonate at the same frequency). For example, for $(2a/d)^2$ between 2 and 3, only the TE$_{011}$ and TM$_{111}$ modes can resonate in the frequency range corresponding to $(2af)^2$ between 16.3×10^8 and 20.4×10^8 (region within dashed rectangle in Fig. 7.11). If the TM$_{111}$ mode is not excited, this frequency range at least can be tuned without spurious resonances from other modes occurring.

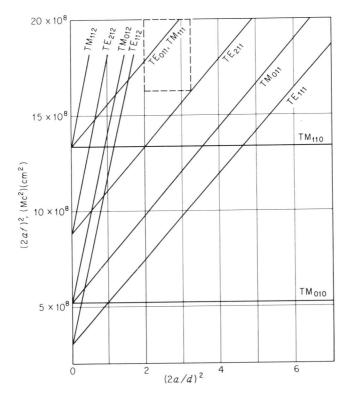

Fig. 7.11 Mode chart for a circular cylindrical cavity.

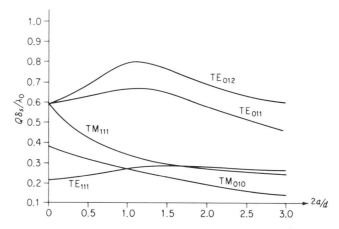

Fig. 7.12 Q for circular-cylindrical-cavity modes.

For the TM modes the Q can be evaluated to give

$$Q\frac{\delta_s}{\lambda_0} = \begin{cases} \dfrac{[p_{nm}^2 + (l\pi a/d)^2]^{\frac{1}{2}}}{2\pi(1 + 2a/d)} & l > 0 \\ \dfrac{p_{nm}}{2\pi(1 + a/d)} & l = 0 \end{cases} \qquad (7.41)$$

Figure 7.12 gives a plot of $Q\delta_s/\lambda_0$ against $2a/d$ for several modes. Note the considerably higher value of Q obtained for the TE_{011} mode relative to that for the TE_{111} mode. Optimum Q occurs for $d \approx 2a$. At $\lambda_0 = 3$ cm, $\delta_s/\lambda_0 = 2.2 \times 10^{-5}$, and hence, from Fig. 7.12, it is apparent that typical values of Q range from 10,000 to 40,000 or more. At $\lambda_0 = 10$ cm, the corresponding values of Q would be $\sqrt{10/3}$ greater.

7.4 Equivalent circuits for cavities

In this section the equivalent circuits of cavities coupled to transmission lines and waveguides are examined. A complete treatment is not given because of the complexity of the problem in general (see Sec. 7.8 for further details). Instead, a number of specific examples are examined in order to indicate the type of results that are obtained.

Aperture-coupled cavity

As an example of an aperture-coupled cavity, consider the rectangular cavity coupled to a rectangular guide by means of a small centered circular hole in the transverse wall at $z = 0$, as illustrated in Fig. 7.13. As indicated earlier, a small circular aperture in a transverse wall behaves as a shunt inductive susceptance with a normalized value given by

(5.25) as

$$\bar{B}_L = \frac{3ab}{8\beta r_0^3} \tag{7.42}$$

where a is the guide width, b is the guide height, r_0 is the aperture radius, and $\beta = [k_0^2 - (\pi/a)^2]^{\frac{1}{2}}$ is the propagation constant for the TE_{10} mode. The equivalent circuit of the aperture-coupled cavity is thus a short-circuited transmission line of length d shunted by a normalized susceptance $\bar{B}_L$.

To analyze this coupled cavity, we shall assume initially that there are no losses. The modifications required when small losses are present are given later. The cavity will exhibit an infinite number of resonances, and the input impedance $\bar{Z}_{\text{in}}$ will have an infinite number of zeros interlaced by an infinite number of poles, this being the general behavior of a distributed-parameter one-port microwave network. If we are interested in a resonance corresponding to a high value of $\bar{Z}_{\text{in}}$, infinite in the case of no loss, we should examine the nature of $\bar{Z}_{\text{in}}$ in the vicinity of one of its poles. This case corresponds to a parallel resonant circuit.

The input impedance is given by the parallel impedance of $j\bar{X}_L$ and $j \tan \beta d$ and is

$$\bar{Z}_{\text{in}} = \frac{-\bar{X}_L \tan \beta d}{j\bar{X}_L + j \tan \beta d} \tag{7.43}$$

where $j\bar{X}_L = (-j\bar{B}_L)^{-1}$. The antiresonances occur when the denominator vanishes, i.e., at the poles of $\bar{Z}_{\text{in}}$, or when

$$\bar{X}_L = -\tan \beta d = \frac{8r_0^3 \beta d}{3abd} \tag{7.44}$$

To solve this equation for the values of β that yield resonances, graphical methods are convenient. By plotting the two sides of (7.44) as functions of βd, the points of intersection yield the solutions for βd. When β is

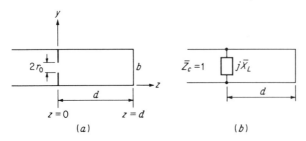

Fig. 7.13 (a) Aperture-coupled rectangular cavity; (b) equivalent transmission-line circuit.

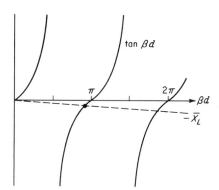

Fig. 7.14 Graphical solution for resonant frequency of aperture-coupled cavity.

known, the resonant frequency may be found from the relation

$$\frac{\omega}{2\pi} = f = \frac{c}{2\pi}\left[\beta^2 + \left(\frac{n}{a}\right)^2\right]^{\frac{1}{2}} \tag{7.45}$$

This graphical solution is illustrated in Fig. 7.14. Note that there are an infinite number of solutions. Normally, $\bar{X}_L$ is very small, so that the value of βd for the fundamental mode is approximately equal to π. The higher-order modes occur for $\beta d = \beta_n d \approx (n - \frac{1}{2})\pi$ when n, an integer, is large. The value of β for the first mode will be denoted by β_1, and the corresponding value of ω by ω_1, as determined by putting $\beta = \beta_1$ in (7.45).

An infinite number of equivalent lumped-parameter networks can be used to represent $\bar{Z}_{\text{in}}$ in the vicinity of ω_1. Usually, the simplest possible network is used. This equivalent network must be chosen so that its input impedance $\bar{Z}$ equals $\bar{Z}_{\text{in}}$ at ω_1. Likewise, for small variations $\Delta\omega$ about ω_1, the two impedances must be equal. A general procedure for specifying this equivalence is obtained by expanding the impedance functions in a power series in $\omega - \omega_1 = \Delta\omega$ about ω_1 and equating these series term by term. Since $\bar{Z}_{\text{in}}$ has a pole at $\omega = \omega_1$, the Taylor expansion cannot be applied directly to $\bar{Z}_{\text{in}}$. However, it may be applied to $(\omega - \omega_1)\bar{Z}_{\text{in}}(\omega)$ to give

$$(\omega - \omega_1)\bar{Z}_{\text{in}}(\omega) = \lim_{\omega \to \omega_1} (\omega - \omega_1)\bar{Z}_{\text{in}}(\omega) + \frac{d}{d\omega}(\omega - \omega_1)\bar{Z}_{\text{in}}\Big|_{\omega_1}(\omega - \omega_1)$$
$$+ \frac{1}{2}\frac{d^2}{d\omega^2}(\omega - \omega_1)\bar{Z}_{\text{in}}\Big|_{\omega_1}(\omega - \omega_1)^2 + \cdots$$

We now obtain

$$\bar{Z}_{\text{in}}(\omega) = \frac{\lim_{\omega \to \omega_1}(\omega - \omega_1)\bar{Z}_{\text{in}}(\omega)}{\omega - \omega_1} + \frac{d}{d\omega}(\omega - \omega_1)\bar{Z}_{\text{in}}\Big|_{\omega_1} + \cdots \tag{7.46}$$

A similar expansion of $\bar{Z}$ gives (note that $\bar{Z}$ must have a pole at ω_1 also)

$$\bar{Z}(\omega) = \frac{\lim_{\omega \to \omega_1} (\omega - \omega_1)\bar{Z}(\omega)}{\omega - \omega_1} + \frac{d}{d\omega}(\omega - \omega_1)\bar{Z}\big|_{\omega_1} + \cdots \tag{7.47}$$

Expansions of this type are called Laurent series expansions, and the coefficient of the $(\omega - \omega_1)^{-1}$ term is called the residue at the pole ω_1. These two series must be made equal term for term up to the highest power in $\omega - \omega_1 = \Delta\omega$ required to represent $\bar{Z}_{\text{in}}$ with sufficient accuracy in the frequency range of interest. For a microwave cavity, the Q is usually so high, and the frequency range $\Delta\omega/\omega_1$ of interest is approximately the range between the two points where $|\bar{Z}_{\text{in}}|$ equals 0.707 of its maximum. This latter fractional frequency band is equal to $1/Q$, and hence $\Delta\omega/\omega$ is so small that normally only the first term in the expansion (7.46) would be required to represent $\bar{Z}_{\text{in}}$ with sufficient accuracy in the vicinity of ω_1. In the present case a simple parallel LC circuit would be sufficient to represent $\bar{Z}_{\text{in}}(\omega)$ around ω_1.

In order to specify the values of L and C, we must evaluate the first terms in (7.46) and (7.47). For the LC circuit we have

$$\bar{Z} = \frac{j\omega L}{1 - \omega^2 LC}$$

We now choose $\omega_1^2 LC = 1$ to produce a pole at ω_1 for $\bar{Z}$. Hence

$$\bar{Z} = \frac{j\omega \omega_1^2 L}{\omega_1^2 - \omega^2}$$

and

$$\lim_{\omega \to \omega_1} (\omega - \omega_1)\bar{Z} = \frac{-j\omega\omega_1^2 L}{\omega + \omega_1}\bigg|_{\omega_1} = \frac{-j\omega_1^2 L}{2}$$

Thus we have

$$\bar{Z}(\omega) = \frac{-j\omega_1^2 L}{2(\omega - \omega_1)} \tag{7.48}$$

for ω near ω_1.

To evaluate the behavior of $\bar{Z}_{\text{in}}$ near ω_1, we can place ω equal to ω_1 in the numerator in (7.43). The denominator is first expanded in a Taylor series in β about β_1 to give

$$\bar{X}_L(\beta) + \tan \beta d \approx \bar{X}_L(\beta_1) + \tan \beta_1 d + \left(\frac{d\bar{X}_L}{d\beta} + \frac{d\tan \beta d}{d\beta}\right)\bigg|_{\beta_1}(\beta - \beta_1)$$

$$= \left(\frac{\bar{X}_{L1}}{\beta_1} + d\sec^2 \beta_1 d\right)(\beta - \beta_1)$$

since $\bar{X}_L(\beta_1) = \bar{X}_{L1} = -\tan \beta_1 d$ and $d\bar{X}_L/d\beta = (1/\beta)\bar{X}_L$. Next we expand β in terms of ω about ω_1 to give

$$\beta \approx \beta_1 + \frac{d\beta}{d\omega}\bigg|_{\omega_1}(\omega - \omega_1)$$

If we denote $d\beta/d\omega$ at ω_1 by β_1', we see that $\bar{Z}_{\text{in}}$ can be expressed as

$$\bar{Z}_{\text{in}} = \frac{j\bar{X}_{L1}\tan\beta_1 d}{[\bar{X}_{L1} + \beta_1 d(1 + \tan^2\beta_1 d)](\beta_1'/\beta_1)(\omega - \omega_1)}$$

upon replacing $\sec^2\beta_1 d$ by $1 + \tan^2\beta_1 d$. Replacing $\tan\beta_1 d$ by $-\bar{X}_{L1}$ now gives

$$\bar{Z}_{\text{in}} = -j\frac{\bar{X}_{L1}^2}{[\bar{X}_{L1} + \beta_1 d(1 + \bar{X}_{L1}^2)](\beta_1'/\beta_1)(\omega - \omega_1)}$$

Normally, $\bar{X}_{L1} \ll 1$, and since $\beta_1 d \approx \pi$, we can make further approximations to obtain (we shall verify that $\bar{X}_{L1} \ll 1$ later)

$$\bar{Z}_{\text{in}} = -j\frac{\bar{X}_{L1}^2}{\beta_1' d(\omega - \omega_1)} \tag{7.49}$$

Comparison with (7.48) shows that we must choose

$$\frac{\omega_1^2 L}{2} = \frac{\bar{X}_{L1}^2}{\beta_1' d}$$

or

$$L = \frac{2\bar{X}_{L1}^2}{\omega_1^2 \beta_1' d} \tag{7.50}$$

The capacitance C is determined by the condition $\omega_1^2 LC = 1$ given earlier.

Up to this point we have neglected the losses in the cavity. For a high-Q cavity these may be accounted for simply by replacing the resonant frequency ω_1 by a complex resonant frequency $\omega_1(1 + j/2Q)$, as indicated in Sec. 7.1. That is, the natural response of a lossy cavity is proportional to $e^{-\delta t + j\omega_1 t}$, and not to $e^{j\omega_1 t}$, where $\delta = \omega_1/2Q$. This is equivalent to having a complex resonant frequency $\omega_1(1 + j/2Q)$. The field in the cavity is, apart from some local fringing because of the presence of the aperture, a TE_{101} mode. Its Q was evaluated in Sec. 7.3, and is given by (7.32). For the lossy case we then have

$$\bar{Z}_{\text{in}} = -j\frac{\bar{X}_{L1}^2}{\beta_1' d(\omega - \omega_1 - j\omega_1/2Q)} \tag{7.51}$$

At resonance ($\omega = \omega_1$), we now obtain a pure resistive impedance $\bar{R}_{\text{in}}$ given by

$$\bar{R}_{\text{in}} = \bar{Z}_{\text{in}} = \frac{2\bar{X}_{L1}^2 Q}{\omega_1 \beta_1' d} \tag{7.52}$$

If we want the cavity to be matched to the waveguide at resonance, we must choose the aperture reactance $\bar{X}_{L1}$ so that $\bar{R}_{in} = 1$; that is,

$$\bar{X}_{L1} = \left(\frac{\omega_1 \beta'_1 d}{2Q}\right)^{\frac{1}{2}} \tag{7.53}$$

This matched condition is referred to as critical coupling. If $\bar{R}_{in}$ is greater than the characteristic impedance of the input line (unity in the case of normalized impedances), the cavity is said to be overcoupled, whereas if $\bar{R}_{in}$ is smaller, the cavity is undercoupled. If $\bar{R}_{in}$ is the normalized input resistance at resonance for a parallel resonant cavity, then $\bar{R}_{in}$ is defined as the coupling parameter K. In the case of series resonance, the coupling parameter equals the input normalized conductance at resonance.

For the rectangular cavity discussed in Sec. 7.3, with $a = b = d = 3$ cm, we found $f_1 = 7{,}070$ Mc and $Q = 12{,}700$. For this cavity

$$\beta'_1 = \frac{\omega_1}{\beta_1 c^2} = 4.7 \times 10^{-11} \text{ sec/cm}$$

and (7.53) gives $\bar{X}_{L1} = 0.0157$ for critical coupling. The corresponding aperture radius r_0 from (7.42) is found to be 0.37 cm. Note that $\bar{X}_{L1} \ll 1$, so that our earlier approximation in neglecting $\bar{X}_{L1}$ compared with unity is justified. Also note that a solution of (7.53) for the required value of $\bar{X}_{L1}$ to give critical coupling must, in general, be carried out simultaneously with the solution of (7.44) for the resonant frequency ω_1. However, for a high-Q cavity, ω_1 may be approximated with negligible error by the frequency corresponding to $\beta d = \pi$ in (7.53). This was done in the above calculation.

For the lossy cavity the equivalent circuit must include a resistance $\bar{R}_{in}$ in parallel with L and C as illustrated in Fig. 7.15a. The reader may readily verify that the input impedance $\bar{Z}$ now becomes, for ω near ω_1,

$$\bar{Z} = -j \frac{\omega_1^2 L}{2(\omega - \omega_1 - j\omega_1/2Q)} \tag{7.54}$$

where $Q = \bar{R}_{in}/\omega_1 L$.

Since the cavity is coupled to an input waveguide, the cavity terminals are loaded by an impedance equal to the impedance seen looking toward the generator from the aperture plane. If the generator is matched to

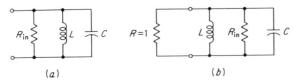

Fig. 7.15 Equivalent circuit for aperture-coupled cavity.

the waveguide, a normalized resistance of unity is connected across the cavity terminals, as in Fig. 7.15b. The external Q_e is given by (Sec. 7.1)

$$Q_e = \frac{1}{\omega_1 L} \tag{7.55a}$$

and the loaded Q_L by

$$Q_L = \left(\frac{1}{Q} + \frac{1}{Q_e}\right)^{-1} = \frac{\bar{R}_{\text{in}}}{(1 + \bar{R}_{\text{in}})\omega_1 L} \tag{7.55b}$$

The loaded and unloaded quality factors are related as follows:

$$\frac{1}{Q_L} = \frac{1}{Q} + \frac{1}{Q_e} = \frac{(1 + \bar{R}_{\text{in}})\omega_1 L}{\bar{R}_{\text{in}}} = \omega_1 L + \frac{\omega_1 L}{\bar{R}_{\text{in}}} = \frac{1}{Q}(1 + K)$$

or

$$Q = (1 + K)Q_L \tag{7.56}$$

In general, the coupling parameter K may be defined as

$$K = \frac{Q}{Q_e} \tag{7.57}$$

For a parallel resonant circuit this is seen to give

$$K = \frac{\bar{R}_{\text{in}}}{\omega_1 L} \frac{\omega_1 L}{1}$$

which agrees with the earlier definition. Likewise, for a series resonant circuit with normalized input conductance $\bar{G}_{\text{in}}$, the unloaded and loaded Q's are given by $Q = \omega_1 L \bar{G}_{\text{in}}$, $Q_e = \omega_1 L$, and hence $K = \bar{G}_{\text{in}} = Q/Q_e$ again. The coupling parameter is a measure of the degree of coupling between the cavity and the input waveguide or transmission line.

The external Q, Q_e, is sometimes called the radiation Q. The reason for this is that the cavity may be considered to radiate power through the aperture into the input waveguide. This power loss by radiation through the aperture is equal to the power lost in the normalized unit resistance connected across the resonator terminals in the equivalent circuit illustrated in Fig. 7.15b.

For the rectangular cavity under discussion, the next-higher frequency at which a resonance occurs corresponds to a series-type resonance at which $|\bar{Z}_{\text{in}}|$ is a minimum. In the loss-free case, $\bar{Z}_{\text{in}} = 0$, and from (7.43) this is seen to correspond to $\tan \beta d = 0$, or $\beta d = \pi$. At a series resonance, $\bar{Y}_{\text{in}}$ has a pole, and consequently an analysis similar to that presented for $\bar{Z}_{\text{in}}$ is applied to $\bar{Y}_{\text{in}}$. It is readily found that in the vicinity of $\omega = \omega_2$, where ω_2 is the value of ω that makes $\beta d = \pi$,

$$\bar{Y}_{\text{in}} = -j\frac{1}{\beta_2'(\omega - \omega_2 - j\omega_2/2Q)}$$

In this case the aperture has no effect, as would be expected, because the standing-wave pattern in the cavity is now such that the transverse electric field is zero at the aperture plane. The input admittance near ω_2 is just that of a short-circuited guide near a half guide wavelength long. This resonance is not of any practical interest since it corresponds to a very loosely coupled cavity.

Loop-coupled cavity

Figure 7.16 illustrates a cavity that is coupled to a coaxial line by means of a small loop. Since the loop is very small, the current in the loop can be considered to be constant. Any mode in the cavity that has a magnetic field with flux lines that thread through the loop will be coupled by the loop. However, at any particular frequency ω, only that mode which is resonant at this frequency will be excited with an appreciable amplitude. The fields excited in the cavity by the current I flowing in the loop can be found by solving for a vector potential arising from the current I. From the vector potential the magnetic field, and hence the flux passing through the loop, may be found. For a unit current let the magnetic flux of the nth mode that threads through the loop be ψ_n. This is then equal to the mutual inductance M_n between the coupling loop and the nth mode. Each mode presents an impedance equivalent to that of a series LCR circuit to the coupling loop. Thus a suitable equivalent circuit is an infinite number of series LCR circuits coupled by mutual inductance to the input coaxial line, as illustrated in Fig. 7.17. The input impedance is thus of the form

$$Z_{in} = j\omega L_0 + j \sum_{n=1}^{\infty} \frac{\omega^3 M_n^2 C_n}{1 - \omega^2 L_n C_n + j\omega C_n R_n} \tag{7.58}$$

where L_0 is the self-inductance of the coupling loop. If we define the resonance frequencies ω_n by $\omega_n^2 L_n C_n = 1$ and the unloaded Q of the nth mode by $Q_n = \omega_n L_n / R_n$, we can rewrite (7.58) as

$$Z_{in} = j\omega L_0 + j \sum_{n=1}^{\infty} \frac{\omega^3 M_n^2}{L_n(\omega_n^2 - \omega^2 + j\omega\omega_n/Q_n)} \tag{7.59}$$

If $\omega \approx \omega_n$, then all terms in the series in (7.59) except the nth term are

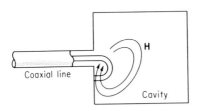

Fig. 7.16 Loop-coupled cavity.

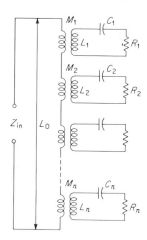

Fig. 7.17 Equivalent circuit of loop-coupled cavity.

small. Thus, in the vicinity of the nth resonance,

$$\begin{aligned}
Z_{\text{in}} &= j\omega L_0 + \frac{j\omega^3 M_n{}^2}{L_n(\omega_n{}^2 - \omega^2 + j\omega\omega_n/Q_n)} \\
&\approx j\omega L_0 - j\frac{\omega_n{}^2 M_n{}^2}{2L_n(\omega - \omega_n - j\omega_n/2Q_n)}
\end{aligned} \qquad (7.60)$$

The equivalent circuit now reduces to a single LCR series circuit mutually coupled to the input line. For efficient excitation of a given mode, the loop should be located at a point where this mode provides a maximum flux linkage.

The preceding results represent a formal solution to the loop-coupled cavity. In order actually to specify the circuit parameters, the boundary-value problem for the fields excited in a cavity by a given loop current must be solved. Also the Q's of the various cavity modes must be determined. For simple cavity shapes these calculations can be carried out with reasonable accuracy. However, they are too lengthy to include here.

7.5 Fabry-Perot resonators

As noted for the expressions for the Q of modes in a cylindrical cavity, the Q decreases as $f^{-\frac{1}{2}}$ as the frequency is raised. This is true for any given mode in any enclosed cavity. Thus, at very short wavelengths, say in the submillimeter range, the Q becomes too small for the usual type of cavity to be a useful device. A further difficulty is the miniature size of the cavity at wavelengths shorter than a few millimeters. This means that in microwave tubes, where electron beams pass through a cavity to interact with the cavity fields, the volume available for interaction becomes so small that oscillations cannot be sustained. To over-

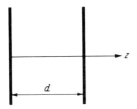

Fig. 7.18 Ideal form of a Fabry-Perot resonator.

come these difficulties in the short-millimeter-wavelength region and throughout the submillimeter (infrared and optical) region, the optical Fabry-Perot interferometer has been adapted as a resonator. A brief treatment of the basic properties of this type of resonator is given in this section. For further details the references given at the end of this chapter may be consulted.

The simplest form of an ideal Fabry-Perot resonator is two infinite parallel plates with a spacing d, as illustrated in Fig. 7.18. The field solutions in a resonator of this type are TEM standing waves. With z measured from the left side plate, the solution can be expressed as

$$E_x = E_0 \sin k_0 z \tag{7.61}$$

$$H_y = jY_0 E_0 \cos k_0 z \tag{7.62}$$

where H_y is obtained from E_x by using Maxwell's curl equation for the electric field. In order for E_x to vanish at both $z = 0$ and d, we must have

$$k_0 d = l\pi$$

or

$$f = \frac{c}{2\pi} k_0 = \frac{cl}{2d} \qquad l = 1, 2, 3, \ldots \tag{7.63}$$

Equation (7.63) determines the resonant frequencies.

The Q of the resonator may be evaluated by computing the energy stored and the power loss in the plates. In a rectangular prism of length d and of unit width and height along x and y, the time-average electric energy stored is

$$W_e = \frac{\epsilon_0}{4} \int_0^d |E_x|^2 \, dz = \frac{\epsilon_0}{4} |E_0|^2 \int_0^d \sin^2 \frac{l\pi z}{d} \, dz$$

$$= \frac{\epsilon_0 |E_0|^2 d}{8} = W_m \tag{7.64}$$

That W_m equals W_e is readily confirmed by evaluating

$$W_m = \frac{\mu_0}{4} \int_0^d |H_y|^2 \, dz$$

The current on each plate is equal to H_y, and hence the power loss per unit area for the two plates is

$$P_l = R_m |Y_0 E_0|^2 \qquad (7.65)$$

Thus the Q of the resonator for the lth mode is

$$Q = \frac{\omega W}{P_l} = \frac{2\omega W_e}{P_l} = \frac{\omega \epsilon_0 Z_0^2 d}{4 R_m} = \frac{\pi \epsilon_0 c l Z_0^2}{4 R_m} = \frac{\pi l Z_0}{4 R_m} \qquad (7.66)$$

upon using (7.63) and replacing c by $(\mu_0 \epsilon_0)^{-\frac{1}{2}}$.

Since the mode index l can be made very large, the Q can also be large. In addition, for a given spacing d, only one mode of oscillation is possible (actually two modes if the other polarization with the electric field along y is counted as a separate mode). In a totally enclosed cavity the Q can also be made large by using a large cavity oscillating in a high-order mode. However, such totally enclosed cavities are difficult to excite in a pure mode because of the close spacing between the resonant frequencies of the high-order modes. As an example of the value of Q obtainable, consider a Fabry-Perot resonator with copper plates spaced 5 cm apart and operating at $\lambda_0 = 0.1$ mm. The mode index is thus

$$l = \frac{2d}{\lambda_0} = 1{,}000$$

The surface resistance is $R_m = (\sigma \delta_s)^{-1} = 0.45$ ohm. Hence (7.66) gives $Q = 650{,}000$, which is a very high Q indeed.

A practical Fabry-Perot resonator must have finite plates and a means of coupling power into the resonator. One method of coupling power into the resonator is to cut a square array of small circular holes (apertures) in the end plate. The resonator can then be excited by illuminating this end plate with electromagnetic radiation from a horn with a lens in the opening for collimating the beam. An arrangement of this sort is illustrated in Fig. 7.19, where it is assumed that both plates are perforated and a similar horn is used as a receiving antenna. The field received by

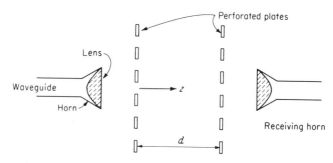

Fig. 7.19 Method of exciting a Fabry-Perot resonator.

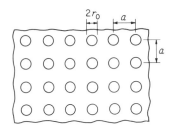

Fig. 7.20 Array of circular holes in end plate of a Fabry-Perot resonator.

the receiving horn will be proportional to the amplitude of the field in the resonator. It will be shown that this field is large only at certain discrete resonant frequencies. Thus transmission through the resonator is negligible except at resonance.

A thin plate (thickness much less than λ_0) perforated by a regular array of small circular apertures as illustrated in Fig. 7.20 acts as a normalized shunt inductive susceptance $-j\bar{B}_L$ to a normally incident TEM wave. This behavior is similar to that for an aperture in a transverse wall in a waveguide, and the cavity is essentially an aperture-coupled one. The normalized susceptance is given by

$$\bar{B}_L = \frac{3a^2}{4k_0 r_0^3} = \frac{3a^2 Z_0}{4\omega\mu_0 r_0^3} \tag{7.67}$$

The reflection coefficient for the infinite plate is given by

$$\Gamma = \frac{1 - \bar{Y}_{\text{in}}}{1 + \bar{Y}_{\text{in}}} = \frac{1 - 1 + j\bar{B}_L}{2 - j\bar{B}_L} = \frac{j\bar{B}_L}{2 - j\bar{B}_L} \tag{7.68}$$

This reflection coefficient takes into account the field radiated through the plate by means of the normalized intrinsic admittance of free space which is included in $\bar{Y}_{\text{in}}$. For finite-size plates, not all the energy coupled into the resonator through one plate will be reflected from the other plate. There will always be some diffraction effects, resulting in some loss of power past the plates at each reflection. If the plates are large in terms of wavelength and the horns illuminate the center portion only, these diffraction losses can be kept quite small.

To analyze the aperture-coupled resonator it is convenient to use an approach based on multiple reflections since this permits diffraction losses to be taken into account simply by replacing the reflection coefficient Γ by an equivalent reflection coefficient Γ_e. The case of no loss is considered first. Let the plane wave coupled into the resonator by the transmitting horn have an electric field with an amplitude E_0 at $z = 0$. This wave arrives at $z = d$ with an amplitude $E_0 e^{-jk_0 d}$. Upon reflection it arrives back at $z = 0$ with an amplitude $\Gamma E_0 e^{-2jk_0 d}$ and is re-reflected. The total field in the resonator may be obtained by summing up all these multiply reflected waves. With reference to Fig. 7.21, we find that, for any z,

$$E_x = E_0 e^{-jk_0 z} + \Gamma E_0 e^{-jk_0(2d-z)} + \Gamma^2 e^{-jk_0(2d+z)} + \cdots$$

$$= E_0 e^{-jk_0 z} \sum_{n=0}^{\infty} \Gamma^{2n} e^{-jk_0 2nd} + \Gamma E_0 e^{-jk_0(2d-z)} \sum_{n=0}^{\infty} \Gamma^{2n} e^{-jk_0 2nd}$$

$$= \frac{E_0(e^{-jk_0 z} + \Gamma e^{-jk_0(2d-z)})}{1 - \Gamma^2 e^{-j2k_0 d}} \tag{7.69}$$

To obtain a high external Q the susceptance $\bar{B}_L$ is made very large. Thus $\Gamma \approx -(1 + 2/j\bar{B}_L)$, and may be replaced by minus unity in the numerator in (7.69). We now obtain

$$E_x = \frac{2jE_0 e^{-jk_0 d} \sin k_0(d-z)}{1 - \Gamma^2 e^{-j2k_0 d}} \tag{7.70}$$

The maximum response of the cavity occurs essentially when the denominator has its minimum value, i.e., when k_0 is such that

$$D = |1 - \Gamma^2 e^{-j2k_0 d}| = \left| 1 - \left(\frac{j\bar{B}_L}{2 - j\bar{B}_L} \right)^2 e^{-j2k_0 d} \right|$$

is a minimum. If we express Γ in the form

$$\Gamma = \rho e^{j\theta}$$

where

$$\rho = \frac{\bar{B}_L}{(4 + \bar{B}_L^2)^{\frac{1}{2}}} \quad \text{and} \quad \theta = \frac{\pi}{2} + \tan^{-1} \frac{\bar{B}_L}{2}$$

then, clearly, D has a minimum when

$$|1 - \rho^2 e^{-j2(k_0 d - \theta)}|$$

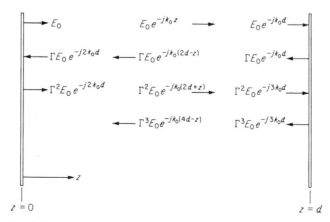

Fig. 7.21 Multiple reflections in a Fabry-Perot resonator.

is a minimum. This occurs when $2(k_0 d - \theta) = 2(l - 1)\pi$, provided we neglect the small variation in ρ with k_0. Substituting for θ, we obtain

$$2k_0 d = 2(l - 1)\pi + 2\theta = (2l - 1)\pi + 2\tan^{-1}\frac{\bar{B}_L}{2} \tag{7.71}$$

Since $\bar{B}_L$ is large, $\theta \approx \pi$; so the resonant frequencies are approximately given by

$$k_0 d = l\pi \quad \text{or} \quad f = \frac{lc}{2d}$$

which agrees with our results for the uncoupled resonator.

The minimum value of the denominator is $1 - \rho^2$. For some value of k_0 slightly removed from the resonance value, the denominator will have a magnitude $\sqrt{2}(1 - \rho^2)$. If the deviation in k_0 is Δk_0, then the external Q of the resonator will be given by

$$Q_e = \frac{k_0}{2\,\Delta k_0} \tag{7.72}$$

as reference to (7.7) shows. To find the change Δk_0, we replace $k_0 d - \theta$ by $(l - 1)\pi + (d - \Delta\theta/\Delta k_0)\,\Delta k_0$ in the expression for D^2 and equate it to $2(1 - \rho^2)^2$; thus

$$\left[1 - \rho^2 \cos 2\left(d - \frac{\Delta\theta}{\Delta k_0}\right)\Delta k_0\right]^2 + \rho^4 \sin^2 2\left(d - \frac{\Delta\theta}{\Delta k_0}\right)\Delta k_0$$

$$= 2(1 - \rho^2)^2 = 1 - 2\rho^2 + \rho^4 + 2\rho^2\left[1 - \cos 2\left(d - \frac{\Delta\theta}{\Delta k_0}\right)\Delta k_0\right]$$

As Δk_0 is small, we may use the approximation $\cos 2(d - \Delta\theta/\Delta k_0)\,\Delta k_0 \approx 1 - 2(\Delta k_0)^2(d - \Delta\theta/\Delta k_0)^2$ to obtain

$$\Delta k_0 = \frac{1 - \rho^2}{2\rho(d - \Delta\theta/\Delta k_0)} \approx \frac{1 - \rho^2}{2(d - \Delta\theta/\Delta k_0)}$$

since ρ is very nearly equal to unity for a high-Q resonator. For $\Delta\theta/\Delta k_0$ we have

$$\frac{\Delta[\tan^{-1}(\bar{B}_L/2)]}{\Delta k_0} = \frac{4}{\bar{B}_L^2 + 4}\frac{\Delta(\bar{B}_L/2)}{\Delta k_0}$$

$$= -\frac{2\bar{B}_L}{(4 + \bar{B}_L^2)k_0} = \frac{-2\rho^2}{k_0\bar{B}_L} \approx \frac{-2}{k_0\bar{B}_L}$$

upon using (7.67). Hence we obtain

$$Q_e = \frac{k_0(d + 2/k_0\bar{B}_L)}{1 - \rho^2} \tag{7.73}$$

for the external Q of the resonator. In a typical resonator $k_0 d = l\pi$ is very large compared with $2/\bar{B}_L$, and consequently the external Q is well approximated by the expression

$$Q_e = \frac{k_0 d}{1 - \rho^2} = \frac{l\pi}{1 - \rho^2} \tag{7.74}$$

The losses in the end plates per unit area are essentially the same as for the unperforated plates, and thus the loaded Q of the resonator is

$$Q_L = \left(\frac{1}{Q_e} + \frac{1}{Q}\right)^{-1} = \frac{\pi l Z_0}{4 R_m + Z_0(1 - \rho^2)} \tag{7.75}$$

for the lth resonant mode. Note that (7.66) was used for Q. For critical coupling, $Q_e = Q$; so

$$4R_m = Z_0(1 - \rho^2) = \frac{4Z_0}{4 + \bar{B}_L{}^2}$$

and the required value of susceptance is

$$\bar{B}_L = \left(\frac{Z_0 - 4R_m}{R_m}\right)^{\frac{1}{2}} \approx \left(\frac{Z_0}{R_m}\right)^{\frac{1}{2}} \tag{7.76}$$

The external Q_e arises from the power radiated by the cavity fields through the perforated plates into the external region. If, in addition to the power radiated through the holes, power is lost by diffraction around the sides of the finite-sized plates, then the effective value of ρ is decreased. This will reduce the external Q, and hence also the loaded Q, Q_L.

At optical frequencies the requirement of low diffraction losses imposes some very stringent requirements on the flatness and parallelism of the plates (mirrors) in a Fabry-Perot resonator. The plates must be flat and parallel to within a fraction of a wavelength. Tolerances of this sort are very difficult to meet. Fortunately, these severe tolerances can be alleviated by replacing the flat mirrors by spherical or paraboloidal mirrors, which have a focusing effect that confines the field to the region near the axis of the resonator.

In the analysis of the Fabry-Perot resonator with finite plates, we have assumed that the field between the mirrors is a TEM standing wave. Such a field is possible only for infinite flat mirrors. For finite mirrors the field must have an amplitude variation with the transverse coordinates such that the amplitude has fallen to a negligible value at the edge of the mirrors. An analysis of the problem shows that there are many field distributions that satisfy this requirement. Hence, for a Fabry-Perot resonator with finite mirrors, many modes of oscillation are possible. Because the transverse dimensions are large in terms of wavelength, these modes are all approximately TEM waves, and therefore they are labeled as TEM_{nml} modes, where n and m refer to the variations in the field amplitude with the transverse coordinates.

The basic method of analysis that has been employed to date is to assume some general expression for the field distribution over one mirror.

This distribution is next regarded as elementary sources, and Huygens' principle is employed to compute the radiation field incident on the second mirror. In order for the original field distribution to correspond to a mode for the resonator, it must reproduce itself at the second mirror apart from some constant complex factor which corresponds to the phase shift and diffraction loss occurring for one transit between the mirrors. Imposition of this constraint leads to an integral equation whose solutions yield the field distribution for the various modes, as well as the phase shift and reduction in amplitude due to diffraction loss per transit between mirrors. A good discussion of this method of analysis and typical results that are obtained is found in the references cited at the end of this chapter.

★7.6 Field expansion in a general cavity

Considerable insight into the general properties of an electromagnetic cavity can be obtained by examining the problem of expanding an arbitrary electromagnetic field into a complete set of modes in a cavity of unspecified shape. Such general modal expansions are required in order to determine the fields excited in a cavity by an arbitrary source. These expansions are also required in the evaluation of the input impedance or admittance of cavities coupled to external transmission lines or waveguides.

A fundamental theorem which is basic to general cavity theory is Helmholtz's theorem.† This theorem states that in a volume V bounded by a closed surface S as in Fig. 7.22 a general vector field $\mathbf{P}(x, y, z)$ is given by

$$\mathbf{P}(x, y, z) = \nabla \left[\int_V \frac{-\nabla' \cdot \mathbf{P}(\mathbf{r}')}{4\pi R} \, dV' + \oint_S \frac{\mathbf{P}(\mathbf{r}') \cdot \mathbf{n}}{4\pi R} \, dS' \right]$$
$$+ \nabla \times \left[\int_V \frac{\nabla' \times \mathbf{P}(\mathbf{r}')}{4\pi R} \, dV' + \oint_S \frac{\mathbf{P}(\mathbf{r}') \times \mathbf{n}}{4\pi R} \, dS' \right] \quad (7.77)$$

† R. Plonsey and R. E. Collin, "Principles and Applications of Electromagnetic Fields," chap. 1, McGraw-Hill Book Company, New York, 1961.

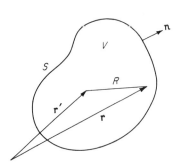

Fig. 7.22 Illustration for Helmholtz's theorem.

where $R = |\mathbf{r} - \mathbf{r}'|$. Thus the volume sources for $\mathbf{P}$ are given by $-\nabla' \cdot \mathbf{P}$ and $\nabla' \times \mathbf{P}$, and the surface sources are given by $\mathbf{P} \cdot \mathbf{n}$ and $\mathbf{P} \times \mathbf{n}$. If, and only if, $\nabla \cdot \mathbf{P} = 0$, $\mathbf{n} \cdot \mathbf{P} = 0$, can $\mathbf{P}$ be derived entirely from the curl of a suitable vector potential. Also, only if $\mathbf{n} \times \mathbf{P} = 0$, $\nabla \times \mathbf{P} = 0$, can $\mathbf{P}$ be derived from the gradient of a scalar potential, as (7.77) shows. If both $\nabla \cdot \mathbf{P} = 0$ and $\nabla \times \mathbf{P} = 0$ in V, then $\mathbf{P}$ is said to be a source-free field in V. In this case $\mathbf{P}$ can always be derived from the gradient of a scalar potential, but this potential must be multivalued if $\mathbf{n} \times \mathbf{P}$ does not equal zero, in some situations. This statement will be clarified later.

In setting up a suitable set of modes in which to expand a vector field $\mathbf{P}$ inside a given volume V, it is necessary to know the boundary conditions that must be imposed on these modes in order that a unique set of modes may be obtained. This uniqueness is needed so that when a solution for the field has been obtained we are assured of the uniqueness of that solution. Since the electromagnetic field satisfies Helmholtz's equation, we are concerned with obtaining a unique solution to this equation.

Consider $\nabla^2 \mathbf{A}_n + k_n^2 \mathbf{A}_n = 0$. We wish to determine the type of boundary conditions that must be imposed on $\mathbf{A}_n$ such that the solution to this equation is unique for a specific eigenvalue k_n. Assume that a second solution $\mathbf{B}_n$ exists such that $\nabla^2 \mathbf{B}_n + k_n^2 \mathbf{B}_n = 0$ also. The difference solution $\mathbf{C}_n = \mathbf{A}_n - \mathbf{B}_n$ satisfies $\nabla^2 \mathbf{C}_n + k_n^2 \mathbf{C}_n = 0$ also. We shall require of $\mathbf{B}_n$ the conditions $\nabla \cdot \mathbf{B}_n = \nabla \cdot \mathbf{A}_n$, $\nabla \times \mathbf{B}_n = \nabla \times \mathbf{A}_n$ in V, so that both $\mathbf{A}_n$ and $\mathbf{B}_n$ have the same volume sources. Then

$$\nabla \cdot \mathbf{C}_n = \nabla \times \mathbf{C}_n = 0$$

in V. Consider

$$k_n^2 \int_V |\mathbf{C}_n|^2 \, dV = -\int_V \mathbf{C}_n \cdot \nabla^2 \mathbf{C}_n \, dV$$

$$= \int_V (\mathbf{C}_n \cdot \nabla \times \nabla \times \mathbf{C}_n - \mathbf{C}_n \cdot \nabla \nabla \cdot \mathbf{C}_n) \, dV$$

Using the relations

$$\nabla \cdot (\mathbf{C}_n \times \nabla \times \mathbf{C}_n) = |\nabla \times \mathbf{C}_n|^2 - \mathbf{C}_n \cdot \nabla \times \nabla \times \mathbf{C}_n$$

and

$$\nabla \cdot (\mathbf{C}_n \nabla \cdot \mathbf{C}_n) = |\nabla \cdot \mathbf{C}_n|^2 + \mathbf{C}_n \cdot \nabla \nabla \cdot \mathbf{C}_n$$

we obtain

$$k_n^2 \int_V |\mathbf{C}_n|^2 \, dV = \int_V |\nabla \times \mathbf{C}_n|^2 \, dV + \int_V |\nabla \cdot \mathbf{C}_n|^2 \, dV$$

$$- \oint_S \mathbf{n} \cdot (\mathbf{C}_n \nabla \cdot \mathbf{C}_n + \mathbf{C}_n \times \nabla \times \mathbf{C}_n) \, dS \quad (7.78)$$

The volume integrals on the right vanish, and to make $\int_V |\mathbf{C}_n|^2 \, dV = 0$, which implies $C_n = 0$, we must make the two surface integrals vanish. This may be accomplished by making $\mathbf{A}_n$ and $\mathbf{B}_n$ satisfy the boundary conditions

$$\nabla \cdot \mathbf{A}_n = \mathbf{n} \times \mathbf{A}_n = 0 \qquad \text{on } S \qquad (7.79a)$$

or

$$\mathbf{n} \cdot \mathbf{A}_n = \mathbf{n} \times \nabla \times \mathbf{A}_n = 0 \qquad \text{on } S \qquad (7.79b)$$

and similarly for $\mathbf{B}_n$. These are the boundary conditions that will be used for the electric- and magnetic-type modes, respectively. Other possibilities are $\mathbf{n} \cdot \mathbf{A}_n = \mathbf{n} \times \mathbf{A}_n = 0$ on S or $\nabla \cdot \mathbf{A}_n = \mathbf{n} \times \nabla \times \mathbf{A}_n = 0$ on S. The imposition of these boundary conditions gives a unique solution. Note that two conditions must be specified for a vector function.

Cavity field expansions in terms of short-circuit modes

The short-circuit modes are those corresponding to the field solutions inside ideal (perfectly conducting walls), totally closed cavities. For these modes the electric field modes $\mathbf{E}_n$ satisfy the boundary condition $\mathbf{n} \times \mathbf{E}_n = 0$ on S, and the magnetic field modes satisfy the boundary condition $\mathbf{n} \cdot \mathbf{H}_n = 0$ on S. In addition, as (7.78) and (7.79) show, we must impose the additional boundary conditions $\nabla \cdot \mathbf{E}_n = 0$ and $\mathbf{n} \times \nabla \times \mathbf{H}_n = 0$ on S in order to obtain unique solutions.

Sometimes it is convenient to consider a cavity as having a surface part of which is perfectly conducting and part of which acts as a perfect open circuit. In this case the boundary conditions for $\mathbf{E}_n$ and $\mathbf{H}_n$ are interchanged on the perfect open-circuit portion of the boundary. In practice it is difficult to solve the cavity problem when the same boundary conditions do not apply on the whole surface. For this reason we restrict our analysis to the use of short-circuit modes, since these are readily found for the common types of cavities encountered.

There are three basic types of cavities to be considered, as illustrated in Fig. 7.23. Type 1 is a simply connected volume with a single surface, whereas type 2 is a simply connected volume with a multiple (double) surface. Finally, type 3 is a multiply connected volume with a single

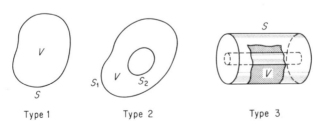

Type 1 Type 2 Type 3

Fig. 7.23 Basic types of cavities.

Sec. 7.6   **Electromagnetic resonators**   347

surface. Examples of the latter type are toroids and short-circuited coaxial lines. Consideration of these basic cavity types is of importance in connection with the existence of zero frequency modes, as will be seen.

Electric field expansion

In general, both $\nabla \times \mathbf{E}$ and $\nabla \cdot \mathbf{E}$ are nonzero; so we require two sets of modes: a solenoidal set with zero divergence and nonzero curl and an irrotational set with zero curl but nonzero divergence. The solenoidal modes are defined by

$$\nabla^2 \mathbf{E}_n + k_n{}^2 \mathbf{E}_n = 0 \qquad \text{in } V$$
$$\mathbf{n} \times \mathbf{E}_n = 0 \qquad \text{on } S \qquad\qquad (7.80)$$
$$\nabla \cdot \mathbf{E}_n = 0 \qquad \text{in } V \text{ and on } S$$

and the irrotational modes by $l_n \mathbf{F}_n = \nabla \Phi_n$, where

$$\nabla^2 \Phi_n + l_n{}^2 \Phi_n = 0 \qquad \text{in } V$$
$$\Phi_n = \text{const} \qquad \text{on } S \qquad \text{that is} \qquad \mathbf{n} \times \nabla \Phi_n = 0 \qquad \text{on } S \qquad (7.81)$$
$$\nabla \times \mathbf{F}_n = 0 \qquad \text{in } V$$

where k_n, l_n are eigenvalues for the problem. The constant is chosen as zero, except for the $n = 0$ mode, which has $l_0 = 0$. The above definition for $\mathbf{F}_n$ is chosen so that

$$\int_V \Phi_n{}^2 \, dV = -\frac{1}{l_n{}^2} \int_V \Phi_n \nabla^2 \Phi_n \, dV = -\frac{1}{l_n{}^2} \int_V (\nabla \cdot \Phi_n \nabla \Phi_n - \nabla \Phi_n \cdot \nabla \Phi_n) \, dV$$

$$= \int_V \mathbf{F}_n \cdot \mathbf{F}_n \, dV - \frac{1}{l_n{}^2} \oint_S \Phi_n \frac{\partial \Phi_n}{\partial n} \, dS = \int_V \mathbf{F}_n \cdot \mathbf{F}_n \, dV$$

Note that only one boundary condition is imposed on the $\mathbf{F}_n$ since scalar functions Φ_n are uniquely determined if either Φ_n or $\partial \Phi_n / \partial n$ is specified on S. The modes are assumed normalized, so that

$$\int_V \Phi_n{}^2 \, dV = \int_V \mathbf{F}_n \cdot \mathbf{F}_n \, dV = \int_V \mathbf{E}_n \cdot \mathbf{E}_n \, dV = 1 \qquad (7.82)$$

For $n = 0$ the eigenvalues l_0, k_0 are assumed to be the zero eigenvalues. The corresponding eigenfunctions $\mathbf{E}_0$, $\mathbf{F}_0$ are the zero frequency modes. We shall show that $\mathbf{E}_0$ cannot be distinguished from $\mathbf{F}_0$; so only the latter will be retained. We have $\nabla^2 \mathbf{E}_0 = 0 = \nabla \nabla \cdot \mathbf{E}_0 - \nabla \times \nabla \times \mathbf{E}_0$, or $\nabla \times \nabla \times \mathbf{E}_0 = 0$, since $\nabla \cdot \mathbf{E}_0 = 0$ by hypothesis. Thus

$$-\int_V \mathbf{E}_0 \cdot \nabla \times \nabla \times \mathbf{E}_0 \, dV = 0$$
$$= \int_V (\nabla \cdot \mathbf{E}_0 \times \nabla \times \mathbf{E}_0 - \nabla \times \mathbf{E}_0 \cdot \nabla \times \mathbf{E}_0) \, dV$$

This gives $\int_V |\nabla \times \mathbf{E}_0|^2 \, dV = \oint_S \mathbf{n} \cdot \mathbf{E}_0 \times \nabla \times \mathbf{E}_0 \, dS = 0$ since $\mathbf{n} \times \mathbf{E}_0 = 0$.

Thus we must have $\nabla \times \mathbf{E}_0 = 0$, which implies $\mathbf{E}_0 = \nabla f$, where f is a scalar function which is constant on the surface S. This latter relation holds for $\mathbf{F}_0$ as well, so that $\mathbf{E}_0$ can be discarded as long as $\mathbf{F}_0$ is retained. The zero frequency mode $\mathbf{F}_0$ will be defined by

$$\mathbf{F}_0 = \nabla \Phi_0$$
$$\nabla \times \mathbf{F}_0 = 0$$
$$\mathbf{n} \times \mathbf{F}_0 = 0 \quad \text{on } S \qquad \text{that is} \qquad \Phi_0 = \text{const} \quad \text{on } S \qquad (7.83)$$
$$\nabla^2 \Phi_0 = 0 \quad \text{since } l_0 = 0$$

For cavity types 1 and 3 a solution for Φ_0 other than a constant does not exist. Hence the $\mathbf{F}_0$ mode is present only for the type 2 cavity, with Φ_0 having different constant values on S_1 and S_2. This mode is just the static electric field that may exist between two conducting bodies at different potentials.

Orthogonality properties

Nondegenerate $\mathbf{E}_n$ and $\mathbf{F}_n$ modes are orthogonal among themselves and with each other. Consider

$$\int_V (\mathbf{E}_m \nabla^2 \mathbf{E}_n - \mathbf{E}_n \nabla^2 \mathbf{E}_m) \, dV = (k_m{}^2 - k_n{}^2) \int_V \mathbf{E}_n \cdot \mathbf{E}_m \, dV$$
$$= \int_V (\mathbf{E}_n \cdot \nabla \times \nabla \times \mathbf{E}_m - \mathbf{E}_m \cdot \nabla \times \nabla \times \mathbf{E}_n) \, dV$$
$$= \int_V [\nabla \times \mathbf{E}_n \cdot \nabla \times \mathbf{E}_m - \nabla \times \mathbf{E}_m \cdot \nabla \times \mathbf{E}_n$$
$$\qquad - \nabla \cdot (\mathbf{E}_n \times \nabla \times \mathbf{E}_m - \mathbf{E}_m \times \nabla \times \mathbf{E}_n)] \, dV$$
$$= \oint_S (\mathbf{n} \times \mathbf{E}_m \cdot \nabla \times \mathbf{E}_n - \mathbf{n} \times \mathbf{E}_n \cdot \nabla \times \mathbf{E}_m) \, dS = 0$$

Hence, if $k_m{}^2 \neq k_n{}^2$,

$$\int_V \mathbf{E}_m \cdot \mathbf{E}_n \, dV = \delta_{nm} \qquad (7.84)$$

where the Kronecker delta $\delta_{nm} = 0$ if $n \neq m$ and equals unity for $n = m$.

For the modes $\mathbf{F}_m$ we show first of all that the Φ_n are orthogonal. We have

$$\int_V (\Phi_n \nabla^2 \Phi_m - \Phi_m \nabla^2 \Phi_n) \, dV = (l_n{}^2 - l_m{}^2) \int_V \Phi_n \Phi_m \, dV$$
$$= \oint_S \left(\Phi_n \frac{\partial \Phi_m}{\partial n} - \Phi_m \frac{\partial \Phi_n}{\partial n} \right) dS = 0$$

so that

$$\int_V \Phi_n \Phi_m \, dV = 0 \qquad \text{when } l_n{}^2 \neq l_m{}^2$$

Next consider

$$\int_V \nabla \cdot \Phi_n \nabla \Phi_m \, dV = \oint_S \Phi_n \frac{\partial \Phi_m}{\partial n} \, dS = 0 = \int_V (\nabla \Phi_n \cdot \nabla \Phi_m + \Phi_n \nabla^2 \Phi_m) \, dV$$

$$= l_n l_m \int_V \mathbf{F}_n \cdot \mathbf{F}_m \, dV - l_m{}^2 \int_V \Phi_n \Phi_m \, dV$$

Since the latter integral is zero, we have

$$\int_V \mathbf{F}_n \cdot \mathbf{F}_m \, dV = \int_V \Phi_n \Phi_m \, dV = \delta_{nm} \tag{7.85}$$

for n and m not both equal to zero. If $m = 0$, $n \neq 0$, the proof of orthogonality still holds. However, the normalization for the $\mathbf{F}_0$ mode will be chosen as

$$\int_V \mathbf{F}_0 \cdot \mathbf{F}_0 \, dV = \int_V |\nabla \Phi_0|^2 \, dV = 1 \tag{7.86}$$

To show that the $\mathbf{E}_n$ and $\mathbf{F}_m$ modes are orthogonal, consider

$$\nabla \cdot \mathbf{F}_m \times \nabla \times \mathbf{E}_n = (\nabla \times \mathbf{F}_m) \cdot (\nabla \times \mathbf{E}_n) - \mathbf{F}_m \cdot \nabla \times \nabla \times \mathbf{E}_n$$

$$= -k_n{}^2 \mathbf{F}_m \cdot \mathbf{E}_n$$

since $\nabla \times \mathbf{F}_m = 0$, $\nabla \times \nabla \times \mathbf{E}_n = \nabla \nabla \cdot \mathbf{E}_n - \nabla^2 \mathbf{E}_n = k_n{}^2 \mathbf{E}_n$. We now obtain

$$k_n{}^2 \int_V \mathbf{F}_m \cdot \mathbf{E}_n \, dV = -\oint_S \mathbf{n} \cdot \mathbf{F}_m \times \nabla \times \mathbf{E}_n \, dS$$

$$= -\oint_S \mathbf{n} \times \mathbf{F}_m \cdot \nabla \times \mathbf{E}_n \, dS = 0 \tag{7.87}$$

For $n = m = 0$, the modes $\mathbf{F}_0$ and $\mathbf{E}_0$ are not orthogonal, being in fact identical. Since the $\mathbf{F}_m$ are orthogonal to the $\mathbf{E}_n$, it follows that both sets are needed, including $\mathbf{F}_0$ in general, to expand an arbitrary electric field.

Magnetic field expansion

To expand an arbitrary magnetic field (we consider the possibility of $\mathbf{n} \cdot \mathbf{H} \neq 0$ over a portion of the surface to exist), we shall set up a dual set of modes analogous to those used to expand the electric field. The solenoidal modes are defined by

$$\begin{aligned}
&\nabla^2 \mathbf{H}_n + k_n{}^2 \mathbf{H}_n = 0 \\
&\nabla \cdot \mathbf{H}_n = 0 \\
&\mathbf{n} \cdot \mathbf{H}_n = 0 \qquad \text{on } S \\
&\mathbf{n} \times \nabla \times \mathbf{H}_n = 0 \qquad \text{on } S
\end{aligned} \tag{7.88}$$

and the irrotational modes are defined by

$$p_n \mathbf{G}_n = \nabla \psi_n$$
$$\Delta^2 \psi_n + p_n^2 \psi_n = 0$$
$$\frac{\partial \psi_n}{\partial n} = 0 \quad \text{on } S \tag{7.89}$$
$$\nabla \times \mathbf{G}_n = 0$$

When $p_n = 0$, we have (assuming this occurs for $n = 0$) $\nabla \times \mathbf{G}_0 = 0$, $\nabla^2 \psi_0 = \nabla \cdot \mathbf{G}_0 = 0$, $\mathbf{n} \cdot \mathbf{G}_0 = 0$. Helmholtz's theorem now states that $\mathbf{G}_0$ can be derived from the curl of a vector potential, say $\mathbf{G}_0 = \nabla \times \mathbf{A}_0$, where $\nabla \times \nabla \times \mathbf{A}_0 = 0$. Since $\mathbf{G}_0$ has also been assumed to be given by $\mathbf{G}_0 = \nabla \psi_0$, it follows that ψ_0 must be multivalued. The function $\mathbf{G}_0$ corresponding to a static magnetic field can exist in the type 3 cavity only. For example, in a short-circuited coaxial line,

$$\mathbf{H} = \frac{I \mathbf{a}_\theta}{2\pi r} = \nabla \frac{I\theta}{2\pi} = \frac{\mathbf{a}_\theta}{r} \frac{\partial}{\partial \theta} \frac{I\theta}{2\pi} = \frac{I \mathbf{a}_\theta}{2\pi r} = -\nabla \times \frac{I \mathbf{a}_z}{2\pi} \ln r$$

Note that the scalar potential $I\theta/2\pi$ is multivalued. If $k_n = 0$ for $n = 0$, then (7.88) gives $\nabla \times \nabla \times \mathbf{H}_0 = 0$. A volume integral of $\mathbf{H}_0 \cdot \nabla \times \nabla \times \mathbf{H}_0$ similar to that employed earlier in connection with $\mathbf{F}_0$ and $\mathbf{E}_0$ shows that $\nabla \times \mathbf{H}_0 = 0$. Hence $\mathbf{H}_0$ has the same properties as $\mathbf{G}_0$; that is, $\nabla \times \mathbf{H}_0 = \nabla \cdot \mathbf{H}_0 = 0$, $\mathbf{n} \cdot \mathbf{H}_0 = 0$ on S. Therefore we shall not retain the $\mathbf{H}_0$ mode, but keep the $\mathbf{G}_0$ mode for the zero frequency mode.

Orthogonality properties

We assume normalization according to the following:

$$\int_V \mathbf{H}_n \cdot \mathbf{H}_n \, dV = \int_V \mathbf{G}_n \cdot \mathbf{G}_n \, dV = \int_V \psi_n^2 \, dV = 1 \tag{7.90a}$$

$$\int_V \mathbf{G}_0 \cdot \mathbf{G}_0 \, dV = \int_V |\nabla \psi_0|^2 \, dV = 1 \tag{7.90b}$$

By methods paralleling that used for the electric-type modes the following orthogonality properties may be derived:

$$\int_V \mathbf{H}_n \cdot \mathbf{H}_m \, dV = \int_V \mathbf{H}_n \cdot \mathbf{G}_m \, dV = \int_V \mathbf{G}_n \cdot \mathbf{G}_m \, dV = \int_V \psi_n \psi_m \, dV = 0 \tag{7.91}$$

for $n \neq m$. For $n = m$ we also have

$$\int_V \mathbf{H}_n \cdot \mathbf{G}_n \, dV = 0 \tag{7.92}$$

Relationship between $\mathbf{E}_n$ and $\mathbf{H}_n$ modes

The eigenvalues for both the $\mathbf{E}_n$, $\mathbf{H}_n$ modes were designated as k_n because they are, in fact, equal. Furthermore, we can show that

$$\nabla \times \mathbf{E}_n = k_n \mathbf{H}_n \qquad \nabla \times \mathbf{H}_n = k_n \mathbf{E}_n \tag{7.93}$$

The curl of the first relation gives

$$\nabla \times \nabla \times \mathbf{E}_n = \nabla \nabla \cdot \mathbf{E}_n - \nabla^2 \mathbf{E}_n = k_n \nabla \times \mathbf{H}_n = k_n{}^2 \mathbf{E}_n = -\nabla^2 \mathbf{E}_n$$

by using the second relation. Similarly,

$$\nabla \times \nabla \times \mathbf{H}_n = -\nabla^2 \mathbf{H}_n = k_n \nabla \times \mathbf{E}_n = k_n{}^2 \mathbf{H}_n$$

Hence (7.93) is consistent with the Helmholtz equation of which the $\mathbf{E}_n$, $\mathbf{H}_n$ are solutions. Furthermore, the boundary condition on $\mathbf{E}_n$ is $\mathbf{n} \times \mathbf{E}_n = 0$ on S. This implies $\mathbf{n} \times \nabla \times \mathbf{H}_n = 0$ on S, which is the boundary condition that has been imposed on the $\mathbf{H}_n$ functions. Also, we have $\mathbf{n} \cdot \mathbf{H}_n = 0$ on S, which implies $\mathbf{n} \cdot \nabla \times \mathbf{E}_n = 0$ on S. Now

$$\mathbf{n} \cdot \nabla \times \mathbf{E}_n = \mathbf{n} \cdot \left[\left(\nabla - \mathbf{n} \frac{\partial}{\partial n} \right) \times (\mathbf{E}_n - \mathbf{nn} \cdot \mathbf{E}_n) \right.$$
$$\left. + \left(\nabla - \mathbf{n} \frac{\partial}{\partial n} \right) \times \mathbf{nn} \cdot \mathbf{E}_n + \mathbf{n} \frac{\partial}{\partial n} \times \mathbf{E}_n \right] = \mathbf{n} \cdot \nabla_t \times \mathbf{E}_{tn} = 0$$

since only the term $\nabla_t \times \mathbf{E}_{tn}$ is in the direction of $\mathbf{n}$. Here t denotes components tangent to the surface S. Since $\mathbf{E}_{tn} = 0$ on S, $\mathbf{n} \cdot \nabla \times \mathbf{E}_n = 0$ on S and is consistent with the boundary condition $\mathbf{n} \cdot \mathbf{H}_n = 0$. Thus (7.93) is the only possible relation between the $\mathbf{E}_n$, $\mathbf{H}_n$ modes. A volume integral of

$$\nabla \cdot \mathbf{E}_n \times \nabla \times \mathbf{E}_n = -\mathbf{E}_n \cdot \nabla \times \nabla \times \mathbf{E}_n + |\nabla \times \mathbf{E}_n|^2 = -k_n{}^2(|\mathbf{E}_n|^2 - |\mathbf{H}_n|^2)$$

shows that the normalization of $\mathbf{H}_n$ is consistent with the normalization of the $\mathbf{E}_n$ also. It should now be apparent that the $\mathbf{E}_n$, $\mathbf{F}_n$ and $\mathbf{H}_n$, $\mathbf{G}_n$ modes have the properties that enable them to represent the electric and magnetic fields, respectively. An arbitrary field would require an infinite sum of these modes for its expansion.

★7.7 Oscillations in a source-free cavity

Consider a type 1 cavity with perfectly conducting walls and free of all currents and charges. We wish to determine the possible modes of oscillation. Let the fields in the cavity be expressed in terms of infinite series of the form

$$\mathbf{E} = \sum_n e_n(t) \mathbf{E}_n + \sum_n f_n(t) \mathbf{F}_n \tag{7.94a}$$

$$\mathbf{H} = \sum_n h_n(t) \mathbf{H}_n + \sum_n g_n(t) \mathbf{G}_n \tag{7.94b}$$

where e_n, f_n, h_n, and g_n are amplitude factors that are functions of time. Since $\mathbf{n} \times \mathbf{E} = \mathbf{n} \cdot \mathbf{H} = 0$ on S and the mode functions satisfy similar boundary conditions, the series are uniformly convergent and may be differentiated term by term. Maxwell's curl equations thus give

$$\nabla \times \mathbf{E} = \sum_n e_n \nabla \times \mathbf{E}_n = \sum_n e_n k_n \mathbf{H}_n = -\mu \sum_n \frac{\partial h_n}{\partial t} \mathbf{H}_n - \mu \sum_n \frac{\partial g_n}{\partial t} \mathbf{G}_n$$

$$\nabla \times \mathbf{H} = \sum_n h_n \nabla \times \mathbf{H}_n = \sum_n h_n k_n \mathbf{E}_n = \epsilon \sum_n \frac{\partial e_n}{\partial t} \mathbf{E}_n + \epsilon \sum_n \frac{\partial f_n}{\partial t} \mathbf{F}_n$$

If the first equation is scalar multiplied by $\mathbf{H}_n$ and $\mathbf{G}_n$ in turn and integrated over the cavity volume, we obtain

$$e_n k_n = -\mu \frac{\partial h_n}{\partial t} \tag{7.95a}$$

$$\frac{\partial g_n}{\partial t} = 0 \tag{7.95b}$$

by virtue of the orthogonality properties (7.91). Similarly, when the second equation is scalar multiplied by $\mathbf{E}_n$ and $\mathbf{F}_n$ in turn and integrated over the cavity volume, we obtain

$$h_n k_n = \epsilon \frac{\partial e_n}{\partial t} \tag{7.96a}$$

$$\frac{\partial f_n}{\partial t} = 0 \tag{7.96b}$$

From (7.95a) and (7.96a) we obtain

$$\frac{\partial^2 e_n}{\partial t^2} + \frac{k_n^2}{\mu \epsilon} e_n = 0 \qquad \frac{\partial^2 h_n}{\partial t^2} + \frac{k_n^2}{\mu \epsilon} h_n = 0$$

The solution for e_n is

$$e_n = e^{j\omega_n t} \tag{7.97a}$$

and from (7.96a) the solution for h_n is then

$$h_n = \frac{jk_n}{\omega_n \mu} e_n = j \sqrt{\frac{\epsilon}{\mu}} e_n = j \sqrt{\frac{\epsilon}{\mu}} e^{j\omega_n t} \tag{7.97b}$$

where $\omega_n = k_n (\mu \epsilon)^{-\frac{1}{2}}$ is the resonant frequency for the nth mode.

In the absence of volume sources, the $\mathbf{F}_n$ and $\mathbf{G}_n$ for $n \neq 0$ do not exist. However, the zero frequency modes $\mathbf{F}_0$ and $\mathbf{G}_0$ may exist in cavity types 2 and 3, respectively. These modes are independent of the $\mathbf{E}_n, \mathbf{H}_n$ modes. The nth free oscillation in the cavity is given by

$$\mathbf{E} = e_n \mathbf{E}_n = \mathbf{E}_n e^{j\omega_n t} \tag{7.98a}$$

$$\mathbf{H} = j\sqrt{\frac{\epsilon}{\mu}} \mathbf{H}_n e^{j\omega_n t} \tag{7.98b}$$

These results are valid if the material in the cavity is lossy as well, provided ϵ and μ are taken as complex quantities. In this case ω_n is complex, with the imaginary part representing a damping of the mode.

Cavity with lossy walls

Consider a cavity with finite conducting walls on which

$$\mathbf{n} \times \mathbf{E} = Z_m \mathbf{n} \times \mathbf{J}_s = Z_m \mathbf{H}_t \tag{7.99}$$

where $\mathbf{H}_t$ is the tangential magnetic field and $\mathbf{n}$ is a unit outward directed normal. The surface impedance $Z_m = (1 + j)/\sigma \delta_s$. Let the fields be expanded as follows:

$$\mathbf{E} e^{j\omega t} = \sum_n e_n \mathbf{E}_n e^{j\omega t} + \sum_n f_n \mathbf{F}_n e^{j\omega t} \tag{7.100a}$$

$$\mathbf{H} e^{j\omega t} = \sum_n h_n \mathbf{H}_n e^{j\omega t} + \sum_n g_n \mathbf{G}_n e^{j\omega t} \tag{7.100b}$$

where e_n, f_n, h_n, and g_n are amplitude constants independent of time. We have assumed a time variation $e^{j\omega t}$ so that the concept of a surface impedance Z_m can be applied to account for the finite conductivity of the walls.

In the present case $\mathbf{n} \times \mathbf{E}$ and $\mathbf{n} \cdot \mathbf{H}$ do not vanish on S, and since the modes in which $\mathbf{E}$ and $\mathbf{H}$ are expanded satisfy the boundary conditions $\mathbf{n} \times \mathbf{E}_n = \mathbf{n} \times \mathbf{F}_n = 0$ on S and $\mathbf{n} \cdot \mathbf{H}_n = \mathbf{n} \cdot \mathbf{G}_n = 0$ on S, the series expansions for $\mathbf{E}$ and $\mathbf{H}$ will not be uniformly convergent at the boundary S. Consequently, the curl of (7.100a) cannot be evaluated term by term. To overcome this difficulty we use the divergence theorem (essentially an integration by parts) to obtain

$$\int_V \nabla \cdot \mathbf{E} \times \mathbf{H}_n \, dV = \int_V (\nabla \times \mathbf{E}) \cdot \mathbf{H}_n \, dV - \int_V \nabla \times \mathbf{H}_n \cdot \mathbf{E} \, dV$$

$$= \oint_S \mathbf{n} \times \mathbf{E} \cdot \mathbf{H}_n \, dS$$

Replacing $\nabla \times \mathbf{H}_n$ by $k_n \mathbf{E}_n$ and $\nabla \times \mathbf{E}$ by $-j\omega\mu \mathbf{H}$ and using the expansions (7.100) and the orthogonal properties of the eigenfunctions now give

$$-j\omega\mu h_n - k_n e_n = \oint_S \mathbf{n} \times \mathbf{E} \cdot \mathbf{H}_n \, dS = Z_m \oint \mathbf{H} \cdot \mathbf{H}_n \, dS \tag{7.101a}$$

Note that $\mathbf{H}_n \cdot \mathbf{H} = \mathbf{H}_n \cdot \mathbf{H}_t$ since $\mathbf{n} \cdot \mathbf{H}_n = 0$. Similarly, we find

$$\int_V \nabla \cdot \mathbf{E}_n \times \mathbf{H} \, dV = \int_V (\nabla \times \mathbf{E}_n \cdot \mathbf{H} - \nabla \times \mathbf{H} \cdot \mathbf{E}_n) \, dV$$

$$= \oint_S \mathbf{n} \times \mathbf{E}_n \cdot \mathbf{H} \, dS = 0$$

Replacing $\nabla \times \mathbf{E}_n$ by $k_n \mathbf{H}_n$ and $\nabla \times \mathbf{H}$ by $j\omega\epsilon \mathbf{E}$ and using the expansion (7.100) yield

$$j\omega\epsilon e_n = k_n h_n \tag{7.101b}$$

This result is the same as that obtained by taking the curl of (7.100b) term by term. However, (7.101a) cannot be obtained by taking the curl of (7.100a) term by term because of its nonuniform convergence.

From the two relations (7.101) separate expressions for e_n and h_n may be obtained. For h_n we find

$$h_n = \frac{j\omega\epsilon}{k^2 - k_n^2} Z_m \oint_S \mathbf{H} \cdot \mathbf{H}_n \, dS \tag{7.102}$$

where $k^2 = \omega^2\mu\epsilon$.

Let us now assume that the field is essentially that of the nth mode and that this mode is not degenerate, i.e., no other mode has the same eigenvalue k_n. Then we have $\mathbf{H} \approx h_n \mathbf{H}_n$, and the surface integral becomes

$$h_n Z_m \oint_S \mathbf{H}_n \cdot \mathbf{H}_n \, dS = h_n Z_m \oint_S |\mathbf{H}_n|^2 \, dS$$

since $\mathbf{H}_n$ is real. The power loss in the walls for the nth mode is

$$P_l = \frac{R_m}{2} \oint_S |\mathbf{H}_n|^2 \, dS$$

and the average stored magnetic energy is

$$W_m = \frac{\mu}{4} \int_V |\mathbf{H}_n|^2 \, dV = \frac{\mu}{4}$$

The Q for the nth mode is

$$Q_n = \frac{2\omega W_m}{P_l}$$

and thus

$$\oint_S |\mathbf{H}_n|^2 \, dS = \frac{2P_l}{R_m} = \frac{4\omega W_m}{Q_n R_m} = \frac{\mu\omega}{Q_n R_m} \tag{7.103}$$

For (7.102) we now get

$$h_n = \frac{j\omega\epsilon(1+j)R_m}{k^2 - k_n^2} \frac{\mu\omega h_n}{Q_n R_m}$$

a relation that can hold only if

$$-\frac{k^2(1-j)}{k^2 - k_n^2} \frac{1}{Q_n} = 1$$

or

$$k = \frac{k_n}{(1 + 1/Q_n - j/Q_n)^{\frac{1}{2}}} \approx k_n \left(1 - \frac{1}{2Q_n} + \frac{j}{2Q_n}\right) \tag{7.104}$$

We thus find that, for the cavity with lossy walls, the resonant frequency differs from the no-loss resonant frequency ω_n by a factor $1 - 1/2Q_n$. In addition, a damping constant $\delta = \omega_n/2Q_n$ is introduced. In terms of

ω and δ, (7.104) gives

$$j\omega - \delta = j\omega_n\left(1 - \frac{1}{2Q_n}\right) - \frac{1}{2Q_n} \tag{7.105}$$

Degenerate modes

The volume orthogonality of the $\mathbf{G}_n$ and $\mathbf{H}_m$ modes holds even if $p_n = k_m$, as an examination of the method used in the proof will show. However, if $k_n = k_m$, then the proof of the volume orthogonality of the $\mathbf{H}_n$ and $\mathbf{H}_m$ modes breaks down. In this case

$$\int_V \mathbf{H}_n \cdot \mathbf{H}_m \, dV$$

may or may not vanish. If the integral does not vanish, the two modes are coupled together, since the average magnetic energy stored in the two modes will contain a nonzero interaction term arising from the above integral.

In addition to volume coupling between degenerate modes, it is also possible to have coupling arising from finite wall losses. The nature of this surface coupling is examined below.

If two modes $\mathbf{H}_n$ and $\mathbf{H}_m$ are degenerate, so that $k_n = k_m$, and if in addition

$$\oint_S \mathbf{H}_n \cdot \mathbf{H}_m \, dS \neq 0 \tag{7.106}$$

then the power loss associated with these two modes will contain a cross-interaction term arising from the above integral. In this case the two modes are said to be coupled together by the finite surface impedance of the cavity walls. It is not possible to have just one of these modes present since the presence of one mode automatically couples the other mode. However, for most practical cavities such coupling does not exist. Nevertheless, the possibility of mode coupling must be kept in mind since, if it exists, both modes must be included in any calculation of energy stored, power loss, and Q. The situation here is much like that occurring for degenerate waveguide modes (Sec. 3.10).

In the case where k_n and k_m are not equal, the surface integral may be shown to vanish for rectangular and cylindrical cavities. Although a general proof is not available, we should anticipate that this is a general property of nondegenerate modes.

The problem of coupled degenerate modes may be circumvented by introducing new modes that are linear combinations of the old degenerate modes in such a fashion that they are uncoupled. If $\mathbf{H}_n$ and $\mathbf{H}_m$ are degenerate coupled modes, choose new modes

$$\mathbf{H}'_n = c_1\mathbf{H}_n + c_2\mathbf{H}_m \tag{7.107a}$$

$$\mathbf{H}'_m = d_1\mathbf{H}_n + d_2\mathbf{H}_m \tag{7.107b}$$

with c_i and d_i chosen so that

$$\oint_S \mathbf{H}'_n \cdot \mathbf{H}'_m \, dS = 0 \tag{7.108a}$$

$$\int_V \mathbf{H}'_n \cdot \mathbf{H}'_m \, dV = 0 \tag{7.108b}$$

$$\int_V |\mathbf{H}'_n|^2 \, dV = \int_V |\mathbf{H}'_m|^2 \, dV = 1 \tag{7.108c}$$

These new modes are uncoupled and can exist independently of each other in the lossy cavity. For these new uncoupled modes the Q may be evaluated for each mode individually, since the cross-coupling term in the expression for power loss has been made equal to zero and, similarly, the cross-coupling term in the expression for stored magnetic energy has been made equal to zero. If more than two modes are degenerate, a similar procedure may be applied to find a new set of uncoupled normalized modes. In a general discussion we may therefore assume that all the degenerate modes have been chosen so that they are uncoupled.

★7.8 Excitation of cavities

In this section we consider the application of the modal expansion of the field in a cavity to the problem of finding the field excited by magnetic and electric dipoles, which may represent a current loop or probe, respectively. In addition, a small aperture may be described in terms of equivalent electric and magnetic dipoles as well. Thus the theory to be developed will be sufficiently general to treat the three common methods of coupling a cavity to an external waveguide or coaxial transmission line.

Let a cavity contain infinitesimal electric and magnetic dipoles

$$\mathbf{P} = \mathbf{P}_0 e^{j\omega t} \delta(\mathbf{r} - \mathbf{r}_0) \tag{7.109a}$$

$$\mathbf{M} = \mathbf{M}_0 e^{j\omega t} \delta(\mathbf{r} - \mathbf{r}_0) \tag{7.109b}$$

at a point whose position is defined by the vector $\mathbf{r}_0$.

The three-dimensional delta function $\delta(\mathbf{r} - \mathbf{r}_0)$ symbolizes that the dipoles are localized at the point $\mathbf{r} = \mathbf{r}_0$. This delta function is defined in such a manner that, for an arbitrary vector $\mathbf{A}$ which is continuous at $\mathbf{r}_0$, we have

$$\int_V \mathbf{A}(\mathbf{r}) \delta(\mathbf{r} - \mathbf{r}_0) \, dV = \mathbf{A}(\mathbf{r}_0)$$

when the point $\mathbf{r}_0$ is included in the volume V (Sec. 2.11).

By analogy with the following equations governing polarization in material bodies,

$$\mathbf{B} = \mu_0(\mathbf{H} + \mathbf{M}) \qquad \mathbf{D} = \epsilon_0 \mathbf{E} + \mathbf{P}$$

Sec. 7.8   Electromagnetic resonators

it is seen that Maxwell's equations become

$$\nabla \times \mathbf{E} = -j\omega \mathbf{B} = -j\omega\mu_0 \mathbf{H} - j\omega\mu_0 \mathbf{M}_0 \delta(\mathbf{r} - \mathbf{r}_0) \tag{7.110a}$$

$$\nabla \times \mathbf{H} = j\omega \mathbf{D} = j\omega\epsilon_0 \mathbf{E} + j\omega \mathbf{P}_0 \delta(\mathbf{r} - \mathbf{r}_0) \tag{7.110b}$$

where a time factor $e^{j\omega t}$ has been suppressed.

We now use the general expansion (7.100) for the fields to give

$$\mathbf{E} = \sum_n e_n \mathbf{E}_n + \sum_n f_n \mathbf{F}_n \tag{7.111a}$$

$$\mathbf{H} = \sum_n h_n \mathbf{H}_n + \sum_n g_n \mathbf{G}_n \tag{7.111b}$$

To find the expansion coefficients e_n, h_n we follow the derivation leading to (7.101) but note that $\nabla \times \mathbf{E}$ and $\nabla \times \mathbf{H}$ are replaced by the right-hand sides of (7.110). Thus

$$\int_V (\nabla \times \mathbf{E}) \cdot \mathbf{H}_n \, dV - \int_V \nabla \times \mathbf{H}_n \cdot \mathbf{E} \, dV = Z_m \oint_S \mathbf{H} \cdot \mathbf{H}_n \, dS$$

$$= \int_V [-j\omega\mu_0 \mathbf{H} - j\omega\mu_0 \mathbf{M}_0 \delta(\mathbf{r} - \mathbf{r}_0)] \cdot \mathbf{H}_n \, dV - k_n \int_V \mathbf{E}_n \cdot \mathbf{E} \, dV$$

$$= -j\omega\mu_0 h_n - k_n e_n - j\omega\mu_0 \mathbf{M}_0 \cdot \mathbf{H}_n(\mathbf{r}_0) \tag{7.112a}$$

and similarly,

$$k_n h_n = j\omega\epsilon_0 e_n + j\omega \mathbf{P}_0 \cdot \mathbf{E}_n(\mathbf{r}_0) \tag{7.112b}$$

To obtain an equation for g_n, consider

$$\int_V \nabla \cdot \mathbf{E} \times \mathbf{G}_n \, dV = \int_V (\mathbf{G}_n \cdot \nabla \times \mathbf{E} - \mathbf{E} \cdot \nabla \times \mathbf{G}_n) \, dV$$

$$= \oint_S \mathbf{n} \times \mathbf{E} \cdot \mathbf{G}_n \, dS = Z_m \oint_S \mathbf{H} \cdot \mathbf{G}_n \, dS$$

Using (7.110a) for $\nabla \times \mathbf{E}$, the expansion (7.111b), and the orthogonal properties of the modes $\mathbf{H}_n$ and $\mathbf{G}_n$ now gives

$$j\omega\mu_0 g_n + j\omega\mu_0 \mathbf{M}_0 \cdot \mathbf{G}_n(\mathbf{r}_0) = -Z_m \oint_S \mathbf{H} \cdot \mathbf{G}_n \, dS \tag{7.113}$$

since $\nabla \times \mathbf{G}_n$ is zero.

In a similar fashion, use of the relation

$$\int_V \nabla \cdot \mathbf{F}_n \times \mathbf{H} \, dV = \int_V (\mathbf{H} \cdot \nabla \times \mathbf{F}_n - \mathbf{F}_n \cdot \nabla \times \mathbf{H}) \, dV$$

$$= - \int_V \mathbf{F}_n \cdot \nabla \times \mathbf{H} \, dV = \oint_S \mathbf{n} \times \mathbf{F}_n \cdot \mathbf{H} \, dV = 0$$

together with (7.110b) and (7.111a) yields

$$j\omega\epsilon_0 f_n = -j\omega \mathbf{P}_0 \cdot \mathbf{F}_n(\mathbf{r}_0) \tag{7.114}$$

We now have the following equations for the expansion coefficients e_n, h_n, f_n, and g_n:

$$j\omega\epsilon_0 e_n = k_n h_n - j\omega \mathbf{P}_0 \cdot \mathbf{E}_n(\mathbf{r}_0) \tag{7.115a}$$

$$h_n = \frac{-1}{k_0^2 - k_n^2}\left(j\omega k_n \mathbf{P}_0 \cdot \mathbf{E}_n + k_0^2 \mathbf{M}_0 \cdot \mathbf{H}_n - j\omega\epsilon_0 Z_m \oint_S \mathbf{H} \cdot \mathbf{H}_n \, dS\right) \tag{7.115b}$$

$$j\omega\epsilon_0 f_n = -j\omega \mathbf{P}_0 \cdot \mathbf{F}_n \tag{7.115c}$$

$$j\omega\mu_0 g_n = -j\omega\mu_0 \mathbf{M}_0 \cdot \mathbf{G}_n - Z_m \oint_S \mathbf{H} \cdot \mathbf{G}_n \, dS \tag{7.115d}$$

where $k_0^2 = \omega^2\mu_0\epsilon_0$ and (7.115a) has been used in order to eliminate e_n and obtain (7.115b).

In many practical problems dealing with cavities the above equations may be simplified. Usually, ω is very nearly equal to a particular resonant frequency ω_n. As (7.115b) shows, all coefficients h_m, $m \neq n$, will then be small compared with h_n. Thus all the coefficients e_m, $m \neq n$, will also be small compared with e_n, and the field is predominantly that described by the $\mathbf{H}_n$, $\mathbf{E}_n$ mode. In the surface integrals, which represent small perturbations from the loss-free solution, we may approximate $\mathbf{H}$ by $h_n\mathbf{H}_n$ without appreciable error. In addition, in the equation for g_n, we may neglect the surface-integral term to a first approximation. We may then use the relation

$$\oint_S \mathbf{H}_n \cdot \mathbf{H}_n \, dS = \frac{2P_l}{R_m} = \frac{\omega\mu_0}{Q_n R_m}$$

derived earlier if the $\mathbf{H}_n$ modes have been chosen so that they are uncoupled. In place of (7.115) we now obtain the following simplified equations:

$$\left[k_0^2\left(1 + \frac{1-j}{Q_n}\right) - k_n^2\right]h_n = -j\omega k_n \mathbf{P}_0 \cdot \mathbf{E}_n - k_0^2 \mathbf{M}_0 \cdot \mathbf{H}_n$$

or

$$h_n = \frac{j\omega k_n \mathbf{P}_0 \cdot \mathbf{E}_n + k_0^2 \mathbf{M}_0 \cdot \mathbf{H}_n}{k_n^2 - k_0^2\left(1 + \dfrac{1-j}{Q_n}\right)} \tag{7.116a}$$

$$e_n = -\frac{j\omega\mu_0(k_n \mathbf{M}_0 \cdot \mathbf{H}_n + j\omega \mathbf{P}_0 \cdot \mathbf{E}_n)}{k_n^2 - k_0^2\left(1 + \dfrac{1-j}{Q_n}\right)} \tag{7.116b}$$

$$\epsilon_0 f_n = -\mathbf{P}_0 \cdot \mathbf{F}_n \tag{7.116c}$$

$$g_n = -\mathbf{M}_0 \cdot \mathbf{G}_n \tag{7.116d}$$

Equations (7.116a) and (7.116b) may be used for all h_m and e_m. Since we have assumed ω to be equal or nearly equal to ω_n, the denominator may be replaced by $k_0{}^2 - k_m{}^2$ for $m \neq n$. For n we may factor the denominator to give

$$\left[k_n - k_0\left(1 + \frac{1-j}{Q_n}\right)^{\frac{1}{2}}\right]\left[k_n + k_0\left(1 + \frac{1-j}{Q_n}\right)^{\frac{1}{2}}\right]$$

$$\approx -2k_n\left[k_0 - k_n\left(1 - \frac{1-j}{2Q_n}\right)\right]$$

since Q_n is very large.

Usually, we are primarily interested in the strength of excitation of the resonant mode $\mathbf{E}_n$, $\mathbf{H}_n$. Its excitation coefficients are given by (7.116a) and (7.116b). The coefficients f_n and g_n describe the local field that exists around the dipole sources. This field is a quasi-static field in its configuration. The excitation of fields in cavities by volume distributions of currents may be solved by the same method outlined above. As an example, consider a cavity with a volume distribution of current $\mathbf{J}(\mathbf{r}_0)e^{j\omega_n t}$. A differential element of current may be considered equivalent to an electric dipole $\mathbf{P}$ with a moment given by $\mathbf{J}/j\omega$. Thus, from a current element $\mathbf{J}(\mathbf{r}_0)\delta(\mathbf{r} - \mathbf{r}_0)$ located at the point $\mathbf{r}_0$, the amplitudes of the nth mode are given by (7.116) divided by $j\omega$. When the current varies with the frequency ω_n of the nth resonant mode, only this mode is excited with a large amplitude. The amplitude of the electric field of the nth mode due to the volume distribution of current is found by superposition, i.e., adding up the contributions from each current element. Thus we have

$$e_n = -j\omega\mu_0 \frac{\int_V \mathbf{J}(\mathbf{r}_0) \cdot \mathbf{E}_n(\mathbf{r}_0)\, dV_0}{k_n{}^2 - k_0{}^2\left(1 + \frac{1-j}{Q_n}\right)} \tag{7.117}$$

where the integration is taken over the volume of the current distribution. A similar expression holds for the amplitude constant h_n.

The fields excited in cavities by volume distributions of currents may also be solved in terms of a vector potential function. A specific application of the use of the vector potential is given in Chap. 9 in connection with the klystron tube; so we do not consider this method here.

Problems

7.1 Show that, on a short-circuited coaxial line one-half wavelength long, the time-average stored electric and magnetic energies are equal. Use the expressions for the fields given by (3.26).

7.2 For the folded coaxial line, show that $b = \sqrt{ad}$ in order for the characteristic impedance of the inner and outer lines to be the same. This is a common form of line for use in high-frequency oscillators. The effective length l is about twice the physical length. At a frequency of 300 Mc, what must $l/2$ be in order that $l = \lambda_0/4$? If

$2d = 5$ cm and $2a = 2$ cm, what must $2b$ equal for equal characteristic impedances? For a copper line ($\sigma = 5.8 \times 10^7$ mhos/m) find the Q and input impedance at resonance. For d fixed, what are the optimum values of a and b that will make Q a maximum?

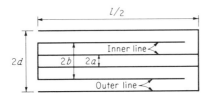

Fig. P 7.2

7.3 Verify that an open-circuited transmission line behaves as a series resonant circuit in the vicinity of the frequency for which it is a quarter wavelength long. Obtain an expression for the input impedance at resonance.

7.4 A short-circuited two-wire line is made of copper. The conductor diameter is 1 cm, the spacing is 3 cm, and the length is 40 cm. Find the antiresonant frequency, the Q, and the input resistance at resonance.

7.5 Find the resonant frequency and Q of a copper rectangular cavity of dimensions $a = b = d = 10$ cm for the TE_{101} mode.

7.6 A cylindrical cavity of radius $a = 2$ cm and a length of 6 cm is filled with a dielectric with permittivity $\epsilon = (2.5 - j0.0001)\epsilon_0$. The cavity is made of copper. Find the resonant frequency and Q for the TE_{111} mode. Note that, in the expression for resonant frequency, the velocity of light in free space, c, must be replaced by

$$c\sqrt{\frac{\epsilon_0}{\epsilon'}}$$

★7.7 Use the results given by (4.25) and (4.26) to show that the Q of a cavity is given by

$$Q = \frac{\omega}{2R}\frac{\partial X}{\partial \omega} = \frac{\omega}{2G}\frac{\partial B}{\partial \omega}$$

where $Z_{in} = R + jX$ and $Y_{in} = G + jB$ are the input impedance and admittance at the terminals. Verify that these formulas give the usual results for a series RLC network and for a parallel RCL network.

7.8 A rectangular cavity of dimensions a, b, d is coupled to a rectangular guide through a capacitive slit. The guide width is a, and the height is b. Obtain an equation for determining the first antiresonant frequency. Find the required slit susceptance for critical coupling. For $a = 2b = 2.5$ cm, $d = 2.5$ cm, and a copper cavity, compute the resonant frequency Q and slit opening t for critical coupling. Use the formula $\bar{B}_c = (2\beta b/\pi) \ln \csc (\pi t/2b)$ for the slit susceptance. What is the loaded Q?

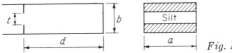

Fig. P 7.8

7.9 For a capacitive diaphragm in a rectangular guide of the dimensions given in Prob. 7.8, obtain an equivalent circuit to represent the susceptance function $\bar{B}_c = (2\beta b/\pi) \ln \csc (\pi t/2b)$ correct to terms up to $\Delta\omega = \omega - \omega_1$ in the vicinity of the frequency ω_1.
Hint: Expand β in a Taylor series about ω_1 and choose a series LC circuit.

7.10 Design a rectangular cavity of length d, height $b = 1.2$ cm, width $a = 2.5$ cm that will resonate at 10,000 Mc. The cavity is critically coupled to a rectangular guide of dimensions a by b. Specify the cavity length d and the radius of the centered circular aperture. Determine the unloaded and loaded Q's if the cavity is made of copper.

7.11 For the aperture-coupled rectangular cavity discussed in the text, let the incident power at resonance be 100 mw. The cavity is critically coupled. Evaluate the peak value of the electric field in the incident wave and in the cavity field. How does the peak amplitude of the cavity field depend on the cavity Q? (See Prob. 7.10 for cavity dimensions.)

7.12 Consider a Fabry-Perot resonator with the plates perforated by a square array of small circular holes. The aperture spacing is 4 mm. The resonator is operated at a wavelength of 8 mm in the 50th resonant mode; that is, $k_0 d \approx 50\pi$. Find the aperture radius r_0 required for critical coupling and the unloaded and loaded Q's. The plates are made of copper with $\sigma = 5.8 \times 10^7$ mhos/m.

7.13 Let the Fabry-Perot resonator of Prob. 7.12 have one plate unperforated. Derive expressions for the resonant frequency, external Q, and loaded Q.

★7.14 A cavity is excited by an impressed electric field $\mathbf{E}_0$ tangent to an aperture surface S_a cut in the cavity wall S. Use the relation $\int_V \nabla \cdot \mathbf{E} \times \mathbf{G}_n \, dV$ to show that the amplitude g_n is given by

$$-j\omega\mu_0 g_n = \oint_S \mathbf{n} \times \mathbf{E} \cdot \mathbf{G}_n \, dS = \int_{S_a} \mathbf{n} \times \mathbf{E}_0 \cdot \mathbf{G}_n \, dS$$

for a cavity with perfectly conducting walls. Using the general expansion (7.100b), show that an alternative expression is

$$-j\omega\mu_0 g_n = -j\omega\mu_0 \int_V \mathbf{H} \cdot \mathbf{G}_n \, dV = \frac{-j\omega\mu_0}{p_n} \oint_S \mathbf{n} \cdot \mathbf{H}\psi_n \, dS$$

upon putting $\mathbf{H} \cdot \nabla\psi_n = \nabla \cdot \mathbf{H}\psi_n$ since $\nabla \cdot \mathbf{H} = 0$ and using the divergence theorem. This last relation shows that the $\mathbf{G}_n$ modes are excited whenever $\mathbf{n} \cdot \mathbf{H}$ does not vanish over S.

★7.15 Show that the two expressions for g_n in Prob. 7.14 are identical.
Hint: Consider

$$\mathbf{n} \cdot \nabla \times \mathbf{E}\psi_n = \psi_n \mathbf{n} \cdot \nabla \times \mathbf{E} - \mathbf{n} \cdot \mathbf{E} \times \nabla\psi_n = -j\omega\mu_0\psi_n \mathbf{n} \cdot \mathbf{H} - \mathbf{n} \times \mathbf{E} \cdot \nabla\psi_n$$

and use Stokes' law to show that

$$\oint_S \mathbf{n} \cdot \nabla \times \mathbf{E}\psi_n \, dS = 0$$

★7.16 Find the eigenfunctions $\mathbf{E}_n$, $\mathbf{H}_n$, $\mathbf{F}_n$, and $\mathbf{G}_n$ for a rectangular cavity of dimensions a, b, c.

★7.17 Obtain a modal expansion similar to (7.111a) and (7.115) for the electromagnetic field in a cavity excited by a volume distribution of electric current $\mathbf{J}(\mathbf{r})$.

References

1. Montgomery, C. G., R. H. Dicke, and E. M. Purcell: "Principles of Microwave Circuits," McGraw-Hill Book Company, New York, 1948.
2. Ragan, G. L. (ed.): "Microwave Transmission Circuits," McGraw-Hill Book Company, New York, 1948.
3. Slater, J. C.: "Microwave Electronics," D. Van Nostrand Company, Inc., Princeton, N.J., 1950.
4. Goubau, G.: "Electromagnetic Waveguides and Cavities," chap. 2, Pergamon Press, New York, 1961.
5. Kurokawa, K.: The Expansions of Electromagnetic Fields in Cavities, *IRE Trans.*, vol. MTT-6, pp. 178–187, April, 1958.
6. Van Bladel, J.: "Electromagnetic Fields," chap. 10, McGraw-Hill Book Company, New York, 1964.

Fabry-Perot resonators

7. Culshaw, W.: High Resolution Millimeter Wave Fabry-Perot Interferometer, *IRE Trans.*, vol. MTT-8, pp. 182–189, March, 1960.
8. Culshaw, W.: Resonators for Millimeter and Submillimeter Wavelengths, *IRE Trans.*, vol. MTT-9, pp. 135–144, March, 1961.
9. Fox, A. G., and T. Li: Resonant Modes in a Maser Interferometer, *Bell System Tech. J.*, vol. 40, pp. 453–488, March, 1961.
10. Boyd, G. D., and J. P. Gordon: Confocal Multimode Resonator for Millimeter through Optical Wavelength Masers, *Bell System Tech. J.*, vol. 40, pp. 489–508, March, 1961.
11. Culshaw, W.: Further Considerations on Fabry-Perot Type Resonators, *IRE Trans.*, vol. MTT-10, pp. 331–339, September, 1962.

8
Periodic structures and filters

Waveguides and transmission lines loaded at periodic intervals with identical obstacles, e.g., a reactive element such as a diaphragm, are referred to as periodic structures. The interest in waveguiding structures of this type arises from two basic properties common to all periodic structures, namely (1) passband-stopband characteristics, and (2) support of waves with phase velocities much less than the velocity of light. The passband-stopband characteristic is the existence of frequency bands throughout which a wave propagates unattenuated (except for incidental conductor losses) along the structure separated by frequency bands throughout which the wave is cut off and does not propagate. The former is called a passband, and the latter is referred to as a stopband. The passband-stopband property is of some interest for its frequency filtering aspects.

The ability of many periodic structures to support a wave having a phase velocity much less than that of light is of basic importance for traveling-wave-tube circuits. In a traveling-wave tube, efficient interaction between the electron beam and the electromagnetic field is obtained only if the phase velocity is equal to the beam velocity. Since the latter is often no greater than 10 to 20 percent of the velocity of light, considerable slowing down of the electromagnetic wave is required. Periodic structures suitable for use in traveling-wave tubes are discussed in this chapter. The actual principles of operation of the tube are covered in Chap. 9.

The last part of the chapter is devoted to an introduction to microwave filter theory. A complete treatment of all aspects of filter theory and design would be much too lengthy to include in this text. However, sufficient material is covered to provide a background so that the technical literature can be read without difficulty.

8.1 Capacitively loaded transmission-line–circuit analysis

To introduce a number of basic concepts, methods of analysis, and typical properties of periodic structures, we shall consider a simple example of a capacitively loaded transmission line. For a physically smooth transmission line, such as a coaxial line, the phase velocity is given by

$$v_p = (LC)^{-\frac{1}{2}} = (\mu_0 \kappa \epsilon_0)^{-\frac{1}{2}} \tag{8.1}$$

where κ is the dielectric constant of the medium surrounding the conductor. A significant reduction in phase velocity can be achieved in a smooth line only by increasing κ. This method has the great disadvantage that the cross-sectional dimensions of the line must also be reduced to avoid the propagation of higher-order modes. The phase velocity cannot be decreased by increasing the shunt capacity C per unit length because any change in the line configuration to increase C automatically decreases the series inductance L per unit length, since $LC = \mu_0 \epsilon$. However, by removing the restriction that the line should be physically smooth, an effective increase in the shunt capacitance per unit length can be achieved without a corresponding decrease in the series inductance L. That is, lumped shunt capacitance may be added at periodic intervals without affecting the value of L. If the spacing between the added lumped capacitors is small compared with the wavelength, it may be anticipated that the line will appear to be electrically smooth, with a phase velocity

$$v_p = \left[\left(C + \frac{C_0}{d}\right)L\right]^{-\frac{1}{2}} \tag{8.2}$$

where C_0/d is the amount of lumped capacitance added per unit length (a capacitor C_0 added at intervals d). The following analysis will verify this conclusion.

One method of obtaining shunt capacitive loading of a coaxial transmission line is to introduce thin circular diaphragms at regular intervals, as in Fig. 8.1. The diaphragms may be machined as an integral part of the center conductor. The fringing electric field in the vicinity of the

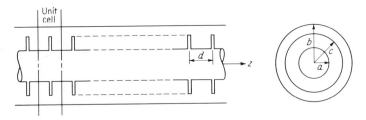

Fig. 8.1 Capacitive loading of a coaxial line by means of thin circular diaphragms.

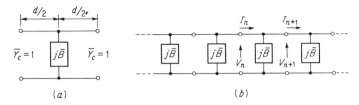

Fig. 8.2 (a) Equivalent circuit for unit cell of loaded coaxial line; (b) cascade connection of basic unit-cell networks.

diaphragm increases the local storage of electric energy and hence may be accounted for, from a circuit viewpoint, by a shunt capacitance. The local field can be described in terms of the incident, reflected, and transmitted dominant TEM mode and a superposition of an infinite number of higher-order E modes. If the cylinder spacing $b - a$ is small compared with the wavelength, the higher-order modes are evanescent and decay to a negligible value in a distance of the order of $b - a$ away from the diaphragm in either direction. An approximate expression for the shunt susceptance of the diaphragm is[†]

$$\bar{B} = \frac{B}{Y_c} = \frac{8(b-c)^2}{\lambda_0 c} \frac{\ln(b/a)}{[\ln(b/c)]^2} \ln \csc \left(\frac{\pi}{2} \frac{b-c}{b-a} \right) \qquad (8.3)$$

where $Y_c = [60 \ln (b/a)]^{-1}$ is the characteristic admittance of an air-filled coaxial line. The expression for $\bar{B}$ is accurate for $b - a \leq 0.1\lambda_0$. In this low-frequency region, $\bar{B}$ has a frequency dependence directly proportional to ω. At higher frequencies $\bar{B}$ will have a more complicated frequency dependence, although the thin diaphragm can still be represented by a shunt susceptance.

The circuit, or network, analysis of a periodic structure involves constructing an equivalent network for a single basic section or unit cell of the structure first. This is followed by an analysis to determine the voltage and current waves that may propagate along the network consisting of the cascade connection of an infinite number of the basic networks. For the structure of Fig. 8.1 an equivalent network of a basic section is a shunt normalized susceptance $\bar{B}$ with a length $d/2$ of transmission line on either side, as in Fig. 8.2a. Figure 8.2b illustrates the voltage-current relationships at the input and output of the nth section in the infinitely long cascade connection.

The relationships between the input variables V_n, I_n and the output variables V_{n+1}, I_{n+1} are readily found by using the $\mathcal{ABCD}$ transmission matrix discussed in Sec. 4.9. The V_n and I_n are the total voltage and current amplitudes, i.e., the sum of the contributions from the incident and reflected TEM waves at the terminal plane. The circuit for a unit

[†] N. Marcuvitz (ed.), "Waveguide Handbook," p. 229, McGraw-Hill Book Company, New York, 1951.

cell may be broken down into three circuits in cascade, namely, a section of transmission line of length $d/2$ (electrical length $\theta/2 = k_0 d/2$), followed by a shunt susceptance $\bar{B}$, which in turn is followed by another length of transmission line. The $\mathcal{ABCD}$ matrix for each of these individual networks is, respectively (Prob. 4.18),

$$\begin{bmatrix} \cos\dfrac{\theta}{2} & j\sin\dfrac{\theta}{2} \\ j\sin\dfrac{\theta}{2} & \cos\dfrac{\theta}{2} \end{bmatrix} \quad \begin{bmatrix} 1 & 0 \\ j\bar{B} & 1 \end{bmatrix} \quad \begin{bmatrix} \cos\dfrac{\theta}{2} & j\sin\dfrac{\theta}{2} \\ j\sin\dfrac{\theta}{2} & \cos\dfrac{\theta}{2} \end{bmatrix}$$

The transmission matrix for the unit cell is obtained by the chain rule [see (4.75)], i.e., the product of the above three matrices, and hence we have

$$\begin{bmatrix} V_n \\ I_n \end{bmatrix} = \begin{bmatrix} \cos\dfrac{\theta}{2} & j\sin\dfrac{\theta}{2} \\ j\sin\dfrac{\theta}{2} & \cos\dfrac{\theta}{2} \end{bmatrix} \begin{bmatrix} 1 & 0 \\ j\bar{B} & 1 \end{bmatrix} \begin{bmatrix} \cos\dfrac{\theta}{2} & j\sin\dfrac{\theta}{2} \\ j\sin\dfrac{\theta}{2} & \cos\dfrac{\theta}{2} \end{bmatrix} \begin{bmatrix} V_{n+1} \\ I_{n+1} \end{bmatrix}$$

$$= \begin{bmatrix} \cos\theta - \dfrac{\bar{B}}{2}\sin\theta & j\left(\dfrac{\bar{B}}{2}\cos\theta + \sin\theta - \dfrac{\bar{B}}{2}\right) \\ j\left(\dfrac{\bar{B}}{2}\cos\theta + \sin\theta + \dfrac{\bar{B}}{2}\right) & \cos\theta - \dfrac{\bar{B}}{2}\sin\theta \end{bmatrix} \begin{bmatrix} V_{n+1} \\ I_{n+1} \end{bmatrix}$$

(8.4)

Note that $\mathcal{A} = \mathcal{D}$, which is always true for a symmetrical network, i.e., a symmetrical unit cell.

If the periodic structure is capable of supporting a propagating wave, it is necessary for the voltage and current at the $(n+1)$st terminal to be equal to the voltage and current at the nth terminal, apart from a phase delay due to a finite propagation time. Thus we assume that

$$V_{n+1} = e^{-\gamma d} V_n \tag{8.5a}$$

$$I_{n+1} = e^{-\gamma d} I_n \tag{8.5b}$$

where $\gamma = j\beta + \alpha$ is the propagation constant for the periodic structure. In terms of the transmission matrix for a unit cell, we now have

$$\begin{bmatrix} V_n \\ I_n \end{bmatrix} = \begin{bmatrix} \mathcal{A} & \mathcal{B} \\ \mathcal{C} & \mathcal{D} \end{bmatrix} \begin{bmatrix} V_{n+1} \\ I_{n+1} \end{bmatrix} = e^{\gamma d} \begin{bmatrix} V_{n+1} \\ I_{n+1} \end{bmatrix}$$

or

$$\left(\begin{bmatrix} \mathcal{A} & \mathcal{B} \\ \mathcal{C} & \mathcal{D} \end{bmatrix} - \begin{bmatrix} e^{\gamma d} & 0 \\ 0 & e^{\gamma d} \end{bmatrix} \right) \begin{bmatrix} V_{n+1} \\ I_{n+1} \end{bmatrix} = 0 \tag{8.6}$$

This equation is a matrix eigenvalue equation for γ. A nontrivial solution for V_{n+1}, I_{n+1} exists only if the determinant vanishes. Hence

$$\begin{vmatrix} \mathcal{A} - e^{\gamma d} & \mathcal{B} \\ \mathcal{C} & \mathcal{D} - e^{\gamma d} \end{vmatrix} = \mathcal{AD} - \mathcal{BC} + e^{2\gamma d} - e^{\gamma d}(\mathcal{A} + \mathcal{D}) = 0 \quad (8.7)$$

For a reciprocal network the determinant $\mathcal{AD} - \mathcal{BC}$ of the transmission matrix equals unity (Sec. 4.9); so we obtain

$$\cosh \gamma d = \frac{\mathcal{A} + \mathcal{D}}{2} \quad (8.8)$$

For the capacitively loaded coaxial line, (8.8), together with (8.4), yields

$$\cosh \gamma d = \cos \theta - \frac{\bar{B}}{2} \sin \theta \quad (8.9)$$

When $|\cos \theta - (\bar{B}/2) \sin \theta| < 1$, we must have $\gamma = j\beta$ and $\alpha = 0$; that is,

$$\cos \beta d = \cos \theta - \frac{\bar{B}}{2} \sin \theta \quad (8.10a)$$

When the right-hand side of (8.9) is greater than unity, $\gamma = \alpha$ and $\beta = 0$; so

$$\cosh \alpha d = \cos \theta - \frac{\bar{B}}{2} \sin \theta > 1 \quad (8.10b)$$

Finally, when the right-hand side of (8.9) is less than -1, we must have $\gamma d = j\pi + \alpha$, so that

$$\cosh \gamma d = \cosh (j\pi + \alpha d) = -\cosh \alpha d$$
$$= \cos \theta - \frac{\bar{B}}{2} \sin \theta < -1 \quad (8.10c)$$

It is apparent, then, that there will be frequency bands for which unattenuated propagation can take place separated by frequency bands in which the wave is attenuated. Note that propagation in both directions is possible since $-\gamma$ is also a solution.

A detailed study of the passband-stopband characteristic is made in Sec. 8.6. For the present we shall confine our attention to the low-frequency limiting value of β. When $d \ll \lambda_0$, $\theta = k_0 d$ is small, and βd will then also be small. Replacing $\cos \theta$ by $1 - \theta^2/2$ and $\sin \theta$ by θ in (8.10a) gives

$$\cos \beta d \approx 1 - \frac{\beta^2 d^2}{2} = 1 - \frac{k_0^2 d^2}{2} - \frac{\bar{B} k_0 d}{2}$$

Using the relations $k_0^2 = \omega^2 \mu_0 \epsilon_0 = \omega^2 LC$ and $\bar{B} = B/Y_c = \omega C_0 (L/C)^{\frac{1}{2}}$, where $\omega C_0 = B$, we obtain

$$\beta^2 = \omega^2 LC + \frac{\omega^2 LC_0}{d}$$

and hence

$$\beta = \omega \sqrt{L\left(C + \frac{C_0}{d}\right)} \tag{8.11}$$

Therefore we find that, at low frequencies where $d \ll \lambda_0$, the loaded line behaves as an electrically smooth line with a shunt capacitance $C + C_0/d$ per unit length. The increase in β results in a reduction of the phase velocity by a factor k_0/β.

Another parameter of importance in connection with periodic structures is the normalized characteristic impedance $\bar{Z}_B$ presented to the voltage and current waves at the reference terminal plane, i.e., input terminals of a unit cell. An expression for $\bar{Z}_B$ may be obtained from (8.6), which may be written as

$$(\mathcal{A} - e^{\gamma d})V_{n+1} = -\mathcal{B} I_{n+1}$$
$$-\mathcal{C} V_{n+1} = (\mathcal{D} - e^{\gamma d}) I_{n+1}$$

Hence

$$\frac{Z_B}{Z_c} = \bar{Z}_B = \frac{V_{n+1}}{I_{n+1}} = \frac{-\mathcal{B}}{\mathcal{A} - e^{\gamma d}} = -\frac{\mathcal{D} - e^{\gamma d}}{\mathcal{C}} \tag{8.12}$$

Replacing $2e^{\gamma d}$ by $\mathcal{A} + \mathcal{D} \pm [(\mathcal{A} + \mathcal{D})^2 - 4]^{\frac{1}{2}}$ from (8.7), we obtain

$$\bar{Z}_B^{\pm} = \frac{2\mathcal{B}}{\mathcal{D} - \mathcal{A} \pm \sqrt{(\mathcal{A} + \mathcal{D})^2 - 4}} \tag{8.13a}$$

where the upper and lower signs refer to propagation in the $+z$ and $-z$ directions, respectively. We are using the convention that the positive directions of V_n and I_n are those indicated in Fig. 8.2, independent of the direction of propagation. For a symmetrical network, $\mathcal{A} = \mathcal{D}$, and since $\mathcal{A}\mathcal{D} - \mathcal{B}\mathcal{C} = 1$, we have $\mathcal{A}^2 - 1 = \mathcal{B}\mathcal{C}$. In this case (8.13a) reduces to

$$\bar{Z}_B^{\pm} = \frac{2\mathcal{B}}{\pm \sqrt{4\mathcal{A}^2 - 4}} = \pm \sqrt{\frac{\mathcal{B}}{\mathcal{C}}} \tag{8.13b}$$

In general, for a lossless structure, $\bar{Z}_B^- = -(\bar{Z}_B^+)^*$ in the passband, since $|\mathcal{A} + \mathcal{D}| < 2$, as (8.8) shows.

If the unit cell is represented by a T network with parameters $\bar{Z}_{11}$, $\bar{Z}_{12}$, and $\bar{Z}_{22}$, then, by using the relations between the $\mathcal{ABCD}$ parameters and the impedance parameters given in Sec. 4.9, we can also show that

$$\cosh \gamma d = \frac{\bar{Z}_{11} + \bar{Z}_{22}}{2\bar{Z}_{12}} \tag{8.14}$$

$$\bar{Z}_B = \frac{\bar{Z}_{11} - \bar{Z}_{22}}{2} \pm \bar{Z}_{12} \sinh \gamma d \tag{8.15}$$

The waves that may propagate along a periodic structure are often called Bloch waves by analogy with the quantum-mechanical electron waves that may propagate through a periodic crystal lattice in a solid. It is for this reason that we have denoted the characteristic impedance as $\bar{Z}_B$ for the Bloch wave. The voltage and current at the nth terminal plane will be denoted by $V_{Bn}{}^\pm$, $I_{Bn}{}^\pm$ for the Bloch waves from now on instead of by the quantities V_n, I_n. The $+$ and $-$ signs refer to Bloch waves propagating in the $+z$ and $-z$ directions. We shall also adopt the convention that the positive direction of current flow for Bloch waves is always in the $+z$ direction; thus $I_B{}^+ = \bar{Y}_B{}^+ V_B{}^+$ and $I_B{}^- = \bar{Y}_B{}^- V_B{}^-$. However, for a symmetrical structure such that $\mathcal{A} = \mathcal{D}$, we shall have $\bar{Y}_B{}^- = -\bar{Y}_B{}^+ = -(\bar{Z}_B{}^+)^{-1}$.

If (8.13) is used, we find that, for the loaded coaxial line,

$$\bar{Z}_B = \sqrt{\frac{\mathcal{B}}{\mathcal{C}}} = \sqrt{\frac{2\sin\theta + \bar{B}\cos\theta - \bar{B}}{2\sin\theta + \bar{B}\cos\theta + \bar{B}}} \tag{8.16}$$

In the low-frequency limit, where we can replace $\sin\theta$ by

$$\theta = k_0 d = \omega d \sqrt{LC}$$

and $\cos\theta$ by 1, we obtain

$$\bar{Z}_B = \sqrt{\frac{2\theta}{2\theta + 2\bar{B}}} = \sqrt{\frac{C}{C + C_0/d}}$$

and thus

$$Z_B = \bar{Z}_B Z_c = \sqrt{\frac{L}{C + C_0/d}} \tag{8.17}$$

Again we see that, in the low-frequency limit, the loaded line is electrically smooth and the characteristic impedance is modified in the anticipated manner by the effective increase in the shunt capacitance per unit length.

The characteristic impedance of a periodic structure is not a unique quantity since it depends on the choice of terminal planes for a unit cell. If the terminal planes are shifted a distance l in the $-z$ direction, the new characteristic impedance becomes

$$\bar{Z}_B' = \frac{\bar{Z}_B + j\tan k_0 l}{1 + j\bar{Z}_B \tan k_0 l} \tag{8.18}$$

8.2 Wave analysis of periodic structures

Periodic structures may be analyzed in terms of the forward- and backward-propagating waves that can exist in each unit cell with about the

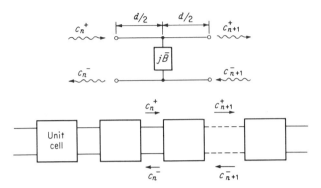

Fig. 8.3 Wave amplitudes in a periodic structure.

same facility as the network approach gives. In the wave approach the wave-amplitude transmission matrix $[A]$ discussed in Sec. 4.9 is used.

With reference to Fig. 8.3, let the amplitudes of the forward- and backward-propagating waves at the nth and $(n + 1)$st terminal plane be c_n^+, c_n^-, c_{n+1}^+, and c_{n+1}^-. The c_{n+1}^+, c_{n+1}^- are related to the c_n^+, c_n^- by the wave-amplitude transmission matrix as follows:

$$\begin{bmatrix} c_n^+ \\ c_n^- \end{bmatrix} = \begin{bmatrix} A_{11} & A_{12} \\ A_{21} & A_{22} \end{bmatrix} \begin{bmatrix} c_{n+1}^+ \\ c_{n+1}^- \end{bmatrix} \tag{8.19}$$

The solution for a Bloch wave requires $c_{n+1}^+ = e^{-\gamma d} c_n^+$ and $c_{n+1}^- = e^{-\gamma d} c_n^-$. Hence (8.19) becomes

$$\begin{bmatrix} A_{11} - e^{\gamma d} & A_{12} \\ A_{21} & A_{22} - e^{\gamma d} \end{bmatrix} \begin{bmatrix} c_{n+1}^+ \\ c_{n+1}^- \end{bmatrix} = 0 \tag{8.20}$$

A nontrivial solution for c_{n+1}^+, c_{n+1}^- is obtained only if the determinant vanishes. Consequently, the eigenvalue equation for γ is

$$A_{11}A_{22} - A_{12}A_{21} + e^{2\gamma d} - e^{\gamma d}(A_{11} + A_{22}) = 0$$

or

$$\cosh \gamma d = \frac{A_{11} + A_{22}}{2} \tag{8.21}$$

since the determinant of the transmission matrix, that is, $A_{11}A_{22} - A_{12}A_{21}$, equals 1 when normalized wave amplitudes are used.

The Bloch wave which can propagate in the periodic structure is made up from forward- and backward-propagating normal transmission-line or waveguide waves that exist between discontinuities. When γ has been determined from (8.21), the ratio of c_n^- to c_n^+ is fixed. This ratio is called the characteristic reflection coefficient Γ_B. Thus the transverse electric field of the Bloch wave will have an amplitude

$$V_{B0} = c_0^+ + c_0^- = c_0^+(1 + \Gamma_B)$$

at the zeroth terminal plane and an amplitude

$$V_{Bn} = c_n^+ + c_n^- = c_n^+(1 + \Gamma_B) = c_0^+(1 + \Gamma_B)e^{-\gamma nd} \qquad (8.22a)$$

at the nth terminal plane. The transverse magnetic field of the Bloch wave will have an amplitude

$$I_{Bn} = c_0^+(1 - \Gamma_B)e^{-\gamma nd} \qquad (8.22b)$$

at the nth terminal plane.

The characteristic reflection coefficient may be found from the pair of equations (8.20) by eliminating $e^{-\gamma d}$ by the use of (8.21). It is usually more convenient to express Γ_B in terms of $\bar{Z}_B$ by using the relation $\bar{Z}_B = (1 + \Gamma_B)/(1 - \Gamma_B)$. Thus we have

$$\Gamma_B^\pm = \frac{\bar{Z}_B^\pm - 1}{\bar{Z}_B^\pm + 1} \qquad (8.23)$$

where the $+$ and $-$ signs refer to Bloch waves propagating in the $+z$ and $-z$ directions, respectively.

The above wave formulation is now applied to the capacitively loaded transmission line discussed earlier. The unit cell is chosen as in Fig. 8.3. The wave-amplitude transmission matrices for the three sections of the unit cell are (Sec. 4.9 and Prob. 8.7)

$$\begin{bmatrix} e^{jk_0 d/2} & 0 \\ 0 & e^{-jk_0 d/2} \end{bmatrix} \begin{bmatrix} \dfrac{2 + j\bar{B}}{2} & j\dfrac{\bar{B}}{2} \\ -j\dfrac{\bar{B}}{2} & \dfrac{4 + \bar{B}^2}{2(2 + j\bar{B})} \end{bmatrix}$$

and another matrix like the first one. The $[A]$ matrix for the unit cell is obtained by multiplying the three component matrices together; thus

$$[A] = \begin{bmatrix} e^{j\theta/2} & 0 \\ 0 & e^{-j\theta/2} \end{bmatrix} \begin{bmatrix} \dfrac{2 + j\bar{B}}{2} & j\dfrac{\bar{B}}{2} \\ -j\dfrac{\bar{B}}{2} & \dfrac{4 + \bar{B}}{2(2 + j\bar{B})} \end{bmatrix} \begin{bmatrix} e^{j\theta/2} & 0 \\ 0 & e^{-j\theta/2} \end{bmatrix}$$

where $\theta = k_0 d$. After multiplication we obtain

$$[A] = \begin{bmatrix} \dfrac{2 + j\bar{B}}{2} e^{j\theta} & j\dfrac{\bar{B}}{2} \\ -j\dfrac{\bar{B}}{2} & \dfrac{4 + \bar{B}^2}{2(2 + j\bar{B})} e^{-j\theta} \end{bmatrix} \qquad (8.24)$$

Making use of (8.21), we find that

$$\cosh \gamma d = \frac{(4 + \bar{B}^2)e^{-j\theta} + (2 + j\bar{B})^2 e^{j\theta}}{4(2 + j\bar{B})} = \cos \theta - \frac{\bar{B}}{2} \sin \theta$$

which is the same as (8.9) obtained earlier.

8.3 Periodic structures composed of unsymmetrical two-port networks

The capacitively loaded coaxial transmission line can be considered as made up of symmetrical sections by choosing terminal planes midway between each diaphragm. For other choices of terminal-plane positions the unit cell would be unsymmetrical, and its equivalent T network would then also be unsymmetrical. Other types of periodic structures are composed of intrinsically unsymmetrical unit cells such that there is no terminal-plane location that will reduce them to a symmetrical structure. Several unsymmetrical structures are illustrated in Fig. 8.4.

For nonsymmetrical structures the Bloch-wave characteristic impedance is given by (8.15), which we rewrite as

$$\bar{Z}_B{}^+ = \zeta + \bar{Z} \tag{8.25a}$$

$$\bar{Z}_B{}^- = \zeta - \bar{Z} \tag{8.25b}$$

where

$$\zeta = \frac{\bar{Z}_{11} - \bar{Z}_{22}}{2} \tag{8.26a}$$

$$\bar{Z} = \pm \bar{Z}_{12} \sinh \gamma d = \pm j \bar{Z}_{12} \sin \beta d \tag{8.26b}$$

and the sign is to be chosen so that $\bar{Z}$ has a positive real part. The phase constant β is given by

$$\cos \beta d = \frac{\bar{Z}_{11} + \bar{Z}_{22}}{2 \bar{Z}_{12}} \tag{8.26c}$$

in the propagation band. The physical length of a unit cell is d. The quantities $\bar{Z}_B{}^\pm$ and βd are often called the iterative parameters of the T network. A consequence of the nonsymmetry of the unit cell is that $\bar{Z}_B{}^+$ is different from $\bar{Z}_B{}^-$.

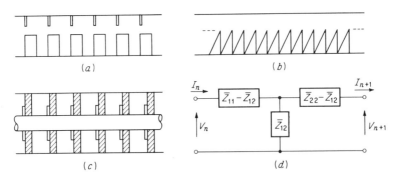

Fig. 8.4 Periodic structures with unsymmetrical unit cells. (*a*, *b*) Rectangular waveguide loaded with thick unsymmetrical diaphragms; (*c*) coaxial line loaded with diaphragms and dielectric rings; (*d*) equivalent T network of a unit cell.

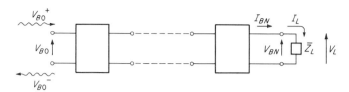

Fig. 8.5 Periodic structure terminated in a load $\bar{Z}_L$.

Let the voltage of the Bloch wave at the zeroth terminal plane be $V_{B0}^\pm$, where the signs $+$ or $-$ refer to Bloch waves propagating in the $+z$ and $-z$ direction, respectively. The corresponding Bloch-wave current is $I_{B0}^\pm = V_{B0}^\pm / \bar{Z}_B^\pm$. At the nth terminal plane the Bloch-wave voltages and currents will be

$$V_{Bn}^\pm = V_{B0}^\pm e^{\mp \gamma nd} \tag{8.27a}$$

$$I_{Bn}^\pm = \frac{V_{B0}^\pm}{\bar{Z}_B^\pm} e^{\mp \gamma nd} \tag{8.27b}$$

Recall that we are taking the positive direction of current flow to be in the $+z$ direction, independent of the direction of propagation for the Bloch waves.

If the restriction is made that the only points at which the voltages and currents will be specified are the terminal planes, the periodic structure has properties similar to any uniform transmission line or waveguide. As such, transmission-line theory can be applied to study the effects of terminating a periodic structure in an arbitrary load impedance, to design matching sections for periodic structures, etc. These applications are discussed in the following two sections.

8.4 Terminated periodic structures

Figure 8.5 illustrates a periodic structure terminated in a load impedance $\bar{Z}_L$ at the Nth terminal plane. The total voltage and current at the nth terminal plane will be a superposition of an incident and reflected Bloch wave; thus

$$V_{Bn} = V_{B0}^+ e^{-j\beta nd} + V_{B0}^- e^{j\beta nd} \tag{8.28a}$$

$$\begin{aligned} I_{Bn} &= I_{B0}^+ e^{-j\beta nd} + I_{B0}^- e^{j\beta nd} \\ &= V_{B0}^+ \bar{Y}_B^+ e^{-j\beta nd} + V_{B0}^- \bar{Y}_B^- e^{j\beta nd} \end{aligned} \tag{8.28b}$$

where $\bar{Y}_B = \bar{Z}_B^{-1}$. At the Nth terminal plane we must have

$$V_L = V_{BN} = \bar{Z}_L I_{BN} = \bar{Z}_L I_L$$

and hence

$$V_{BN}^+ + V_{BN}^- = \bar{Z}_L (\bar{Y}_B^+ V_{BN}^+ + \bar{Y}_B^- V_{BN}^-) \tag{8.29}$$

The reflection coefficient Γ_L of the load for Bloch waves is, from (8.29),

$$\Gamma_L = \frac{V_{BN}^-}{V_{BN}^+} = -\frac{\bar{Z}_L \bar{Y}_B^+ - 1}{\bar{Z}_L \bar{Y}_B^- - 1} = -\frac{\bar{Z}_B^-}{\bar{Z}_B^+} \frac{\bar{Z}_L - \bar{Z}_B^+}{\bar{Z}_L - \bar{Z}_B^-}$$

$$= \frac{\bar{Z} - \zeta}{\bar{Z} + \zeta} \frac{\bar{Z}_L - \bar{Z} - \zeta}{\bar{Z}_L + \bar{Z} - \zeta} \tag{8.30}$$

For a symmetrical structure $\zeta = 0$ and the expression for Γ_L reduces to the usual form.

The Bloch-wave reflection coefficient at the nth terminal plane is

$$\Gamma_n = \frac{V_{Bn}^-}{V_{Bn}^+} = \frac{V_{BN}^- e^{-j(N-n)\beta d}}{V_{BN}^+ e^{j(N-n)\beta d}} = \Gamma_L e^{-j2(N-n)\beta d} \tag{8.31}$$

The input impedance at the nth terminal plane is

$$\bar{Z}_n = \frac{V_{Bn}^+ + V_{Bn}^-}{I_{Bn}^+ + I_{Bn}^-} = \frac{V_{Bn}^+(1 + \Gamma_n)}{V_{Bn}^+ \bar{Y}_B^+ + V_{Bn}^- \bar{Y}_B^-}$$

$$= \frac{1 + \Gamma_n}{\bar{Y}_B^+ + \Gamma_n \bar{Y}_B^-} = \frac{\bar{Z}_B^+ \bar{Z}_B^-(1 + \Gamma_n)}{\bar{Z}_B^- + \bar{Z}_B^+ \Gamma_n} \tag{8.32}$$

From (8.32) we can also obtain the alternative expressions

$$\bar{Z}_n - \zeta = \bar{Z} \frac{\bar{Z}_B^- - \bar{Z}_B^+ \Gamma_n}{\bar{Z}_B^- + \bar{Z}_B^+ \Gamma_n} \tag{8.33a}$$

$$\Gamma_n = -\frac{\bar{Z}_B^-}{\bar{Z}_B^+} \frac{(\bar{Z}_n - \zeta) - \bar{Z}}{(\bar{Z}_n - \zeta) + \bar{Z}} \tag{8.33b}$$

If (8.31) is used to express Γ_n in terms of Γ_L and (8.30) is used to express Γ_L in terms of $\bar{Z}_L$, we find that (8.33a) gives

$$\bar{Z}_n - \zeta = \bar{Z} \frac{\bar{Z}_L - \zeta + j\bar{Z} \tan(N-n)\beta d}{\bar{Z} + j(\bar{Z}_L - \zeta) \tan(N-n)\beta d} \tag{8.34}$$

This equation gives the transformation of impedance along a periodic structure. It differs somewhat from the usual transmission-line formula when the unit cell is unsymmetrical, so that $\zeta \neq 0$.

For a Bloch wave propagating in the $+z$ direction, the periodic structure must be terminated in a load $\bar{Z}_L = \bar{Z}_B^+ = \zeta + \bar{Z}$ to avoid a reflected wave. Similarly, the matched-load termination for a Bloch wave propagating in the $-z$ direction is $-\bar{Z}_B^- = \bar{Z} - \zeta$. The two characteristic Bloch-wave impedances are the iterative impedances for the T network of the unit cell. With voltages and currents chosen as in Fig. 8.6, it is readily shown that an impedance $\bar{Z}_B^+$ connected at terminals 2 is transformed into itself at terminals 1. Similarly, an impedance $\bar{Z}_B^-$ connected at terminals 1 is transformed into itself at terminals 2. It is for this reason that $\bar{Z}_B^\pm$ is called an iterative (repeating) impedance.

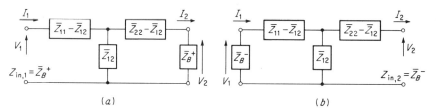

Fig. 8.6 Iterative impedance properties of a T network.

For a lossless T network, $\bar\zeta$ is pure imaginary and $\bar Z$ is pure real in the propagation band. Ambiguity in the sign of $\bar Z$ as given by (8.15) or (8.26b) may be avoided by noting that, in a passband, $\bar Z$ is real and positive, in order to be consistent with our choice of positive direction for current. We must have positive real power transmission, and hence

$$P = \operatorname{Re} \tfrac{1}{2} V_B{}^+(I_B{}^+)^* = \operatorname{Re} \tfrac{1}{2}|I_B{}^+|^2 Z_B{}^+ = \tfrac{1}{2}|I_B{}^+|^2 \bar Z > 0 \tag{8.35}$$

Another criterion that may be used is the one requiring the reactive part of $\bar Z_B{}^+$ to have a positive derivative with respect to ω (Sec. 4.3). There is also reactive power in a Bloch wave in a passband, and this is given by

$$P_{\text{reactive}} = \tfrac{1}{2}|I_B{}^+|^2 \bar\zeta \tag{8.36}$$

Complex power for a Bloch wave propagating in the $-z$ direction is given by $-\tfrac{1}{2} V_B{}^-(I_B{}^-)^*$ because of our choice of direction for positive current.

8.5 Matching of periodic structures

If a periodic structure is connected to a smooth transmission line or waveguide, some means of matching the periodic structure to the input waveguide must be provided to avoid reflection of the incident power. A situation encountered quite frequently is the one where the periodic structure is identical with the input waveguide apart from the periodic loading. One way of providing a matched transition from the unloaded to the loaded waveguide is to use a tapered intermediate section. The matching taper section is similar to the loaded waveguide except that the periodic loading is gradually reduced to zero over a distance of about a wavelength. Figure 8.7a illustrates a tapered transition in a rectangular waveguide connected to a similar guide periodically loaded with diaphragms.

Any of the matching networks discussed in Chap. 5 may also be used to match a periodically loaded guide to an unloaded guide. For example, at some distance $d'/2$ in front of the first terminal plane for the periodic structure, the characteristic admittance $\bar Y_B$ of the periodic structure is transformed into an admittance $1 - j\bar B'$, so that placing a shunt sus-

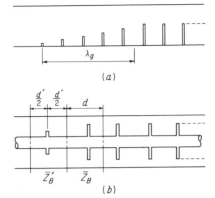

Fig. 8.7 (a) Tapered transition matching section for a diaphragm-loaded rectangular guide; (b) quarter-wave transformer matching of a capacitively loaded coaxial line.

ceptance $j\bar{B}'$ at this point provides a matched transition. Matching by means of a shunt susceptance may be viewed as an application of the quarter-wave transformer matching technique. The unit cell consisting of the shunt susceptance $j\bar{B}'$ plus a length $d'/2$ of transmission line (or waveguide) on either side, as in Fig. 8.7b, may be considered as part of an infinite periodic structure with a propagation phase constant β' and a normalized characteristic impedance $\bar{Z}'_B$. If the parameters $\bar{B}'$ and d' are chosen so that $\beta'd' = \pi/2$, $\bar{Z}'_B = \bar{Z}_B^{\frac{1}{2}}$, then at the input terminal to the matching section, (8.34) gives

$$\bar{Z}_{in} = \frac{(\bar{Z}'_B)^2}{\bar{Z}_B} = 1 \tag{8.37}$$

Note that the matching section is a symmetrical structure, so that $\bar{Z}'_B = \bar{Z}'$ and $\varsigma' = 0$. We also require that $\bar{Z}_B$ be real in order for (8.37) to have a real solution for $\bar{Z}'_B$. When these conditions are met, we see that the matching section behaves essentially as a quarter-wave transformer. For symmetrical structures the required values of $\bar{B}'$ and d' may be found from (8.8) and (8.13b); thus, as $\cos \beta'd' = \cos(\pi/2) = 0$, we have

$$\mathcal{Q}' = \mathfrak{D}' = 0 \tag{8.38a}$$

$$(\bar{Z}'_B)^2 = \frac{\mathcal{B}'}{\mathcal{C}'} = \bar{Z}_B \tag{8.38b}$$

For the capacitively loaded transmission line we obtain, by using (8.4) and (8.16) applied to the matching section,

$$2 \cot k_0 d' = \bar{B}' \tag{8.39a}$$

$$(\bar{Z}'_B)^2 = \frac{2 \sin k_0 d' + \bar{B}' \cos k_0 d' - \bar{B}'}{2 \sin k_0 d' + \bar{B}' \cos k_0 d' + \bar{B}'} = \tan^2 \frac{k_0 d'}{2} = \bar{Z}_B \tag{8.39b}$$

when (8.39a) is used to eliminate $\bar{B}'$. The above results are the equivalent of those derived in Chap. 5, i.e., given by (5.8) and (5.9).

To obtain a match over a wide frequency band, more elaborate matching networks must be used since a single shunt susceptance usually does not provide a match over a wide frequency band. Broadband matching is complicated by the fact that the characteristic impedance $\bar{Z}_B$ of a periodic structure is a function of frequency. No general technique exists for designing broadband matching networks because of the general nature of $\bar{Z}_B$. Each periodic structure must be considered by itself so that the frequency variation in $\bar{Z}_B$ can be incorporated into the design. For this reason, the matching problem is not discussed any further.

8.6 k_0-β diagram

We now return to a detailed study of the passband-stopband characteristics of the capacitively loaded coaxial transmission line discussed in the earlier sections. The information contained in the eigenvalue equation for the propagation constant β in a periodic structure is usually plotted on a k_0-β (or ω-β) plane. The curves of β versus k_0 show immediately the frequency bands for propagation and also the stopbands in which no propagation takes place. The resultant plot is called the k_0-β diagram, or the Brillouin diagram.†

For the capacitively loaded coaxial line, (8.9) gave

$$\cos \beta d = \cos k_0 d - \frac{\bar{B}}{2} \sin k_0 d = \cos k_0 d - K k_0 d \sin k_0 d \qquad (8.40)$$

where $\bar{B}/2 = \omega C_0/2Y_c$ has been expressed as $Kk_0 d$. Curves of $k_0 d$ versus βd are sketched in Fig. 8.8 for $K = 1$, that is, for $\bar{B} = 2k_0 d$. A low-frequency passband exists for $0 < k_0 d < 0.416\pi$. This passband is followed by a stopband and further alternating passbands and stopbands. As $k_0 d$ becomes large, the loading is increased, since $\bar{B}$ increases with k_0. This has the effect of decreasing the width of the passbands in terms of frequency.

The edges of the bands occur when the magnitude of the right-hand side of (8.40) exceeds unity. The lower edge of the first passband occurs when $0 < k_0 d < \pi$ and

$$\cos k_0 d - K k_0 d \sin k_0 d = -1$$

This equation may be solved for $k_0 d$ to give

$$\cot \frac{k_0 d}{2} = K k_0 d \qquad (8.41a)$$

$$\cos \frac{k_0 d}{2} = 0 \qquad (8.41b)$$

† Named after Brillouin, who used diagrams of this sort to illustrate the energy-band structure in periodic crystalline media.

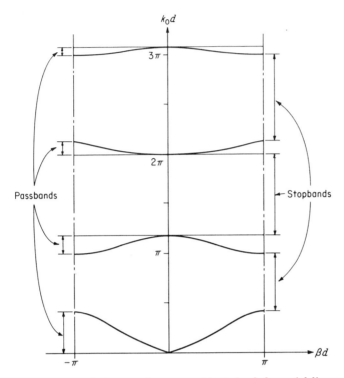

Fig. 8.8 k_0d-βd diagram for a capacitively loaded coaxial line, $\bar{B} = 2k_0d$.

The corresponding principal value of βd is π, and the values of k_0d obtained from (8.41) mark the edges of all the bands for this value of βd. The edges of the bands where $\beta d = 0$ are obtained by equating (8.40) to unity, in which case we obtain

$$\tan \frac{k_0d}{2} = -Kk_0d \tag{8.42a}$$

$$\sin \frac{k_0d}{2} = 0 \tag{8.42b}$$

One edge of the passband always occurs when the spacing between discontinuities equals one-half wavelength in the unloaded waveguide, in the present case, when k_0d is a multiple of π. When the spacing between discontinuities equals one-half wavelength, they may all be lumped together, with the result that the line becomes effectively loaded at a single point by an infinite susceptance (or reactance). Clearly, power transmission along the periodic structure must reduce to zero at this frequency.

Only the principal value of βd is plotted in Fig. 8.8. In addition,

$\beta d + 2n\pi$, where n is an arbitrary integer, are solutions. These other solutions are the propagation constants of the spatial harmonics into which the Bloch wave may be expanded. The spatial harmonics are discussed in Sec. 8.8.

The k_0-β diagram for other types of periodic structures exhibit features similar to those in Fig. 8.8. For example, if the capacitive loading is replaced by inductive shunt loading, the relative locations of the passbands and stopbands are interchanged. The zero-frequency region will be a stopband since the shunt inductors will short-circuit the line at zero frequency.

8.7 Group velocity and energy flow

The phase velocity for a Bloch wave in a periodic structure is given by

$$v_p = \frac{\omega}{\beta} = \frac{k_0}{\beta} c = \frac{k_0 d}{\beta d} c \qquad (8.43)$$

With reference to Fig. 8.9, it is seen that $k_0 d/\beta d$ is the slope of the line from the origin to a point P on the $k_0 d$-βd diagram. Since β is a function of ω, the periodic structure has frequency dispersion. The group velocity v_g as given by (Sec. 3.11)

$$v_g = \frac{d\omega}{d\beta} = c \frac{d(k_0 d)}{d(\beta d)} \qquad (8.44)$$

is therefore different from the phase velocity. Again referring to Fig. 8.9, it is seen that the group velocity is equal to the slope of the tangent to the curve of k_0 versus β multiplied by the velocity of light c. Thus we have

$$v_p = c \tan \phi_p \qquad v_g = c \tan \phi_g$$

where ϕ_p and ϕ_g are the angles given in Fig. 8.9.

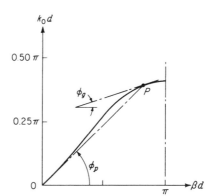

Fig. 8.9 Enlarged drawing of first passband for a capacitively loaded transmission line, $\bar{B} = 2k_0 d$.

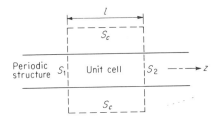

Fig. 8.10 A unit cell of a periodic structure.

For the capacitively loaded coaxial line, use of the eigenvalue equation for β, that is, (8.10a), enables us to obtain

$$v_g = c\frac{dk_0}{d\beta} = c\frac{d\theta}{d(\beta d)} = \frac{c \sin \beta d}{\left(\dfrac{\bar{B}}{2k_0 d} + 1\right) \sin k_0 d + \dfrac{\bar{B}}{2} \cos k_0 d} \quad (8.45)$$

This expression shows that the group velocity becomes zero when $\beta d = 0$ or π, except when k_0 also equals zero. Thus, as the edges of the passbands are approached, the group velocity goes to zero.

The group velocity is also the signal velocity for any signal consisting of a sufficiently narrow band of frequencies such that β can be approximated by a linear function of ω throughout the band. The signal delay τ for propagation through a unit cell is given by

$$\tau = \frac{d}{v_g} \quad (8.46)$$

In Sec. 3.11 it was shown that for a waveguide, which is a dispersive medium, the velocity of energy flow in a propagating wave was equal to the group velocity. The same result will be shown to hold for a lossless periodic structure also.

Consider a unit cell of a lossless periodic structure as in Fig. 8.10. A surface S is chosen to consist of surfaces S_1 and S_2 at the input and output terminal planes plus a cylindrical surface S_c surrounding the structure. If the periodic structure is enclosed by a perfectly conducting waveguide, the surface S_c coincides with the waveguide wall. If the periodic structure is an open-boundary structure, the surface S_c is that of a cylinder with infinite radius. In both cases $\mathbf{n} \times \mathbf{E}$ vanishes on S_c, so that the Poynting vector is zero over this surface. For generality we shall let the unit cell contain regions with frequency-dispersive material, i.e., material with parameters μ and ϵ that are functions of ω.

In the derivation of Foster's reactance theorem in Sec. 4.3 it was shown that [see (4.24a)]

$$\oint_S \left(\mathbf{E} \times \frac{\partial \mathbf{H}^*}{\partial \omega} + \frac{\partial \mathbf{E}^*}{\partial \omega} \times \mathbf{H}\right) \cdot d\mathbf{S} = -j\int_V \left(\mathbf{H} \cdot \mathbf{H}^* \frac{\partial \omega \mu}{\partial \omega} + \mathbf{E} \cdot \mathbf{E}^* \frac{\partial \omega \epsilon}{\partial \omega}\right)$$

$$= -4j(W_m + W_e) \quad (8.47a)$$

since the latter integral is equal to four times the time-average energy stored in the volume bounded by S. Since the Poynting vector is zero on S_c and $d\mathbf{S}$ is directed inward, we have

$$\int_{S_1} \left(\mathbf{E}_1 \times \frac{\partial \mathbf{H}_1^*}{\partial \omega} + \frac{\partial \mathbf{E}_1^*}{\partial \omega} \times \mathbf{H}_1 \right) \cdot \mathbf{a}_z \, dS$$

$$- \int_{S_2} \left(\mathbf{E}_2 \times \frac{\partial \mathbf{H}_2^*}{\partial \omega} + \frac{\partial \mathbf{E}_2^*}{\partial \omega} \times \mathbf{H}_2 \right) \cdot \mathbf{a}_z \, dS = -4j(W_m + W_e) \quad (8.47b)$$

where $\mathbf{E}_1$, $\mathbf{H}_1$ are the fields at terminal plane 1 and $\mathbf{E}_2$, $\mathbf{H}_2$ are the fields at terminal plane 2. For a Bloch wave, $\mathbf{E}_2 = \mathbf{E}_1 e^{-j\beta l}$, where βl is the phase shift through a unit cell of length l. We thus find that

$$\mathbf{E}_2 \times \frac{\partial \mathbf{H}_2^*}{\partial \omega} + \frac{\partial \mathbf{E}_2^*}{\partial \omega} \times \mathbf{H}_2 = \mathbf{E}_1 \times \frac{\partial \mathbf{H}_1^*}{\partial \omega} + jl \frac{d\beta}{d\omega} \mathbf{E}_1 \times \mathbf{H}_1^*$$

$$+ \frac{\partial \mathbf{E}_1^*}{\partial \omega} \times \mathbf{H}_1 + jl \frac{d\beta}{d\omega} \mathbf{E}_1^* \times \mathbf{H}_1$$

Consequently, (8.47b) gives (note that the integral over S_2 can be evaluated as an integral over S_1)

$$-2jl \frac{d\beta}{d\omega} \operatorname{Re} \int_{S_1} \mathbf{E}_1 \times \mathbf{H}_1^* \cdot \mathbf{a}_z \, dS = -4jl \frac{d\beta}{d\omega} P = -4j(W_m + W_e)$$

where $P = \frac{1}{2} \operatorname{Re} \int_{S_1} \mathbf{E}_1 \times \mathbf{H}_1^* \cdot \mathbf{a}_z \, dS$ is the power transmitted across a terminal plane. We now see that

$$v_g = \frac{d\omega}{d\beta} = \frac{P}{(W_m + W_e)/l} \quad (8.48)$$

But the energy density $(W_m + W_e)/l$ in a unit cell multiplied by the velocity of energy flow is equal to the power P, and therefore the group velocity is the velocity of energy flow.

8.8 Floquet's theorem and spatial harmonics

It has been noted that in an infinite periodic structure the field of a Bloch wave repeats at every terminal plane except for a propagation factor $e^{-\gamma d}$, where d is the length of a unit cell. Since the choice of location of a terminal plane within a unit cell is arbitrary, we see that the field at any point in a unit cell will take on exactly the same value at a similar point in any other unit cell except for a propagation factor $e^{-\gamma d}$ from one cell to the next. Thus, if the field in the unit cell between $0 \leq z \leq d$ is $\mathbf{E}(x, y, z)$, $\mathbf{H}(x, y, z)$, the field in the unit cell located in the region $d \leq z \leq 2d$ must be

$$e^{-\gamma d}\mathbf{E}(x, y, z - d), \quad e^{-\gamma d}\mathbf{H}(x, y, z - d)$$

Consequently, the field in a periodic structure is described by a solution of the form

$$\mathbf{E}(x, y, z) = e^{-\gamma z}\mathbf{E}_p(x, y, z) \tag{8.49a}$$

$$\mathbf{H}(x, y, z) = e^{-\gamma z}\mathbf{H}_p(x, y, z) \tag{8.49b}$$

where $\mathbf{E}_p$ and $\mathbf{H}_p$ are periodic functions of z with period d; for example,

$$\mathbf{E}_p(x, y, z + nd) = \mathbf{E}_p(x, y, z) \tag{8.49c}$$

The possibility of expressing the field in a periodic structure in the form given by (8.49) is often referred to as Floquet's theorem.† From (8.49a) we see that the electric field at $z_1 + d$ is related to the field at z_1 as follows:

$$\begin{aligned}\mathbf{E}(x, y, z_1 + d) &= e^{-\gamma(z_1+d)}\mathbf{E}_p(x, y, z_1 + d) \\ &= e^{-\gamma(z_1+d)}\mathbf{E}_p(x, y, z_1) = e^{-\gamma d}\mathbf{E}(x, y, z_1)\end{aligned}$$

which has, indeed, the correct repetitive properties of a Bloch wave.

Any periodic function such as $\mathbf{E}_p(x, y, z)$ may be expanded into an infinite Fourier series; thus

$$\mathbf{E}_p(x, y, z) = \sum_{n=-\infty}^{\infty} \mathbf{E}_{pn}(x, y)e^{-j2n\pi z/d} \tag{8.50}$$

where $\mathbf{E}_{pn}$ are vector functions of x and y. Multiplying both sides by $e^{j2m\pi z/d}$ and integrating over a unit cell, i.e., from $z = 0$ to d, give

$$\mathbf{E}_{pm}(x, y) = \frac{1}{d}\int_0^d \mathbf{E}_p(x, y, z)e^{j2m\pi z/d}\, dz \tag{8.51}$$

since the exponential functions form a complete orthogonal set; i.e.,

$$\int_0^d e^{-j2n\pi z/d}e^{j2m\pi z/d}\, dz = \begin{cases} 0 & m \neq n \\ d & m = n \end{cases}$$

The field in a periodic structure can now be represented as

$$\begin{aligned}\mathbf{E}(x, y, z) &= \sum_{n=-\infty}^{\infty} \mathbf{E}_{pn}(x, y)e^{-j\beta z - j2n\pi z/d} \\ &= \sum_{n=-\infty}^{\infty} \mathbf{E}_{pn}(x, y)e^{-j\beta_n z}\end{aligned} \tag{8.52}$$

where $\gamma = j\beta$ and $\beta_n = \beta + 2n\pi/d$. Each term in this expansion is called a spatial harmonic (or a Hartree harmonic) and has a propagation phase constant β_n. Some of the β_n will be negative whenever the integer n is sufficiently negative. The corresponding phase velocity of the nth

---

† Actually, Floquet's work dealt with differential equations with periodic coefficients. The case of periodic boundary conditions is an extension of that work.

spatial harmonic is

$$v_{pn} = \frac{\omega}{\beta_n} = \frac{\omega}{\beta + 2n\pi/d} \tag{8.53}$$

and will be negative whenever β_n is negative. The group velocity of the nth harmonic is

$$v_{gn} = \frac{d\omega}{d\beta_n} = \left(\frac{d\beta_n}{d\omega}\right)^{-1} = \left(\frac{d\beta}{d\omega}\right)^{-1} = v_g \tag{8.54}$$

and is the same for all harmonics. From the above relations it is seen that some of the spatial harmonics (approximately one-half) have phase and group velocities that are directed in opposite directions. This property is made use of in the backward-wave traveling-wave-tube oscillator. The term backward wave, or reverse wave, is often used to refer to a wave with oppositely directed phase and group velocities. The voltage and current waves can, of course, also be expanded into an infinite set of spatial harmonics (Prob. 8.10).

Although a Bloch wave can be expanded into an infinite set of spatial harmonics, all the spatial harmonics must be simultaneously present in order that the total field may satisfy all the boundary conditions. The eigenvalue equation for β for a periodic structure always yields solutions $\beta_n = \beta + 2n\pi/d$, in addition to the fundamental solution. These other possible solutions are clearly the propagation constants of the spatial harmonics. A complete $k_0 d$-βd diagram thus exhibits $k_0 d$ as a periodic function of βd; that is, the βd curve is continued periodically outside the range $-\pi \leq \beta d \leq \pi$. The slope of the line from the origin to any point on the curve still gives the phase velocity, and the slope of the tangent to the curve gives the group velocity, when multiplied by c.

8.9 Periodic structures for traveling-wave tubes

Traveling-wave tubes require a structure capable of supporting an electromagnetic wave with a phase velocity equal to the velocity of the electron beam. Since the latter is usually much smaller than the velocity of light, the required structure is commonly referred to as a slow-wave structure. A common type of slow-wave circuit used in traveling-wave tubes is the helix. The helix is treated in the following two sections, and hence this section is restricted to a discussion of some of the other types of slow-wave periodic structures suitable for use in traveling-wave tubes.

A periodic slow-wave structure often used for the linear magnetron tube is the vane-type structure illustrated in Fig. 8.11. It consists essentially of a corrugated plane with thick teeth. It will be instructive to apply Floquet's theorem and carry out an analysis of this structure in order to illustrate the general techniques employed. Edge effects at $x = \pm a/2$ will be neglected for simplicity; i.e., we shall treat the struc-

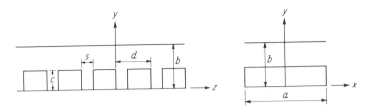

Fig. 8.11 Vane-type, or corrugated-plane, periodic structure.

ture as being infinitely wide. If a is large compared with the spacing b, and this in turn is small compared with λ_0, the edge effects will not produce a significant change in the characteristics of the ideal structure. For use in a magnetron, a strong axial electric field is required, and hence we shall examine the possibility of having TM- or E-type modes.

For TM modes having no variation with x, the field components may be expressed in terms of the single magnetic field component H_x that is present. We have

$$\nabla \times \mathbf{H} = -\mathbf{a}_x \times \nabla H_x = j\omega\epsilon_0 \mathbf{E}$$

and so

$$E_y = -j\frac{Z_0}{k_0}\frac{\partial H_x}{\partial z} \qquad (8.55a)$$

$$E_z = j\frac{Z_0}{k_0}\frac{\partial H_x}{\partial y} \qquad (8.55b)$$

The field H_x is a solution of

$$\left(\frac{\partial^2}{\partial y^2} + \frac{\partial^2}{\partial z^2} + k_0^2\right)H_x = 0 \qquad (8.56)$$

According to Floquet's theorem, the field H_x can be expressed in the form $e^{-j\beta z}\psi(y, z)$, where $\psi(y, z)$ is periodic in z with a period d. Hence we shall assume that

$$H_x = \sum_{n=-\infty}^{\infty} f_n(y)e^{-j\beta_n z}$$

where $\beta_n = \beta + 2n\pi/d$ and the $f_n(y)$ are functions of y to be determined. The substitution of this series into (8.56) shows that the $f_n(y)$ are solutions of

$$\frac{d^2 f_n(y)}{dy^2} - (\beta_n^2 - k_0^2)f_n(y) = 0 \qquad (8.57)$$

Above the corrugations, i.e., in the region $c \leq y \leq b$, we must choose the f_n so that E_z will vanish on the perfectly conducting wall at $y = b$. Thus we require $df_n/dy = 0$ at $y = b$. Since solutions to (8.57) are $\sinh h_n y$

and $\cosh h_n y$, where $h_n = (\beta_n^2 - k_0^2)^{1/2}$, we choose

$$f_n(y) = a_n \cosh h_n(b - y)$$

where a_n is a constant. At $y = b$, this function has a zero derivative. For the fields H_x and E_z in the region above the corrugations, we now have

$$H_x = \sum_{n=-\infty}^{\infty} a_n \cosh h_n(b - y) \, e^{-j\beta_n z} \tag{8.58a}$$

$$E_z = -j\frac{Z_0}{k_0} \sum_{n=-\infty}^{\infty} a_n h_n \sinh h_n(b - y) \, e^{-j\beta_n z} \tag{8.58b}$$

upon using (8.55b).

As a next step we must obtain a suitable expansion for H_x in each corrugation, or slot. If $H_1(y, z)$ is the field in the slot extending from $z = 0$ to $z = s$ and for $0 \leq y \leq c$, then the field in the nth slot beginning at $z = nd$ will be $e^{-j\beta_n d} H_1(y, z - nd)$ according to Floquet's theorem. Therefore we need to concentrate on one slot only. We must determine H_x so that E_y will vanish at $z = 0$ and s and also so that E_z will vanish at $y = 0$. A suitable expansion to use is

$$H_x = \sum_{m=0}^{\infty} g_m(y) \cos \frac{m\pi z}{s}$$

since $d[\cos(m\pi z/s)]/dz$ vanishes at $z = 0$ and s. If this expansion is substituted into (8.56), we find that

$$\frac{d^2 g_m(y)}{dy^2} - \left[\left(\frac{m\pi}{s}\right)^2 - k_0^2\right] g_m(y) = 0$$

Normally, $s \ll \lambda_0$, so that $m\pi/s > k_0$ for $m \neq 0$. Thus appropriate solutions that have a zero derivative at $y = 0$ are

$$g_m(y) = b_m \cosh l_m y$$

where $l_m = [(m\pi/s)^2 - k_0^2]^{1/2}$ and b_m is a constant. For $m = 0$, the solution is $g_0(y) = b_0 \cos k_0 y$, and this part of the solution corresponds to a TEM standing wave in the slot. This mode has $E_y = 0$. In the first slot we can thus write

$$H_x = \sum_{m=0}^{\infty} b_m \cosh l_m y \cos \frac{m\pi z}{s} \tag{8.59a}$$

$$E_z = j\frac{Z_0}{k_0} \sum_{m=0}^{\infty} b_m l_m \sinh l_m y \cos \frac{m\pi z}{s} \tag{8.59b}$$

The final step in the analysis is to determine the expansion coefficients a_n and b_m by imposing boundary conditions at the plane $y = c$ separating the two regions. We require the tangential electric and magnetic fields to be continuous across the gap $y = c$, $0 \leq z \leq s$. In addition, we require the tangential electric field to vanish on the upper faces of the teeth, i.e., at $y = c$ for $s \leq z \leq d$, or $nd + s \leq z \leq (n + 1)d$ in general. Using (8.58) and (8.59), we see that the boundary conditions require

$$\sum_{n=-\infty}^{\infty} a_n e^{-j\beta_n z} \cosh h_n(b - c) = \sum_{m=0}^{\infty} b_m \cosh l_m c \cos \frac{m\pi z}{s} \qquad 0 \leq z \leq s \tag{8.60a}$$

$$\sum_{n=-\infty}^{\infty} a_n h_n e^{-j\beta_n z} \sinh h_n(b - c)$$

$$= \begin{cases} -\sum_{m=0}^{\infty} b_m l_m \sinh l_m c \cos \dfrac{m\pi z}{s} & 0 \leq z \leq s \\ 0 & s \leq z \leq d \end{cases} \tag{8.60b}$$

If we multiply (8.60b) by $e^{j\beta z}$, we obtain

$$\sum_{n=-\infty}^{\infty} a_n h_n \sinh h_n(b - c)\, e^{-j2n\pi z/d}$$

$$= \begin{cases} -\sum_{m=0}^{\infty} b_m e^{j\beta z} l_m \sinh l_m c \cos \dfrac{m\pi z}{s} & 0 \leq z \leq s \\ 0 & s \leq z \leq d \end{cases} \tag{8.60c}$$

Now the coefficients in a Fourier series are uniquely determined only if the function which the series is to represent is specified for the complete interval over which the series is orthogonal. The functions $e^{-j2n\pi z/d}$ are orthogonal over the range 0 to d, and thus, since the left-hand side of (8.60c) is specified for all z over one period, we can obtain unique expressions for the a_n in terms of the b_m from (8.60c). Note that this is not true for (8.60a), which holds only in the region $0 \leq z \leq s$. If we multiply (8.60c) on both sides by $e^{j2r\pi z/d}$ and integrate from 0 to d, we obtain (r is an integer)

$$d a_r h_r \sinh h_r(b - c) = -\sum_{m=0}^{\infty} b_m l_m \sinh l_m c \int_0^s e^{j(\beta + 2r\pi/d)z} \cos \frac{m\pi z}{s}\, dz$$

$$= \sum_{m=0}^{\infty} b_m l_m \sinh l_m c \, \frac{j(\beta + 2\pi r/d)[(-1)^m e^{j\beta_r s} - 1]}{(\beta + 2\pi r/d)^2 - (m\pi/s)^2} \tag{8.61}$$

since $\int_0^d e^{(-j2\pi z/d)(n-r)}\, dz = 0$ for $n \neq r$ and equals d for $n = r$. The above represents an infinite set of equations, i.e., one for each value of r. Although (8.60a) is not a unique equation for the a_n, it does specify

Sec. 8.9   Periodic structures and filters

the b_m uniquely in terms of the a_n. The a_n have already been expressed in terms of the b_m; so we may regard them as known. Multiplying (8.60a) by $\cos(r\pi z/s)$ and integrating from 0 to s give

$$\frac{s}{\epsilon_{0r}} b_r \cosh l_r c = \sum_{n=-\infty}^{\infty} a_n \cosh h_n(b-c) \int_0^s e^{-j\beta_n z} \cos \frac{r\pi z}{s} dz$$

$$= -j \sum_{n=-\infty}^{\infty} a_n \cosh h_n(b-c) \frac{\left(\beta + \dfrac{2n\pi}{d}\right)[1 - (-1)^r e^{-j(\beta + 2n\pi/d)s}]}{(\beta + 2n\pi/d)^2 - (r\pi/s)^2} \quad (8.62)$$

where the Neumann factor $\epsilon_{0r} = 1$ for $r = 0$ and equals 2 for $r > 0$. This is also an infinite set of equations since $r = 0, 1, 2, \ldots, \infty$. Equations (8.61) and (8.62) constitute two linear systems of equations involving the a_n and b_m. If the solutions for the a_n as given by (8.61) are substituted into (8.62), the result is a homogeneous set of equations for the b_m. For a nontrivial solution, the determinant of this homogeneous set of equations must vanish. Setting the determinant equal to zero yields the eigenvalue equation for β. However, the sets of equations are of infinite order, so that, in practice, an exact solution is not possible. Therefore we shall find only a first approximation to the exact eigenvalue equation.

If the slot spacing s is small compared with λ_0, it seems reasonable to expect that the field in the slot can be approximated by the TEM standing-wave field alone. Thus we shall take all b_m except b_0 equal to zero. If we lump all the constants in (8.61) together and replace r by n, the equation is of the form

$$a_n = \sum_{m=0}^{\infty} b_m R_{mn} \qquad n = 0, \pm 1, \ldots \quad (8.63a)$$

Likewise, (8.62) is an equation of the form

$$b_m = \sum_{n=-\infty}^{\infty} a_n T_{nm} \qquad m = 0, 1, 2, \ldots \quad (8.63b)$$

Replacing a_n by (8.63a) gives

$$b_m = \sum_{n=-\infty}^{\infty} \sum_{m=0}^{\infty} b_m R_{mn} T_{nm} \qquad m = 0, 1, 2, \ldots \quad (8.64)$$

The determinant of this infinite set of homogeneous equations, when equated to zero, gives the exact eigenvalue equation for β. When we take all b_m except b_0 equal to zero, we obtain instead

$$b_0 = \sum_{n=-\infty}^{\infty} b_0 R_{0n} T_{n0}$$

or

$$1 - \sum_{n=-\infty}^{\infty} R_{0n} T_{n0} = 0 \tag{8.65}$$

for a first approximation to the eigenvalue equation. Now R_{0n} are all the constants in (8.61), multiplying b_0 when the equation is solved for a_n and with r replaced by n. Likewise, T_{n0} is the constant relating b_0 to the a_n in (8.62). When these values for R_{0n} and T_{0n} are substituted into (8.65), we obtain

$$\frac{1}{k_0 d \tan k_0 c} = \frac{s}{d} \sum_{n=-\infty}^{\infty} \left[\frac{\sin (\beta_n s/2)}{\beta_n s/2} \right]^2 \frac{1}{h_n d \tanh h_n (b-c)} \tag{8.66}$$

For slow waves β is much larger than k_0, and hence h_n can be replaced by β_n in this equation with negligible error. In this case the right-hand side is not dependent on k_0. By evaluating the right-hand side for a range of assumed values for β, the corresponding value of k_0 may be found by solving (8.66). The numerical work is straightforward but tedious. Typical results are given by Hutter.†

A reasonably accurate description of the dispersion curve relating β to k_0 may also be obtained from a simple transmission-line analysis. The region above the corrugations is essentially a parallel-plate transmission line (strip line) with a characteristic impedance $Z_1 = Z_0(b-c)$ per unit width. The slots are short-circuited transmission-line stubs connected in series with the main line at periodic intervals d. The stubs present a reactance

$$jX = jZ_0 s \tan k_0 c$$

to the main line. The equivalent circuit of the structure is therefore of the form shown in Fig. 8.12. This periodic circuit may be analyzed in the same way that the capacitively loaded transmission line was. It is

† R. G. E. Hutter, "Beam and Wave Electronics in Microwave Tubes," sec. 7.4, D. Van Nostrand Company, Inc., Princeton, N.J., 1960.

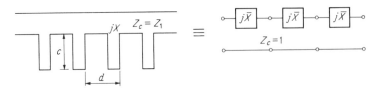

Fig. 8.12 Simplified equivalent circuit for the vane structure of Fig. 8.11.

Sec. 8.9 Periodic structures and filters 389

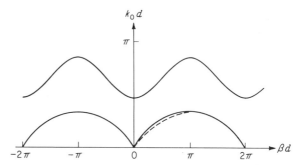

Fig. 8.13 k_0d-βd diagram for a stub-loaded transmission line.

readily found that the eigenvalue equation for β is

$$\cos \beta d = \cos k_0 d - \frac{X}{2Z_1} \sin k_0 d$$

$$= \cos k_0 d - \frac{s}{2(b-c)} \tan k_0 c \sin k_0 d \qquad (8.67)$$

This equation is quite accurate as long as $s \ll d$ and also much smaller than λ_0.

For frequencies such that $0 < k_0 c < \pi/2$, the loading is inductive, and for $\pi/2 < k_0 c < \pi$, it is capacitive, etc. A typical dispersion curve for the case $s = 2(b - c)$ and $d = 0.83c$ is given in Fig. 8.13. For these dimensions the phase velocity is reduced by a factor of about 3 only. Much greater reduction factors are obtained by making $s/(b - c)$ larger so as to increase the normalized characteristic impedance of the stubs. For comparison, the results from (8.66) for $d = 2s$ are plotted also (broken curve), verifying the accuracy of (8.67). The first cutoff occurs approximately when the stubs become resonant, i.e., for $k_0 c = \pi/2$. Increasing c will therefore also reduce the phase velocity in the first passband, since βd will equal π for a smaller value of k_0.

The foregoing analysis is typical for periodic structures that cannot be represented by simple transmission-line circuits. The essential steps to be followed are summarized:

1. Obtain suitable field expansions in each region of the periodic structure. This involves solution of the Helmholtz equation and the use of Floquet's theorem.

2. Impose appropriate boundary conditions on the fields at all common boundaries separating the different regions. In general, it will be found that both E and H modes may be required in order to satisfy the boundary conditions.

3. By Fourier analysis convert the boundary conditions into algebraic equations for the amplitude constants.

4. The system of algebraic equations can be written in the form of a homogeneous set of equations. Equating the determinant to zero gives the eigenvalue equation for β. Since the equations are usually infinite in order, some assumption must be made as regards the number of nonzero amplitude constants that will be chosen. Equating the higher-order amplitude constants to zero results in an approximate eigenvalue equation.

Periodic structures for millimeter-wave traveling-wave tubes

At millimeter wavelengths a helix has too small a diameter to be a useful slow-wave structure. Various forms of tape ladder lines, interdigital tape lines, and meander tape lines are preferred. Illustrations of these structures are given in Fig. 8.14. A discussion of these structures, together with typical k_0-β curves, is given in a paper by Harvey,† which also provides references to the original analysis of these structures.

The two structures shown in Fig. 8.14d and e are complementary, or dual, structures. That is, the meander line is obtained by interchanging the open region and the conducting region in the interdigital line. For a complementary structure of this type we can show that the field is also a dual solution and hence both structures have exactly the same k_0-β

† A. F. Harvey, Periodic and Guiding Structures at Microwave Frequencies, *IRE Trans.*, vol. MTT-8, pp. 30–61, January, 1960.

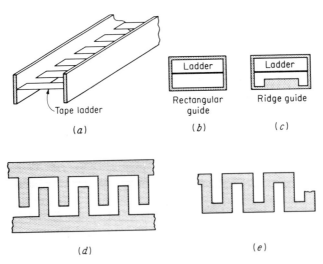

Fig. 8.14 (a–c) Tape ladder lines; (d) interdigital tape line; (e) meander tape line.

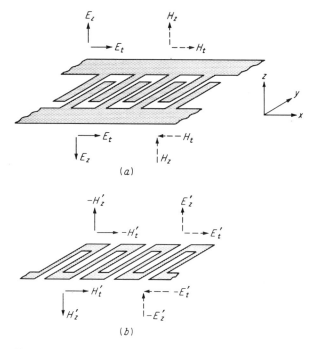

Fig. 8.15 Illustration of dual properties of interdigital and meander lines.

dispersion curve. A detailed discussion of the dual properties is given below.

Let the interdigital line be located in the xy plane as in Fig. 8.15. Let us consider a mode of propagation having an electric field for which the transverse (x and y components) field $\mathbf{E}_t$ is an even function of z, that is, the same on the upper and lower sides of the structure. Since $\nabla \cdot \mathbf{E} = 0$, we now have $\partial E_z / \partial z = -\nabla_t \cdot \mathbf{E}_t$, and hence $\partial E_z / \partial z$ is an even function of z, and E_z must then be an odd function of z. From the curl equation $\nabla \times \mathbf{E} = -j\omega\mu_0 \mathbf{H}$, we can readily conclude that $\mathbf{H}_t$ must be an odd function of z and H_z an even function of z. The field structure is illustrated in Fig. 8.15a. The field $\mathbf{E}_t$ will vanish on the conducting surface, and since $\mathbf{H}_t$ is an odd function of z, it must vanish on the open part of the xy plane. At the conductor surface, $|\mathbf{H}_t|$ will equal one-half the total current density in the line since the total change in $|\mathbf{H}_t|$ across the conductor must equal the total current density.

A dual field $\mathbf{E}'$, $\mathbf{H}'$ given by

$$\mathbf{E}' = \pm Z_0 \mathbf{H} \qquad \mathbf{H}' = \mp Y_0 \mathbf{E}$$

is easily shown to satisfy Maxwell's equations

$$\nabla \times \mathbf{E}' = -j\omega\mu_0 \mathbf{H}' \qquad \nabla \times \mathbf{H}' = j\omega\epsilon_0 \mathbf{E}'$$

if the field **E**, **H** does. The dual field is a solution to the meander-line problem (Fig. 8.15b), provided we choose the dual solution

$$\mathbf{E}' = Z_0\mathbf{H} \qquad \mathbf{H}' = -Y_0\mathbf{E} \tag{8.68a}$$

above the meander-line plane and the solution

$$\mathbf{E}' = -Z_0\mathbf{H} \qquad \mathbf{H}' = Y_0\mathbf{E} \tag{8.68b}$$

below the structure. In both regions the primed fields satisfy Maxwell's equations. The field $\mathbf{E}'_t$ will vanish on the conducting portions of the meander line since the field $\mathbf{H}_t$ was zero in the open regions of the interdigital line. Similarly, $\mathbf{H}'_t$ vanishes over the open regions of the meander-line plane since the field $\mathbf{E}_t$ was zero on the conducting surfaces of the interdigital line. All boundary conditions being satisfied, the solution is complete. It may now be concluded that both structures must have the same k_0-β dispersion curve. It should be noted, however, that duality applies only if the two structures are exact complements; that is, superimposing the two structures must result in the whole xy plane being a single conducting sheet. The sides of the interdigital line must therefore extend to $y = \pm\infty$, and the line must be infinitely long. However, in practice, the field is confined to the vicinity near the cuts, so that the sides do not have to extend much beyond the toothed region before they can be terminated with negligible disturbance of the field. The duality principle used above is often referred to as Babinet's principle.

8.10 Sheath helix

The sheath helix is an approximate model of a tape helix. The tape helix, illustrated in Fig. 8.16a, consists of a thin ribbon, or tape, wound

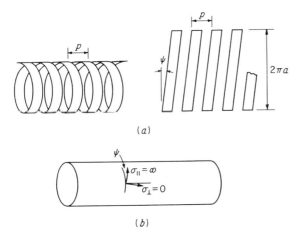

Fig. 8.16 (a) A tape helix; (b) sheath helix.

into a helical structure. The pitch is denoted by p, and the pitch angle by ψ. If the spacing between turns and the ribbon width are made to approach zero, the resultant structure becomes electrically smooth. At the boundary surface $r = a$, the boundary conditions for the electric field may be approximated by the conditions that the conductivity in the direction along the tape (direction of current) is infinite, whereas that in the direction perpendicular to the tape is zero. The use of these boundary conditions permits a solution for the electromagnetic field guided by the helix to be obtained with relative ease. This anisotropic conducting cylinder model of a tape helix is called the sheath helix, illustrated in Fig. 8.16b. The field solution, derived below, will show that the sheath helix supports a slow wave with a phase velocity $v_p = c \sin \psi$. The wave may be considered to propagate along the helical conductor with a velocity c, and hence progresses along the axial direction z with a phase velocity $c \sin \psi$. The sheath-helix model is valid at low frequencies, where p is much smaller than λ_0. At higher frequencies a more realistic model must be used, and the existence of spatial harmonics then becomes apparent, as shown in Sec. 8.11.

The field solution for the helix consists of both E and H modes since these are coupled together by the boundary conditions at $r = a$. Along the direction of the tape, the tangential electric field must vanish, since the conductivity in this direction is taken as infinite; thus

$$E_{\phi 1} \cos \psi + E_{z1} \sin \psi = E_{\phi 2} \cos \psi + E_{z2} \sin \psi = 0 \tag{8.69a}$$

where the subscripts 1 and 2 refer to the field components in the two regions $r \leq a$ and $r \geq a$. The component of electric field on the cylindrical surface $r = a$ and perpendicular to the tape must be continuous since the conductivity is taken as zero in this direction. Hence

$$E_{z1} \cos \psi - E_{\phi 1} \sin \psi = E_{z2} \cos \psi - E_{\phi 2} \sin \psi \tag{8.69b}$$

The component of **H** tangent to the tape must also be continuous since no current flows perpendicular to the tape; so a third boundary condition is

$$H_{z1} \sin \psi + H_{\phi 1} \cos \psi = H_{z2} \sin \psi + H_{\phi 2} \cos \psi \tag{8.69c}$$

Expansions for the E and H modes in the two regions $r \lessgtr a$ may be obtained in terms of the axial field components E_z and H_z, as shown in Secs. 3.1 and 3.7. The axial fields $E_z = e_z(r, \phi)e^{-j\beta z}$, $H_z = h_z(r, \phi)e^{-j\beta z}$ are solutions of

$$\nabla_t^2 e_z + (k_0^2 - \beta^2)e_z = 0$$

Since we anticipate slow-wave solutions for which $\beta^2 > k_0^2$, the solutions involve Bessel functions with imaginary arguments, that is, $J_n(r\sqrt{k_0^2 - \beta^2})$ and $Y_n(r\sqrt{k_0^2 - \beta^2})$. In place of these functions, the modified Bessel functions $I_n(r\sqrt{\beta^2 - k_0^2})$, $K_n(r\sqrt{\beta^2 - k_0^2})$ are more convenient to use. These functions are related to the J_n and Y_n functions as follows:

$$I_n(x) = j^{-n}J_n(jx) \tag{8.70a}$$

$$K_n(x) = \frac{\pi}{2}j^{n+1}[J_n(jx) + jY_n(jx)] \tag{8.70b}$$

For small values of x, the K_n functions approach infinity in a logarithmic fashion, and hence only the I_n functions are used in the region $r < a$. For r large, the asymptotic forms

$$I_n(x) \sim \frac{1}{\sqrt{2\pi x}} e^x \tag{8.71a}$$

$$K_n(x) \sim \sqrt{\frac{\pi}{2x}} e^{-x} \tag{8.71b}$$

are valid. Since we require a field that decays for large r, only the functions K_n are employed in the region $r > a$. Suitable expansions for e_z and h_z in the two regions are now seen to be

$$e_z = \begin{cases} \sum_{n=-\infty}^{\infty} a_n e^{-jn\phi} I_n(hr) & r \le a \\ \sum_{n=-\infty}^{\infty} b_n e^{-jn\phi} K_n(hr) & r \ge a \end{cases}$$

$$h_z = \begin{cases} \sum_{n=-\infty}^{\infty} c_n e^{-jn\phi} I_n(hr) & r \le a \\ \sum_{n=-\infty}^{\infty} d_n e^{-jn\phi} K_n(hr) & r \ge a \end{cases}$$

where $h = (\beta^2 - k_0^2)^{1/2}$ and a_n, b_n, c_n, and d_n are unknown amplitude constants.

For the sheath-helix model it is possible to find a solution for a field that satisfies the boundary conditions (8.69) for each integer n. We are primarily interested in the solution $n = 0$, which has circular symmetry. If we make use of (3.12), (3.13), and (3.17), together with the relations

$$\frac{dI_0(hr)}{d(hr)} = I_1(hr) \qquad \frac{dK_0(hr)}{d(hr)} = -K_1(hr)$$

we find that the field in the two regions can be expressed as follows:

For $r < a$,

$$E_z = a_0 I_0(hr) e^{-j\beta z} \qquad E_r = \frac{j\beta}{h} a_0 I_1(hr) e^{-j\beta z} \qquad E_\phi = -\frac{j\omega\mu_0}{h} c_0 I_1(hr) e^{-j\beta z} \tag{8.72a}$$

$$H_z = c_0 I_0(hr) e^{-j\beta z} \qquad H_r = \frac{j\beta}{h} c_0 I_1(hr) e^{-j\beta z} \qquad H_\phi = \frac{j\omega\epsilon_0}{h} a_0 I_1(hr) e^{-j\beta z}$$

For $r > a$,

$$E_z = b_0 K_0(hr) e^{-j\beta z} \qquad E_r = -\frac{j\beta}{h} b_0 K_1(hr) e^{-j\beta z} \qquad E_\phi = \frac{j\omega\mu_0}{h} d_0 K_1(hr) e^{-j\beta z}$$
(8.72b)

$$H_z = d_0 K_0(hr) e^{-j\beta z} \qquad H_r = -\frac{j\beta}{h} d_0 K_1(hr) e^{-j\beta z}$$

$$H_\phi = -\frac{j\omega\epsilon_0}{h} b_0 K_1(hr) e^{-j\beta z}$$

for the $n = 0$ mode.

If the above expressions for the fields are substituted into the boundary conditions (8.69), the result is four homogeneous equations for the constants a_0, b_0, c_0, and d_0. A nontrivial solution exists only if the determinant vanishes. Equating the determinant to zero results in the eigenvalue equation for β, which is

$$\frac{K_1(ha) I_1(ha)}{K_0(ha) I_0(ha)} = \frac{(ha)^2 \tan^2 \psi}{(k_0 a)^2}$$
(8.73)

For ha greater than 10, the ratio $K_1 I_1 / K_0 I_0$ rapidly approaches unity. In this region (8.73) gives $h = k_0 \cot \psi$, from which we obtain

$$\beta = (k_0^2 + h^2)^{\frac{1}{2}} = k_0 \csc \psi$$
(8.74)

The resultant phase velocity v_p is

$$v_p = \frac{\omega}{\beta} = \frac{k_0}{\beta} c = c \sin \psi$$
(8.75)

and is reduced by the factor $\sin \psi$. A plot of v_p/c as a function of $k_0 a$ is given in Fig. 8.17 as determined by the solution of (8.73) with ψ equal to 10°. For $k_0 a$ greater than 0.25, the phase velocity is well approximated by (8.75). In the frequency range where $v_p = c \sin \psi$, the group velocity v_g is also equal to $c \sin \psi$, and there is no frequency dispersion.

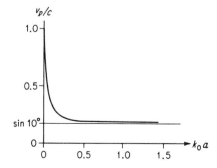

Fig. 8.17 Phase-velocity reduction factor for a sheath helix with pitch angle $\psi = 10°$.

8.11 Some general properties of a helix

The tape helix consists of a thin ribbon of metal wound into a helical structure, as shown in Fig. 8.16a. A helix may also be constructed by the use of a round wire. The parameters describing the helix are the pitch p, or turn-to-turn spacing, the diameter $2a$, and the pitch angle ψ. These parameters are given in Fig. 8.16a, which shows a developed view of a tape helix.

The helix is a periodic structure with respect to translation by a distance p along the axis and also with respect to rotation through an arbitrary angle θ, followed by a translation $p\theta/2\pi$ along the axial direction. In other words, an infinitely long helix translated along the z axis by a distance p or rotated by an angle θ and then shifted by a distance $p\theta/2\pi$ along z will coincide with itself.

The above periodic, or symmetry, properties of the helix place certain restrictions on the nature of the field solutions. If $\mathbf{E}_1(r, \phi, z)$, where r and ϕ are cylindrical coordinates, is a solution for the electric field, then $\mathbf{E}_1(r, \phi + \theta, z + p\theta/2\pi)$ multiplied by a propagation factor $e^{-j\beta(p\theta/2\pi)}$ is another solution, since the point $r, \phi + \theta, z + p\theta/2\pi$ is indistinguishable from the point r, ϕ, z. The solution $\mathbf{E}_1(r, \phi, z)$ must be periodic in ϕ and, apart from a propagation factor $e^{-j\beta z}$, must also be periodic in z with period p. Hence $\mathbf{E}_1$ may be expanded in the double Fourier series

$$\mathbf{E}_1(r, \phi, z) = \sum_{m=-\infty}^{\infty} \sum_{n=-\infty}^{\infty} \mathbf{E}_{1,mn}(r) e^{-jm\phi - j2n\pi z/p} e^{-j\beta z} \qquad (8.76)$$

where $\mathbf{E}_{1,mn}(r)$ are vector functions of r corresponding to the usual amplitude constants in a Fourier series. The relationship between translation and rotation noted above requires that $e^{j\beta z}\mathbf{E}_1(r, \phi, z)$ does not change when ϕ, z are replaced by $\phi + \theta, z + p\theta/2\pi$. Thus, in (8.76), we require

$$e^{-jm(\phi+\theta)-j2n\pi(z+p\theta/2\pi)/p} = e^{-jm(\phi+\theta)-jn\theta-j2n\pi z/p}$$
$$= e^{-jm\phi - j2n\pi z/p}$$

This condition will hold only if $m = -n$. Consequently, for a helix, the double Fourier series expansion for the electric field reduces to a single series of the form

$$\mathbf{E}_1(r, \phi, z) = \sum_{n=-\infty}^{\infty} \mathbf{E}_{1,n}(r) e^{-jn(2\pi z/p - \phi)} e^{-j\beta z} \qquad (8.77)$$

The solution for a helix proceeds by expanding the field in the two regions $r < a$ and $r > a$ into an infinite series of E and H modes expressed in cylindrical coordinates. The boundary conditions at $r = a$ will couple the E and H modes together, so that pure E or H modes cannot exist independently. For the nth term in (8.77), the radial dependence in the region $r > a$ will accord with the modified Bessel function of the second kind, that is, $K_n(h_n r)$, of order n and with an argument

$$h_n r = \left[\left(\beta + \frac{2n\pi}{p}\right)^2 - k_0^2\right]^{\frac{1}{2}} r$$

The K_n functions are asymptotic to $(\pi/2h_n r)^{\frac{1}{2}} e^{-h_n r}$ for r large; so the field will decay exponentially as long as all h_n are real, i.e., for all $(\beta + 2n\pi/p)^2$ greater than k_0^2. When the field decays exponentially it corresponds to a surface-wave mode guided by the helix.

At a given frequency only certain discrete values of β, say β_m, are possible solutions. For each value of β_m, corresponding to a particular mode of propagation, the field is given by a Fourier series of the form

$$\mathbf{E}_m(r, \phi, z) = \sum_{n=-\infty}^{\infty} \mathbf{E}_{m,n}(r) e^{-jn(2\pi z/p - \phi) - j\beta_m z} \tag{8.78}$$

Each term in this expansion is called a spatial harmonic, and has a propagation phase constant $\beta_m + 2n\pi/p$. On a $k_0 p$-βp diagram the region above the lines $k_0 = \pm \beta$ is a forbidden region, as shown in Fig. 8.18, since it corresponds to a situation where $h_0 = (\beta^2 - k_0^2)^{\frac{1}{2}}$ is imaginary and the $n = 0$ spatial harmonic does not decay in the radial direction. Since we also require $|\beta + 2n\pi/p|$ to be greater than k_0, all possible allowed values of β, corresponding to bound surface-wave modes, are further restricted to lie in the unshaded triangular regions in Fig. 8.18. The boundaries of these regions are marked by the lines

$$k_0 = \pm \left(\beta \pm \frac{2n\pi}{p}\right)$$

where n is an integer. In the forbidden regions the propagation constant turns out to be complex rather than pure real, a feature which is different from that of a normal cutoff mode.

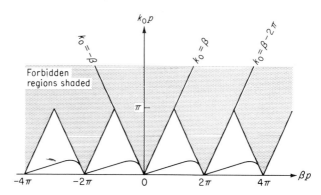

Fig. 8.18 Illustration of allowed and forbidden regions in the k_0-β diagram for a helix.

A first approximation to the solution for a tape helix is obtained by assuming that the current is directed along the direction of the tape only, is uniform across the width of the tape, and has a propagation factor $e^{-j\beta z}$. A typical k_0-β curve obtained on this basis and with $\psi = 10°$ is shown in Fig. 8.18. For further results the paper by Sensiper or the book by Watkins may be consulted.†

8.12 Introduction to microwave filters

The ideal filter network is a network that provides perfect transmission for all frequencies in certain passband regions and infinite attenuation in the stopband regions. Such ideal characteristics cannot be obtained, and the goal of filter design is to approximate the ideal requirements to within an acceptable tolerance. Filters are used in all frequency ranges to provide as nearly perfect transmission as possible for signals falling within desired passband frequency ranges, together with rejection of those signals and noise outside the desired frequency bands. Filters fall into three main categories, namely (1) low-pass filters that transmit all signals between zero frequency and some upper limit ω_c and attenuate all frequencies above the cutoff value ω_c, (2) high-pass filters that pass all frequencies above a lower cutoff value ω_c and reject all frequencies below ω_c, (3) bandpass filters that pass all frequencies in a range ω_1 to ω_2 and reject frequencies outside this range. The complement to the bandpass filter, i.e., the band-rejection filter, which attenuates frequencies in the range ω_1 to ω_2, is also of interest in certain applications.

At low frequencies the "building blocks" for filters are ideal inductors and capacitors. These elements have very simple frequency characteristics, and a very general and complete synthesis procedure has been developed for the design of filters utilizing them. It is possible to synthesize directly filters with a wide variety of prescribed frequency characteristics. The filter design problem at microwave frequencies where distributed parameter elements must be used is much more complicated, and no complete theory or synthesis procedure exists. The complex frequency behavior of microwave circuit elements makes it virtually impossible to develop a general and complete synthesis procedure. However, in spite of these added complications at microwave frequencies, a number of useful techniques have been developed for the design of microwave filters. The case of narrowband filters is particularly straightforward since many microwave elements will have frequency characteristics essentially like those of an ideal inductive or capacitive reactance over a limited frequency range. In this case a low-frequency prototype filter

† S. Sensiper, Electromagnetic Wave Propagation on Helical Structures, *Proc. IRE*, vol. 43, pp. 149-161, February, 1955.

D. A. Watkins, "Topics in Electromagnetic Theory," John Wiley & Sons, Inc., New York, 1958.

may be used as a model. The microwave filter is realized by replacing all inductors and capacitors by suitable microwave circuit elements that have similar frequency characteristics over the frequency range of interest. For this reason a good deal of the effort in microwave filter design has been based directly on the application of low-frequency filter-synthesis techniques.

There are essentially two low-frequency filter-synthesis techniques in common use. These are referred to as the image-parameter method (and variations thereof, such as the constant-k and m-derived filters) and the insertion-loss method. The image-parameter method provides a filter design having the required passband and stopband characteristics, but does not specify the exact frequency characteristics over each region. The insertion-loss method begins with a complete specification of a physically realizable frequency characteristic, and from this a suitable filter network is synthesized. The image-parameter method suffers from the shortcomings that a good deal of cut-and-try procedures must often be resorted to in order to obtain an acceptable overall frequency characteristic. For this reason the insertion-loss method is preferable and is the only method considered in detail in the following sections. The image-parameter method is only briefly outlined, in order to show its relationship to the properties of periodic networks as already discussed.

The labor involved in filter synthesis is largely obviated by the use of certain frequency transformations and element normalizations. These enable high-pass and bandpass filters operating over arbitrary frequency bands and between arbitrary resistive load terminations to be obtained from a basic low-pass filter design. The characteristics of any filter will, of course, be modified by the losses that are present in all physical network elements. To incorporate the effect of lossy elements into the synthesis procedure makes the synthesis theory a great deal more involved; so this is usually not done. At microwave frequencies losses can be kept reasonably small, to the extent that most filter designs based on the use of lossless elements do perform satisfactorily.

The aim of the following sections is to present the essential features of low-frequency filter synthesis, frequency transformations, normalized filter design, and the applications of these techniques to microwave filter design. A number of typical microwave filters are also discussed. An extensive account of all aspects of microwave filter design is beyond the scope of this text. However, a number of selected references are given where further details may be found.

8.13 Image-parameter method of filter design

Filters designed by the image-parameter method have many features in common with those of periodic structures. As noted in the previous sections, a cascade connection of lossless two-port networks behaves similar to a transmission line. For unsymmetrical networks two characteristic impedances $\bar{Z}_B^{\pm} = \pm\bar{Z} + \zeta$ occur, and each section has a propagation factor $e^{\pm\gamma d}$. A periodic structure of this form has passband and stopband characteristics and is therefore a bandpass filter. However, the proper load termination to prevent reflections is $\bar{Z}_B$ and is complex when $\zeta \neq 0$. Usually, a filter must operate between resistive load terminations, and it would not be possible to have matched input and output terminations in this case unless ζ were zero, i.e., unless symmetrical networks were used or unless matching sections were used at the input and output. For this reason the image-parameter method of filter design is based on considerations somewhat different from those which have been discussed for periodic structures.

Consider a single two-port network with parameters $\mathcal{A}$, $\mathcal{B}$, $\mathcal{C}$, and $\mathcal{D}$. Let the output be terminated in a load Z_{i2}, and let the input be terminated in a load Z_{i1}, as in Fig. 8.19. For particular values of Z_{i1} and Z_{i2}, known as the image impedances, the input impedance at port 1 equals Z_{i1} and that at port 2 equals Z_{i2}. These impedances then provide matched terminations for the two-port network, and if they are real, they also provide a maximum power transfer when the generator has an internal impedance equal to the image impedance. The governing equations for the two-port network are

$$V_1 = \mathcal{A}V_2 + \mathcal{B}I_2 \qquad I_1 = \mathcal{C}V_2 + \mathcal{D}I_2$$

and hence

$$\frac{V_1}{I_1} = Z_{\text{in},1} = \frac{\mathcal{A}V_2 + \mathcal{B}I_2}{\mathcal{C}V_2 + \mathcal{D}I_2} = \frac{\mathcal{A}Z_{i2} + \mathcal{B}}{\mathcal{C}Z_{i2} + \mathcal{D}} \tag{8.79}$$

If we solve for V_2 and I_2 in terms of V_1 and I_1, we obtain

$$V_2 = \mathcal{D}V_1 - \mathcal{B}I_1 \qquad I_2 = -\mathcal{C}V_1 + \mathcal{A}I_1$$

We thus have

$$-\frac{V_2}{I_2} = Z_{\text{in},2} = -\frac{\mathcal{D}V_1 - \mathcal{B}I_1}{-\mathcal{C}V_1 + \mathcal{A}I_1} = \frac{\mathcal{D}Z_{i1} + \mathcal{B}}{\mathcal{C}Z_{i1} + \mathcal{A}} \tag{8.80}$$

The requirement that $Z_{i1} = Z_{\text{in},1}$ and $Z_{i2} = Z_{\text{in},2}$ gives

$$Z_{i1}(\mathcal{C}Z_{i2} + \mathcal{D}) = \mathcal{A}Z_{i2} + \mathcal{B} \qquad Z_{i2}(\mathcal{C}Z_{i1} + \mathcal{A}) = \mathcal{D}Z_{i1} + \mathcal{B}$$

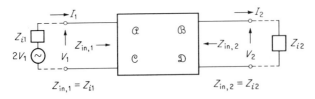

Fig. 8.19 Image parameters for a two-port network.

A simultaneous solution of these equations gives

$$Z_{i1} = \sqrt{\frac{\mathcal{AB}}{\mathcal{CD}}} \qquad (8.81a)$$

$$Z_{i2} = \sqrt{\frac{\mathcal{DB}}{\mathcal{AC}}} \qquad (8.81b)$$

Also we find that $Z_{i2} = (\mathcal{D}/\mathcal{A})Z_{i1}$.

If a generator with internal impedance Z_{i1} is connected at port 1 and the output port 2 is terminated in a load Z_{i2}, the voltage and current transfer ratios are readily found from the relations

$$V_2 = \mathcal{D}V_1 - \mathcal{B}I_1 = \left(\mathcal{D} - \frac{\mathcal{B}}{Z_{i1}}\right)V_1$$

$$I_2 = -\mathcal{C}V_1 + \mathcal{A}I_1 = (-\mathcal{C}Z_{i1} + \mathcal{A})I_1$$

where V_1 is the voltage across the network terminals at port 1 (the generator voltage is $2V_1$). Thus we find that

$$\frac{V_2}{V_1} = \sqrt{\frac{\mathcal{D}}{\mathcal{A}}}\,(\sqrt{\mathcal{AD}} - \sqrt{\mathcal{BC}}) \qquad (8.82a)$$

$$\frac{I_2}{I_1} = \sqrt{\frac{\mathcal{A}}{\mathcal{D}}}\,(\sqrt{\mathcal{AD}} - \sqrt{\mathcal{BC}}) \qquad (8.82b)$$

In a similar manner the transfer constants from port 2 to port 1 are found to be [or from (8.82)]

$$\frac{V_1}{V_2} = \sqrt{\frac{\mathcal{A}}{\mathcal{D}}}\,(\sqrt{\mathcal{AD}} + \sqrt{\mathcal{BC}}) \qquad (8.83a)$$

$$\frac{I_1}{I_2} = \sqrt{\frac{\mathcal{D}}{\mathcal{A}}}\,(\sqrt{\mathcal{AD}} + \sqrt{\mathcal{BC}}) \qquad (8.83b)$$

The image propagation factor $e^{-\gamma}$ is defined as

$$e^{-\gamma} = \sqrt{\mathcal{AD}} - \sqrt{\mathcal{BC}} \qquad (8.84a)$$

whence it is found that

$$e^{\gamma} = \sqrt{\mathcal{AD}} + \sqrt{\mathcal{BC}} \qquad (8.84b)$$

and

$$\cosh \gamma = \sqrt{\mathcal{AD}} \qquad (8.84c)$$
$$\sinh \gamma = \sqrt{\mathcal{BC}} \qquad (8.84d)$$

The factor $(\sqrt{\mathcal{A}/\mathcal{D}})^2$ is interpreted as an impedance transformation ratio and may be viewed as an ideal transformer of turns ratio $\sqrt{\mathcal{A}/\mathcal{D}}$.

For a lossless network, $\mathcal{A}$ and $\mathcal{D}$ are real and $\mathcal{B}$ and $\mathcal{C}$ are imaginary. In the passband of a filter, γ is pure imaginary and equal to $j\beta$, and this occurs for $|\mathcal{AD}| < 1$, as (8.84c) shows. Also, in the passband, the image impedances are pure real, whereas in a stopband they are pure imaginary, as the following considerations show. In a passband, $\mathcal{B}$ and $\mathcal{C}$ must be of the same sign, so that $\mathcal{BC} = j|\mathcal{B}|j|\mathcal{C}| = -|\mathcal{BC}|$ will make $\sinh \gamma$ in (8.84d) pure imaginary; that is, $\gamma = j\beta$. Thus, in (8.81), the quantity under the square root will be real and positive since $\mathcal{AD}$ must be positive to give a real solution for $\cosh \gamma$. Hence the image impedances are real in a passband.

If N two-port networks are connected in cascade and these have propagation constants γ_n, $n = 1, 2, \ldots, N$, and voltage transformation ratios

$$T_1 = \sqrt{\frac{\mathcal{D}_1}{\mathcal{A}_1}} \qquad T_2, \ldots, T_n, \ldots, T_N$$

and the output section is terminated in an impedance equal to its output image impedance, the overall voltage transfer ratio is

$$\frac{V_N}{V_1} = \frac{V_N}{V_{N-1}} \cdots \frac{V_2}{V_1} = T_1 T_2 \cdots T_N e^{-\gamma_1 - \gamma_2 \cdots -\gamma_N} = \prod_{n=1}^{N} T_n e^{-\gamma_n} \qquad (8.85)$$

provided also that the output image impedance of any one section is equal to the input image impedance of the adjacent section. With this filter network terminated in a load impedance Z_{iN} equal to the image impedance of the output section, and with the generator at the input having an internal impedance Z_{i1}, the overall network is matched for maximum power transfer. The filter operates between impedance levels of Z_{iN} and Z_{i1}, which provide an overall impedance-ratio change of amount

$$\frac{Z_{iN}}{Z_{i1}} = \frac{Z_{iN}}{Z_{iN-1}} \cdots \frac{Z_{i2}}{Z_{i1}} = \prod_{n=1}^{N} T_n^2 \qquad (8.86)$$

If symmetrical two ports are used, $\mathcal{A} = \mathcal{D}$ and $Z_{i1} = Z_{i2}$, and both are equal to the Bloch-wave characteristic impedance Z_B^+. For a symmetrical network no transformation or change in impedance level is obtained. The filter consisting of N symmetrical sections terminated in load impedances equal to the image impedance Z_i behaves exactly like an infinite periodic structure, with its characteristic passband and stopband features.

In the image-parameter method of filter design, the two-port parameters $\mathcal{A}$, $\mathcal{B}$, $\mathcal{C}$, $\mathcal{D}$ are chosen to provide the required passbands and stopbands. In addition, the image parameters are chosen equal to the terminating impedances at the center of the passband. The shortcomings of the filter are now apparent, because the image impedances are functions of frequency and normally do not remain equal to the terminating impedances over the whole desired passband. This results in some loss in transmission (loss due to mismatch) within the passband, an amount that cannot be prescribed or determined before the filter has been designed. In addition, there is no means available for controlling the rate at which the attenuation builds up with frequency beyond the edges of the passband, apart from increasing the number of filter sections. Nevertheless, many useful microwave filters have been designed on this basis.†

8.14 Filter design by insertion-loss method

The power loss ratio of a network was defined in Sec. 5.11 as the available, or incident, power divided by the actual power delivered to the load; thus

$$P_{LR} = \frac{1}{1 - \Gamma\Gamma^*} = \frac{1}{1 - \rho^2} \qquad (8.87)$$

where Γ is the input reflection coefficient for a lossless network terminated in a resistive load impedance $Z_L = R_L$. The insertion loss, measured in decibels, is

$$L = 10 \log P_{LR} \qquad (8.88)$$

when the terminating resistive load impedance equals the internal impedance of the generator at the input end. In general, the insertion loss is defined as the ratio of the power delivered to the load when connected directly to the generator to the power delivered when the filter is inserted.

The insertion-loss method of filter design begins by specifying the power loss ratio P_{LR} or the magnitude of the reflection coefficient $|\Gamma| = \rho$ as a function of ω. A network that will give the desired power loss ratio is then synthesized. This procedure is seen to be essentially the same as was followed in the synthesis of quarter-wave transformers in Secs. 5.9 and 5.10. Indeed, the multisection quarter-wave transformer may be considered a particular type of bandpass filter. It must be kept in mind, however, that a completely arbitrary $\Gamma(\omega)$ as a function of ω cannot be chosen since it may not correspond to a physical network. The restrictions to be imposed on Γ are known as the conditions for physical realizability, and some of these are discussed below.

For a passive network it is clear that the reflected power cannot exceed the incident power, and hence one restriction on $\Gamma(\omega)$ is

$$|\Gamma(\omega)| \leq 1 \qquad (8.89)$$

If the normalized input impedance of the network is

$$\bar{Z}(\omega) = \bar{R}(\omega) + j\bar{X}(\omega)$$

† S. B. Cohn, chaps. 26 and 27 in Radio Research Laboratory Staff, "Very High Frequency Techniques," vol. 2, McGraw-Hill Book Company, New York, 1947.

For a discussion of image-parameter methods at low frequencies, see E. A. Guillemin, "Communication Networks," vol. 2, John Wiley & Sons, Inc., New York, 1935.

we have

$$\Gamma(\omega) = \frac{\bar{Z}_{in} - 1}{\bar{Z}_{in} + 1} = \frac{\bar{R}(\omega) - 1 + j\bar{X}(\omega)}{\bar{R}(\omega) + 1 + j\bar{X}(\omega)}$$

As shown in Sec. 4.4, $\bar{R}$ is an even function of ω and $\bar{X}$ is an odd function of ω. Hence

$$\Gamma(-\omega) = \frac{\bar{R}(\omega) - 1 - j\bar{X}(\omega)}{\bar{R}(\omega) + 1 - j\bar{X}(\omega)} = \Gamma^*(\omega)$$

and thus

$$|\Gamma(\omega)|^2 = \rho^2(\omega) = \Gamma\Gamma^* = \Gamma(\omega)\Gamma(-\omega) \tag{8.90}$$

It is apparent from this relation that $\rho^2(\omega) = \rho^2(-\omega)$ is an even function of ω and must therefore contain only even powers of ω. Now any low-frequency impedance function (impedance of a network made up of resistors, capacitors, and inductors) can be expressed as the ratio of two polynomials in ω. Consequently, Γ can also be expressed as the ratio of two polynomials. It follows that $\rho^2(\omega)$ can then be expressed in the form

$$\rho^2(\omega) = \frac{M(\omega^2)}{M(\omega^2) + N(\omega^2)} = \frac{(\bar{R} - 1)^2 + \bar{X}^2}{(\bar{R} + 1)^2 + \bar{X}^2} \tag{8.91}$$

where M and N are real and nonnegative polynomials in ω^2. The power loss ratio can now be expressed as

$$P_{LR} = 1 + \frac{M(\omega^2)}{N(\omega^2)} = 1 + \frac{[\bar{R}(\omega) - 1]^2 + [\bar{X}(\omega)]^2}{4\bar{R}(\omega)} \tag{8.92}$$

The last result in (8.92) shows that $N(\omega^2)$ must be an even polynomial in ω since it equals $4\bar{R}(\omega)$. Hence we write $N(\omega^2) = Q^2(\omega)$, which is clearly an even polynomial in ω. If we denote $M(\omega^2)$ by the even polynomial $P(\omega^2)$ instead, we have

$$P_{LR} = 1 + \frac{P(\omega^2)}{Q^2(\omega)} \tag{8.93}$$

The conditions specified on P_{LR} up to this point are necessary conditions in order that the network may be physically realizable. It may be shown that the condition that the power loss ratio P_{LR} be expressible in the form (8.93) is also a sufficient condition for the network to be realizable.[†] In succeeding sections we consider suitable forms for the polynomials P and Q and the types of networks required to yield the corresponding power loss ratio.

[†] G. L. Ragan (ed.), "Microwave Transmission Circuits," sec. 9.13, McGraw-Hill Book Company, New York, 1948.

8.15 Specification of power loss ratio

There are virtually an unlimited number of different forms that could be specified for the power loss ratio and be realized as a physical network. However, many of these networks could be anticipated to be very complex and hence of little practical utility. The power loss ratios that have been found most useful for microwave filter design are those that give a maximally flat passband response and those that give an equal-ripple, or Chebyshev, response in the passband. Such passband-response characteristics correspond to those of the binomial and Chebyshev multisection quarter-wave transformers discussed in Chap. 5. The maximally flat filter (commonly called a Butterworth filter) and the Chebyshev filter are described below for the low-pass case only. In a following section it is shown how high-pass and bandpass filter characteristics may be obtained from the low-pass filter response by suitable frequency transformations, or mappings.

Maximally flat filter characteristic

The power loss ratio for a maximally flat low-pass filter is obtained by choosing the polynomial Q equal to unity and choosing $P(\omega^2)$ equal to $k^2(\omega/\omega_c)^{2N}$. Hence we have

$$P_{LR} = 1 + k^2 \left(\frac{\omega}{\omega_c}\right)^{2N} \tag{8.94}$$

The passband is the region from $\omega = 0$ to the cutoff value ω_c. The maximum value of P_{LR} in the passband is $1 + k^2$, and for this reason k^2 is called the passband tolerance. For $\omega > \omega_c$ the power loss ratio increases indefinitely at a rate dependent on the exponent $2N$, which in turn is related to the number of filter sections employed. A typical filter characteristic is illustrated in Fig. 8.20 for $N = 2$.

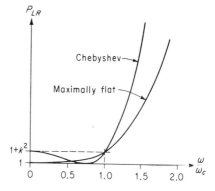

Fig. 8.20 Low-pass-filter response for maximally flat and Chebyshev filters for $N = 2$.

Chebyshev filter

The power loss ratio for the equal-ripple, or Chebyshev, filter is chosen as

$$P_{LR} = 1 + k^2 T_N^2 \left(\frac{\omega}{\omega_c}\right) \tag{8.95}$$

where $T_N(\omega/\omega_c)$ is the Chebyshev polynomial of degree N discussed in Sec. 5.10. Recall that

$$T_N\left(\frac{\omega}{\omega_c}\right) = \cos\left(N \cos^{-1} \frac{\omega}{\omega_c}\right)$$

and thus $T_N(\omega/\omega_c)$ oscillates between ± 1 for $|\omega/\omega_c| \leq 1$ and increases monotonically for ω/ω_c greater than unity. The power loss ratio will oscillate between 1 and $1 + k^2$ in the passband, will equal k^2 at the cutoff frequency, and will increase monotonically for $\omega > \omega_c$. A typical response curve is shown in Fig. 8.20 for $N = 2$. One particularly striking feature of the Chebyshev response curve compared with the maximally flat curve is its much greater rate of rise beyond the cutoff point. This means that the corresponding filter has a much sharper cutoff region separating the passband and stopband, which is usually a desired characteristic.

For ω/ω_c large, the power loss ratio for the Chebyshev filter approaches

$$P_{LR} \approx \frac{k^2}{4} \left(\frac{2\omega}{\omega_c}\right)^{2N} \tag{8.96}$$

Compared with the maximally flat response characteristic this is larger by a factor 2^{2N-2}. In fact, no other polynomial $P(\omega^2)$ yielding a passband tolerance of k^2 can yield a rate of increase of P_{LR} greater than that given by (8.96). Conversely, for a specified rate of increase in the power loss ratio beyond the cutoff frequency, the Chebyshev polynomial gives the smallest passband tolerance. In this sense the Chebyshev filter represents an optimum design. The proof is similar to that used in Sec. 5.11 to prove the optimum properties of the Chebyshev quarter-wave transformer.

When the power loss ratio is equal to $1 + k^2$, the magnitude of the reflection coefficient at the input is

$$\rho = \frac{k}{(1 + k^2)^{\frac{1}{2}}}$$

The input voltage standing-wave ratio is given by

$$s = \frac{1 + \rho}{1 - \rho} = \frac{(1 + k^2)^{\frac{1}{2}} + k}{(1 + k^2)^{\frac{1}{2}} - k}$$

If P is chosen as unity and Q is set equal to $k^2 T_N^2(\omega/\omega_c)$, a high-pass

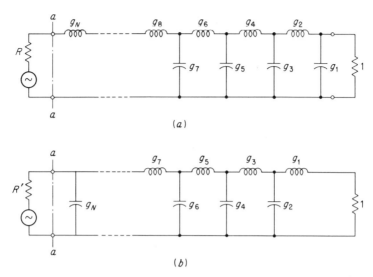

Fig. 8.21 Low-pass ladder-prototype-filter networks.

filter having Chebyshev behavior in the attenuation or stopband is obtained. It is also possible to choose P and Q so as to give Chebyshev behavior in both the passband and the stopband. The required network is, however, usually too complex to be realized in a satisfactory manner with microwave circuit elements.

8.16 Some low-pass-filter designs

The maximally flat and Chebyshev low-pass-filter power loss ratios discussed in the preceding section can be realized by means of a ladder network of capacitors and inductors in the form illustrated in Fig. 8.21. The load impedance is chosen equal to 1 ohm, and the generator impedance as R. The circuit in Fig. 8.21b is the dual of the circuit in Fig. 8.21a. Both circuits can be designed to give the same power loss ratio. For maximally flat or Chebyshev response in the passband the ladder network is symmetric for an odd number of elements. This is also true for N even, in the case of the maximally flat filter. The element values are denoted by g_k, and are the same in both circuits. However, the required generator impedance R' in the network of Fig. 8.21b is equal to $1/R$. For N odd both R and R' equal unity.

If we let Z_{in} be the input impedance at the plane aa in Fig. 8.21a, the reflection coefficient will be

$$\Gamma = \frac{Z_{\text{in}} - R}{Z_{\text{in}} + R}$$

In terms of Z_{in} and R, the power loss ratio is readily computed to be

$$P_{LR} = 1 + \frac{|Z_{in} - R|^2}{2R(Z_{in} + Z_{in}^*)} \tag{8.97}$$

At $\omega = 0$, all capacitors appear as infinite impedances and all inductors as zero impedances, and hence $Z_{in} = 1$. For a maximally flat filter or a Chebyshev filter with N odd, we must have unity power loss ratio at $\omega = 0$. This requires that we choose the generator impedance equal to unity. For a Chebyshev filter with N even, we have $P_{LR} = 1 + k^2$ at $\omega = 0$, and hence

$$1 + k^2 = 1 + \frac{(1 - R)^2}{4R}$$

or

$$R = 2k^2 + 1 - \sqrt{4k^2(1 + k^2)} = \frac{1}{R'} \tag{8.98}$$

The required values of the elements g_k in the ladder network can be obtained very readily by solving for Z_{in} and equating the power loss ratio as given by (8.97) to the desired power loss ratio for N up to 3 or 4. As an example, consider the case of $N = 2$ for the circuit in Fig. 8.21a. We readily find that

$$Z_{in} = j\omega L + \frac{1}{1 + j\omega C}$$

where $C = g_1$ and $L = g_2$. Using (8.97) gives

$$P_{LR} = 1 + \frac{(1 - R)^2 + \omega^2(L^2 + C^2R^2 - 2LC) + \omega^4 L^2 C^2}{4R}$$

To make P_{LR} equal unity at $\omega = 0$ and to obtain maximally flat response, we must choose $R = 1$ and $L^2 + C^2 - 2LC = 0$. If we specify cutoff to occur at $\omega = 1$ with a passband tolerance of k^2, we also have

$$1 + k^2 = 1 + \frac{L^2 C^2}{4}$$

or $LC = 2k$. Hence we find $L = C = (2k)^{1/2}$. For Chebyshev response we equate P_{LR} to

$$1 + k^2 T_2^2\left(\frac{\omega}{\omega_c}\right) = 1 + k^2\left[2\left(\frac{\omega}{\omega_c}\right)^2 - 1\right]^2$$

in order to determine L and C.

For large values of N the above procedure is very laborious to carry out. In place of this direct substitution scheme general solutions have been worked out.† In addition, tables of element values, i.e., values for

† V. Belevitch, Chebyshev Filters and Amplifier Networks, *Wireless Eng.*, vol. 29, pp. 106–107, April, 1952.

the g_k, have been prepared by a number of people.†

For the maximally flat network with a power loss ratio

$$P_{LR} = 1 + \omega^{2N} \tag{8.99}$$

the element values are given by

$$R = 1 \tag{8.100a}$$

$$g_k = 2 \sin \frac{2k-1}{2N} \pi \qquad k = 1, 2, \ldots, N \tag{8.100b}$$

where g_k is the value either of inductance in henrys or of capacitance in farads.

† L. Weinberg, Network Design by Use of Modern Synthesis Techniques and Tables, *Proc. Natl. Electron. Conf.*, vol. 12, 1956.

S. Cohn, Direct Coupled Resonator Filters, *Proc. IRE*, vol. 45, pp. 187–196, February, 1957.

See also "The Microwave Engineers' Handbook," Horizon House, Inc., 1963–1964.

Table 8.1 Values of g_k for maximally flat filter

k	N=2	N=3	N=4	N=5
1	1.414	1.00	0.7654	0.6180
2	1.414	2.00	1.848	1.618
3		1.00	1.848	2.000
4			0.7654	1.618
5				0.6180

Table 8.2 Values of g_k for Chebyshev filter with $k^2 = 0.0233$

k	N=2	N=3	N=4	N=5
1	0.8430	1.0315	1.1088	1.1468
2	0.6220	1.1474	1.3061	1.3712
3		1.0315	1.7703	1.9750
4			0.8180	1.3712
5				1.1468

For a Chebyshev low-pass filter with $\omega_c = 1$, the element values are given by

$$R = \begin{cases} 1 & N \text{ odd} \quad (8.101a) \\ 2k^2 + 1 - 2k\sqrt{1+k^2} & N \text{ even} \quad (8.101b) \end{cases}$$

$$g_k = \frac{4a_{k-1}a_k}{b_{k-1}g_{k-1}} \quad (8.101c)$$

where

$$a_k = \sin\frac{2k-1}{2N}\pi$$

$$b_k = \sinh^2\frac{\beta}{2N} + \sin^2\frac{k\pi}{N}$$

$$\beta = \ln\frac{\sqrt{1+k^2}-1}{\sqrt{1+k^2}+1}$$

Numerical values for the g_k are given in Tables 8.1 and 8.2 for N up to 5, $\omega_c = 1$, and a passband tolerance k^2 equal to 0.0233 (a 0.1-db ripple in the passband) for the Chebyshev filter. More extensive tables are given in the references cited.

8.17 Frequency transformations

The low-pass filter with cutoff at $\omega_c = 1$ and terminated in a 1-ohm load impedance may be used as a basis for the design of high-pass and bandpass filters with arbitrary resistive load termination. For this reason it is referred to as a prototype filter. For the purpose of this section it is convenient to denote the frequency variable for the low-pass prototype filter by ω'. The power loss ratio may be expressed in the form

$$P_{LR} = 1 + P(\omega'^2) \quad (8.102)$$

for maximally flat and Chebyshev responses.

If it is desirable to have a load termination R_L different from 1 ohm, the required filter is obtained by multiplying all other reactances and the generator resistance by a factor R_L. That is, the prototype-filter reactances can be viewed as normalized with respect to R_L. The new values for the inductances and capacitances are

$$L'_k = R_L L_k \quad (8.103a)$$

$$C'_k = \frac{C_k}{R_L} \quad (8.103b)$$

$$R' = R_L R \quad (8.103c)$$

where R' denotes the new value of R here, and not the generator resistance in the dual circuit of Fig. 8.21b. In the discussion of frequency transformations below, we assume a 1-ohm termination since the change in impedance level of the filter can be made as a last step in the design process.

If we replace the frequency variable ω' by a new variable ω according to

$$\omega' = f(\omega)$$

the power loss ratio becomes

$$P_{LR} = 1 + P(\omega'^2) = 1 + P[f^2(\omega)] \tag{8.104}$$

As a function of ω' or f, this power loss ratio is that of the low-pass prototype filter, but as a function of ω, it has a different characteristic, depending on how $f(\omega)$ is chosen. A number of different frequency transformations, or mappings, are considered below.

Frequency expansion

If it is desirable to change the cutoff frequency from unity to some other value ω_c, we choose

$$f(\omega) = \frac{\omega}{\omega_c}$$

and thus

$$P_{LR} = 1 + P\left(\frac{\omega^2}{\omega_c^2}\right) \tag{8.105}$$

Cutoff occurs when the argument ω/ω_c equals unity or when $\omega = \omega_c$. The series reactances and shunt susceptances in the prototype filter must be replaced by new reactances and susceptances

$$jX'_k = j\left(\frac{\omega}{\omega_c}\right)L_k \qquad jB'_k = j\left(\frac{\omega}{\omega_c}\right)C_k$$

when ω' is changed to ω/ω_c in order to yield the power loss ratio given by (8.105). Examination of the latter equations shows that the new values of the L_k and C_k must be

$$L'_k = \frac{L_k}{\omega_c} \tag{8.106a}$$

$$C'_k = \frac{C_k}{\omega_c} \tag{8.106b}$$

The power loss ratios in terms of ω' and ω are compared in Fig. 8.22a and b.

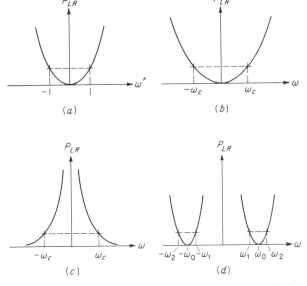

Fig. 8.22 (a) Low-pass prototype-filter response; (b) frequency expansion; (c) low-pass to high-pass transformation; (d) low-pass to bandpass transformation.

Low-pass to high-pass transformation

A high-pass filter is obtained by choosing $f(\omega)$ equal to $-\omega_c/\omega$ to yield a power loss ratio

$$P_{\text{LR}} = 1 + P\left(\frac{\omega_c^2}{\omega^2}\right) \qquad (8.107)$$

This frequency transformation maps the point $\omega' = 0$ into the points $\omega = \pm\infty$, the points $\omega' = \pm 1$ into $\omega = \mp\omega_c$, and the points $\omega' = \pm\infty$ into $\omega = 0$. The effect is to interchange the passband and stopband regions, as illustrated by Fig. 8.22a and c. To obtain a power loss ratio of the form (8.107), the series reactances and shunt susceptances in the new filter must be chosen as follows:

$$jX'_k = -j\frac{\omega_c}{\omega}L_k \qquad jB'_k = -j\frac{\omega_c}{\omega}C_k$$

From these equations it is seen that all inductances L_k must be replaced by capacitances C'_k and the C_k must be replaced by inductances L'_k in the following manner:

$$C'_k = \frac{1}{\omega_c L_k} \qquad (8.108a)$$

$$L'_k = \frac{1}{\omega_c C_k} \qquad (8.108b)$$

Low-pass to bandpass transformation

To obtain a bandpass filter consider a change of variable according to

$$\omega' = f(\omega) = \frac{\omega_0}{\omega_2 - \omega_1}\left(\frac{\omega}{\omega_0} - \frac{\omega_0}{\omega}\right) \tag{8.109}$$

This equation may be solved for ω to give

$$\omega = \omega'\frac{\omega_2 - \omega_1}{2} \pm \sqrt{\omega'^2\left(\frac{\omega_2 - \omega_1}{2}\right)^2 + \omega_0^2}$$

If we also choose $\omega_0^2 = \omega_1\omega_2$, we obtain

$$\omega = \omega'\frac{\omega_2 - \omega_1}{2} \pm \frac{1}{2}\sqrt{\omega'^2(\omega_2 - \omega_1)^2 + 4\omega_1\omega_2} \tag{8.110}$$

The point $\omega' = 0$ is seen to map into the points $\omega = \pm\omega_0$, and $\omega' = \pm 1$ maps into the four points $\pm(\omega_2 - \omega_1)/2 \pm (\omega_2 + \omega_1)/2 = \pm\omega_2$ and $\pm\omega_1$. Thus the prototype-filter passband between ± 1 maps into passbands extending from ω_1 to ω_2 and $-\omega_1$ to $-\omega_2$, which represent bandpass filters with band centers at $\pm\omega_0$ equal to the geometric mean of ω_1 and ω_2, as in Fig. 8.22d.

The required filter elements may be deduced by considering the frequency behavior of series and parallel connections of L and C, as in Fig. 8.23. For the series circuit we have

$$jX' = j\omega L + \frac{1}{j\omega C} = j\sqrt{\frac{L}{C}}\left(\omega\sqrt{LC} - \frac{1}{\omega\sqrt{LC}}\right)$$

and for the parallel circuit we have

$$jB' = j\omega C + \frac{1}{j\omega L} = j\sqrt{\frac{C}{L}}\left(\omega\sqrt{LC} - \frac{1}{\omega\sqrt{LC}}\right)$$

If we make $\sqrt{LC}$ equal to ω_0^{-1}, we obtain

$$jX' = j\sqrt{\frac{L}{C}}\left(\frac{\omega}{\omega_0} - \frac{\omega_0}{\omega}\right) \qquad jB' = j\sqrt{\frac{C}{L}}\left(\frac{\omega}{\omega_0} - \frac{\omega_0}{\omega}\right)$$

The required frequency transformation may now be seen to be obtained if we replace the reactance $jX_k = j\omega L_k$ and the susceptance $jB_k = j\omega C_k$ in the prototype filter by series and parallel tuned circuits such that

$$jX'_k = j\sqrt{\frac{L'_k}{C'_k}}\left(\frac{\omega}{\omega_0} - \frac{\omega_0}{\omega}\right) = jL_k\frac{\omega_0}{\omega_2 - \omega_1}\left(\frac{\omega}{\omega_0} - \frac{\omega_0}{\omega}\right)$$

$$jB'_k = j\sqrt{\frac{C'_k}{L'_k}}\left(\frac{\omega}{\omega_0} - \frac{\omega_0}{\omega}\right) = jC_k\frac{\omega_0}{\omega_2 - \omega_1}\left(\frac{\omega}{\omega_0} - \frac{\omega_0}{\omega}\right)$$

Fig. 8.23 Series and parallel tuned circuits.

This requires that we choose

$$L'_k C'_k = \frac{1}{\omega_0{}^2} = \frac{1}{\omega_1 \omega_2} \tag{8.111a}$$

$$\sqrt{\frac{L'_k}{C'_k}} = \omega_0 L'_k = \frac{\omega_0 L_k}{\omega_2 - \omega_1}$$

or

$$L'_k = \frac{L_k}{\omega_2 - \omega_1} \tag{8.111b}$$

for the series circuit. For the parallel circuit we must choose

$$L'_k C'_k = \frac{1}{\omega_0{}^2} = \frac{1}{\omega_1 \omega_2} \tag{8.112a}$$

$$\sqrt{\frac{C'_k}{L'_k}} = \omega_0 C'_k = \frac{\omega_0 C_k}{\omega_2 - \omega_1}$$

or

$$C'_k = \frac{C_k}{\omega_2 - \omega_1} \tag{8.112b}$$

The resultant filter network is illustrated in Fig. 8.24.

Periodic bandpass mapping

The general concept of frequency mapping may be applied in a variety of other ways as well. All that is necessary is to replace the reactances $j\omega' L_k$ and susceptances $j\omega' C_k$ in the prototype filter by other reactance and susceptance functions having ω' replaced by a new function $f(\omega)$ of ω. As a further example we shall consider the effect of replacing all L_k by short-circuited transmission-line stubs of length l and characteristic impedance Z_k and the capacitors C_k by open-circuited stubs of length l and characteristic admittance Y_k. The new reactance and susceptance functions become

$$jX'_k = jZ_k \tan\left(\frac{\omega}{c} l\right) = jZ_k \tan \theta = jg_k \tan \theta$$

$$jB'_k = jY_k \tan \theta = jg_k \tan \theta$$

This filter is illustrated in Fig. 8.25 and has a power loss ratio given by

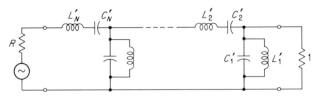

Fig. 8.24 A bandpass-filter network.

Sec. 8.18 Periodic structures and filters 415

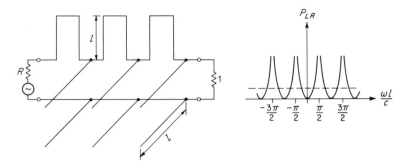

Fig. 8.25 A transmission-line filter and its response as obtained from the low-pass prototype filter by a frequency transformation.

$$P_{LR} = 1 + P(\tan^2 \theta) \quad (8.113)$$

The frequency transformation $\omega' = \tan(\omega l/c) = \tan\theta$ maps the whole ω' axis periodically into intervals of length π in the θ domain or of length $\pi c/l$ in the ω domain. This filter would be satisfactory at low frequencies where junction effects at the points where the stubs are connected to the main line are negligible and where the separation between stubs could be kept very small in terms of wavelength.†

At microwave frequencies the alternative occurrence of series and parallel tuned circuits in the filter derived from the low-pass prototype filter by a frequency transformation is an undesirable feature. It is difficult to construct a filter of this type using microwave elements. However, it is possible to convert the filter with series tuned elements into an equivalent filter containing only parallel tuned circuits in a cascade connection, or to convert the circuit into one containing only series tuned circuits. The desired transformation may be obtained by the use of impedance and admittance inverters or quarter-wave transformers, as discussed in the next section.

8.18 Impedance inverters

An impedance inverter is an ideal quarter-wave transformer. A load impedance connected at one end is seen as an impedance that has been inverted with respect to the characteristic impedance squared at the input. Impedance inverters may be used to convert a bandpass-filter network of the type shown in Fig. 8.24 into a network containing only series tuned circuits.

† If a filter is made up of transmission lines of commensurate length and resistors, the only frequency variable occurring is $\tan(\omega l/c)$. In this case the frequency transformation $\omega' = \tan(\omega l/c)$ permits the use of conventional low-frequency-network synthesis techniques to be applied. See Refs. 7 to 9 and 15 given at the end of this chapter.

Consider a bandpass filter designed from a low-pass prototype according to the methods presented in Sec. 8.17. We shall assume also that the impedance level has been adjusted so that the load termination is R_L instead of 1 ohm. The generator impedance is denoted by R_g, where $R_g = RR_L$ and equals R_L for N odd, and also for N even for the maximally flat filter. The parameters of this network are denoted by L'_k and C'_k. The network is illustrated in Fig. 8.26a. Figure 8.26b illustrates an equivalent network obtained by the use of impedance inverters. The inverter between elements C_{0k}, L_{0k} and C_{0k+1}, L_{0k+1} has a characteristic impedance $K_{k+1,k}$. All the capacitances and inductances are chosen so that

$$L'_k C'_k = L_{0k} C_{0k} = \omega_0^{-2} \tag{8.114}$$

However, for flexibility in design, we allow arbitrary change in impedance level throughout the network, and thus all the parameters R_{0L}, R_{0g}, and L_{0k}, $k = 1, 2, \ldots, N$, may be arbitrarily chosen. The characteristic impedances of the inverters are thereby fixed, as the following analysis shows.

For the network in Fig. 8.26a a straightforward calculation shows that the input impedance at the plane aa looking toward R_L is

$$\begin{aligned}
Z_{\text{in}} &= j\omega L'_2 + \frac{1}{j\omega C'_2} + \frac{j\omega L'_1}{1 - (\omega/\omega_0)^2 + j\omega L'_1/R_L} \\
&= \sqrt{\frac{L'_2}{C'_2}} \left[j\left(\frac{\omega}{\omega_0} - \frac{\omega_0}{\omega}\right) + \sqrt{\frac{C'_2}{L'_2}} \frac{j\omega L'_1}{1 - (\omega/\omega_0)^2 + j\omega L'_1/R_L} \right]
\end{aligned} \tag{8.115}$$

For the circuit of Fig. 8.26b, the input impedance at the input to the first inverter is K_{10}^2/R_{0L}. Adding this to the impedance $j\omega L_{01} + 1/j\omega C_{01}$,

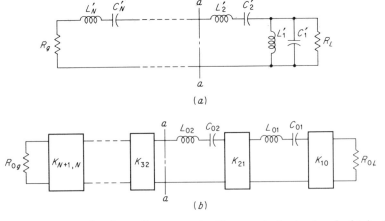

Fig. 8.26 (a) A bandpass-filter network; (b) an equivalent network obtained by use of impedance inverters.

inverting with respect to $K_{21}{}^2$, and combining with the impedance $j\omega L_{02} + 1/j\omega C_{02}$, we find that the input impedance at the plane aa is

$$Z_{0,\text{in}} = \sqrt{\frac{L_{02}}{C_{02}}}\left[j\left(\frac{\omega}{\omega_0} - \frac{\omega_0}{\omega}\right) + \sqrt{\frac{C_{02}}{L_{02}}}\frac{j\omega K_{21}{}^2 C_{01}}{1 - (\omega/\omega_0)^2 + j\omega C_{01} K_{10}{}^2/R_{0L}}\right] \tag{8.116}$$

When we compare (8.116) with (8.115), we see that they will be the same functions of ω apart from a constant $\sqrt{L_{02}/C_{02}}$ in place of $\sqrt{L_2'/C_2'}$, which represents a change in impedance level, if we make the quantity inside the brackets the same. Thus let

$$\frac{K_{10}{}^2 C_{01}}{R_{0L}} = \frac{L_1'}{R_L}$$

or

$$K_{10} = \sqrt{\frac{L_1' R_{0L}}{C_{01} R_L}} = \sqrt{\frac{L_{01} R_{01}}{C_1' R_L}} \tag{8.117}$$

since $L_1'/C_{01} = L_{01}/C_1'$, because $L_1' C_1' = L_{01} C_{01} = \omega_0{}^{-2}$. Similarly, let

$$\sqrt{\frac{C_{02}}{L_{02}}} K_{21}{}^2 C_{01} = \sqrt{\frac{C_2'}{L_2'}} L_1'$$

which gives

$$K_{21} = \sqrt{\frac{L_{02} L_{01}}{C_1' L_2'}} \tag{8.118}$$

With the above choice for K_{10} and K_{21}, both networks have the same input impedance function, apart from a change in impedance level. This impedance will be denoted by $\sqrt{L_2'/C_2'}\, Z$ for the network in Fig. 8.26a and will then be equal to $\sqrt{L_{02}/C_{02}}\, Z$ for the network in Fig. 8.26b. If the foregoing analysis is repeated to find the input impedances at the junction between the fourth and fifth elements, similar results are obtained. The required impedances are readily found by replacing L_1', C_1' by L_3', C_3'; L_2', C_2' by L_4', C_4'; etc., and replacing R_L by $\sqrt{L_2'/C_2'}\, Z$ and R_{0L} by $\sqrt{L_{02}/C_{02}}\, Z$. The required values of K_{32} and K_{43} to maintain electrical equivalence between the two networks may also be found by making the above changes in the expressions for K_{10} and K_{21}. Thus, from (8.117), we obtain

$$K_{32} = \sqrt{\frac{L_{03}}{C_3'}\left(\frac{L_{02}}{C_{02}}\frac{C_2'}{L_2'}\right)^{\frac{1}{2}}} = \sqrt{\frac{L_{02} L_{03}}{L_2' C_3'}} \tag{8.119}$$

$$K_{43} = \sqrt{\frac{L_{03} L_{04}}{C_3' L_4'}} \tag{8.120}$$

By induction, or by repeating the above process, we find that, in general, for k up to N,

$$K_{k,k-1} = \begin{cases} \sqrt{\dfrac{L_{0k}L_{0k-1}}{L'_k C'_{k-1}}} & k \text{ even} & (8.121a) \\[1em] \sqrt{\dfrac{L_{0k}L_{0k-1}}{C'_k L'_{k-1}}} & k \text{ odd} & (8.121b) \end{cases}$$

For the last inverter consider first the case where N is even, so that the last element in the filter of Fig. 8.26a is a series branch. With reference to Fig. 8.27 and by analogy with (8.115) and (8.116), the respective input impedances at the planes aa are

$$Z_\text{in} = \sqrt{\frac{L'_N}{C'_N}}\, Z$$

$$Z_{0,\text{in}} = \sqrt{\frac{L_{0N}}{C_{0N}}}\, Z$$

where Z is the same for both networks. The network in Fig. 8.27a is properly terminated in an impedance R_g. The proper termination for the network in Fig. 8.27b, if the last inverter is not used, will be R_g multiplied by the ratio of the impedance levels in the two networks; that is,

$$R'_{0g} = \frac{R_g \sqrt{L_{0N}/C_{0N}}}{\sqrt{L'_N/C'_N}} = R_g \sqrt{\frac{L_{0N} C'_N}{L'_N C_{0N}}}$$

However, if an additional impedance inverter is used, we can choose an arbitrary termination R_{0g}, provided the characteristic impedance $K_{N+1,N}$ is chosen so as to make R_{0g} appear as a load R'_{0g} at the plane aa in Fig. 8.27b. Thus we require $K_{N+1,N} = \sqrt{R_{0g} R'_{0g}}$. Noting that

$$\frac{L_{0N} C'_N}{L'_N C_{0N}} = \frac{L_{0N}^2}{L'_N{}^2}$$

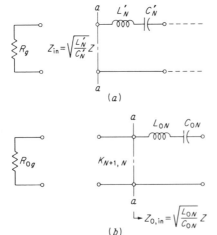

Fig. 8.27 Last section of filter for N even.

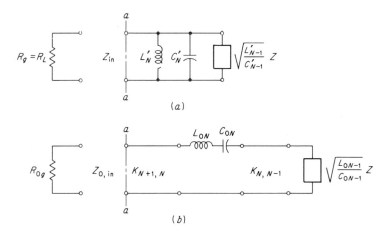

Fig. 8.28 Last section of filter for N odd.

and using the earlier expression for R'_{0g}, we obtain

$$K_{N+1,N} = \sqrt{\frac{R_{0g}R_g L_{0N}}{L'_N}} \qquad N \text{ even} \tag{8.122}$$

When N is odd, the last section in the prototype filter is a shunt branch. In this case we have the situation illustrated in Fig. 8.28. The input impedances at the planes aa are, respectively,

$$Z_{\text{in}} = \left(j\omega C'_N + \frac{1}{j\omega L'_N} + \sqrt{\frac{C'_{N-1}}{L'_{N-1}}}\frac{1}{Z}\right)^{-1}$$

$$= \sqrt{\frac{L'_N}{C'_N}}\left[j\left(\frac{\omega}{\omega_0} - \frac{\omega_0}{\omega}\right) + \frac{1}{Z}\sqrt{\frac{L'_N C'_{N-1}}{C'_N L'_{N-1}}}\right]^{-1}$$

$$Z_{0,\text{in}} = K^2_{N+1,N}\left(j\omega L_{0N} + \frac{1}{j\omega C_{0N}} + \frac{K^2_{N,N-1}}{Z}\sqrt{\frac{C_{0N-1}}{L_{0N-1}}}\right)^{-1}$$

$$= K^2_{N+1,N}\sqrt{\frac{C_{0N}}{L_{0N}}}\left[j\left(\frac{\omega}{\omega_0} - \frac{\omega_0}{\omega}\right) + \frac{K^2_{N,N-1}}{Z}\sqrt{\frac{C_{0N}C_{0N-1}}{L_{0N}L_{0N-1}}}\right]^{-1}$$

Since Z is the same for both networks, the quantity inside the brackets is the same for $Z_{0,\text{in}}$ as for Z_{in}, in view of the way in which $K_{N,N-1}$ is chosen [see (8.121b)]. The relative impedance levels of the two networks are thus $\sqrt{L'_N/C'_N}$ and $K^2_{N+1,N}\sqrt{C_{0N}/L_{0N}}$. The two terminating impedances $R_g = R_L$ and R_{0g} must be proportional to the respective impedance levels, and hence

$$\frac{R_{0g}}{R_L} = K^2_{N+1,N}\sqrt{\frac{C_{0N}C'_N}{L_{0N}L'_N}}$$

and thus

$$K_{N+1,N} = \sqrt{\frac{R_{0g}L_{0N}}{R_L C'_N}} \quad N \text{ odd} \tag{8.123}$$

All the required characteristic impedances of the inverters are now specified.

The network in Fig. 8.26a may also be converted into an equivalent network containing only parallel tuned circuits by the use of admittance inverters. The equivalent circuit is illustrated in Fig. 8.29. Again we must have $L_{0k}C_{0k} = \omega_0^{-2}$, but all L_{0k} and the resistive terminations R_{0L} and R_{0g} may be arbitrarily chosen provided the characteristic admittances $J_{k,k-1}$ of the inverters satisfy the following relations:

$$J_{10} = \sqrt{\frac{C_{01}}{C'_1 R_{0L} R_L}} \tag{8.124a}$$

$$J_{k,k-1} = \sqrt{\frac{C_{0k}C_{0k-1}}{C'_k L'_{k-1}}} \quad k \text{ odd} \tag{8.124b}$$

$$J_{k,k-1} = \sqrt{\frac{C_{0k}C_{0k-1}}{L'_k C'_{k-1}}} \quad k \text{ even} \tag{8.124c}$$

$$J_{N+1,N} = \sqrt{\frac{C_{0N}}{C'_N R_g R_{0g}}} \quad N \text{ odd} \tag{8.124d}$$

$$J_{N+1,N} = \sqrt{\frac{C_{0N}R_g}{L'_N R_{0g}}} \quad N \text{ even} \tag{8.124e}$$

Note that for N odd we also have $R_g = R_L$ in the prototype circuit. The derivation of these expressions parallels that used for the impedance inverters.

The impedance and admittance inverters used in the preceding analysis were assumed to be ideal frequency-independent quarter-wave transformers of electrical length $\pi/2$ independent of frequency. Such ideal inverters do not, of course, exist. Nevertheless, the theory can be applied in practice. For very narrowband bandpass filters, say bandwidths of 1 or 2 percent, a quarter-wave length of transmission line or waveguide does not depart appreciably from the ideal inverter having a $\pi/2$ electrical length. A filter designed on the basis of ideal inverters would have a response very close to the theoretical response in this case. For greater bandwidths the departure of a quarter-wave transformer

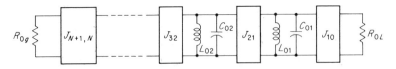

Fig. 8.29 Equivalent filter network obtained by use of admittance inverters.

from the ideal can be incorporated into the design by splitting the transformer into an ideal one with two additional short lengths of transmission line on either side to account for the excess or deficit in phase shift from the ideal phase shift of 90°. For example, a transformer of length $0.3\lambda_0$ at a wavelength of λ_0 can be treated as a transformer of

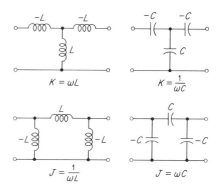

Fig. 8.30 Lumped-element inverters.

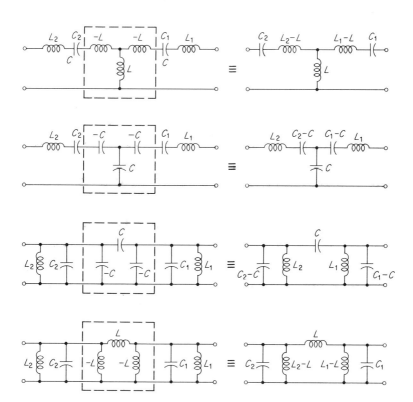

Fig. 8.31 Reactance-coupled resonator-type circuits obtained by use of lumped-element inverters.

length $0.25\lambda_0$ plus sections of line of length $0.025\lambda_0$ on both sides. The excess length of line may be incorporated into and made part of the resonant circuit on either side of the inverter.

Lumped-element circuits that act as impedance inverters and admittance inverters are illustrated in Fig. 8.30 along with their equivalent characteristic impedance or admittance. These circuits involve negative elements and are frequency-dependent. However, the negative elements may be absorbed into the elements of the adjacent resonant circuits to eliminate them from the overall network. The resultant filter then consists of tuned circuits coupled by single reactive elements, as illustrated in Fig. 8.31. Application of these techniques to microwave filter design is discussed in the following sections.

8.19 A transmission-line filter

The first application of the foregoing design formulas is made to the design of a transmission-line filter using quarter-wave-coupled short-circuited shunt stubs. A three-section filter of this type is illustrated in Fig. 8.32. We shall consider a narrowband filter (fractional bandwidth of 2 percent) so that the quarter-wave sections of transmission line between the stubs may be treated as an ideal transformer with 90° phase shift throughout the passband. If the short-circuited stubs are a quarter wave long at the frequency ω_0, then, in the vicinity of ω_0, the normalized input admittance to the stub is [see Eq. (7.20)]

$$\bar{Y}_{in} = j\bar{Y}(\omega - \omega_0)\frac{l}{c} = j\bar{Y}\frac{\pi}{2}\frac{\omega - \omega_0}{\omega_0} \qquad (8.125)$$

where l is the stub length, $\bar{Y}$ is its normalized characteristic admittance, and $\omega_0 l/c = \pi/2$. The input admittance of a parallel tuned circuit is

$$Y_{in} = j\omega C + \frac{1}{j\omega L} = j\sqrt{\frac{C}{L}}\left(\frac{\omega}{\omega_0} - \frac{\omega_0}{\omega}\right)$$

$$\approx j2\sqrt{\frac{C}{L}}\frac{\omega - \omega_0}{\omega_0} = j2C(\omega - \omega_0) \qquad (8.126)$$

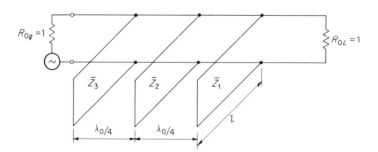

Fig. 8.32 Transmission-line filter using quarter-wave-coupled stubs.

Hence the short-circuited stub behaves as a parallel tuned circuit with $\sqrt{C/L} = \tilde{Y}\pi/4$ in the vicinity of the frequency for which it is a quarter wave long. These shunt stubs may be used for the parallel resonant circuits in the filter network of Fig. 8.29.

We shall specify that the filter terminations are both equal and also equal to the characteristic impedance of the main line. The characteristic impedances of the stubs will be denoted by $\bar{Z}_1, \bar{Z}_2$, etc., when normalized with respect to the terminating impedances. In the present case the impedance level throughout the filter is maintained constant, and we must have all $J_{n,n-1}$ equal to unity.

For the three-element filter we shall specify the following requirements:

$$f_1 = \frac{\omega_1}{2\pi} = 1{,}000 \text{ Mc} \qquad f_2 = \frac{\omega_2}{2\pi} = 1{,}020 \text{ Mc}$$

$$f_0 = \frac{\omega_0}{2\pi} = \frac{\sqrt{\omega_1 \omega_2}}{2\pi} = 1{,}009.95 \text{ Mc} \qquad k^2 = 0.0233$$

For a Chebyshev response the element values for the normalized low-pass prototype filter are given in Table 8.2 as $g_1 = 1.0315 = g_3$, $g_2 = 1.1474$. For the bandpass prototype filter of Fig. 8.24 the required element values are given by (8.111) and (8.112), and are

$$C_1' = \frac{g_1}{\omega_2 - \omega_1} = \frac{1.0315}{2\pi(2 \times 10^7)} = C_3' = 5.1575 \times 10^{-8} \times (2\pi)^{-1}$$

$$L_2' = \frac{g_2}{\omega_2 - \omega_1} = \frac{1.1474}{2\pi(2 \times 10^7)} = 5.737 \times 10^{-8} \times (2\pi)^{-1}$$

In addition, we require $L_k' C_k' = \omega_0^{-2}$ for all values of k. To find the required element values in the filter network of Fig. 8.29 we make use of (8.124). With $R_{0L} = R_L = 1$, we find from (8.124a) that $C_{01} = C_1'$ in order that $J_{10} = 1$. From (8.124c) we obtain $C_{02}C_{01} = L_2' C_1'$, or $C_{02} = L_2'$, since J_{21} must equal unity. Similarly, for $J_{32} = 1$, we obtain $C_{03}C_{02} = C_3' L_2'$, or $C_{03} = C_3'$, from (8.124b). We also require $C_{0k}L_{0k} = \omega_0^{-2}$. These relations permit us to calculate the ratios

$$\left(\frac{C_{0k}}{L_{0k}}\right)^{\frac{1}{2}} = \omega_0 C_{0k}$$

and we find that

$$\omega_0 C_{01} = \omega_0 C_1' = 5.1575 \times 10^{-8} \frac{\omega_0}{2\pi} = \omega_0 C_{03}$$

$$\omega_0 C_{02} = \omega_0 L_2' = 5.737 \times 10^{-8} \frac{\omega_0}{2\pi}$$

From (8.125) and (8.126) we find that the normalized admittances of the stubs are given by

$$\bar{Y}_k = \frac{4}{\pi}\sqrt{\frac{C_{0k}}{L_{0k}}} = \frac{4\omega_0}{\pi} C_{0k} = \frac{2c}{l} C_{0k} \tag{8.127}$$

Hence

$$\bar{Y}_1 = \frac{4}{\pi}\sqrt{\frac{C_{01}}{L_{01}}} = \frac{4}{\pi}\omega_0 C_{01} = \frac{4}{\pi}\frac{\omega_0 g_1}{\omega_2 - \omega_1} = \bar{Y}_3 = 66.5$$

$$\bar{Y}_2 = \frac{4}{\pi}\frac{\omega_0 g_2}{\omega_2 - \omega_1} = 73.6$$

The required normalized characteristic admittances are quite large and may be difficult to obtain in practice. If the stubs are loaded with dielectric, the admittance is increased by a factor $\kappa^{\frac{1}{2}}$, where κ is the dielectric constant. Use of dielectric filling will permit larger spacing between the conductors in the stubs. In addition, the stubs could be made $3\lambda_0/4$ long, which has the effect of making $\bar{Y}_{\text{in}}$ in (8.125) equal to $j\bar{Y}\frac{3\pi}{2}\frac{\omega - \omega_0}{\omega_0}$, an increase by a factor of 3. This will reduce the required stub admittances by a factor of 3. When the required values of the $\bar{Y}_k$ are known, the dimensions of the stubs can be determined and the mechanical design of the filter carried out.

For bandwidths greater than about 5 percent, the above procedure would not be very accurate because of the neglect of the frequency dependence of the quarter-wave admittance inverters. The frequency behavior of the shunt stubs can be approximated by that of a parallel LC circuit over a band much greater than 5 percent; so the main correction to be made is in the departure from 90° phase shift for the quarter-wave coupling sections. One method of incorporating the frequency sensitivity of the quarter-wave sections into the design is discussed in a paper by Matthaei, to which the interested reader is referred.†

8.20 Quarter-wave-coupled cavity filters

Quarter-wave-coupled cavity filters are similar to the filter discussed in the preceding section except that the transmission-line stubs are replaced by cavities. The filter is realized in practice by placing diaphragms in a waveguide. To understand the basis for design in the narrowband case, we must first consider the equivalent circuit of a section of waveguide loaded with two identical diaphragms a distance l_k apart, as shown in Fig. 8.33.

For a waveguide, the important frequency variable is not ω but rather $(\beta/k_0)\omega = \beta c$, since waveguide diaphragms have susceptances that vary very nearly as β or β^{-1} and the electrical length of a section of guide is

† G. L. Matthaei, Design of Wide-band and Narrow-band Band-pass Filters on the Insertion Loss Basis, *IRE Trans.*, vol. MTT-8, pp. 580–593, November, 1960.

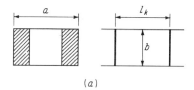

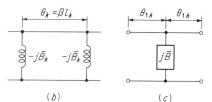

Fig. 8.33 (a) Rectangular waveguide loaded with two inductive diaphragms to form a cavity; (b) exact equivalent circuit; (c) approximate equivalent circuit

proportional to β. The normalized frequency variable $\omega/\omega_0 = \lambda_0/\lambda$ is therefore replaced by $\lambda_{g0}/\lambda_g = \beta/\beta_0$, where λ_{g0} is the guide wavelength at $\omega = \omega_0$ and λ_g is the corresponding value at any ω. Consequently, in all design formulas, we replace ω by βc, where c is the velocity of light.

The exact equivalent circuit for a waveguide loaded with two identical inductive diaphragms with normalized susceptance $-j\bar{B}_k$ is shown in Fig. 8.33b. For filter design according to the methods developed in preceding sections we must replace the exact equivalent circuit by an approximate shunt circuit. Mumford has shown that an equivalent circuit of the form illustrated in Fig. 8.33c has the same frequency characteristics as the exact equivalent circuit has over a narrow band of frequencies around ω_0.[†] The results obtained by Mumford are presented here without derivation. The derivation is straightforward, and may be found in Mumford's paper. The shunt susceptance $\bar{B}$ is expressed in the form

$$\bar{B} = \sqrt{\frac{C}{L}} \left(\frac{\beta}{\beta_0} - \frac{\beta_0}{\beta} \right) \approx 2 \sqrt{\frac{C}{L}} \frac{\Delta\beta}{\beta_0}$$

where $\Delta\beta = \beta - \beta_0$ is small. When a resonant circuit of this type is connected across a transmission line, it is loaded by a shunt conductance of normalized value unity on each side. The loaded Q of the circuit is thus

$$Q_k = \tfrac{1}{2}(\beta_0 c)C = \frac{1}{2}\sqrt{\frac{C}{L}} \tag{8.128}$$

since $\beta_0 c = (LC)^{-\frac{1}{2}}$. Hence we may express $\bar{B}$ in terms of the loaded Q; thus

$$\bar{B} = 4Q_k \frac{\Delta\beta}{\beta_0} \tag{8.129}$$

[†] W. W. Mumford, Maximally Flat Filters in Waveguides, *Bell System Tech. J.*, vol. 27, pp. 684–714, October, 1948.

The value obtained for Q_k by Mumford is

$$Q_k = \frac{\tan^{-1}(2/\bar{B}_k)}{2\sin^{-1}\dfrac{2}{(\bar{B}_k^4 + 4\bar{B}_k^2)^{\frac{1}{2}}}}$$

$$\approx \frac{(\bar{B}_k^4 + 4\bar{B}_k^2)^{\frac{1}{2}}}{4}\tan^{-1}\frac{2}{\bar{B}_k} \qquad (8.130)$$

since $\bar{B}_k$ is large compared with unity for a narrowband (high-Q) filter. The required diaphragm spacing l_k to give perfect transmission through the cavity at $\omega = \omega_0$ is given by

$$\tan \beta_0 l_k = -\frac{2}{\bar{B}_k} \qquad (8.131)$$

The two sections of line with electrical length θ_{1k} in the circuit of Fig. 8.33c are to be chosen so that

$$\beta_0 l_k + 2\theta_{1k} = \theta_k + 2\theta_{1k} = \pi \qquad (8.132)$$

at the frequency ω_0. These additional lengths of line in the equivalent circuit of a single cavity are absorbed into and made part of the quarter-wave coupling lines in the filter.

The design of maximally flat and Chebyshev filters with N odd is straightforward. If the prototype circuit of Fig. 8.29 is used, it is only necessary to make

$$Q_k = \frac{1}{2}\sqrt{\frac{C_{0k}}{L_{0k}}} \qquad (8.133)$$

and to choose C_{0k}, L_{0k} so that $C_{0k}L_{0k} = (\beta_0 c)^{-2}$ and all $J_{k,k-1}$ equal unity. The procedure is the same as that discussed in Sec. 8.19, dealing with the transmission-line filter. The section of waveguide between cavity k and $k+1$ has an electrical length equal to $\pi/2$. Since this includes θ_{1k+1} and θ_{1k} from the adjacent cavities, the physical length of the quarter-wave coupling line between cavities k and $k+1$ will be

$$l_{k,k+1} = \frac{1}{\beta_0}\left(\frac{\pi}{2} - \theta_{1k} - \theta_{1k+1}\right)$$

$$= \frac{\lambda_{g0}}{2\pi}\left(\frac{\theta_k + \theta_{k+1}}{2} - \frac{\pi}{2}\right) = \frac{l_k + l_{k+1}}{2} - \frac{\lambda_{g0}}{4} \qquad (8.134)$$

upon using (8.132). A schematic illustration of the filter is given in Fig. 8.34. Formulas for the required diaphragm dimensions to yield the specified value of $\bar{B}_k$ are given in Sec. 5.5. The power loss ratio for the filter is obtained by replacing ω/ω_0 by β/β_0. For a Chebyshev filter it is given by [see (8.104) and (8.109)]

$$P_{LR} = 1 + k^2 T_N{}^2\left[\frac{\beta_0}{\beta_2 - \beta_1}\left(\frac{\beta}{\beta_0} - \frac{\beta_0}{\beta}\right)\right] \qquad (8.135)$$

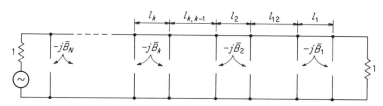

Fig. 8.34 Quarter-wave-coupled waveguide-cavity filter.

where β_2 and β_1 are the values of β at the edges of the passband. If β_1 and β_2 are specified, then

$$\beta_0 = \sqrt{\beta_1 \beta_2} \tag{8.136}$$

It should also be noted that for the waveguide filter, ω_0, ω_1, and ω_2 in the design formulas (8.111) and (8.112) must be replaced by $\beta_0 c$, $\beta_1 c$, and $\beta_2 c$, where c is the velocity of light; i.e., replace $k_0 = \omega/c$ by β. To illustrate the procedure we shall evaluate the required susceptance $-j\bar{B}_1$ for the first cavity in the filter shown in Fig. 8.34.

Assume that a five-element filter will be needed. The response is to be of the Chebyshev type with a passband tolerance $k^2 = 0.0233$. The waveguide to be used has a width of $a = 0.9$ in. The passband is to extend from $f_1 = 10{,}000$ Mc to $f_2 = 10{,}400$ Mc. The corresponding values of $k_0 = \omega/c$ are 2.1 and 2.18 rad/cm. The values of β_1 and β_2 are thus

$$\beta_1 = \left[(2.1)^2 - \left(\frac{\pi}{a}\right)^2\right]^{\frac{1}{2}} = (4.4 - 1.89)^{\frac{1}{2}} = 1.59$$

$$\beta_2 = [(2.18)^2 - 1.89]^{\frac{1}{2}} = 1.7$$

The center of the band occurs at $\beta_0 = (\beta_1 \beta_2)^{\frac{1}{2}} = 1.64$, which gives $f_0 = 10{,}200$ Mc. From Table 8.2 we find $g_1 = 1.1468$. Using (8.112b) gives

$$C_1' = \frac{\beta_0}{\beta_2 - \beta_1} \frac{C_1}{\beta_0 c} = \frac{\beta_0}{\beta_2 - \beta_1} \frac{g_1}{\beta_0 c}$$

Since J_{10} is to equal unity and $R_L = R_{0L} = 1$, we have $C_{01} = C_1'$ from (8.124a). Using (8.133), we obtain

$$Q_1 = \frac{1}{2}\sqrt{\frac{C_{01}}{L_{01}}} = \tfrac{1}{2}(\beta_0 c) C_{01} = \frac{\beta_0}{\beta_2 - \beta_1} \frac{g_1}{2}$$

For the kth resonator we should obtain

$$Q_k = \frac{\beta_0}{\beta_2 - \beta_1} \frac{g_k}{2} \tag{8.137}$$

For Q_1 we obtain 8.56, and from this result we can determine $\bar{B}_1$ by using (8.130). For $\bar{B}_k$ large, we can replace $\tan^{-1}(2/\bar{B}_k)$ by $2/\bar{B}_k$, and we then

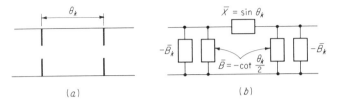

Fig. 8.35 A waveguide cavity and its equivalent circuit.

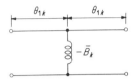

Fig. 8.36 An impedance inverter.

find that

$$\bar{B}_k = 2(Q_k^2 - 1)^{\frac{1}{2}} \tag{8.138}$$

Thus we find that $\bar{B}_1 = 17$. The required diaphragm dimensions can now be determined, and also the cavity length from (8.131). The above procedure has to be repeated for each cavity in the filter chain.

8.21 Direct-coupled cavity filters

Direct-coupled cavity filters have the advantage that the physical structure is more compact than the corresponding quarter-wave-coupled cavity filter. A design procedure for direct-coupled cavity filters that is accurate for bandwidths up to 20 percent has been developed by Cohn.[†] Cohn's design method is based on the use of the network in Fig. 8.26 as a prototype. The design formulas will be presented here without the detailed derivation.

The waveguide cavity and its equivalent circuit shown in Fig. 8.33a and b may also be represented by a Π network shunted with inductive susceptances at each end, as in Fig. 8.35. The two shunt susceptances $\bar{B} = -\cot(\theta_k/2)$ may be neglected compared with $\bar{B}_k$ since $\bar{B}_k$ will be large, and θ_k is nearly equal to π; so $\bar{B}$ is small compared with unity. The series arm $\bar{X}$ is thus used as the series resonant circuit in the prototype filter.

For impedance inverters Cohn uses the shunt inductive reactance plus two short sections of waveguide (equivalent transmission lines), as shown in Fig. 8.36. For this circuit the impedance inverting properties

[†] S. B. Cohn, Direct Coupled Resonator Filters, *Proc. IRE*, vol. 45, pp. 187–196, February, 1957.

are obtained if

$$\theta_{1k} = -\tfrac{1}{2}\tan^{-1}\frac{2}{\bar{B}_k} \tag{8.139a}$$

$$\bar{B}_k = \frac{1 - K^2}{K} \tag{8.139b}$$

where K is the characteristic impedance of the quarter-wave impedance inverter. With θ_{1k} and $\bar{B}_k$ determined at a frequency ω_0, it is found that the inverter does not depart appreciably from its ideal characteristics over a 20 percent band.

In the vicinity of $\omega = \omega_0$, where $\theta_k = \pi$, the series reactance $\bar{X}$ behaves as

$$\bar{X} = \sin\theta_k = \sin(\theta_k - \pi + \pi)$$

$$\approx -(\theta_k - \pi) = -(\beta - \beta_0)l = \frac{\beta_0 - \beta}{\beta_0}\pi \approx -\frac{\pi}{2}\left(\frac{\beta}{\beta_0} - \frac{\beta_0}{\beta}\right)$$

where $\beta_0 l = \pi$. This frequency behavior is similar (apart from the sign, which is immaterial) to that for a series resonant circuit for which $X = \sqrt{L/C}\,(\omega/\omega_0 - \omega_0/\omega) \approx 2\sqrt{L/C}\,(\omega - \omega_0)/\omega_0$ if ω/ω_0 is replaced by the new frequency variable β/β_0.

When the negative line lengths of the impedance inverters are absorbed as part of the cavity length, the physical length of the kth cavity becomes

$$l_k = \frac{\lambda_{g0}}{2} + \frac{\lambda_{g0}}{2\pi}(\theta_{1k} + \theta_{1k+1}) \tag{8.140}$$

In the prototype circuit of Fig. 8.26 we must choose all $\sqrt{L_{0k}/C_{0k}}$ equal to $\pi/2$ to obtain a correspondence with the type of series resonant circuit employed here. In addition, we choose

$$C_{0k}L_{0k} = (\beta_0 c)^{-2} \tag{8.141}$$

The impedance-inverter parameters as given by (8.117), (8.121), (8.122), and (8.123) thus become known in terms of the C'_k and L'_k, which are related to the g_k in the low-pass prototype. From the known values of the $Z_{k+1,k}$ the shunt susceptances $\bar{B}_k$ may be found. The filter is illustrated schematically in Fig. 8.37. The design formulas obtained as out-

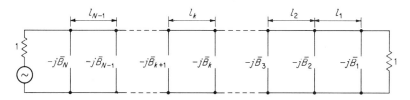

Fig. 8.37 A direct-coupled waveguide-cavity filter.

lined above are

$$\bar{B}_1 = \frac{1 - w/g_1}{\sqrt{w/g_1}} \tag{8.142a}$$

$$\bar{B}_2 = \frac{1}{w}\left(1 - \frac{w^2}{g_1 g_2}\right)\sqrt{g_1 g_2} \tag{8.142b}$$

$$\bar{B}_k = \frac{1}{w}\left(1 - \frac{w^2}{g_k g_{k-1}}\right)\sqrt{g_k g_{k-1}} \tag{8.142c}$$

$$\bar{B}_N = \frac{1 - wR/g_N}{\sqrt{wR/g_N}} \qquad R = 1 \text{ for } N \text{ odd} \tag{8.142d}$$

where $w = \dfrac{\pi}{2}\dfrac{\beta_2 - \beta_1}{\beta_0}$ and the g_k are the element values from the low-pass prototype filter. Note that $R = 1$ for N odd, and also for N even in the case of maximally flat filters; otherwise R is given by (8.101b). The length of the kth cavity at $\beta = \beta_0$ is

$$l_k = \frac{\lambda_{g0}}{2} - \frac{\lambda_{g0}}{4\pi}\left(\tan^{-1}\frac{2}{\bar{B}_{k+1}} + \tan^{-1}\frac{2}{\bar{B}_k}\right) \tag{8.143}$$

The power loss ratio is given by substituting $\dfrac{\beta_0}{\beta_2 - \beta_1}\left(\dfrac{\beta}{\beta_0} - \dfrac{\beta_0}{\beta}\right)$ for ω' in the low-pass prototype filter response.† Note also that $\beta_0^2 = \beta_1\beta_2$.

We shall terminate the discussion on filters at this point even though we have not covered a number of other important topics, such as the phase characteristics and related signal delay and distortion problems and the effect of losses on filter characteristics. For a discussion of these topics the references given at the end of this chapter should be consulted.

Problems

8.1 Find the 𝒶ℬ𝒸𝒟 matrix for the following networks: (a) a shunt susceptance $j\bar{B}$, (b) a series reactance $j\bar{X}$, (c) a shunt reactance $j\bar{X}_1$ followed by a series reactance $j\bar{X}_2$.

8.2 Derive the relations (8.14) and (8.15).

8.3 Consider a T network terminated in a load Z. Evaluate the input impedance Z_{in} and show that the condition that Z transforms into Z_{in}, that is, $Z_{\text{in}} = Z$, leads to the characteristic values $Z_B^{\pm}$ for the periodic structure.

8.4 Show that the eigenvalue equation for the propagation constant of a Bloch wave on a transmission line loaded at intervals d with a series reactance $j\bar{X}$ is $\cosh \gamma d = \cos k_0 d - (\bar{X}/2) \sin k_0 d$.

8.5 Show that (8.21) may be expressed in the form

$$\cosh \gamma d = \frac{\cos \phi}{|S_{12}|}$$

† Cohn uses a somewhat different frequency variable, which, however, for the bandwidths considered, is essentially the same as we have used.

where ϕ is the phase angle of S_{12}, and S_{12} is the scattering-matrix off-diagonal element for the unit cell (Sec. 4.8).

8.6 Express Γ_B in terms of the A_{ij} by noting the similarity between (8.20) and (8.6) and that between Γ_B and $\bar{Z}_B^{-1}$.

8.7 Show that the wave-amplitude transmission matrix for a shunt susceptance $j\bar{B}$ is

$$[A] = \begin{bmatrix} \dfrac{2+j\bar{B}}{2} & j\dfrac{\bar{B}}{2} \\ -j\dfrac{\bar{B}}{2} & \dfrac{4+\bar{B}^2}{2(2+j\bar{B})} \end{bmatrix}$$

8.8 A load $\bar{Z}_L$ on an ordinary transmission line gives a reflection coefficient $\Gamma'_L = (\bar{Z}_L - 1)/(\bar{Z}_L + 1)$. Show that (8.30), giving Γ_L for a Bloch wave, may be expressed as

$$\Gamma_L = \frac{1 + \Gamma_B^-}{1 + \Gamma_B^+} \frac{\Gamma_B^+ - \Gamma'_L}{\Gamma'_L - \Gamma_B^-}$$

where $\Gamma_B^\pm$ are the characteristic reflection coefficients of the component waves making up the Bloch wave.

8.9 For Bloch waves in the capacitively loaded coaxial line, show that the TEM voltage waves between any two consecutive diaphragms are given by

$$V = V_n^+ e^{-jk_0(z-nd)} + V_n^- e^{jk_0(z-nd)} = V_n^+ e^{-jk_0(z-nd)} + \Gamma_B^+ V_n^+ e^{jk_0(z-nd)}$$

for the Bloch wave propagating in the $+z$ direction, and

$$V = V_n^+ e^{-jk_0(z-nd)} + \Gamma_B^- V_n^+ e^{jk_0(z-nd)}$$

for the Bloch wave propagating in the $-z$ direction. The 0'th terminal plane has been chosen as $z = 0$ and $V_B^+ = V_n^+(1 + \Gamma_B^+)$, $V_B^- = V_n^+(1 + \Gamma_B^-)$.

8.10 Consider an infinite transmission line loaded with shunt capacitive susceptances $j\bar{B}$ at $z = nd$, $n = -\infty$ to ∞. Show that the current and voltage waves that make up a Bloch wave are given by

$$V_B = V^+ e^{-jk_0 z} + V^- e^{jk_0 z}$$
$$I_B = I^+ e^{-jk_0 z} + I^- e^{jk_0 z} = V^+ e^{-jk_0 z} - V^- e^{jk_0 z}$$

where $V^- = -V^+(1 - e^{-j\theta+j\beta d})/(1 - e^{j\theta+j\beta d})$ and $\theta = k_0 d$. Let $V_B(z)$ equal $V_p(z)e^{-j\beta z}$ and expand $V_p(z)$ into an infinite series of spatial harmonics. Note that the relation between V^+ and V^- may be obtained by using the condition that $V_B(z = d) = e^{-j\beta d} V_B(0)$ and that β is given by (8.10a).

8.11 For the sheath helix show that the eigenvalue equation for the nth mode is

$$\frac{(h^2 a^2 + n\beta a \cot \psi)^2}{(k_0 ha^2 \cot \psi)^2} = \frac{K'_n(ha) I'_n(ha)}{K_n(ha) I_n(ha)}$$

8.12 Consider an N-section filter made up of a capacitively loaded coaxial line of N unit cells. The filter is terminated in a resistive load $\bar{R}$ equal to the image impedance at zero frequency, i.e., equal to 1. The generator at the input has an internal resistance equal to $\bar{R}$. Show that the power delivered to the load is given by

$$P = \frac{V_g^2}{|\bar{R} + \bar{Z}_{\text{in}}|^2} \operatorname{Re} \bar{Z}_{\text{in}}$$

where V_g is the generator voltage and $\bar{Z}_{\text{in}}$ is given by

$$\bar{Z}_{\text{in}} = \bar{Z}_i \frac{\bar{R} + \bar{Z}_i \tanh \gamma Nd}{\bar{Z}_i + \bar{R} \tanh \gamma Nd}$$

and $\bar{Z}_i$, γd are the image parameters at any frequency. In the passband where Z_i is real and $\tanh \gamma d = j \tan \beta d$, verify that

$$P = \frac{V_g^2 Z_i^2 R(1+t^2)}{4R^2 Z_i^2 + (R^2 + Z_i^2)t^2} = \frac{V_g^2(1+t^2)}{4R(1+t^2) + R\left(\dfrac{R^2 - Z_i^2}{RZ_i}\right)^2 t^2} \qquad t = \tan N\beta d$$

Thus show that the power loss ratio becomes

$$P_{\text{LR}} = \frac{V_g^2/4R}{P} = 1 + \left(\frac{R^2 - Z_i^2}{2RZ_i}\right)^2 \sin^2 N\beta d$$

Plot P_{LR} as a function of ω in the passband for the case where $\bar{B} = 10\omega$. See (8.10a) and (8.16) for expressions giving $\bar{Z}_i = \bar{Z}_B$ and βd. Verify that in the stopband the power loss ratio is given by

$$P_{\text{LR}} = 1 + \frac{1}{4}\left|\frac{R}{Z_i} - \frac{Z_i}{R}\right|^2 \sinh^2 N\alpha d$$

8.13 Design a three-cavity quarter-wave-coupled filter with the following specifications: waveguide width $a = 0.9$ in., band edges at $f_1 = 10{,}000$ Mc, $f_2 = 10{,}400$ Mc, passband tolerance $k^2 = 0.0233$, Chebyshev response. Inductive diaphragms with circular holes are to be used. Determine the hole radii and diaphragm spacings.

8.14 Design a four-cavity direct-coupled cavity filter having Chebyshev response. The passband tolerance is $k^2 = 0.0233$, band edges occur at $f_1 = 9{,}500$ Mc, $f_2 = 10{,}500$ Mc, and the guide width is 0.9 in. Specify the diaphragm dimensions and spacing. Use any convenient inductive diaphragm.

8.15 Design a four-cavity direct-coupled maximally flat waveguide filter with the specifications given in Prob. 8.14. Note that the maximally flat filter designed from the low-pass prototype has a passband tolerance of 1. To obtain a passband tolerance of k^2 between β_1 and β_2, the design must be carried out for a wider bandwidth, say β_1' to β_2'. Thus we should have

$$\left[\frac{\beta_0}{\beta_2' - \beta_1'}\left(\frac{\beta}{\beta_0} - \frac{\beta_0}{\beta}\right)\right]^{10} = 1 \qquad \text{for } \beta = \beta_1', \beta_2'$$

Also $\beta_1'\beta_2' = \beta_0^2$. Determine β_2' and β_1' so that

$$\left[\frac{\beta_0}{\beta_2' - \beta_1'}\left(\frac{\beta}{\beta_0} - \frac{\beta_0}{\beta}\right)\right]^{10} = k^2 \qquad \text{when } \beta = \beta_1 \text{ and } \beta_2$$

If the design is carried out using these values of β_1' and β_2', the required passband tolerance of k^2 will be maintained in the passband between β_1 and β_2. Show that, in general, $\beta_1'\beta_2' = \beta_1\beta_2 = \beta_0^2$ and $\beta_2' - \beta_1' = (\beta_2 - \beta_1)k^{-1/N}$.

8.16 For the circuits in Fig. 8.33b and c show that the normalized image impedances are given by

$$(1 - \bar{B}_k^2 - 2\bar{B}_k \cot \theta_k)^{-\frac{1}{2}} \qquad \text{and} \qquad \left(\frac{2\sin^2 \theta_{1k} + \cos 2\theta_{1k} - \bar{B}}{2\sin 2\theta_{1k} + \cos 2\theta_{1k} + \bar{B}}\right)^{\frac{1}{2}}$$

When $\omega = \omega_0$, $\bar{B} = 0$, show that $\bar{B}_k$ must be related to θ_k by (8.131) to make Z_i equal to unity. Show that the image phase constants ϕ for the two circuits are given by

$$\cos \phi = \cos \theta_k + \bar{B}_k \sin \theta_k \quad \text{and} \quad \cos \phi = \cos 2\theta_{1k} - \frac{\bar{B}}{2} \sin 2\theta_{1k}$$

and will be equal at ω_0 if $\theta_k + 2\theta_{1k} = \pi$.

References

Periodic structures

1. Brillouin, L.: "Wave Propagation in Periodic Structures," 2d ed., Dover Publications, Inc., New York, 1953.
2. Slater, J. C.: "Microwave Electronics," D. Van Nostrand Company, Inc., Princeton, N.J., 1950.
3. Watkins, D. A.: "Topics in Electromagnetic Theory," John Wiley & Sons, Inc., New York, 1958.
4. Collin, R. E.: "Field Theory of Guided Waves," McGraw-Hill Book Company, New York, 1960.
5. Bevensee, R. M.: "Electromagnetic Slow Wave Systems," John Wiley & Sons, Inc., New York, 1964.

Microwave filters

The following references supplement those given in the text:

6. Riblet, H. J.: Synthesis of Narrow Band Direct Coupled Filters, *Proc. IRE*, vol. 40, pp. 1219–1223, October, 1952.
7. Jones, E. M. T.: Synthesis of Wide Band Microwave Filters to Have Prescribed Insertion Loss, *IRE Conv. Record*, pt. 5, pp. 119–128, 1956.
8. Ozaki, H., and J. Ishii: Synthesis of a Class of Strip Line Filters, *IRE Trans.*, vol. CT-5, pp. 104–109, June, 1958.
9. Ozaki, H., and J. Ishii: Synthesis of Transmission Line Networks and the Design of UHF Filters, *IRE Trans.*, vol. CT-2, pp. 325–336, December, 1955.
10. Young, L.: The Quarter-wave Transformer Prototype Circuit, *IRE Trans.*, vol. MTT-8, pp. 483–489, September, 1960.
11. Ghose, R. N.: "Microwave Circuit Theory and Analysis," McGraw-Hill Book Company, New York, 1963.
12. Riblet, H. J.: A Unified Discussion of High Q Waveguide Filter Design Theory, *IRE Trans.*, vol. MTT-6, pp. 359–368, October, 1958.
13. Jones, E. M. T., and J. T. Bolljahn: Coupled Strip Transmission Line Filters and Directional Couplers, *IRE Trans.*, vol. MTT-4, pp. 75–81, April, 1956.
14. Bradley, E. H.: Design and Development of Strip Line Filters, *IRE Trans.*, vol. MTT-4, pp. 86–93, April, 1956.
15. Wenzel, R. J.: Exact Design of TEM Microwave Networks Using Quarter-wave Lines, *IEEE Trans.*, vol. MTT-12, pp. 94–111, June, 1964.
16. Matthaei, G. L., L. Young, and E. M. T. Jones: "Microwave Filters, Impedance-matching Networks, and Coupling Structures," McGraw-Hill Book Company, New York, 1964.

9
Microwave tubes

9.1 Introduction

Conventional low-frequency tubes, such as triodes, fail to operate at microwave frequencies because the electron transit time from the cathode to the grid becomes an appreciable fraction of the period of the sinusoidal signal to be amplified. In other words, propagation times become significant, and the same limitations that are inherent in low-frequency circuits are present in low-frequency tubes also. Microwave tubes must be designed to utilize the wave-propagation phenomena to best advantage.

Broadly speaking, there are two basic types of microwave tubes, those that employ electromagnetic cavities (klystrons and some magnetrons) and those that employ slow-wave circuits (traveling-wave tubes). Both types of tubes utilize an electron beam on which space-charge waves and cyclotron waves can be excited. The space-charge waves are primarily longitudinal oscillations of the electrons and interact with the electromagnetic fields in cavities and slow-wave circuits to give amplification. The properties of cavities and slow-wave circuits have already been discussed. What remains to be done is to examine the propagation of space-charge waves on electron beams and then to consider the interactions that take place between electron beams and the fields in cavities and slow-wave circuits.

The purpose of this chapter is to examine the nature of electron beams and the space-charge waves that they can support. In addition, the interaction of the beam with a microwave cavity or slow-wave circuit is to be studied in order to explain the operating principles of a number of different microwave tubes. Space does not permit a detailed treatment of the many different varieties of microwave tubes in existence. We shall concentrate on fundamentals that form, more or less, the basic operating principles of all microwave tubes.

Two approaches may be used in analyzing the dynamic behavior of the electron beam. The earliest approach used was the ballistic, or lagrangian, approach. In this method the motion of an individual electron is studied in detail, and it is assumed that all other electrons behave in a similar way. The ballistic approach has the advantage of permitting certain nonlinear, or large-signal, effects to be treated fairly easily.

The other approach is the field approach, sometimes called the eulerian, or hydrodynamical, approach. In this method the electron beam is essentially treated as a charged fluid. Field variables that describe the velocity, charge density, a-c current, etc., at an arbitrary point as a function of time are introduced. However, no attempt is made to follow the motion of a single electron. The field approach, which leads to the space-charge waves, is more unifying and lends itself to the treatment of all different types of microwave tubes within the same general mathematical framework. Therefore only the field approach is used in this text.

An exact analysis of a microwave tube would be very difficult and laborious to carry out. As in any other physical problem, it is necessary to introduce a number of simplifying assumptions in order to arrive at a mathematical model that can be analyzed without too many complications. The success of a simplified theory must then be judged by the extent to which it predicts and agrees with experimental results.

The first few sections of this chapter discuss a number of models used for the electron beam and the propagation of space-charge waves on these beams. The governing equations are Maxwell's equations and Newton's laws, together with the Lorentz force equation. The equation of motion for a charge element is a nonlinear equation, but may be linearized by assuming small-signal conditions; i.e., all a-c quantities are small compared with d-c quantities. We shall consider only the small-signal situation since this will suffice to develop the operating principles of microwave tubes. Large-signal analysis is a great deal more difficult, and the theory, in general, is not fully developed.

After treating the dynamics of the electron beam, the klystron and traveling-wave tube are examined in detail. A number of other tubes are also discussed, but in a more qualitative way.

9.2 Electron beams with d-c conditions

By means of a suitable electron gun consisting of a cathode, accelerating electrodes, and focusing electrodes, a beam of electrons with essentially a uniform velocity v_0 can be produced.† Figure 9.1 is a schematic illus-

---

† The design of electron guns is not treated in this text. For a discussion of these see:
    J. R. Pierce, "Theory and Design of Electron Beams," D. Van Nostrand Company, Inc., Princeton, N.J., 1950.
    K. R. Spangenberg, "Vacuum Tubes," McGraw-Hill Book Company, New York, 1948.

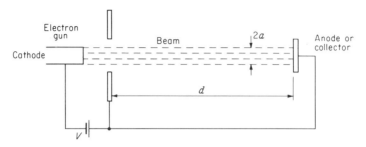

Fig. 9.1 A cylindrical electron beam.

tration of a cylindrical electron beam with a radius a. If the potential difference through which the electron is accelerated is V, the velocity v_0 is given by

$$v_0 = \left(\frac{2Ve}{m}\right)^{1/2} = 5.93 \times 10^5 V^{1/2} \quad \text{m/sec} \tag{9.1}$$

where $-e$ is the electron charge and m is the mass of the electron. For $V = 1,000$ volts, $v_0 = 1.87 \times 10^7$ m/sec $= 0.0625c$. The beam perveance is defined by the quantity $IV^{-3/2}$, where I is the total beam current.

The coulomb repulsive force, or d-c space-charge force, will tend to cause the electron beam to disperse, i.e., cause outward radial motion of the electrons. The space-charge force will be proportional to the density of the beam, i.e., to the number of electrons per unit volume. For the usual density of beams employed in microwave tubes (10^{12} to 10^{15} electrons per cubic meter), the dispersion of the beam due to space-charge forces is negligible if the drift space is short (d is small in Fig. 9.1). This condition exists in many klystrons, but in traveling-wave tubes the beam must travel over distances which are so long that considerable dispersion may take place unless some means of keeping the beam together or focused is employed. The means by which the d-c space-charge forces are counteracted leads to three commonly used beam models. These models are discussed below.

Ion-neutralized beam

Even with the high vacuum employed in a microwave tube, a great many neutral gas particles are still present. Many of these gas molecules become ionized by means of collisions with the relatively high energy electrons. The presence of positive ions will tend to neutralize the negative space charge of the electron beam. The positive ions, however, need not be considered in the interaction of a high-frequency electromagnetic field with the beam because their mass is at least 1,800 times greater than the electron's mass, and hence the a-c motion of the ions is negligible by comparison with that of the electrons.

Although all electron beams are ion-neutralized to some extent, complete electron space-charge neutralization is rarely achieved. However, for the purpose of mathematical analysis, a completely ion-neutralized electron beam is sometimes postulated as a model. Beam spreading due to space-charge forces is discussed in Spangenberg's book.

Beam with axially confined flow

If a very large static magnetic field B_0 in the direction of the beam velocity is applied, the effect is to constrain the electrons from moving in the radial direction. The space-charge forces tend to impart a radial velocity to the electron. The magnetic field B_0 produces a force $-e\mathbf{v}_r \times \mathbf{B}_0$, which causes the electrons to execute circular motion about the magnetic field lines and thus prevents the beam from dispersing in the radial direction.

In the magnetically focused beam the field B_0 has its flux lines threading through the cathode surface, as in Fig. 9.2a. Some electron diffusion across the magnetic field lines will occur, but if B_0 is made large enough, the amount of beam dispersion can be kept small.

For the purpose of mathematical analysis it is convenient to assume that B_0 is made infinite since in this case no electron motion in a transverse direction can take place. The analysis of the behavior of the beam under a-c conditions is thereby greatly simplified since electron motion can now occur only in the axial direction (one-dimensional motion). The

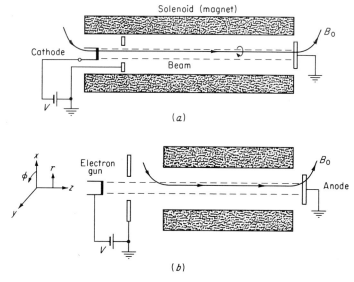

Fig. 9.2 (a) Magnetic focusing for axially confined flow; (b) magnetic focusing for Brillouin flow.

axially confined flow model is commonly used in the treatment of traveling-wave tubes.

Brillouin flow

In Brillouin flow (to be described), the axial magnetic field B_0 is not permitted to thread through the cathode surface. Since the field lines are continuous, they must move away from the beam region in the radial direction near the cathode, as shown in Fig. 9.2b. When the beam enters the magnetic field region, it is given a uniform rotation at the Larmor frequency $\omega_l = eB_0/2m$ by the magnetic field. In cylindrical coordinates r, ϕ, z, the equation of motion for an electron,

$$m \frac{d\mathbf{v}}{dt} = -e(\mathbf{E} + \mathbf{v} \times \mathbf{B})$$

may be written in component form as

$$\frac{d^2 r}{dt^2} - r\left(\frac{d\phi}{dt}\right)^2 = -\frac{e}{m}\left(E_r + rB_0 \frac{d\phi}{dt}\right) \quad (9.2a)$$

$$r \frac{d^2\phi}{dt^2} + 2 \frac{dr}{dt}\frac{d\phi}{dt} = \frac{e}{m} B_0 \frac{dr}{dt} \quad (9.2b)$$

$$\frac{d^2 z}{dt^2} = 0 \qquad \frac{dz}{dt} = v_0 \quad (9.2c)$$

in the region where $\mathbf{B} = B_0 \mathbf{a}_z$ and is uniform. It is assumed that $E_\phi = E_z = 0$. The radial electric field may be found by using Gauss' law. If the d-c charge density of the beam is $-\rho_0$, then $2\pi r D_r = -\pi r^2 \rho_0$, or $E_r = -r\rho_0/2\epsilon_0$. The radial space-charge force on an electron is thus $-eE_r = re\rho_0/2\epsilon_0$. If $d^2\phi/dt^2 = 0$, we find from (9.2b) that

$$\frac{d\phi}{dt} = \omega_l = \frac{eB_0}{2m} \quad (9.3)$$

If this solution is to satisfy (9.2a) and also make $d^2 r/dt^2$ vanish, we require

$$r\omega_l^2 = \frac{e}{m}\left(-\frac{r\rho_0}{2\epsilon_0} + rB_0 \omega_l\right)$$

or

$$\omega_l^2 = \frac{e\rho_0}{2m\epsilon_0} = \frac{\omega_p^2}{2} \quad (9.4)$$

where $\omega_p = (e\rho_0/m\epsilon_0)^{1/2}$ is called the plasma frequency. Typical values of ω_p for beams used in microwave tubes range from 10^7 to 10^9. If the focusing field B_0 is chosen to satisfy (9.4), there will be no radial acceleration of the electrons. The equilibrium condition in the radial direction is actually a balance of outward radial forces $-eE_r$ due to space charge and $m\omega_l^2 r$ due to centrifugal acceleration against the inward magnetic radial force $e\omega_l r B_0$. Electron-beam flow under these conditions is referred to as Brillouin flow.

Although we have given the conditions for steady-state Brillouin flow within the uniform $\mathbf{B}_0$ field region, we did not show that a beam leaving a cathode with a velocity $v_0\mathbf{a}_z$ will assume Brillouin-flow characteristics as it enters into the uniform $\mathbf{B}_0$ field region through the nonuniform field region in front of the cathode. To show this requires demonstrating that the change in angular momentum of the beam from an initial value of zero to its final value for Brillouin flow is equal to the time integral of the torque $ev_0 B_{0r}$ produced by the radial magnetic field component in the nonuniform region. The reader is referred to Brillouin's original paper for the derivation.†

The conditions required for Brillouin flow can be achieved in practice. Even if the beam is partially ion-neutralized, as long as ρ_0 is not zero, a value for B_0 such that (9.4) holds can be found. However, the behavior of a beam with Brillouin flow under a-c conditions is more difficult to treat since transverse motion of the electrons is permitted. For this reason the ideal axially confined flow model is more commonly used.

In the magnetron-type (M-type) traveling-wave tube a planar sheet beam is used. For this type of beam an analogous flow, referred to as planar Brillouin flow, can take place. The properties of sheet beams are discussed in Sec. 9.11, dealing with M-type traveling-wave tubes, and hence are not covered in this section.

9.3 Space-charge waves on beams with confined flow

This section is devoted to an analysis of space-charge waves on an axially confined electron beam inside a cylindrical waveguide of radius b. The radius of the beam is a, as in Fig. 9.3. Small-signal conditions are assumed.

The beam is considered to be uniform in density in a cross-sectional plane. The d-c charge density is $-\rho_0$, and the axial velocity is v_0. The d-c current density in the z direction is $J_0 = -\rho_0 v_0$. The d-c parameters ρ_0, v_0, and J_0 are independent of space and time coordinates. Under a-c

† L. Brillouin, A Theorem of Larmor and Its Importance for Electrons in Magnetic Fields, *Phys. Rev.*, vol. 67, p. 260, 1945.
W. G. Dow, Nonuniform D.C. Electron Flow in Magnetically Focused Cylindrical Beams, *Advan. Electron. Electron Phys.*, vol. 10, 1958.
Pierce, *op. cit.*

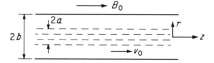

Fig. 9.3 Electron beam inside a cylindrical waveguide.

conditions with time dependence $e^{j\omega t}$, there will be a-c components of charge density, velocity, and current that vary with time and the spatial coordinates. These a-c components are denoted by ρ, $\mathbf{v}$, and $\mathbf{J}$ without the time factor $e^{j\omega t}$. The a-c fluctuation in electron density from the d-c value N will be denoted by n.

The electromagnetic field satisfies the equations

$$\nabla \times \mathbf{E} = -j\omega\mu_0\mathbf{H} \tag{9.5a}$$

$$\nabla \times \mathbf{H} = j\omega\epsilon_0\mathbf{E} + \mathbf{J} \tag{9.5b}$$

$$\nabla \cdot \mathbf{E} = \frac{\rho}{\epsilon_0} \tag{9.5c}$$

$$\nabla \cdot \mathbf{B} = 0 \tag{9.5d}$$

$$\nabla \cdot \mathbf{J} = -\frac{\partial \rho}{\partial t} \tag{9.5e}$$

A unit volume of the beam with charge density $-\rho_0 + \rho$ and charge-mass ratio η equal to that for electrons, that is, $\eta = e/m$, has a motion governed by the equation

$$(N + n)m\frac{d\mathbf{v}_t}{dt} = (\rho - \rho_0)(\mathbf{E} + \mathbf{v}_t \times \mathbf{B} + \mathbf{v}_t \times \mathbf{B}_0) \tag{9.5f}$$

where $N + n$ is the number of electrons per unit volume and $\mathbf{v}_t = \mathbf{v}_0 + \mathbf{v}$ is the total velocity. For small-signal conditions, $|\mathbf{B}| \ll |\mathbf{B}_0|$; so the force term $\mathbf{v}_t \times \mathbf{B}$ can be neglected compared with $\mathbf{v}_t \times \mathbf{B}_0$. In addition, since $B = \mu_0 H \approx \mu_0 Y_0 E = E/c$, we see that $|\mathbf{v}_t \times \mathbf{B}|$ is smaller than $|\mathbf{E}|$ by a factor v_t/c. Hence a-c magnetic forces are negligible. The total velocity $\mathbf{v}_t$ is a function of x, y, z, and t. In addition, the position x, y, z of a charge element is a function of time. Consequently,

$$\frac{d\mathbf{v}_t}{dt} = \frac{\partial \mathbf{v}_t}{\partial t} + \frac{\partial \mathbf{v}_t}{\partial x}\frac{dx}{dt} + \frac{\partial \mathbf{v}_t}{\partial y}\frac{dy}{dt} + \frac{\partial \mathbf{v}_t}{\partial z}\frac{dz}{dt} = \frac{\partial \mathbf{v}_t}{\partial t} + (\mathbf{v}_t \cdot \nabla)\mathbf{v}_t \tag{9.6}$$

since $\mathbf{v}_t = \mathbf{a}_x \, dx/dt + \mathbf{a}_y \, dy/dt + \mathbf{a}_z \, dz/dt$. Thus expansion of $d\mathbf{v}_t/dt$ leads to a nonlinear term $(\mathbf{v}_t \cdot \nabla)\mathbf{v}_t$, depending on v_t^2. However, for an a-c velocity $\mathbf{v}$ that is small compared with the d-c velocity $\mathbf{v}_0$, we have

$$[(\mathbf{v} + \mathbf{v}_0) \cdot \nabla](\mathbf{v} + \mathbf{v}_0) = (\mathbf{v} + \mathbf{v}_0) \cdot \nabla \mathbf{v} \approx (\mathbf{v}_0 \cdot \nabla)\mathbf{v} \tag{9.7}$$

since $\mathbf{v}_0$ is constant and the second-order term $(\mathbf{v} \cdot \nabla)\mathbf{v}$ is negligible and may be dropped. Thus we obtain

$$(N + n)m\left[\frac{\partial \mathbf{v}}{\partial t} + (\mathbf{v}_0 \cdot \nabla)\mathbf{v}\right] = (\rho - \rho_0)(\mathbf{E} + \mathbf{v} \times \mathbf{B}_0)$$

But the terms involving n and ρ are products of two a-c quantities and may be dropped for small-signal conditions. Hence the first-order linearized equation of motion becomes

$$\frac{\partial \mathbf{v}}{\partial t} + (\mathbf{v}_0 \cdot \nabla)\mathbf{v} = -\eta(\mathbf{E} + \mathbf{v} \times \mathbf{B}_0) \tag{9.8}$$

since $Ne = \rho_0$ and $e/m = \eta$.

For the cylindrical beam under consideration we also have $\mathbf{v}_0 = \mathbf{a}_z v_0$ and a time dependence $e^{j\omega t}$. If we let $\mathbf{B}_0$ approach infinity, the transverse components of $\mathbf{v}$ must vanish, so that the term $\mathbf{v} \times \mathbf{B}_0$ in (9.8) will vanish. Thus $\mathbf{v}$ has a component in the z direction only, and (9.8) gives

$$j\omega v + v_0 \frac{\partial v}{\partial z} = -\eta E_z \tag{9.9}$$

When $\mathbf{v}$ has only a z component, the a-c current density $\mathbf{J}$ has only a z component since the total current is

$$\mathbf{J}_0 + \mathbf{J} = (-\rho_0 + \rho)(\mathbf{v}_0 + \mathbf{v}) = -\rho_0 \mathbf{v}_0 + (\rho \mathbf{v}_0 - \rho_0 \mathbf{v}) + \rho \mathbf{v}$$
$$\approx -\rho_0 \mathbf{v}_0 + (\rho \mathbf{v}_0 - \rho_0 \mathbf{v}) \tag{9.10}$$

after dropping the second-order term $\rho \mathbf{v}$, which is the product of two small a-c quantities. The d-c and a-c currents are thus

$$\mathbf{J}_0 = -\rho_0 \mathbf{v}_0 \tag{9.11a}$$
$$\mathbf{J} = \rho \mathbf{v}_0 - \rho_0 \mathbf{v} \tag{9.11b}$$

From the continuity equation (9.5e) we obtain

$$\frac{\partial J}{\partial z} = -j\omega \rho \tag{9.12}$$

Equations (9.9), (9.11b), and (9.12) permit us to express J as a function of E_z. Maxwell's equations (9.5a) and (9.5b) may then be solved in the usual manner to obtain wave solutions.

Since we are looking for wave solutions, we may assume that all a-c quantities have a z dependence $e^{-j\beta z}$. In this case (9.9) and (9.12) give

$$(j\omega - j\beta v_0)v = -\eta E_z \tag{9.13a}$$
$$j\beta J = j\omega \rho \tag{9.13b}$$

For convenience, ω/v_0 will be denoted by β_0, which may be interpreted as the d-c propagation constant for the beam. Using (9.11b) and (9.13), we find that

$$J = -j \frac{\omega_p^2}{\omega} \frac{\beta_0^2 \epsilon_0 E_z}{(\beta_0 - \beta)^2} \tag{9.14}$$

where $\omega_p^2 = \rho_0 \eta / \epsilon_0$ is the plasma frequency squared, and $-\rho_0$ is the d-c electron charge density of the beam.

To solve Maxwell's equations for the beam inside a cylindrical guide it will be convenient to introduce the vector potential. For a mode having azimuthal symmetry (no ϕ dependence), all boundary conditions can be satisfied by a vector potential having only a z component $A_z(r,z) = \psi(r)e^{-j\beta z}$. The equation satisfied by A_z is

$$\nabla^2 A_z + k_0^2 A_z = -\mu_0 J$$

From A_z we obtain

$$E_z = -j\omega A_z + \left.\frac{\nabla\nabla\cdot\mathbf{A}_z}{j\omega\mu_0\epsilon_0}\right|_z = \frac{k_0^2 + \partial^2 A_z/\partial z^2}{j\omega\mu_0\epsilon_0} = \frac{k_0^2 - \beta^2}{j\omega\mu_0\epsilon_0}A_z$$

Using (9.14) to express E_z in terms of J gives

$$\mu_0 J = -\left(\frac{\omega_p}{\omega}\right)^2 \beta_0^2 \frac{k_0^2 - \beta^2}{(\beta_0 - \beta)^2} A_z \tag{9.15}$$

The Helmholtz equation for A_z now becomes

$$\nabla_t^2 A_z + p^2 A_z = 0 \qquad 0 \leq r \leq a \tag{9.16a}$$
$$\nabla_t^2 A_z - h^2 A_z = 0 \qquad a \leq r \leq b \tag{9.16b}$$

where we have replaced ∇^2 by $\nabla_t^2 - \beta^2$ and put

$$p^2 = -\beta^2 + k_0^2 + \left(\frac{\omega_p}{\omega}\right)^2 \beta_0^2 \frac{\beta^2 - k_0^2}{(\beta_0 - \beta)^2}$$

$$= -(\beta^2 - k_0^2)\left[1 - \left(\frac{\omega_p}{\omega}\right)^2\left(\frac{\beta_0}{\beta_0 - \beta}\right)^2\right] \tag{9.17a}$$

$$h^2 = \beta^2 - k_0^2 \tag{9.17b}$$

after using (9.15) to express J in terms of A_z.

The analysis, when completed, will show that the space-charge waves are slow waves, with $\beta \approx \beta_0 \gg k_0$, and hence p and h will be real. With no ϕ variation, (9.16) reduces to

$$\frac{d^2\psi}{dr^2} + \frac{1}{r}\frac{d\psi}{dr} + \left\{\begin{array}{c} p^2 \\ -h^2 \end{array}\right\}\psi = 0 \tag{9.18}$$

where $A_z = \psi(r)e^{-j\beta z}$. The equation for ψ is Bessel's equation of order zero, and the solutions are $J_0(pr)$, $Y_0(pr)$, $J_0(jhr)$, and $Y_0(jhr)$. Instead of using the Bessel functions with imaginary argument, we use the modified Bessel functions $I_0(hr)$, $K_0(hr)$. In the region $0 \leq r \leq a$ we cannot use Y_0 since it becomes infinite. Therefore we let

$$\psi(r) = C_1 J_0(pr) \qquad 0 \leq r \leq a$$
$$\psi(r) = C_2 I_0(hr) + C_3 K_0(hr) \qquad a \leq r \leq b$$

where C_1, C_2, C_3 are arbitrary constants. The axial electric field must vanish at $r = b$ and must be continuous at $r = a$. These conditions hold for A_z, and hence for ψ also. Thus

$$C_1 J_0(pa) = C_2 I_0(ha) + C_3 K_0(ha) \tag{9.19a}$$
$$0 = C_2 I_0(hb) + C_3 K_0(hb) \tag{9.19b}$$

Besides E_z, the only other field components present are E_r and H_ϕ. These are given in terms of A_z by

$$E_r = -\frac{\beta}{\omega\mu_0\epsilon_0}\frac{\partial A_z}{\partial r} = -\frac{\beta c}{k_0}\frac{\partial A_z}{\partial r}$$

$$H_\phi = -\frac{1}{\mu_0}\frac{\partial A_z}{\partial r} = \frac{k_0}{\beta} Y_0 E_r$$

Continuity of H_ϕ at $r = a$ requires

$$C_1 p J_0'(pa) = C_2 h I_0'(ha) + C_3 h K_0'(ha) \tag{9.20}$$

where the prime denotes differentiation with respect to the argument pa or ha. In order for (9.19) and (9.20) to have a nontrivial solution for C_1, C_2, and C_3, the determinant of the coefficients must vanish. Thus we find that

$$p\frac{J_0'(pa)}{J_0(pa)} = h\frac{K_0(hb)I_0'(ha) - K_0'(ha)I_0(hb)}{K_0(hb)I_0(ha) - K_0(ha)I_0(hb)} \tag{9.21}$$

This transcendental equation, together with the relations (9.17), which give

$$p^2 = -h^2 + \left(\frac{\omega_p}{\omega}\right)^2 \frac{\beta_0^2 h^2}{(\beta_0 - \sqrt{h^2 + k_0^2})^2} \tag{9.22}$$

determines the propagation constant β.

Two special cases are now examined. First consider the case where b and a are made very large. Then, since

$$I_0(x) \sim \frac{e^x}{\sqrt{2\pi x}} \qquad K_0(x) \sim \sqrt{\frac{\pi}{2x}} e^{-x}$$

for large x, we find that (9.21) gives $p \tan(pa - \pi/4) = h$. But since we are letting a go to infinity, the only possible solution independent of a is $p = h = 0$. From (9.17) we then obtain the following solutions for β:

$$\beta = \pm k_0 \tag{9.23a}$$

$$\beta = \beta_0\left(1 \pm \frac{\omega_p}{\omega}\right) \tag{9.23b}$$

But with $p = h = 0$, $\beta = \pm k_0$, all field components vanish as reference to the equations given earlier for E_z, H_ϕ, and E_r shows. Thus this is a trivial solution. The other solutions $\beta = \beta_0(1 \pm \omega_p/\omega)$ correspond to the space-charge waves. The wave velocities are

$$\frac{\omega}{\beta} = \frac{v_0}{1 \pm \omega_p/\omega} \approx v_0\left(1 \mp \frac{\omega_p}{\omega}\right)$$

since $\omega_p \ll \omega$ for conditions that are typical in microwave tubes. The wave velocities are slightly greater and slightly smaller than the d-c beam velocity v_0. The two waves are called the fast and slow space-charge waves. For $p = 0$, both E_r and H_ϕ vanish but E_z remains finite. The space-charge waves may thus be viewed as a longitudinal oscillation of the electrons in the beam. When $\omega = \omega_p$, one solution corresponds to $\beta = 0$, that is, no propagation. Thus it is seen that the plasma frequency is a natural frequency of oscillation for electrons in an infinite beam. Since the transverse fields are zero, the space-charge waves are not changed even if B_0 is finite as long as the beam has infinite radius.

As a second special case consider the situation $b = a$ so that the beam fills the waveguide. From (9.21) we now see that the right-hand side becomes infinite since the denominator vanishes. Thus we require $J_0(pa) = 0$. Hence pa takes on values typical of those for TM$_{0m}$ modes in a circular guide. The lowest-order solution is (Sec. 3.7) $pa = 2.405$, or in general $pa = p_{0m}$, where the p_{0m} are given in Table 3.3. Using (9.22), we now find that

$$\left(\frac{p_{0m}}{a}\right)^2 = -h^2\left[1 - \left(\frac{\omega_p}{\omega}\right)^2 \frac{\beta_0^2}{(\beta_0 - \sqrt{h^2 + k_0^2})^2}\right]$$

$$= (k_0^2 - \beta^2)\left[1 - \left(\frac{\omega_p}{\omega}\right)^2 \frac{\beta_0^2}{(\beta_0 - \beta)^2}\right] \quad (9.24)$$

For the field waves we expect β to be approximately equal to k_0. Then, since $\omega_p \ll \omega$ and $\beta_0 \gg k_0$, an approximate solution of (9.24) is

$$\beta^2 = k_0^2 - \left(\frac{p_{0m}}{a}\right)^2 \quad (9.25)$$

This is the unperturbed propagation constant for TM$_{0m}$ modes in a cylindrical guide. A correction to β may be obtained by using the solution given by (9.25) in the term multiplied by ω_p^2 in (9.24).

Of greater interest are the space-charge waves for which $\beta \approx \beta_0$. For these $k_0^2 \ll \beta^2$; so (9.24) may be approximated by

$$\left(\frac{p_{0m}}{a}\right)^2 = -\beta^2\left[1 - \left(\frac{\omega_p}{\omega}\right)^2 \frac{\beta_0^2}{(\beta_0 - \beta)^2}\right]$$

which is a quadratic equation in β^2. To obtain an approximate solution, let $\beta = \beta_0(1 + \delta)$, where δ will be small. Then we obtain

$$\left(\frac{p_{0m}}{a}\right)^2 = -\beta_0^2\left[1 - \left(\frac{\omega_p}{\omega}\right)^2 \frac{1}{\delta^2}\right]$$

which gives

$$\delta = \frac{\pm(\omega_p/\omega)\beta_0}{[\beta_0^2 + (p_{0m}/a)^2]^{1/2}}$$

and hence

$$\beta = \beta_0\left[1 \pm \frac{\omega_p}{\omega}\left(1 + \frac{p_{0m}^2}{\beta_0^2 a^2}\right)^{-1/2}\right] \quad (9.26)$$

Note that δ is small, which justifies the approximations made. We may express (9.26) in the same form as (9.23b) by introducing an effective plasma frequency ω_q given by

$$\omega_q = \omega_p \left(1 + \frac{p_{0m}^2}{\beta_0^2 a^2}\right)^{-\frac{1}{2}} = F\omega_p \qquad (9.27)$$

where F is called the space-charge reduction factor; for example, $\omega_q^2 = F^2 \rho_0 \eta/\epsilon_0$; so the effective space charge is $F^2 \rho_0$. Hence we can write

$$\beta = \beta_0 \left(1 \pm \frac{\omega_q}{\omega}\right) \qquad (9.28)$$

For the beam completely filling the guide, there are again a slow and a fast space-charge wave, with velocities slightly greater and slightly smaller than the beam velocity v_0. However, the effective plasma frequency is reduced because of transverse variations in the field. Nevertheless, the space-charge waves have very small transverse field components E_r, H_ϕ. Only the axial electric field E_z is large.

In the general case, when $a \neq b$, the solution for β is tedious. The final results may be expressed in the form (9.28) for the space-charge waves by introducing the effective plasma frequency or the space-charge reduction factor. Some typical results computed from curves given by Branch and Mihran are shown in Fig. 9.4.†

† G. M. Branch and T. G. Mihran, Plasma Frequency Reduction Factors in Electron Beams, *IRE Trans.*, vol. ED-2, pp. 3–11, 1955.

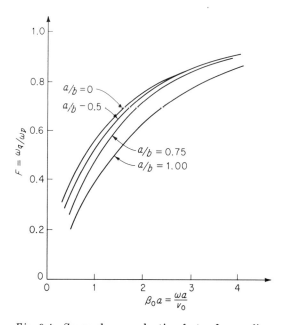

Fig. 9.4 Space-charge reduction factor for a cylindrical electron beam of radius a, velocity v_0, inside a circular guide of radius b. The data apply to the dominant TE_{01} space-charge mode.

The space-charge-wave theory was first developed by Hahn and Ramo in 1939. Since that time space-charge waves under a variety of conditions have been studied. The references cited at the end of this chapter will provide an introduction to the literature on this topic. For the analysis of the ordinary, or O-type, traveling-wave tube, the model discussed above is accurate enough to describe the main operating characteristics.

9.4 Space-charge waves on unfocused beams

Many low-power klystrons employ electron beams without magnetic field focusing when the distance (drift-space length) the beam must travel is short. The propagation of space-charge waves on this type of beam is therefore of interest. We shall consider a beam of radius a and with d-c parameters $-\rho_0, v_0 \mathbf{a}_z$. For space-charge waves with axial symmetry, the only field components present are E_r, E_z, and H_ϕ. The governing equations are Maxwell's equations (9.5) and the force equation (9.8), with $\mathbf{B}_0$ equated to zero.†

For space-charge waves we may assume a z dependence $e^{-j\beta z}$. In component form (9.5a) to (9.5c) and (9.8) are

$$j\beta E_r + \frac{\partial E_z}{\partial r} = j\omega\mu_0 H_\phi \tag{9.29a}$$

$$j\beta H_\phi = j\omega\epsilon_0 E_r + J_r \tag{9.29b}$$

$$\frac{1}{r}\frac{\partial}{\partial r} rH_\phi = j\omega\epsilon_0 E_z + J_z \tag{9.29c}$$

$$\frac{1}{r}\frac{\partial}{\partial r} rE_r - j\beta E_z = \frac{\rho}{\epsilon_0} \tag{9.29d}$$

$$v_r = \frac{j\eta E_r}{\omega - \beta v_0} \tag{9.29e}$$

$$v_z = \frac{j\eta E_z}{\omega - \beta v_0} \tag{9.29f}$$

In addition, we have the relation

$$\mathbf{J} = \rho \mathbf{v}_0 - \rho_0 \mathbf{v}$$

which gives

$$J_r = -\rho_0 v_r = \sigma E_r \tag{9.30a}$$

$$J_z = -\rho_0 v_z + \rho v_0 = \sigma E_z + \rho v_0 \tag{9.30b}$$

† Static space-charge forces are assumed to be negligible, which is a valid assumption for a low-density beam and a short drift space. Alternatively, the beam may be assumed to be ion-neutralized.

where (9.29e) and (9.29f) have been used, and the effective conductivity σ has been introduced as follows:

$$\sigma = \frac{-j\eta\rho_0}{\omega - \beta v_0} = \frac{-j\epsilon_0\omega_p{}^2}{\omega - \beta v_0} \tag{9.31}$$

When we make use of (9.30), the continuity equation $\nabla \cdot \mathbf{J} = -j\omega\rho$ is found to give

$$\sigma\left(\frac{1}{r}\frac{\partial}{\partial r}rE_r - j\beta E_z\right) - j\beta v_0\rho = -j\omega\rho$$

If (9.29d) is used to replace the term in parentheses, we obtain

$$\left(\frac{\sigma}{\epsilon_0} - j\beta v_0 + j\omega\right)\rho = 0 \tag{9.32}$$

If the a-c charge density ρ does not vanish, we must have

$$\frac{\sigma}{\epsilon_0} - j\beta v_0 + j\omega = 0$$

This requires that β be given by

$$\beta = \frac{\omega \pm \omega_p}{v_0} = \beta_0\left(1 \pm \frac{\omega_p}{\omega}\right)$$

The corresponding wave solutions are the space-charge waves in an infinite beam for which $E_r = H_\phi = 0$. For a beam with finite radius the boundary conditions at $r = a$ cannot be satisfied with these values of β. Consequently, the space-charge waves that we are looking for must have a different value of β and, in addition, must have a zero a-c space-charge density ρ, so that (9.32) will hold. For these waves, (9.30) now gives $\mathbf{J} = \sigma\mathbf{E}$, or

$$\mathbf{J} = j\omega\epsilon_0\frac{\sigma}{j\omega\epsilon_0}\mathbf{E}$$

Maxwell's curl equation for $\mathbf{H}$ becomes

$$\nabla \times \mathbf{H} = j\omega\epsilon_0\mathbf{E} + \mathbf{J} = j\omega\epsilon_0\left(1 + \frac{\sigma}{j\omega\epsilon_0}\right)\mathbf{E} = j\omega\epsilon\mathbf{E}$$

where the effective permittivity ϵ of the beam is given by

$$\epsilon = \epsilon_0\left(1 + \frac{\sigma}{j\omega\epsilon_0}\right) = \epsilon_0\left[1 - \frac{\omega_p{}^2}{\omega(\omega - \beta v_0)}\right] \tag{9.33}$$

By using the effective permittivity ϵ, the beam may be treated as a dielectric cylinder. The equation satisfied by E_z is thus

$$\nabla^2 E_z + k^2 E_z = \nabla_t^2 E_z + (k^2 - \beta^2)E_z = 0 \tag{9.34}$$

where

$$k^2 = \omega^2\mu_0\epsilon = k_0^2\left[1 - \frac{\omega_p^2}{\omega(\omega - \beta v_0)}\right]$$

For space-charge waves we anticipate slow waves, for which $\beta \gg k_0$. Hence a suitable solution for E_z is

$$E_z = C_1 I_0(pr) \qquad 0 \leq r \leq a \tag{9.35a}$$

$$E_z = C_2 K_0(hr) \qquad r \geq a \tag{9.35b}$$

where $p = (\beta^2 - k^2)^{1/2}$, $h = (\beta^2 - k_0^2)^{1/2}$, and C_1, C_2 are amplitude constants to be determined.

The boundary conditions at $r = a$ are different from those for a beam with confined flow, for the following reasons.† The electrons have a radial a-c velocity, and hence the boundary of the beam does not remain at its d-c position $r = a$ since electrons will oscillate back and forth about the d-c boundary. The resultant boundary thus becomes rippled, as shown in Fig. 9.5a. The positive charge shown in this figure represents a deficit of negative charge. The effect of the rippled boundary on the radial electric field under small-signal conditions may be accounted for by replacing the rippled boundary by a layer of surface charge. This is, of course, exactly what is done in the case of a dielectric boundary in which polarization charge oscillates back and forth about a mean boundary surface. The surface charge is given in terms of the dielectric polarization **P** by $\mathbf{P} \cdot \mathbf{n} = (\mathbf{D} - \epsilon_0\mathbf{E}) \cdot \mathbf{n}$. For the electron beam the corresponding surface-charge density ρ_s arises from two causes, namely, charge flowing toward the boundary because of the radial current J_r, and charge carried to a given point z owing to surface charge moving with the beam. That is, the rate of increase of surface charge is (Fig. 9.5b)

$$\frac{\partial \rho_s}{\partial t} = j\omega\rho_s = J_r - v_0\frac{\partial \rho_s}{\partial z} = J_r + j\beta v_0\rho_s$$

The term $-v_0\,\partial\rho_s/\partial z$ arises as follows: Let the surface charge density at a point z be ρ_s. Then at the point $z - dz$ the charge density is approximately $\rho_s - (\partial\rho_s/\partial z)\,dz$. The rate at which charge is carried away from the point z on the boundary is $v_0\rho_s$, and the rate at which charge from the adjacent point $z - dz$ flows toward z is $v_0[\rho_s - (\partial\rho_s/\partial z)\,dz]$. The net rate of accumulation of charge in an interval dz, due to the finite beam velocity v_0, is thus $-v_0(\partial\rho_s/\partial z)\,dz = j\beta v_0\rho_s\,dz$, since the z dependence is $e^{-j\beta z}$. The charge density is obtained by dividing by dz. Our final expression for the equivalent surface charge density is

$$\rho_s = \frac{J_r}{j(\omega - \beta v_0)} = \frac{\sigma E_r}{j(\omega - \beta v_0)} = \frac{-\eta\rho_0 E_r}{(\omega - \beta v_0)^2} \tag{9.36}$$

† See also W. C. Hahn, Small Signal Theory of Velocity Modulated Electron Beams, *Gen. Elec. Rev.*, vol. 42, pp. 258–270, 1939.

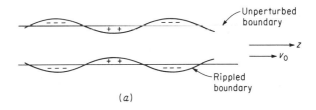
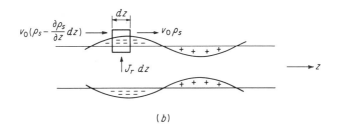
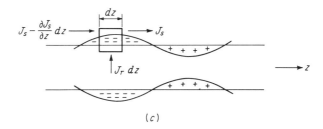

Fig. 9.5 Boundary conditions at the surface of a rippled beam.

The amount of charge which has crossed the unperturbed boundary is $-\rho_0 r$, where r is the a-c displacement of a unit volume of charge. Since

$$\frac{dr}{dt} = v_r = \frac{\partial r}{\partial t} + (\mathbf{v}_0 \cdot \nabla) r = j(\omega - \beta v_0) r$$

we see that

$$\rho_s = -\rho_0 r = \frac{-\rho_0 v_r}{j(\omega - \beta v_0)} = \frac{J_r}{j(\omega - \beta v_0)}$$

which is an alternative derivation of (9.36).

The boundary condition to be applied to the radial electric field is

$$E_{2r} - E_{1r} = \frac{\rho_s}{\epsilon_0} = \frac{-\eta \rho_0 E_{1r}}{\epsilon_0 (\omega - \beta v_0)^2} = \frac{-\omega_p{}^2 E_{1r}}{(\omega - \beta v_0)^2} \tag{9.37}$$

where the subscripts 1 and 2 refer to the fields in the regions $r < a$ and $r > a$, respectively. The above boundary condition may also be expressed in the form

$$\epsilon_0 E_{2r} = \left[\epsilon_0 + \frac{\sigma}{j(\omega - \beta v_0)}\right] E_{1r}$$

This latter result is similar to that which holds at the boundary of a dielectric cylinder except that ω is replaced by $\omega - \beta v_0$ because of the uniform motion of the cylinder, in the z direction, with velocity v_0.

Associated with the equivalent surface charge is an equivalent surface current of density $\mathbf{J}_s = J_s \mathbf{a}_z$. To obtain an expression for J_s, consider Fig. 9.5c. The total rate at which charge flows into a small region of length dz on the boundary is

$$J_r \, dz - \frac{\partial J_s}{\partial z} dz$$

and must equal the rate of increase $j\omega\rho_s \, dz$ of surface charge density in an interval dz. Consequently,

$$-\frac{\partial J_s}{\partial z} = j\beta J_s = j\omega\rho_s - J_r = j\beta v_0 \rho_s$$

or

$$J_s = \rho_s v_0 \tag{9.38}$$

after replacing J_r by $j(\omega - \beta v_0)\rho_s$ from (9.36). For $\omega \gg \omega_p$, the total surface current is usually much larger than the total volume current, and it is therefore very important to include it.

The boundary condition to be applied to H_ϕ is

$$H_{2\phi} - H_{1\phi} = J_s = \rho_s v_0 \tag{9.39}$$

When H_ϕ satisfies this discontinuity relation, the boundary condition on the radial electric field is also satisfied. With the above boundary conditions we are now able to complete the solution to our beam problem.

When we combine (9.29a) and (9.29b), we obtain

$$H_\phi = \frac{j\omega\epsilon_0 + \sigma}{\beta^2 - k_0^2 + j\omega\mu_0\sigma} \frac{\partial E_z}{\partial r} = \frac{j\omega\epsilon}{\beta^2 - k^2} \frac{\partial E_z}{\partial r}$$

Referring to (9.35), we thus find that

$$H_\phi = H_{1\phi} = \frac{j\omega\epsilon}{p} C_1 I_0'(pr) = \frac{j\omega\epsilon}{p} C_1 I_1(pr) \qquad r < a \tag{9.40a}$$

$$H_\phi = H_{2\phi} = \frac{j\omega\epsilon_0}{h} C_2 K_0'(hr) = -\frac{j\omega\epsilon_0}{h} C_2 K_1(hr) \qquad r > a \tag{9.40b}$$

We require E_z to be continuous at $r = a$ and H_ϕ to satisfy the condition (9.39). Therefore

$$C_1 I_0(pa) = C_2 K_0(ha) \tag{9.41a}$$

$$-\frac{j\omega\epsilon_0}{h} C_2 K_1(ha) - \frac{j\omega\epsilon}{p} C_1 I_1(pa) = \rho_s v_0 = \frac{\sigma v_0}{j(\omega - \beta v_0)} E_{1r}$$

From (9.29b) we find $E_{1r}(j\omega\epsilon_0 + \sigma) = j\beta H_{1\phi}$, and hence the second boundary condition becomes

$$\frac{C_2 K_1(ha)}{h} = -\frac{C_1 I_1(pa)}{p}\left[1 - \frac{\omega_p^2}{(\omega - \beta v_0)^2}\right] \tag{9.41b}$$

Dividing (9.41b) by (9.41a) gives

$$\frac{I_1(pa)}{p I_0(pa)}\left[1 - \frac{\omega_p^2}{(\omega - \beta v_0)^2}\right] = -\frac{K_1(ha)}{h K_0(ha)} \tag{9.42}$$

The propagation constant β is determined by a solution of this equation, together with the relation

$$\beta^2 = p^2 + k^2 = p^2 + k_0^2\left[1 - \frac{\omega_p^2}{\omega(\omega - \beta v_0)}\right] = h^2 + k_0^2 \tag{9.43}$$

For space-charge waves, $\beta \gg k_0$, so that h and p can be replaced by β, an approximation that may be used to simplify (9.42) and (9.43).

The results obtained by solving (9.42) and (9.43) may be expressed in the form (9.28):

$$\beta = \beta_0\left(1 \pm \frac{\omega_q}{\omega}\right)$$

At microwave frequencies, ω_q/ω is usually in the range 0.01 to 0.1; so β differs from β_0 by only a few percent or less. In Fig. 9.6 the plasma-frequency reduction factor $F = \omega_q/\omega_p$ is plotted as a function of βa. If p and h are replaced by β in (9.42), that equation may be put into the form

$$\beta a = \beta_0 a\left[1 \pm \frac{\omega_p}{\omega}\left(\frac{K_1 I_0}{K_0 I_1} + 1\right)^{-\frac{1}{2}}\right]$$

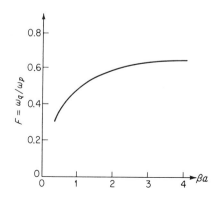

Fig. 9.6 Plasma-frequency reduction factor F for an unfocused cylindrical electron beam as a function of βa.

from which it is seen that

$$F = \frac{\omega_q}{\omega_p} = \left[1 + \frac{K_1(\beta a)I_0(\beta a)}{K_0(\beta a)I_1(\beta a)}\right]^{-\frac{1}{2}} \quad (9.44)$$

This expression was evaluated to obtain F, as given in Fig. 9.6. The data in Fig. 9.6 may be used to solve for the corresponding values of $\beta_0 a$; that is,

$$\beta_0 a = \frac{\beta a}{1 \pm F(\omega_p/\omega)} \quad (9.45)$$

In a beam with confined flow only axial a-c convection currents are permitted. For this reason an a-c charge density ρ exists since the current J_z is a function of z and must have associated with it space-charge-density fluctuations. For the unfocused beam, a-c radial convection currents are also present, and this makes it possible for the total current $\mathbf{J} + j\omega\epsilon_0\mathbf{E}$ to be solenoidal, i.e., to form continuous closed flow lines without terminating in a-c space charge.

9.5 A-c power relations

The a-c power associated with the space-charge waves on an electron beam is of importance for understanding the gain mechanism of traveling-wave tubes and, of course, for power calculations. The small-signal power theorem for beams with confined flow was first derived by Chu.[†] Further extensions have been made by Haus and Bobroff[‡] and Klüver.[§] The derivation is straightforward, but the interpretation of the various terms that enter in is not always clear, depending on the type of beam model being considered.

Maxwell's equations for the fields associated with the beam are

$$\nabla \times \mathbf{E} = -j\omega\mu_0\mathbf{H} \quad \nabla \times \mathbf{H} = j\omega\epsilon_0\mathbf{E} + \mathbf{J}$$

where the small-signal a-c current density is given by (the surface current is included for generality)

$$\mathbf{J} = -\rho_0\mathbf{v} + \rho\mathbf{v}_0 + \rho_s\mathbf{v}_0$$

In addition, we have the equation of motion

$$j\omega\mathbf{v} + (\mathbf{v}_0 \cdot \nabla)\mathbf{v} = -\eta(\mathbf{E} + \mathbf{v} \times \mathbf{B}_0)$$

[†] L. J. Chu, A Kinetic Power Theorem, paper presented at the IRE-PGED Electron Tube Research Conference, Durham, N.H., June, 1951.

[‡] H. A. Haus and D. Bobroff, Small Signal Power Theorem for Electron Beams, *J. Appl. Phys.*, vol. 28, pp. 694–703, June, 1957.

[§] J. W. Klüver, Small Signal Power Conservation Theorem for Irrotational Electron Beams, *J. Appl. Phys.*, vol. 29, pp. 618–622, April, 1958.

Expanding the following expression and using the above equations, we obtain

$$\nabla \cdot (\mathbf{E} \times \mathbf{H}^*) = \mathbf{H}^* \cdot \nabla \times \mathbf{E} - \mathbf{E} \cdot \nabla \times \mathbf{H}^*$$
$$= -j\omega\mu_0 \mathbf{H} \cdot \mathbf{H}^* + j\omega\epsilon_0 \mathbf{E} \cdot \mathbf{E}^* - \mathbf{E} \cdot \mathbf{J}^* \quad (9.46)$$

The continuity equation in the interior of the beam may be written as

$$\mathbf{v}_0 \nabla \cdot \mathbf{J} = -j\omega\rho \mathbf{v}_0 = -j\omega(\rho \mathbf{v}_0 - \rho_0 \mathbf{v}) - j\omega\rho_0 \mathbf{v}$$
$$= -j\omega \mathbf{J} - j\omega\rho_0 \mathbf{v}$$

If we multiply the complex conjugate of this equation by $\mathbf{v}/\eta$, we obtain

$$\frac{\mathbf{v} \cdot \mathbf{v}_0}{\eta} \nabla \cdot \mathbf{J}^* = \frac{j\omega}{\eta} \mathbf{v} \cdot \mathbf{J}^* + \frac{j\omega\rho_0}{\eta} \mathbf{v} \cdot \mathbf{v}^*$$

Multiplying the equation of motion by $\mathbf{J}^*/\eta$ gives

$$\mathbf{J}^* \cdot \mathbf{E} + \mathbf{J}^* \cdot \mathbf{v} \times \mathbf{B}_0 = -\frac{j\omega}{\eta} \mathbf{v} \cdot \mathbf{J}^* - \frac{\mathbf{J}^*}{\eta} \cdot (\mathbf{v}_0 \cdot \nabla)\mathbf{v}$$

Since the transverse components of $\mathbf{J}$ and $\mathbf{v}$ are in the same direction, the term

$$\mathbf{J}^* \cdot \mathbf{v} \times \mathbf{B}_0 = \mathbf{J}^* \times \mathbf{v} \cdot \mathbf{B}_0 = 0$$

The addition of the above two equations thus gives

$$\mathbf{J}^* \cdot \mathbf{E} - \frac{j\omega\rho_0}{\eta} \mathbf{v} \cdot \mathbf{v}^* = -\frac{\mathbf{v} \cdot \mathbf{v}_0}{\eta} \nabla \cdot \mathbf{J}^* - \frac{\mathbf{J}^*}{\eta} \cdot (\mathbf{v}_0 \cdot \nabla)\mathbf{v}$$

If we introduce the term $\nabla \cdot (\mathbf{v} \cdot \mathbf{v}_0 \mathbf{J}^*) = \mathbf{v} \cdot \mathbf{v}_0 \nabla \cdot \mathbf{J}^* + \mathbf{J}^* \cdot \nabla(\mathbf{v} \cdot \mathbf{v}_0)$, we obtain for the right-hand side

$$-\nabla \cdot \left(\frac{\mathbf{v} \cdot \mathbf{v}_0}{\eta} \mathbf{J}^* \right) - \frac{1}{\eta} [\mathbf{J}^* \cdot (\mathbf{v}_0 \cdot \nabla)\mathbf{v} - \mathbf{J}^* \cdot \nabla\mathbf{v} \cdot \mathbf{v}_0]$$

By expanding the bracketed term in rectangular coordinates, it is found to reduce to

$$\frac{v_0}{\eta} \mathbf{a}_z \times \mathbf{J}^* \cdot \nabla \times \mathbf{v}$$

when $\mathbf{v}_0 = v_0 \mathbf{a}_z$. Hence we have

$$\mathbf{J}^* \cdot \mathbf{E} - \frac{j\omega\rho_0}{\eta} \mathbf{v} \cdot \mathbf{v}^* = -\nabla \cdot \left(\frac{\mathbf{v} \cdot \mathbf{v}_0}{\eta} \mathbf{J}^* \right) - \frac{v_0}{\eta} \mathbf{a}_z \times \mathbf{J}^* \cdot \nabla \times \mathbf{v} \quad (9.47)$$

The a-c kinetic-power theorem is obtained by adding (9.47) to (9.46) to obtain

$$\nabla \cdot \left(\mathbf{E} \times \mathbf{H}^* - \frac{\mathbf{v} \cdot \mathbf{v}_0}{\eta} \mathbf{J}^* \right) = -j\omega\mu_0 \mathbf{H} \cdot \mathbf{H}^* - j\omega \frac{\rho_0}{\eta} \mathbf{v} \cdot \mathbf{v}^*$$
$$+ j\omega\epsilon_0 \mathbf{E} \cdot \mathbf{E}^* + \frac{v_0}{\eta} \mathbf{a}_z \times \mathbf{J}^* \cdot \nabla \times \mathbf{v} \quad (9.48)$$

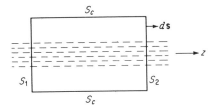

Fig. 9.7 A section of an electron beam.

The term $(\rho_0/2\eta)\mathbf{v} \cdot \mathbf{v}^*$ is the a-c kinetic-energy density in the beam since $\rho_0/\eta = (\rho_0/e)m$ is the mass density per unit volume.

Let us now specialize to the case of a beam with confined flow for which $\mathbf{J} = \mathbf{a}_z J_z$ and the last term on the right-hand side of (9.48) vanishes. The real part of the above equation then gives

$$\text{Re } \nabla \cdot \left(\mathbf{E} \times \mathbf{H}^* - \frac{\mathbf{v} \cdot \mathbf{v}_0}{\eta} \mathbf{J}^* \right) = 0 \tag{9.49}$$

The volume integral over a volume enclosing the beam between cross sections S_1 and S_2 as in Fig. 9.7 may be converted into a surface integral over the surface $S_1 + S_2 + S_c$. Thus the following power-conservation theorem is obtained:

$$\text{Re } \oint_S \tfrac{1}{2} \mathbf{E} \times \mathbf{H}^* \cdot d\mathbf{S} = -\text{Re } \int_{S_2} \frac{-\mathbf{v} \cdot \mathbf{v}_0}{2\eta} \mathbf{J}^* \cdot \mathbf{a}_z \, dS$$
$$+ \text{Re } \int_{S_1} \frac{-\mathbf{v} \cdot \mathbf{v}_0}{2\eta} \mathbf{J}^* \cdot \mathbf{a}_z \, dS \tag{9.50}$$

The term $(-\mathbf{v} \cdot \mathbf{v}_0)/\eta = -(m/e)\mathbf{v} \cdot \mathbf{v}_0 = V_k$ has the dimensions of a voltage, and is called the kinetic voltage. It is the term that gives the conversion of kinetic energy of the beam into electromagnetic energy. In order that a net amount of electromagnetic energy may flow out of the surface S, the quantity

$$\text{Re } \oint_S \tfrac{1}{2} V_k \mathbf{J}^* \cdot d\mathbf{S}$$

must be negative.

A fuller appreciation of the above relations may be obtained by considering the application to space-charge waves in an axially confined infinite cross-section electron beam. For the slow and fast space-charge waves, the a-c kinetic power may be obtained by using (9.13a), (9.14), and (9.49). It is found that

$$\text{Re } (\tfrac{1}{2} V_{kf} J_f^*) = \text{Re } \frac{-v_0 v_f J_f^*}{2\eta} = -\frac{\omega \omega_p^2 \epsilon_0}{2v_0^2 (\beta_f - \beta_0)^3} |E_{zf}|^2$$

$$\text{Re } (\tfrac{1}{2} V_{ks} J_s^*) = -\frac{\omega \omega_p^2 \epsilon_0}{2v_0^2 (\beta_s - \beta_0)^3} |E_{zs}|^2$$

Since $\beta_s > \beta_0 > \beta_f$, the slow space-charge wave has a negative a-c kinetic power whereas the fast space-charge wave has a positive a-c kinetic power. If the beam is excited, with only the slow space-charge wave, the significance of the negative a-c kinetic power is that some of the d-c flow energy of the beam has been extracted and converted into negative a-c energy by the excitation process. The negative a-c kinetic energy must in turn be converted into electromagnetic energy flow in order to maintain power conservation. In the discussion of the traveling-wave tube it will be seen that the slow space-charge wave is the one that produces amplification.

9.6 Velocity modulation

The preceding sections have established the existence of space-charge waves on electron beams. We must now examine the problem of exciting these waves, i.e., producing an a-c velocity modulation on the beam. In klystron tubes velocity modulation is commonly produced by passing the beam through two closely spaced grids located at the center of a cylindrical reentrant cavity, as in Fig. 9.8. The particular form of cavity used is chosen in order to satisfy the requirement of high a-c electric field strength across the grids (which requires small grid spacing) and yet maintain a high cavity Q. The latter requires a large volume–surface area ratio. If we let $\text{Re}\,(E_g e^{j\omega t}) = E_g \cos \omega t$ be the cavity electric field across the gap (often referred to as the buncher gap), those electrons entering the gap when $E_g \cos \omega t$ is directed in the negative z direction will be accelerated and will leave with a velocity greater than v_0. Electrons entering the gap region when $E_g \cos \omega t$ is directed in the positive z direction are slowed down and will leave with a velocity less than v_0. It is apparent, then, that an applied a-c electric field between two parallel

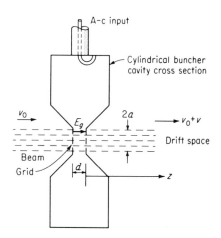

Fig. 9.8 Velocity modulation of an electron beam.

grids will velocity-modulate an electron beam. The analytical details of the modulation process are discussed below.†

We shall consider an unfocused electron beam of the type discussed in Sec. 9.4. Let the cavity field in the gap region be $E_g \cos \omega t$. Electrons will traverse the gap with essentially the entrance velocity v_0. If the time at which a particular electron passes the midplane $z = -d/2$ is t_1, then the field in the cavity at time t when this electron is at a position $z = -d/2 + v_0(t - t_1)$ is

$$E_g \cos \omega t = E_g \cos \frac{\omega}{v_0}\left(z + \frac{d}{2} + v_0 t_1\right)$$

The work done by the cavity field on the electron during its transit through the gap is

$$W = \int_{-d}^{0} -eE_g \cos \beta_0 \left(z + \frac{d}{2} + v_0 t_1\right) dz$$
$$= -eE_g d \frac{\sin (\beta_0 d/2)}{\beta_0 d/2} \cos \omega t_1 \quad (9.51)$$

where $\beta_0 = \omega/v_0$ is the d-c propagation constant for the beam. The beam-coupling parameter M is defined to be

$$M = \frac{\sin (\beta_0 d/2)}{\beta_0 d/2} \quad (9.52)$$

For an electron passing the midplane at time t, the work done on the electron is clearly given by

$$-eE_g \, dM \cos \omega t$$

The work done on an electron results in an increase in its kinetic energy. If the exit velocity from the buncher cavity is $v_0 + v_z$, we have

$$\tfrac{1}{2}m(v_0 + v_z)^2 - \tfrac{1}{2}mv_0^2 = \tfrac{1}{2}m(2v_0 v_z + v_z^2) \approx mv_0 v_z = -eE_g \, dM \cos \omega t \quad (9.53)$$

since for small-signal conditions $v_z \ll v_0$. In complex form (9.53) may be written as

$$v_z e^{j\omega t} = \frac{-\eta}{v_0} E_g \, dM \, e^{j\omega t}$$

Thus the axial a-c beam velocity at the exit grid has a value

$$v_z = \frac{-\eta}{v_0} M(E_g d) \quad (9.54)$$

The foregoing first-order analysis predicts that there will be zero

† This analysis is based on a ballistic formulation, and not on a field approach, since the former is more straightforward.

average work done in bunching the electron beam since the average of (9.51) over one period from t_1 to $t_1 + 1/f$ is zero. This result is not correct, and in actual fact, a net amount of average work is required to velocity-modulate the beam. To determine the average work done, a second-order analysis must be performed.[†] The principal effect of requiring a finite amount of work to velocity-modulate the beam can be represented by an equivalent shunt conductance loading the buncher cavity (beam loading of the buncher cavity). The magnitude of this shunt conductance is typically such as to reduce the unloaded Q of the buncher cavity by a factor of 2 or so. However, even though the first-order analysis given above is not sufficiently accurate to give the beam-loading equivalent conductance, it does give a satisfactory answer for the velocity modulation of the beam, which was the information we were interested in obtaining from the analysis.

Since we now know the velocity modulation of the beam at the exit grid of the buncher cavity, we are in a position to evaluate the amplitudes of the space-charge waves that will be excited on the beam in the drift space beyond the buncher cavity. We shall treat the case of an unfocused beam in detail. The case of a beam with confined flow is somewhat easier to analyze and is the model usually assumed in the analysis of the klystron, even though it is not the type of beam used in most klystrons. However, it turns out that the results for the unfocused beam and the beam with confined flow are essentially equivalent, the main difference being that, for the unfocused beam, the major contribution to the a-c current comes from the equivalent surface current on the beam, whereas for the beam with confined flow, the a-c current is a volume current distributed over the cross section of the beam. During the course of the analysis the results for the case of the beam with confined flow will be given for comparative purposes.

In the drift space $z > 0$, space-charge waves will be launched because of the a-c velocity modulation of the entering beam. At the plane of the exit grid, the radial electric field of the space-charge waves is short-circuited and must be zero. This condition can be met by a suitable combination of the fast and slow space-charge waves. If we let E_{rs} and E_{rf} be the radial electric field of the slow and fast space-charge waves, we require $E_{rs} = -E_{rf}$ at $z = 0$. But E_{rs} and E_{rf} depend on r according to the first-order modified Bessel function $I_1(pr)$, where p is different for the fast and slow waves. However, for typical beams, $p \approx \beta \approx \beta_0$; so the radial dependence can be taken as $I_1(\beta_0 r)$ with negligible error, in which case the boundary conditions at $z = 0$ can be satisfied without introducing higher-order space-charge modes.

[†] K. R. Spangenberg, "Vacuum Tubes," chap. 17, McGraw-Hill Book Company, New York, 1948.
M. Chodorow and C. Suskind, "Fundamentals of Microwave Electronics," chap. 3, McGraw-Hill Book Company, New York, 1964.

The required boundary conditions on E_r can be met if we choose the two space-charge waves to have the same amplitude and to combine in phase for the a-c axial velocity v_z at $z = 0$. Thus let v_{zs} and v_{zf} be the amplitudes of the slow and fast space-charge-wave axial velocities, so that we may write

$$v_z = v_{zs}e^{-j\beta_s z} + v_{zf}e^{-j\beta_f z} = v_{zf}(e^{-j\beta_f z} + e^{-j\beta_s z})$$

where $\beta_s = \beta_0(1 + \omega_q/\omega)$ and $\beta_f = \beta_0(1 - \omega_q/\omega)$, and ω_q is the effective plasma frequency equal to $F\omega_p$. Introducing these expressions gives

$$v_z = 2v_{zf} \cos \beta_q z \, e^{-j\beta_0 z} \tag{9.55}$$

where $\beta_q = \beta_0 \omega_q/\omega = \omega_q/v_0$. Note that v_{zf} is a function of r according to $I_0(\beta_0 r)$. However, $\beta_0 a$ is small, so that v_{zf} is almost constant. Thus we may equate $2v_{zf}$ at $r = 0$ to v_z as given by (9.54) to obtain

$$v_z = \frac{-\eta}{v_0} MV_g \cos \beta_q z \, I_0(\beta_0 r) e^{-j\beta_0 z} \tag{9.56}$$

for the a-c axial velocity at any point in the drift space $z > 0$. In (9.56) we have put V_g for the exciting gap voltage $E_g d$.

For the unfocused beam the a-c space charge density ρ is zero, and from (9.30b) we find $J_z = -\rho_0 v_z$. Consequently, the a-c axial current density in the drift space is

$$J_z = \frac{\eta \rho_0}{v_0} MV_g \cos \beta_q z \, I_0(\beta_0 r) e^{-j\beta_0 z} \tag{9.57}$$

The surface current J_s will be evaluated later, and will turn out to be more important than the volume current.

To find E_z we use (9.29f) to obtain

$$E_{zs} = \frac{\omega - \beta_s v_0}{j\eta} v_{zs} = -\frac{\omega_q}{j\eta} v_{zs} = -\frac{\omega_q}{j\eta} v_{zf}$$

$$E_{zf} = \frac{\omega - \beta_f v_0}{j\eta} v_{zf} = \frac{\omega_q}{j\eta} v_{zf}$$

Hence the axial electric field is given by

$$E_z = \frac{\omega_q}{j\eta} v_{zf}(e^{-j\beta_f z} - e^{-j\beta_s z}) = \frac{2\omega_q}{\eta} v_{zf} e^{-j\beta_0 z} \sin \beta_q z$$

Introducing the earlier expression for v_{zf} gives

$$E_z = -\beta_q MV_g I_0(\beta_0 r) \sin \beta_q z \, e^{-j\beta_0 z} \tag{9.58}$$

Note that E_z vanishes at the exit grid, where $z = 0$.

Sec. 9.6

From (9.29d) we have

$$\frac{1}{r}\frac{\partial}{\partial r}rE_{rs} = j\beta_s E_{zs} = -(\beta_0 + \beta_q)\frac{\omega_q}{\eta}v_{zf} \approx -\frac{\omega_q}{\eta}\beta_0 v_{zf}$$

$$\frac{1}{r}\frac{\partial}{\partial r}rE_{rf} = j\beta_f E_{zf} = (\beta_0 - \beta_q)\frac{\omega_q}{\eta}v_{zf} \approx \frac{\omega_q}{\eta}\beta_0 v_{zf}$$

$$\frac{1}{r}\frac{\partial}{\partial r}r(E_{rs} + E_{rf}) = -2\beta_q\frac{\omega_q}{\eta}v_{zf} \approx 0$$

since $\beta_0 \gg \beta_q$. The boundary conditions on E_r are consequently satisfied to a very good degree of approximation. If desired, a small adjustment of the amplitudes of the two space-charge waves could make $E_{rs} + E_{rf}$ vanish exactly at a particular value of r. However, since we have already approximated $I_1(pr)$ by $I_1(\beta_0 r)$ for both waves, the present approximation of dropping β_q relative to β_0 is consistent with our earlier assumptions.

If we examine the expressions for J_z and v_z, we see that, because of the beating or interference between the two space-charge waves, the a-c current and velocity vary according to $\cos\beta_q z$ in the drift space. Maximum volume current density occurs when

$$z = n\frac{\lambda_q}{2} \tag{9.59}$$

where $\lambda_q = 2\pi/\beta_q$ is the space-charge wavelength and n is an integer.

If we had considered a beam with confined flow as discussed in Sec. 9.3, we should have

$$J_z = -\frac{\omega\rho_0 v_z}{\omega - \beta v_0}$$

by using (9.11b) and (9.12). In this case

$$J_{zs} = \frac{\omega\rho_0 v_{zs}}{\omega_q} \qquad J_{zf} = -\frac{\omega\rho_0 v_{zf}}{\omega_q}$$

If v_z varies according to $\cos\beta_q z$, then J_z will vary as $\sin\beta_q z$ for a beam with confined flow. The different behavior of the two beam models arises because of the zero a-c volume space charge density in the unfocused beam, a condition that can exist because radial a-c motion of the electrons is permitted. The axial current for the two beam models is given by

$$J_z = -\rho_0 v_z + \rho v_0 \qquad J_z = -\rho_0 v_z$$

The space charge ρ changes the relationship between J_z and v_z from $J_z = -\rho_0 v_z$ to $J_z = -\rho_0 v_z \omega/(\omega - \beta v_0)$, which in turn produces the difference in z variation of the current amplitude. In some klystrons where a high power and long drift spaces are used, magnetic focusing is employed. In this case the confined-flow-beam model would be the appropriate one to use. For the beam with confined flow, the axial current density at $r = 0$ would be found to be

$$J_z = \frac{j\eta\rho_0}{v_0} \frac{\omega}{\omega_q} MV_g \sin \beta_q z \, e^{-j\beta_0 z} \tag{9.60}$$

The peak current density is a factor ω/ω_q greater than the maximum current density for the unfocused beam as given by (9.57). This suggests that the confined-flow beam is superior. This is usually not the case because the surface current J_s for the unfocused beam may contribute in a very substantial way to the total a-c axial current, as the following analysis will show.

The surface current density is given by (9.36) and (9.38). For the two space-charge waves we obtain

$$J_{ss} = -\frac{\eta\rho_0 v_0}{(\omega - \beta_s v_0)^2} E_{rs} = -\frac{\eta\rho_0 v_0}{\omega_q^2} E_{rs}$$

$$J_{sf} = -\frac{\eta\rho_0 v_0}{\omega_q^2} E_{rf}$$

When we combine (9.29a) and (9.29b) we find

$$E_r = \frac{j\beta}{p^2} \frac{\partial E_z}{\partial r} = \frac{j\beta}{p} C_1 I_1(pr) \approx jC_1 I_1(\beta_0 r)$$

where $C_1 I_0(pr) = E_z$. Consequently,

$$J_s = -\frac{j\eta\rho_0 v_0}{\omega_q^2} I_1(\beta_0 a)(C_{1f} e^{-j\beta_f z} + C_{1s} e^{-j\beta_s z})$$

Now $E_{zs} = -(\omega_q/j\eta)v_{zf}$ and $E_{zf} = (\omega_q/j\eta)v_{zf}$; so we have $C_{1s} = -(\omega_q/j\eta)v_{zf}$, $C_{1f} = (\omega_q/j\eta)v_{zf}$, where $2v_{zf} = -(\eta/v_0)MV_g$. The final expression for J_s becomes

$$J_s = \frac{j\eta\rho_0}{\omega_q} MV_g I_1(\beta_0 a) \sin \beta_q z \, e^{-j\beta_0 z} \tag{9.61}$$

We shall now compare the relative contributions to the total axial current. The total surface current is given by

$$I_s = 2\pi a \frac{j\eta\rho_0}{\omega_q} MV_g I_1(\beta_0 a) \sin \beta_q z \, e^{-j\beta_0 z} \tag{9.62a}$$

and the total volume current flowing in the axial direction is

$$I_z = \pi a \frac{2\eta\rho_0}{\beta_0 v_0} MV_g I_1(\beta_0 a) \cos \beta_q z \, e^{-j\beta_0 z} \tag{9.62b}$$

where we have used the result

$$\int_0^a I_0(\beta_0 r) 2\pi r \, dr = \frac{2\pi a}{\beta_0} I_1(\beta_0 a)$$

The ratio of the peak amplitudes is

$$\frac{(I_s)_{\max}}{(I_z)_{\max}} = \frac{\omega}{\omega_q} \tag{9.63}$$

This gives the very interesting result that the total surface current in the case of an unfocused beam is a factor ω/ω_q larger than the total volume current. In fact, the total surface current given by (9.62a) is equal to the total volume current in a small-radius confined-flow beam, that is, $\pi a^2 J_z$, where J_z as given by (9.60) for confined flow has a peak amplitude equal to that of I_s if we replace $I_1(\beta_0 a)$ by $\beta_0 a/2$. We therefore conclude that both types of beams are about equally efficient, at least for short drift spaces, where beam spreading would not be important. It is apparent that a-c space-charge bunching is an important mechanism in the production of high-density a-c currents in the velocity-modulated beam. For confined flow the a-c space-charge bunches are formed within the beam, and for the unfocused beam the a-c space-charge bunches appear on the beam surface in the form of a rippled boundary. A sketch of the electric field associated with the two beam models is given in Fig. 9.9.

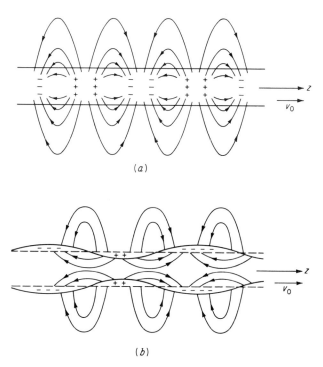

Fig. 9.9 Electric field lines associated with a-c space-charge bunching in (a) a beam with confined flow, (b) an unfocused beam.

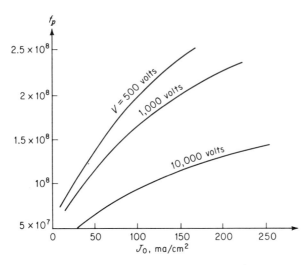

Fig. 9.10 Plasma frequency as a function of beam current density.

The positive charge shown is only an equivalent charge that accounts for a net migration of electrons out of the region, leaving a net negative charge density less than ρ_0, which can be viewed as a superposition of a small positive charge density on the constant d-c background density $-\rho_0$.

In judging the relative amplitudes of the volume current and surface current in an unfocused beam, the ratio ω_q/ω must be known. Usually ω_q does not differ by more than a factor of $\frac{1}{2}$ or so from the plasma radian frequency ω_p. In Fig. 9.10, $f_p = \omega_p/2\pi$ is plotted as a function of the beam current density J_0 in milliamperes per square centimeter for several values of beam-accelerating voltage V. Note that f_p is proportional to $J_0^{\frac{1}{2}}/V^{\frac{1}{4}}$.

9.7 Two-cavity klystron

A schematic illustration of a typical two-cavity klystron amplifier is shown in Fig. 9.11. The first cavity is excited by the input signal, which can be coupled to the cavity by a coaxial-line loop or a waveguide aperture. The first cavity acts as the buncher and velocity-modulates the beam. The second cavity is separated from the buncher by a drift space of length l, which should ideally be chosen so that the a-c current at the second (sometimes called the catcher) cavity is a maximum. The second cavity is thus excited by the a-c signal impressed on the beam in the form of a velocity modulation with a resultant production of an a-c current. The a-c current on the beam is such that the level of excitation of the second cavity is much greater than that in the buncher cavity, and hence amplification takes place. The output signal is taken from the second cavity. If desired, a portion of the amplified output can be fed back to the buncher cavity in a regenerative manner to obtain self-sustained oscillations.

One form of klystron analysis begins with an assumed lumped-parameter equivalent circuit for the output cavity and evaluates the current flowing in this equivalent circuit from an electron beam passing through the cavity by calculating the rate of change with time of the charge induced on the grids at the center of the cavity. This analysis gives a correct picture of klystron behavior, but it fails to illustrate the mechanism of the electromagnetic field interaction with the beam as it actually takes place within the cavity. A more satisfactory approach is to begin with a field analysis that will eventually lead to an equivalent circuit and the basis for a circuit-type analysis of the problem. This is the approach presented below. We first evaluate the field set up in a cavity by the passage of an a-c current through the cavity. This leads to an equivalent circuit for the cavity in the vicinity of one of the resonant frequencies. The next step is to evaluate the response of the cavity (or its equivalent circuit) to the passage of a velocity-modulated electron beam on which the a-c current is in the form of a propagating current wave. This leads to a definition of the beam coupling coefficient, which is a measure of how effective the modulated electron beam is in exciting a response in the cavity. The third step, which we have presented in Sec. 9.6, is the evaluation of the a-c current produced on an initially unmodulated electron beam when it passes through a cavity in which an oscillating electric field exists. These three phases of the analysis substantially provide the complete picture of the operation of a klystron.

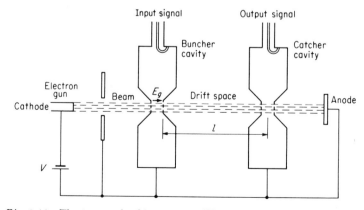

Fig. 9.11 The two-cavity klystron amplifier.

Excitation of a cylindrical cavity

In a klystron cavity it is desirable to have a small grid spacing in order to make the beam coupling parameter $M = [\sin(\beta_0 d/2)]/(\beta_0 d/2)$ close to unity. The transit angle $\beta_0 d = \omega d/v_0$ should be kept small. In addition, a high cavity Q is desired, and this leads to the use of a reentrant cavity. The analysis of the modes in a cavity of this configuration is difficult, and therefore we shall consider instead an ordinary cylindrical cavity. The principal features involved in the excitation of the latter type of cavity by a modulated electron beam are the same as for the reentrant-type cavity.

The cylindrical cavity to be studied is illustrated in Fig. 9.12. The cavity radius is b, and the cavity length is d. Two small cylindrical holes are cut in the center and replaced by grids to allow an electron beam to pass through. The beam radius is a, and is considered very small compared with the cavity radius b. We shall first study the excitation of this cavity by an axial a-c current of the form

$$J_z = Je^{j\omega t} \qquad 0 \leq r \leq a \tag{9.64}$$

Later on we shall consider a traveling-wave current $Je^{j\omega t - j\beta_0 z}$ of the type existing on a velocity-modulated beam and shall find that, for the latter, the cavity response is modified by the beam coupling parameter M.

In view of the uniformity of the current in the z direction and the axial symmetry, only TM_{0m0} modes are excited. These have E_r, E_ϕ, H_r, and H_z equal to zero. It is convenient to introduce the vector potential A_z, which is a solution of

$$\nabla^2 A_z + k_0^2 A_z = \begin{cases} -\mu_0 J & 0 \leq r \leq a \\ 0 & r > a \end{cases} \tag{9.65}$$

At $r = b$ we have $A_z = 0$, so that E_z will vanish on the boundary. From

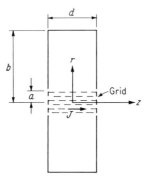

Fig. 9.12 Cylindrical cavity excited by an axial current.

A_z we obtain

$$E_z = -j\omega A_z \tag{9.66a}$$

$$H_\phi = -\frac{1}{\mu_0}\frac{\partial A_z}{\partial r} \tag{9.66b}$$

since there is no z or ϕ variation.

The natural modes of the cavity are solutions of the equation

$$\nabla^2 A_{z,0m0} + k_{0m0}^2 A_{z,0m0} = 0 \tag{9.67}$$

where $k_{0m0} = \omega_{0m0}(\mu_0\epsilon_0)^{\frac{1}{2}}$ and ω_{0m0} is the resonant frequency of the TM$_{0m0}$ mode. With no z or ϕ variation, ∇^2 becomes

$$\nabla^2 = \frac{1}{r}\frac{\partial}{\partial r}r\frac{\partial}{\partial r}$$

and the solutions to (9.67) are Bessel functions. That is,

$$A_{z,0m0} = C_m J_0\left(\frac{p_{0m}r}{b}\right) \tag{9.68}$$

where p_{0m} is chosen so that $J_0(p_{0m}) = 0$, C_m is an arbitrary constant, and for no axial variation $k_{0m0} = p_{0m}/b$ (Sec. 7.3).

For a solution to (9.65) we may choose

$$A_z = \sum_{m=1}^{\infty} C_m J_0\left(\frac{p_{0m}r}{b}\right) \tag{9.69}$$

since the Bessel functions are analogous to the sine and cosine functions and may be used as such in a Fourier series expansion of the vector potential. When we substitute (9.69) into (9.65) we obtain

$$\sum_{m=1}^{\infty} C_m(k_{0m0}^2 - k_0^2) J_0\left(\frac{p_{0m}r}{b}\right) = \mu_0 J \quad 0 \leq r \leq a \tag{9.70}$$

since (9.69) is a solution of (9.67).

The following orthogonality property holds:

$$\int_0^b J_0\left(\frac{p_{0m}r}{b}\right) J_0\left(\frac{p_{0n}r}{b}\right) r\, dr = \begin{cases} 0 & n \neq m \\ \dfrac{b^2}{2} J_1^2(p_{0m}) & n = m \end{cases} \tag{9.71}$$

Thus, if we multiply both sides of (9.70) by $rJ_0(p_{0m}r/b)$ and integrate, we obtain

$$C_m = \frac{2\mu_0}{(k_{0m0}^2 - k_0^2)b^2 J_1^2(p_{0m})} \int_0^a J J_0\left(\frac{p_{0m}r}{b}\right) r\, dr$$

$$= \frac{2\mu_0 J a J_1(p_{0m}a/b)c^2}{p_{0m}b J_1^2(p_{0m})(\omega_{0m0} - \omega)(\omega_{0m0} + \omega)} \tag{9.72}$$

after replacing k_0 by ω/c, and similarly for k_{0m0}. This equation holds for all values of m. We see immediately from this expression that only for those modes for which $\omega \approx \omega_{0m0}$ will the excitation amplitude be large. In addition, we see that, if $\omega = \omega_{0m0}$ for a particular value of m, C_m becomes infinite. An infinite response occurs because the cavity is ideal and is driven at one of its natural resonant frequencies. A practical cavity has a finite Q and will not respond with an infinite amplitude. As shown in Secs. 7.1 to 7.3, the effect of a finite Q is to replace the resonant frequency by

$$\omega_{0m0}\left(1 + \frac{j}{2Q_{0m0}}\right)$$

where Q_{0m0} is the Q of the TM_{0m0} mode. The unloaded cavity Q is given by (7.41) as

$$Q_{0m0} = \frac{p_{0m}c}{(1 + b/d)\omega_{0m0}\delta_s} \tag{9.73}$$

Thus the excitation amplitudes for a cavity with finite Q are given by

$$C_m = -\frac{2\mu_0 a J c^2 J_1(p_{0m}a/b)}{p_{0m}b J_1^2(p_{0m})(\omega_{0m0} + \omega)(\omega - \omega_{0m0} - j\omega_{0m0}/2Q_{0m0})} \tag{9.74}$$

If we choose $\omega = \omega_{010}$, then C_1 will be large and all the other C_m will be small. In this case

$$A_z = C_1 J_0\left(\frac{p_{01}r}{b}\right) = -\frac{2j\mu_0 a J c^3 J_1(p_{01}a/b)Q}{b^2 J_1^2(p_{01})\omega^3} J_0\left(\frac{p_{01}r}{b}\right) \tag{9.75}$$

where ω and Q now refer to the TM_{010} mode. Note that A_z is proportional to the Q, and hence a high Q is desirable. If the cavity is coupled to an external load, we must replace Q by the loaded Q, Q_L.

The total a-c current is $I = \pi a^2 J$. Also, since $a \ll b$, we can replace $J_1(p_{01}a/b)$ by $p_{01}a/2b$. In the vicinity of the resonant frequency ω_{010}, that is, for $\omega = \omega_{010} + \Delta\omega$, we find that the electric field E_z is given by

$$E_z = -j\omega A_z = \frac{jI J_0(p_{01}r/b)}{2\pi b^2 \epsilon_0 J_1^2(p_{01})(\Delta\omega - j\omega_{010}/2Q)} \tag{9.76}$$

If we introduce an equivalent voltage V_0 as the line integral of E_z across the cavity gap at $r = 0$, that is, $V_0 = -E_z d$, we may define an admittance Y_c for the cavity as follows:

$$Y_c = \frac{I}{V_0} = \frac{2\pi b^2}{d}\epsilon_0 J_1^2(p_{01})\left(j\Delta\omega + \frac{\omega_{010}}{2Q}\right) \tag{9.77}$$

For a lumped-parameter LCG_0 circuit with resonant frequency

$$\omega_{010} = (LC)^{-\frac{1}{2}}$$

Fig. 9.13 Equivalent circuit of excited cavity with no loading.

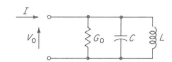

as in Fig. 9.13, we have

$$Y_{\text{in}} = G_0 + 2jC\,\Delta\omega = G_0\left(1 + j\frac{2\,\Delta\omega}{\omega_{010}}Q\right)$$

where $Q = \omega_{010}C/G_0$. Comparison with (9.77) shows that the equivalent cavity conductance G_0 is given by (note that $\epsilon_0 = Y_0/c$)

$$G_0 = \pi b^2 Y_0 J_1{}^2(p_{01})\frac{\omega_{010}}{dQc} \tag{9.78}$$

The above equivalent circuit thus seems a possible circuit to represent the cavity in the vicinity of the resonant frequency ω_{010} of the TM$_{010}$ mode. However, we must show that it correctly accounts for the properties of the cavity even in the absence of the beam current I.

For the TM$_{010}$ mode, the energy stored in the electric field is given by [we assume $E_z = J_0(p_{01}r/b)$]

$$W_e = \frac{\epsilon_0}{4}\int_0^d\int_0^{2\pi}\int_0^b J_0{}^2\left(\frac{p_{01}r}{b}\right)r\,dr\,d\phi\,dz = \frac{\pi\epsilon_0}{4}b^2\,dJ_1{}^2(p_{01})$$

The corresponding voltage across the cavity at $r = 0$ is $V = E_z d = d$. Since $Q = 2\omega W_e/P_l$, the power loss in the cavity is given by

$$P_l = \frac{2\omega W_e}{Q}$$

If we define a conductance G so that $P_l = \tfrac{1}{2}V^2 G$, we find that

$$G = \frac{2P_l}{V^2} = \frac{4\omega W_e}{V^2 Q} = \frac{\pi\epsilon_0 b^2 \omega_{010} J_1{}^2(p_{01})}{dQ} = G_0$$

since $\omega = \omega_{010}$. Thus the two definitions for the cavity conductance lead to consistent results.

If required, C and L are given by $\omega_{010}C = QG_0$ and $\omega_{010}^2 LC = 1$. Thus L and C can be found from the known values of Q and G_0.

Cavity excitation by a velocity-modulated beam

In a velocity-modulated unfocused beam the a-c current is predominantly the a-c beam surface current $I_s = 2\pi a J_s$, where J_s is given by (9.61) since $\omega \gg \omega_q$ for typical beams. The z dependence of the current is $\sin\beta_q z\, e^{-j\beta_0 z}$ when z is measured from the output grid of the buncher cavity. The output cavity should be located a distance $l = (n + \tfrac{1}{2})(\lambda_q/2)$ from the input cavity so that $\sin\beta_q z$ will equal unity and I_s will have a maximum. Because $\beta_q \ll \beta_0$, we have $\beta_q d \ll 1$ and the a-c current variation with z can be taken as $e^{-j\beta_0 l}e^{-j\beta_0 z}$ in the output cavity, where we have replaced

z by $l + z$, so that the new origin is at the center of the output cavity. Thus the cavity is excited by a traveling-wave current.

When the a-c current has a z dependence, all the TM_{0mn} modes are excited. The solutions for A_z are thus of the form

$$A_z = C_{mn} J_0\left(\frac{p_{0m}r}{b}\right) \cos \frac{n\pi z}{d}$$

We then have

$$\sum_{m=1}^{\infty} \sum_{n=0}^{\infty} C_{mn} \left[\left(\frac{p_{0m}}{b}\right)^2 + \left(\frac{n\pi}{d}\right)^2 - k_0^2\right] J_0\left(\frac{p_{0m}r}{b}\right) \cos \frac{n\pi z}{d} = \mu_0 J_z(r, z) \tag{9.79}$$

in place of (9.70). However, if $\omega \approx \omega_{010}$, only the $m = 1$, $n = 0$, or TM_{010} mode is excited with a large amplitude. To find C_{10}, we multiply (9.79) by $rJ_0(p_{0m}r/b) \cos(n\pi z/d)$ with $m = 1$, $n = 0$ and integrate over r and z to obtain

$$C_{10} = \frac{2\mu_0}{(k_{010}^2 - k_0^2)b^2 J_1^2(p_{01})} \int_0^a J_z(r) r J_0\left(\frac{p_{01}r}{b}\right) dr \int_{-d/2}^{d/2} \frac{1}{d} e^{-j\beta_0 z} \, dz$$

This latter result shows that the excitation amplitude is modified by the factor

$$\int_{-d/2}^{d/2} \frac{1}{d} e^{-j\beta_0 z} \, dz = \frac{\sin(\beta_0 d/2)}{\beta_0 d/2} = M$$

which is the beam coupling factor. The integral over r may be replaced by aJ_s without the factor $e^{-j\beta_0 z}$ since $J_z = 0$ except at $r = a$ and $J_0(p_{01}a/b) \approx 1$ for $a \ll b$. Hence, for $\omega = \omega_{010} + \Delta\omega$, we obtain for E_z, at $r = 0$,

$$E_z = \frac{jM(jI_1)e^{-j\beta_0 l}}{2\pi b^2 \epsilon_0 J_1^2(p_{01})(\Delta\omega - j\omega_{010}/2Q)} \tag{9.80}$$

where I_1 is obtained from (9.62a) and is

$$I_1 = 2\pi a \frac{\eta \rho_0}{\omega_q} M I_1(\beta_0 a) V_g \approx \pi a^2 \epsilon_0 \frac{\omega_p^2}{\omega_q} \beta_0 M V_g \tag{9.81}$$

since for $\beta_0 a \ll 1$ we have $I_1(\beta_0 a) \approx \beta_0 a/2$ and are assuming that an optimum length l of drift space is employed so that $\sin \beta_q l$ equals unity. If we compare (9.80) with (9.76), we find that the only change in E_z is replacing the current I by I_1 and multiplying by the beam coupling coefficient M and an irrelevant phase factor $e^{-j\beta_0 l}$.

The voltage developed across the cavity is modified by the factor M. Consequently, the effective current that flows in the equivalent circuit to produce the voltage V is

$$I_e = MI_1 \tag{9.82}$$

from which it is seen that the term beam coupling coefficient is clearly

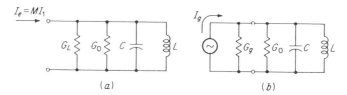

Fig. 9.14 (a) Equivalent circuit of output cavity; (b) equivalent circuit of input cavity.

appropriate. For good coupling between the beam and the cavity a short transit time $\beta_0 d$ is required.

When an external load is coupled to the output cavity, it can be represented by an additional conductance G_L in shunt with G_0. The equivalent circuit of the output cavity is shown in Fig. 9.14a, where $I_e = MI_1$ is the equivalent current that flows in the circuit. The power supplied to the external load G_L is

$$P_0 = \frac{1}{2}\frac{|I_e|^2 G_L}{(G_0 + G_L)^2} \tag{9.83}$$

A similar equivalent circuit may be assumed for the input cavity, as in Fig. 9.14b. For identical cavities the cavity conductance is G_0 for both. If we assume that the beam produces negligible loading on the input cavity, the total conductance in the equivalent circuit of the input cavity is $G_0 + G_g$, where G_g is the equivalent conductance of the signal source.†

The input power to the buncher cavity at resonance is

$$P_{\text{in}} = \frac{1}{2}\frac{|I_g|^2 G_0}{(G_g + G_0)^2} \tag{9.84}$$

which results in a voltage

$$V_g = \frac{I_g}{G_0 + G_g} = \left(\frac{2P_{\text{in}}}{G_0}\right)^{\frac{1}{2}} \tag{9.85}$$

developed across G_0. This is also the velocity-modulating voltage. We can now evaluate the power gain, or amplification, of the klystron. Combining (9.81), (9.83), and (9.85), we obtain

$$A = \frac{P_0}{P_{\text{in}}} = \frac{|I_e|^2 G_L}{|I_g|^2 G_0}\left(\frac{G_0 + G_g}{G_0 + G_L}\right)^2$$

$$= \left(\frac{\pi a^2 \epsilon_0 \beta_0 \omega_p{}^2}{\omega_q}\right)^2 \frac{M^4 G_L}{G_0(G_0 + G_L)^2}$$

$$= \left(\pi a^2 Y_0 \beta_0 \frac{\omega_p}{c}\frac{\omega_p}{\omega_q}\right)^2 \frac{M^4 G_L}{G_0(G_0 + G_L)^2} \tag{9.86}$$

† The conductance representing the beam loading on the input cavity is given by $G_b = 1/2(I_0/V)M[M - \cos(\beta_0 d/2)]$ and is easily taken into account by adding it to $G_0 + G_g$ when it is not negligible. Note that I_0 is the d-c beam current and V the d-c accelerating voltage. See Spangenberg, *op. cit.*, chap. 17.

As an example consider $G_L = G_0^{-1}$ and a beam radius $a = 0.2$ cm. If we also assume a beam current density of 100 ma/cm² and an accelerating voltage of 1,000 volts, we find, from Fig. 9.10, that $\omega_p = 1.02 \times 10^9$. For $f = 10^{10}$, $\omega = 6.28 \times 10^{10}$ and $\beta_0 = \omega/v_0 = 33.6$ rad/cm. Hence $\beta_0 a = 6.72$. Since $\omega_p \ll \omega$, we have $\beta a \approx \beta_0 a$, and Fig. 9.6 then shows that $\omega_q \approx 0.7 \omega_p$. Using these data, the power amplification A is found to be $0.094(Y_0/G_0)^2 M^4$. To evaluate G_0 we make use of (9.78) and (9.73). We have $p_{01} = 2.405$, and since $p_{01} = k_{010}b$, we get

$$b = \frac{2.405c}{6.28 \times 10^{10}} = 1.15 \text{ cm}$$

To keep $\beta_0 d$ small, we must choose d very small. If we take $d = 0.05$ cm, $\beta_0 d = 1.68$ and $M^4 = 0.62$. It would be desirable to make d even smaller, but then the Q, and hence G_0, become small for the type of cavity we are considering. For a copper cavity we find from (9.73) that $Q = 785$, which is not very large. If a reentrant-type cavity were used, a Q about ten times larger could be obtained. Using (9.78) gives $G_0 \approx 0.06 Y_0$. The power amplification is thus 16.2, or 12 db. Considerably higher gain would be obtained by using a reentrant-type cavity since d could then be made smaller and still a high unloaded Q maintained. However, even with the nonoptimum cavity that we have considered, the gain is quite good. For the particular example we have evaluated, the beam loading conductance G_b is very small compared with the cavity conductance G_0. For a more efficient cavity with a much higher unloaded Q, the cavity conductance G_0 would be much smaller and the beam loading conductance G_b might not then be negligible.

In order to obtain greater power gain than can be obtained from a two-cavity klystron, multicavity tubes are used. The gain increases exponentially with the number of cavities employed. In the multicavity klystron the first cavity is again used to provide the initial velocity modulation of the beam. The last cavity in the chain is used as the output cavity. The intermediate cavities are kept unloaded by any external circuits and used to increase the modulation and hence the a-c current on the beam. Power gains of 50 to 60 db can be achieved with multicavity klystrons.

9.8 Reflex klystron

The reflex klystron is an oscillator tube with a built-in feedback mechanism. It uses the same cavity for bunching and for the output cavity. A sketch of the reflex klystron is given in Fig. 9.15. The operation is as follows: If we assume an initial a-c field in the cavity, the beam will be velocity-modulated as it passes through the cavity. Upon entering the drift space, the beam is decelerated and reversed (reflected) by the large d-c field set up by the repeller or reflector electrode at potential $-V_r$.

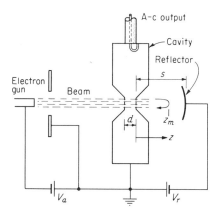

Fig. 9.15 The reflex klystron.

Thus the beam is made to pass through the cavity again, but in the opposite direction. By proper choice of the reflector voltage V_r, the beam can be made to pass through the cavity on its return flight when the a-c current phase angle is such that the field excited in the cavity by the returning beam adds in phase with the initial modulating field. The feedback is then positive, and oscillations will build up in amplitude until the system losses and nonlinear effects prevent further buildup.

When the velocity-modulated beam enters the drift space, it is subjected to a constant decelerating field V_r/s, where s is the cavity output grid–reflector spacing. As a result, the beam propagation constant $\beta_0 = \omega/v_0$ is gradually reduced to zero, and then increased back up to $-\beta_0$. The total phase change undergone by the a-c current on the beam will be given by

$$\theta = 2 \int_0^{z_m} \beta_0(z)\, dz$$

where z_m is the maximum distance an electron can penetrate into the drift space. We can evaluate θ in terms of the transit time T for an electron to return to the cavity. We have, in the drift space,

$$\frac{dv_0(z)}{dt} = -\eta \frac{V_r}{s}$$

which integrates to $v_0(z) = v_0 - \eta t V_r/s$. Hence

$$v_0(z_m) = v_0 - \eta \frac{V_r}{s} \frac{T}{2} = 0$$

which gives $T/2 = v_0 s/\eta V_r$. The return time is equal to $T/2$ also, so that

$$\theta = \omega T = \frac{2 v_0 s \omega}{\eta V_r} \tag{9.87}$$

If we let V_g be the accelerating-gap a-c voltage, the a-c beam current reflected back through the cavity is given by (9.62a) when $\beta_0 z$ is replaced by θ and $\beta_q z$ by $\beta_q \theta/\beta_0 = \omega_q \theta/\omega$. Thus

$$I_s = \frac{j\pi a^2 \eta \rho_0 \beta_0 M V_g}{\omega_q} \sin \frac{\omega_q \theta}{\omega} e^{-j\theta}$$

where we have approximated $I_1(\beta_0 a)$ by $\beta_0 a/2$. In a reflex klystron θ is usually quite small, so that $\sin(\omega_q \theta/\omega)$ may be replaced by $\omega_q \theta/\omega$. The effective current for excitation of the cavity is $I_e = I_s M$, and is given by

$$I_e = \frac{\pi a^2 \eta \rho_0 M^2 V_g}{v_0} \theta e^{j(\pi/2-\theta)} = \frac{I_0}{V_a} \frac{V_g}{2} M^2 \theta(\sin \theta + j \cos \theta) \qquad (9.88)$$

where I_0 is the total d-c beam current $\pi a^2 \rho_0 v_0$ and V_a is the accelerating voltage from which $v_0^2 = 2\eta V_a$.

The a-c electronic admittance of the beam is defined by

$$Y_e = \frac{I_e}{V_g} = \frac{I_0}{V_a} \frac{M^2}{2} \theta(\sin \theta + j \cos \theta) \qquad (9.89)$$

The equivalent circuit of the reflex klystron consists of the electronic admittance Y_e in shunt with the equivalent circuit of the loaded cavity, as in Fig. 9.16. Oscillations can take place when the net conductance is less than zero, or more specifically when

$$Y_e + (G_L + G_0)\left(1 + j\frac{2\,\Delta\omega}{\omega_{010}} Q_L\right) = 0 \qquad (9.90)$$

where Q_L is the loaded Q of the cavity. Since θ is a function of the reflector voltage, as given by (9.87), oscillations depend on an appropriate choice of V_r. In Fig. 9.17 we have plotted the admittance Y_e in polar

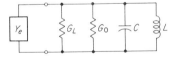

Fig. 9.16 Equivalent circuit for a reflex klystron.

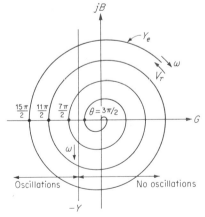

Fig. 9.17 Admittance diagram for a reflex klystron.

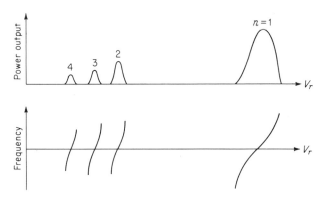

Fig. 9.18 Reflex-klystron tuning curves.

form as a function of θ. Note that $|Y_e|$ increases with θ. On the same plane we have plotted the negative cavity admittance

$$-Y = -(G_0 + G_L)\left(1 + j\frac{2\,\Delta\omega}{\omega_{010}}Q_L\right)$$

which is a straight line parallel to the jB axis at $G = -(G_0 + G_L)$, provided we assume G_0, G_L, Q_L independent of ω in the vicinity of the resonant frequency ω_{010}. The construction shows that oscillations are possible for θ in the vicinity of $3\pi/2, 7\pi/2$, etc., since in this region $G_e + G_0 + G_L < 0$. In addition, oscillations will take place for a range of values of V_r about the points that make $\theta = 3\pi/2 + 2n\pi$. Each value of n gives a mode of oscillation. In typical klystrons as many as seven or more modes of oscillation can be obtained. For stable oscillations $Y_e + Y = 0$, and consequently the frequency of oscillation varies as V_r is varied to tune across a given mode. Typical tuning curves giving power output and frequency as a function of reflector voltage are shown in Fig. 9.18. Physically, the various modes arise because of the increased transit time for electrons into the drift space when V_r is reduced. Oscillations occur when the transit time T equals $(\frac{3}{4} + n)f^{-1}$ or $\frac{3}{4} + n$ a-c periods since the a-c current has the proper phase under these conditions.

Commercially available reflex klystrons range from small-size units producing 100 mw of power up to units capable of delivering several watts of power under continuous operation. Klystron amplifiers employing two or more cavities are available in a size range from units capable of handling a few hundred milliwatts up to several hundred kilowatts of amplified output power.

9.9 Magnetron

This section is devoted to a qualitative description of the magnetron oscillator. The basic structure of a magnetron is a number of identical

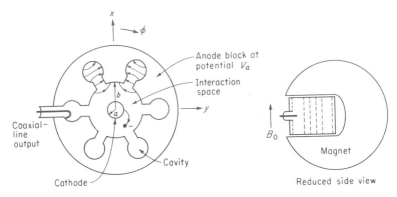

Fig. 9.19 A multicavity magnetron.

resonators arranged in a cylindrical pattern around a cylindrical cathode, as shown in Fig. 9.19. A permanent magnet is used to produce a strong magnetic field normal to the cross section. The anode is kept at a high positive voltage V_a relative to the cathode. Electrons emitted from the cathode are accelerated toward the anode block, but the presence of the magnetic field B_0 produces a force $-ev_r B_0$ in the azimuthal direction which causes the electron trajectory to be deflected in the same direction. If the cathode radius is a and the anode radius is b, the potential at any radius r is $V(r) = V_a[\ln (r/a)]/[\ln (b/a)]$. The velocity of an electron at this radius is given by

$$v(r) = [2\eta V(r)]^{\frac{1}{2}}$$

The electron can execute circular motion, at the radius r, about the cathode if the outward centrifugal force mv^2/r and the radial electric field force $-eE_r = eV_a/[r \ln (b/a)]$ are exactly balanced by the inward magnetic force $ev(r)B_0$. For circular motion at radius r, we therefore have

$$\frac{mv^2}{r} + \frac{eV_a}{r \ln (b/a)} = evB_0 \tag{9.91a}$$

or since $v = \omega_e r$, where ω_e is the electron's angular velocity,

$$\omega_e^2 - \eta B_0 \omega_e + \frac{\eta V_a}{r^2 \ln (b/a)} = 0 \tag{9.91b}$$

For later reference, we solve (9.91) for the cathode-anode accelerating voltage V_a:

$$V_a = \left(\omega_e r^2 \ln \frac{b}{a}\right)\left(B_0 - \frac{\omega_e}{\eta}\right) \tag{9.92}$$

This value of V_a will permit an electron to execute circular motion at a radius r and with an angular frequency ω_e. If now there is present an a-c electromagnetic field that propagates in the azimuthal direction with a phase velocity equal to the electron velocity $\omega_e r$, strong interaction between the field and the circulating electron cloud can take place. The possibility of this type of electromagnetic field being present is discussed below.

The multicavity magnetron is a periodic structure in the azimuthal, or ϕ, direction. If there are N cavities, the period in ϕ is $2\pi/N$. According to Floquet's theorem, each field component can be expanded in the form

$$\psi(r, \phi) = \sum_{n=-\infty}^{\infty} e^{-j\beta\phi - j2n\pi\phi/p} \psi_n(r)$$

$$= \sum_{n=-\infty}^{\infty} \psi_n(r) e^{-j\beta\phi - jnN\phi} \qquad (9.93)$$

where the period $p = 2\pi/N$. But since the structure closes on itself, $\psi(r, 2\pi) = \psi(r, 0)$. The only possible values of β that will make $\beta 2\pi$ equal a multiple of 2π are

$$\beta_m = m \qquad m = 0, \pm 1, \pm 2, \ldots \qquad (9.94)$$

With the value of β specified, a corresponding frequency ω becomes specified, say ω_m, which is the resonant frequency for the mth mode. In other words, when $\omega = \omega_m$, we obtain a value m for β_m. Thus a typical field component will have the form

$$\psi_m(r, \phi) e^{j\omega_m t} = \sum_{n=-\infty}^{\infty} \psi_n(r) e^{-j(m+nN)\phi + j\omega_m t}$$

The phase velocity in the azimuthal direction ϕ for the nth spatial harmonic of the mth resonant mode is

$$v_{p,nm} = \frac{\omega_m r}{\beta_{mn}} = \frac{\omega_m r}{m + nN} \qquad (9.95)$$

at the radius r; that is, angular phase velocity is ω_m/β_{mn}.

The usual mode employed in a magnetron oscillator is the π mode, where the phase change between adjacent cavities is π rad, or 180°. Each cavity with its input gap acts as a short-circuited transmission line a quarter wavelength long, and hence has a maximum electric field across the gap. For the π mode the field is oppositely directed at adjacent cavities. A sketch of the electric field lines in two cavities is given in Fig. 9.19. For the π mode, $\beta_m \phi = m\phi$ must equal π for a change in ϕ equal to one period $2\pi/N$. Hence $m = N/2$, and the phase velocity for the nth spatial harmonic becomes

$$v_{p,nN/2} = \frac{2\omega_{N/2} r}{N(1 + 2n)} \qquad (9.96)$$

In order to obtain interaction between the electron cloud and one of the spatial harmonics at a particular radius r, we must choose V_a so that $\omega_e r = v(r) = v_{p,nN/2}$, or

$$\omega_e = \frac{2\omega_{N/2}}{N(1+2n)} \tag{9.97}$$

The required voltage V_a to obtain synchronism between the electron cloud and the a-c field may be found from (9.92). If we choose a value of r midway between the cathode and anode, that is, $r = (b+a)/2$, and note that in typical magnetrons b/a is small enough so that $\ln(b/a) \approx 2(b-a)/(b+a)$, we obtain

$$V_a = \frac{\omega_m}{m+nN} \frac{b^2 - a^2}{2} \left(B_0 - \frac{\omega_m/\eta}{m+nN} \right) \tag{9.98a}$$

in general, and

$$V_a = \frac{2\omega_{N/2}}{N(1+2n)} \frac{b^2 - a^2}{2} \left[B_0 - \frac{2\omega_{N/2}}{\eta N(1+2n)} \right] \tag{9.98b}$$

for the π mode, where $m = N/2$.

From a physical viewpoint the synchronism between the electron cloud and the nth spatial harmonic of the a-c field implies that those electrons located in the field where E_ϕ acts to slow down the electrons will give up energy to the field. As the electrons slow down they move radially outward [see (9.91)], and eventually are intercepted by the anode. Electrons that are accelerated by the a-c field move in toward the cathode until they get into a proper phase relationship such as to give up energy to the field. When the latter happens, they begin to slow down and spiral out toward the anode. Thus the only electrons that are lost from the interaction space are those that have given up a net amount of energy to the a-c field.

The a-c power may be coupled out from one of the cavities by a coaxial-line loop as shown in Fig. 9.19 or by means of a waveguide.

9.10 O-type traveling-wave tube

The ordinary, or O-type, traveling-wave tube employs a magnetically focused electron beam and a slow-wave structure such as a helix, discussed in Chap. 8. The electron-beam velocity is adjusted to be approximately equal to the phase velocity for an electromagnetic wave propagating along the helix. Under these conditions a strong interaction between the beam and the field can take place. From another viewpoint we can consider the presence of the slow-wave circuit to modify the space-charge wave-propagation constant in such a manner that it becomes complex and represents a growing wave. We shall present a more satisfactory picture of the gain mechanism after we have analyzed a particular tube

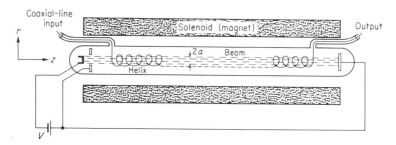

Fig. 9.20 O-type traveling-wave tube.

configuration in detail. A full appreciation of the physical principles involved is somewhat difficult to obtain without a detailed study.

For simplicity we shall use the sheath-helix model discussed in Sec. 8.10 and an axially confined flow beam model (B_0 infinite) of the type treated in Sec. 9.3. In addition, we shall assume that the beam completely fills the region interior to the helix. This assumption is not true in practice, but we make it, nevertheless, in order to simplify the analysis. The basic principle of operation of the tube is not changed by this assumption. The traveling-wave tube is operated in an axially symmetric mode; so all field quantities will be independent of the angle ϕ.

Figure 9.20 illustrates the construction of a typical traveling-wave tube. The main components are an electron gun, a helix, a solenoid to produce the focusing field B_0, and suitable input and output a-c coupling to the helix. The helix is taken to have a radius a and a pitch angle ψ. It is approximated by a cylindrical sheath with infinite conductivity along the direction of the winding and zero conductivity in the perpendicular direction.

In Sec. 8.10 it was shown that both TM and TE modes were required in order to satisfy the boundary conditions at $r = a$. However, for a beam with axially confined flow, where only a z component of a-c velocity is permitted, the TE modes are not affected by the beam since these have $E_z = 0$. Hence the field components for the TE mode for $n = 0$ are those given by (8.72) in Sec. 8.10. Similarly, for $r > a$, that is, outside the helix, the field components for the TM mode are those given by (8.72) in Sec. 8.10. Inside the helix region the TM field in the presence of the beam is that given in Sec. 9.3. However, the pertinent equations are repeated here for convenience. The vector potential A_z is a solution of (9.16a),

$$\nabla_t^2 A_z + p^2 A_z = 0$$

where

$$p^2 = (k_0^2 - \beta^2)\left[1 - \left(\frac{\omega_p}{\omega}\right)^2 \left(\frac{\beta_0}{\beta_0 - \beta}\right)^2\right] = -g^2 \tag{9.99}$$

For the present problem p^2 will turn out to be negative; so we shall replace p^2 by $-g^2$. The solution for A_z is then proportional to $I_0(gr)$. Since E_z is proportional to A_z, we can choose

$$E_z = a_0 I_0(gr) e^{-j\beta z}$$

where a_0 is an amplitude constant. The field components E_r and H_ϕ are readily found from Maxwell's equations; i.e.,

$$E_r = \frac{j\beta}{\beta^2 - k_0^2} \frac{\partial E_z}{\partial r} \qquad H_\phi = \frac{k_0}{\beta} Y_0 E_r$$

Thus we can write the following expressions for the fields in the two regions for the $n = 0$, or axially symmetric, case:

For TE modes,

$$H_z = c_0 I_0(hr) e^{-j\beta z}$$

$$H_r = \frac{j\beta}{h} c_0 I_1(hr) e^{-j\beta z} \qquad r < a$$

$$E_\phi = -\frac{j\omega\mu_0}{h} c_0 I_1(hr) e^{-j\beta z}$$

$$H_z = d_0 K_0(hr) e^{-j\beta z}$$

$$H_r = -\frac{j\beta}{h} d_0 K_1(hr) e^{-j\beta z} \qquad r > a$$

$$E_\phi = \frac{j\omega\mu_0}{h} d_0 K_1(hr) e^{-j\beta z}$$

For the TM mode,

$$E_z = a_0 I_0(gr) e^{-j\beta z}$$

$$E_r = \frac{j\beta g}{h^2} a_0 I_1(gr) e^{-j\beta z} \qquad r < a$$

$$H_\phi = \frac{j\omega\epsilon_0 g}{h^2} a_0 I_1(gr) e^{-j\beta z}$$

$$E_z = b_0 K_0(hr) e^{-j\beta z}$$

$$E_r = -\frac{j\beta}{h} b_0 K_1(hr) e^{-j\beta z} \qquad r > a$$

$$H_\phi = -\frac{j\omega\epsilon_0}{h} b_0 K_1(hr) e^{-j\beta z}$$

where $h^2 = \beta^2 - k_0^2$.

The boundary conditions at $r = a$ for the sheath helix are given by

(8.69). For the present problem they yield

$$\frac{-j\omega\mu_0}{h} c_0 I_1(ha) \cos\psi + a_0 I_0(ga) \sin\psi = 0$$

$$\frac{j\omega\mu_0}{h} d_0 K_1(ha) \cos\psi + b_0 K_0(ha) \sin\psi = 0$$

$$a_0 I_0(ga) \cos\psi + \frac{j\omega\mu_0}{h} c_0 I_1(ha) \sin\psi$$
$$= b_0 K_0(ha) \cos\psi - \frac{j\omega\mu_0}{h} d_0 K_1(ha) \sin\psi$$

and

$$c_0 I_0(ha) \sin\psi + \frac{j\omega\epsilon_0 g}{h^2} a_0 I_1(ga) \cos\psi$$
$$= d_0 K_0(ha) \sin\psi - \frac{j\omega\epsilon_0}{h} b_0 K_1(ha) \cos\psi$$

If we solve for c_0 and d_0 from the first two equations and substitute into the latter two equations, we obtain two homogeneous equations for a_0 and b_0. For a nontrivial solution the determinant must vanish. Equating the determinant to zero gives

$$g \frac{I_1(ga)}{I_0(ga)} = \frac{h^3 \tan^2\psi}{k_0^2} \left[\frac{I_0(ha)}{I_1(ha)} + \frac{K_0(ha)}{K_1(ha)} \right] - h \frac{K_1(ha)}{K_0(ha)} \qquad (9.100)$$

For most traveling-wave tubes the parameters are such that ga and ha are large. In this case the ratio of the Bessel functions in (9.100) approaches unity, and we obtain

$$g = 2 \frac{h^3}{k_0^2} \tan^2\psi - h \qquad (9.101a)$$

From (9.99) we have

$$g^2 = h^2 \left[1 - \left(\frac{\omega_p}{\omega}\right)^2 \left(\frac{\beta_0}{\beta_0 - \beta}\right)^2 \right]$$

and hence

$$\left[1 - \left(\frac{\omega_p}{\omega}\right)^2 \left(\frac{\beta_0}{\beta_0 - \beta}\right)^2 \right]^{\frac{1}{2}} = 2 \frac{h^2}{k_0^2} \tan^2\psi - 1 \qquad (9.101b)$$

where $h^2 = \beta^2 - k_0^2$. The above is a sixth-degree equation in β, and cannot be solved exactly.

Since we are dealing with a slow-wave system, β^2 will be large compared with k_0^2 and $h^2 \approx \beta^2$. In addition, we can equate $k_0 \cot\psi$ to β_0 since the phase velocity of the helix in the absence of a beam is chosen equal to the beam velocity v_0. That is, $k_0 \csc\psi$ is the propagation constant for the helix, and for ψ small, $\sin \psi$ can be replaced by $\tan \psi$. We thus obtain

$$1 - \left(\frac{\omega_p}{\omega}\right)^2 \left(\frac{\beta_0}{\beta_0 - \beta}\right)^2 = \left(\frac{2\beta^2}{\beta_0^2} - 1\right)^2$$

We now assume that $\beta = \beta_0(1 + \delta)$, where δ is small. With this substitution we get

$$\delta^2 - \left(\frac{\omega_p}{\omega}\right)^2 = \delta^2(1 + 4\delta + 2\delta^2)^2$$
$$= 4\delta^6 + 16\delta^5 + 20\delta^4 + 8\delta^3 + \delta^2 \quad (9.102)$$

For δ small, we can drop all but the term involving the lowest power of δ. This is the cubic term, and thus

$$\delta^3 = -\frac{1}{8}\left(\frac{\omega_p}{\omega}\right)^2$$

The three solutions for δ are $\frac{1}{2}(\omega_p/\omega)^{\frac{2}{3}}$ multiplied by the cube roots of -1, which are -1 and $(1 \pm j\sqrt{3})/2$. Hence

$$\delta_1 = -\frac{1}{2}\left(\frac{\omega_p}{\omega}\right)^{\frac{2}{3}} \quad (9.103a)$$

$$\delta_2 = \frac{1}{4}\left(\frac{\omega_p}{\omega}\right)^{\frac{2}{3}}(1 - j\sqrt{3}) \quad (9.103b)$$

$$\delta_3 = \frac{1}{4}\left(\frac{\omega_p}{\omega}\right)^{\frac{2}{3}}(1 + j\sqrt{3}) \quad (9.103c)$$

Since ω_p/ω is small, the assumption that δ was small is justified. The corresponding propagation constants are

$$j\beta_1 = j\beta_0\left[1 - \frac{1}{2}\left(\frac{\omega_p}{\omega}\right)^{\frac{2}{3}}\right] \quad (9.104a)$$

$$j\beta_2 = j\beta_0\left[1 + \frac{1}{4}\left(\frac{\omega_p}{\omega}\right)^{\frac{2}{3}}(1 - j\sqrt{3})\right] \quad (9.104b)$$

$$j\beta_3 = j\beta_0\left[1 + \frac{1}{4}\left(\frac{\omega_p}{\omega}\right)^{\frac{2}{3}}(1 + j\sqrt{3})\right] \quad (9.104c)$$

The first solution corresponds to a wave with a phase velocity slightly greater than the beam velocity. The other two solutions have phase velocities slightly less than the beam velocity, and in addition $j\beta_2$ corresponds to a decaying wave whereas $j\beta_3$ corresponds to a growing wave. The growth constant α_g is

$$\alpha_g = \beta_0 \frac{\sqrt{3}}{4}\left(\frac{\omega_p}{\omega}\right)^{\frac{2}{3}} \quad (9.105)$$

If all three waves are present at the input, only the latter wave will predominate at the output.

There are additional solutions to the eigenvalue equation (9.101b).

We should expect a wave propagating in the $-z$ direction, with $\beta \approx -k_0 \csc \psi \approx -\beta_0$, which is not significantly perturbed by the beam. We therefore assume that $\beta = -\beta_0(1 + \delta)$ and consider δ small. Substituting into (9.101b) and retaining the smallest power term in δ give

$$\delta = -\frac{1}{32}\left(\frac{\omega_p}{\omega}\right)^2$$

Hence a fourth solution is

$$j\beta_4 = -j\beta_0\left[1 - \frac{1}{32}\left(\frac{\omega_p}{\omega}\right)^2\right] \approx -j\beta_0 \quad (9.106)$$

The remaining two solutions of (9.101b) give values for β approximately equal to $\pm k_0$. However, the eigenvalue equation (9.101b) is an approximation to the true eigenvalue equation (9.100), obtained by assuming that ga and ha are large and that β is large compared with k_0. Therefore the two solutions $\beta = \pm k_0$ to the sixth-degree equation (9.101b) are not solutions of (9.100) and do not correspond to physical waves.

The a-c current and velocity are given by (9.14) and (9.13a) as

$$v = \frac{j\eta E_z}{v_0(\beta_0 - \beta)} \quad (9.107)$$

$$J = -j\frac{\omega_p^2}{\omega}\frac{\beta_0^2}{(\beta_0 - \beta)^2}\epsilon_0 E_z \quad (9.108)$$

These equations show that v and J are negligible for the three waves for which β is significantly different from β_0. Thus v and J arise from the first three slow waves discussed. The fourth wave can be excited by reflection at the output end of the tube. If it is reflected at the input end also, it will be amplified and, with continued reflection and amplification, will result in oscillations. To avoid this undesirable feature, an attenuating resistive vane or an integral ferrite isolator is built into the traveling-wave tube.

At the input end we must have the total a-c current and velocity associated with the three forward slow waves vanish. Thus the initial conditions at the input $z = 0$ are

$$J_1 + J_2 + J_3 = 0 \qquad v_1 + v_2 + v_3 = 0$$

When we assume that

$$E_z = I_0(\beta_0 r)(C_1 e^{-j\beta_1 z} + C_2 e^{-j\beta_2 z} + C_3 e^{-j\beta_3 z})$$

and make use of (9.108) and the initial conditions, we find that

$$\frac{J_2}{J_1} = e^{j2\pi/3} \qquad \frac{J_3}{J_1} = e^{-j2\pi/3}$$

Consequently, all three waves at the input have equal magnitudes; that is, we find that $C_1 = C_2 = C_3$. The growing wave will have an amplitude equal to one-third that of the input signal. Therefore the amplitude gain of a traveling-wave tube is

$$\frac{E_0}{E_i} = \tfrac{1}{3}e^{\alpha_g l}$$

where α_g is given by (9.105) and l is the tube length. The power gain in decibels is

$$A = 20 \log 0.333 + 20\alpha_g l \log e$$
$$= -9.54 + 3.75\beta_0 l \left(\frac{\omega_p}{\omega}\right)^{\frac{2}{3}} \qquad (9.109)$$

With the aid of the preceding results we can now describe the physical mechanism of the gain. We note that the growing wave has a phase velocity slightly less than the beam velocity. This growing wave is the perturbed slow space-charge wave. The a-c kinetic-power density of the fast and slow space-charge waves are [see (9.49), (9.13a), and (9.14)]

$$\operatorname{Re} \frac{-v_0 v_f J_f^*}{2\eta} = -\frac{\omega \omega_p^2 \epsilon_0}{2v_0^2(\beta_f - \beta_0)^3} |E_{zf}|^2$$

$$\operatorname{Re} \frac{-v_0 v_s J_s^*}{2\eta} = -\frac{\omega \omega_p^2 \epsilon_0}{2v_0^2(\beta_s - \beta_0)^3} |E_{zs}|^2$$

The slow space-charge wave has $\beta_s > \beta_0$ and hence has a negative a-c kinetic-power density, whereas the fast space-charge wave has a positive a-c kinetic-power density. Since the slow wave grows, it therefore loses energy, and the conservation theorem (9.49) then requires that the electromagnetic power increase. The a-c current of the slow wave will have a phase angle relative to E_z such that $\operatorname{Re}(E_{zs}J_s^*)$ is negative and the current continually gives up energy to the field. This may be verified by substituting β_3 for β in (9.14) to obtain

$$\operatorname{Re}(E_{zs}J_s^*) = -2\sqrt{3}\left(\frac{\omega_p}{\omega}\right)^{\frac{2}{3}} \beta_0^2 \epsilon_0 |E_{zs}|^2$$

As a further aid to the understanding of the traveling-wave tube, it may be noted that it can be viewed as a large number of closely spaced cavity gaps operating as a multicavity klystron. The adjacent turns of the helix are then considered as constituting a gap.

The main advantage of the traveling-wave tube over the klystron is its relatively broad frequency band of operation. Typical units provide gains of 30 to 50 db over an octave or more in frequency. Power-handling capability ranges from milliwatts to megawatts.

9.11 M-type traveling-wave tube

The magnetron-type (M-type) traveling-wave tube is a linear version of the cylindrical magnetron. Figure 9.21 is a schematic illustration of an

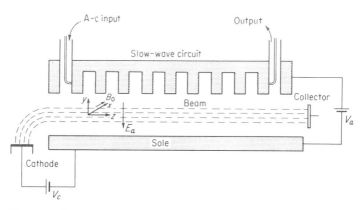

Fig. 9.21 M-type traveling-wave tube.

M-type tube using a corrugated, or comblike, slow-wave circuit. The electron beam is much wider than it is thick and approximates a sheet beam. A potential V_a is applied between the sole and the anode block. A large static magnetic field is applied in a direction perpendicular to the beam velocity $v_0\mathbf{a}_z$, and the static electric field $-E_a\mathbf{a}_y$ arises from the anode to sole potential V_a. The electrons moving upward from the cathode at potential V_c are deflected by the magnetic field into a beam moving in the positive z direction. The desired type of flow is the one where there is only a z-directed velocity $v_0(y)$, which in general is a function of y. Electron flow takes place in a crossed **E** and **B** field, which is typical of magnetron-type tubes.

For stable flow, $v_0(y)\mathbf{a}_z$ does not vary with z. If we denote by $V(y)$ the potential at an arbitrary value of y between the sole and anode block, we must have a balance between the magnetic force

$$-ev_0(y)\mathbf{a}_z \times \mathbf{a}_x B_0 = -eB_0 v_0(y)\mathbf{a}_y$$

and the electric field force $\mathbf{a}_y e\, \partial V/\partial y$. Thus

$$\frac{\partial V}{\partial y} = v_0(y)B_0 \tag{9.110}$$

The velocity $v_0(y)$ may be found from the energy equation

$$\tfrac{1}{2}mv_0^2(y) = e(V - V_c)$$

The derivative with respect to y gives

$$v_0(y)\frac{\partial v_0}{\partial y} = \eta \frac{\partial V}{\partial y} \tag{9.111}$$

The potential $V(y)$ arises from the applied potential V_a and from the d-c space charge within the beam. Under equilibrium conditions the force

$-e(\mathbf{E} + \mathbf{v} \times \mathbf{B}_0)$ acting on an electron is zero. The divergence of this equation thus gives

$$\nabla \cdot \mathbf{E} + \nabla \cdot \mathbf{v} \times \mathbf{B}_0 = 0 = -\frac{\rho_0 s}{\epsilon_0} + B_0 \frac{\partial v_0}{\partial y} \quad (9.112)$$

since $\mathbf{v} = v_0 \mathbf{a}_z$ and $\mathbf{B}_0 = B_0 \mathbf{a}_x$. In this equation $-\rho_0$ is the d-c negative space charge density and s is a factor giving the fraction of the negative space charge which is not neutralized by positive ions. For no positive ions present, $s = 1$. If we assume that $s = 1$, the set of relations (9.110) to (9.112) can hold only if

$$\omega_c{}^2 = \eta^2 B_0{}^2 = \frac{\rho_0 \eta}{\epsilon_0} = \omega_p{}^2$$

or

$$\omega_c = \omega_p \quad (9.113)$$

as can be determined by eliminating $\partial V/\partial y$ and $\partial v_0/\partial y$. When this condition holds, the flow is referred to as planar Brillouin flow.

With the above model for the beam, it is possible to solve for space-charge waves that can propagate on the beam. In the presence of a slow-wave structure, the propagation constants become perturbed and a growing wave is produced similar to that in the O-type tube. For a detailed analysis the reader is referred to the citations given at the end of this chapter. The principles involved are not sufficiently different from those already discussed to warrant inclusion in this text.

9.12 Other types of microwave tubes

In addition to the main types of microwave tubes already discussed, there are a variety of others as well. In one form of traveling-wave tube, the resistance-wall amplifier, the helix is replaced by a circular guide lined with a resistive material. The resistive lining enables a slow wave to propagate in the guide, a wave that is highly attenuated in the absence of a beam. If an electron beam is present, amplification takes place with a growth constant α_g large enough to offset the attenuation due to the resistive lining. Thus a net overall amplification is obtained.

In another form of traveling-wave tube, the double-stream amplifier, two parallel electron beams are used. In this tube one of the beams provides the slow-wave structure, or circuit, for the other beam.

It is also possible to amplify the space-charge waves directly by passing the beam through a succession of accelerating and decelerating regions. This type of tube is called a velocity-jump amplifier because the beam velocity v_0 is periodically changed, or jumped, to new values.

For both the O-type and M-type traveling-wave tubes it is possible to adjust the beam velocity so that it is equal to the phase velocity of any one of the spatial harmonics making up the Bloch wave that can propagate along the periodic structure used for the slow-wave circuit. In particular, interaction between the beam and one of the backward-propagating spatial harmonics is possible. Consider a Bloch wave propagating in the $-z$ direction. For this wave, E_z has the expansion

$$E_z(r, z) = \sum_{n=-\infty}^{\infty} E_n(r) e^{j\beta z + j2n\pi z/p}$$

where p is the period of the periodic structure in the z direction. If we want interaction between the beam and the $n = -1$ spatial harmonic, it is only necessary to choose

$$v_0 = v_p = \frac{\omega}{-(\beta - 2\pi/p)} = \frac{\omega}{2\pi/p - \beta}$$

If the period p is small enough, the $n = -1$ spatial harmonic has a phase velocity directed in the $+z$ direction and its group velocity is in the $-z$ direction. If the $n = -1$ spatial harmonic is amplified, all the other spatial harmonics are also amplified, since they must all be present with very definite amplitudes in order that the boundary conditions may be satisfied. The amplification of the noninteracting spatial harmonics comes about because of the increasing surface current and charge induced on the metallic boundaries by the amplified spatial harmonic that interacts with the beam. Tubes employing interaction with a backward spatial harmonic are usually used as oscillators and are called backward-wave oscillators, or carcinotrons. They have their output coupling at the cathode end.

There are still other forms of microwave tubes, and no doubt more will be developed. For more extensive discussion the cited references at the end of this chapter should be consulted.

9.13 Noise in microwave tubes

The noise properties of a microwave amplifier are important since they will ultimately set the lower limit to the useful signal power in a communication link. Signals below some lower threshold will be completely masked by noise arising from the microwave amplifier, and cannot be detected. In this section we present a number of useful results for dealing with noise problems. We do not give any theoretical development of the noise generated by a microwave tube because of space limitations. A detailed discussion of the theory of noise in electron tubes and other devices is given in a recent book edited by Smullin and Haus, to which the reader is referred.†

† L. D. Smullin and H. A. Haus (eds.), "Noise in Electron Devices," John Wiley & Sons, Inc., New York, 1959.

The main sources of noises in electron tubes are (1) the random emission of electrons from the cathode, which results in a random variation of the d-c charge density $-\rho_0$ of the beam; (2) random variation in the velocity with which electrons are emitted, which results in a random variation of the d-c beam velocity v_0; (3) nonuniform emission over the cathode surface, with a resultant increase in the spread of the random velocity component of the beam; (4) random interception of electrons by grids, the slow-wave circuit, e.g., helix, and other electrodes; and (5) secondary emission of electrons from the collector. Noise manifests itself as a rapid and random fluctuation of the field amplitude at the tube output. It has a frequency spectrum that is essentially constant from zero frequency to a frequency well above the microwave range. If the random fluctuations are larger than the amplified output signal, the latter will be completely masked. The two parameters normally used to describe the noise properties of an amplifier are the noise figure F and the effective noise temperature T_A. The significance and meaning of these parameters are taken up following discussion of Johnson noise.

Johnson noise

The thermal noise caused by random motion of electrons in a resistor R at an absolute temperature T is called Johnson noise (sometimes referred to as Nyquist noise also). If the noise voltage appearing across a resistor R were amplified by an ideal noise-free amplifier and presented on an oscilloscope, it would have the appearance shown in Fig. 9.22. The voltage is positive and negative with equal probability, so that the mean, or average, value is zero. The mean-square value is of course not zero, and is given by

$$\overline{e_n{}^2} = \lim_{T \to \infty} \frac{1}{2T} \int_{-T}^{T} e^2(t)\, dt$$

The mean-square value can be calculated from basic physical laws. In practice, it is desirable to work with the noise power or root-mean-square value of the noise voltage associated with the noise power developed in a small-frequency interval Δf. The latter is given by

$$e_n = \left(\frac{4hfR\,\Delta f}{e^{hf/kT} - 1}\right)^{\frac{1}{2}} \tag{9.114}$$

where h is Planck's constant, k is Boltzmann's constant, and T is the absolute temperature in degrees Kelvin. For the temperatures and fre-

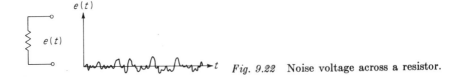

Fig. 9.22 Noise voltage across a resistor.

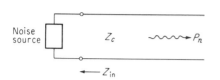
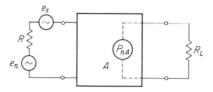

Fig. 9.23 An arbitrary noise source connected to a transmission line.

Fig. 9.24 An amplifier with internal noise sources.

quencies involved in radio and microwave communications, $hf \ll kT$, so that (9.114) may be approximated by

$$e_n = (4kTR\,\Delta f)^{\frac{1}{2}} \tag{9.115}$$

Let us now consider a resistor R at temperature T connected to a lossless transmission line with a characteristic impedance $Z_c = R$. The resistor will then cause a noise power of amount

$$P_n = \left(\frac{e_n}{R + Z_c}\right)^2 Z_c = \frac{e_n{}^2}{4R} = kT\,\Delta f \tag{9.116}$$

to be propagated away in the form of a TEM wave. The quantity $kT\,\Delta f$ is thus the maximum available noise power that can be obtained from a resistor at temperature T.

If a resistor is connected at the input of an amplifier with input impedance $Z_{\text{in}} = R$, so that maximum power transfer occurs, the noise-power input to the amplifier is $kT\,\Delta f$ in the frequency interval Δf. If the amplifier power gain is A, the amplified output noise is $kTA\,\Delta f$.

Equivalent noise temperature of a noise source

Consider an arbitrary noise source, e.g., a gas-discharge tube, connected to an infinite lossless transmission line. Also let the noise source be matched to the transmission line in the frequency interval Δf. The source is matched if an incident wave in the frequency interval Δf about the frequency f under consideration is completely absorbed by the noise source terminating the transmission line (Fig. 9.23). Let the noise power delivered by the noise source to the transmission line be P_n, in the frequency interval Δf centered about the frequency f. The equivalent noise temperature T_e of the noise source is defined to be the temperature a resistor R equal to Z_c must have to deliver the same amount of noise power; thus

$$T_e = \frac{P_n}{k\,\Delta f} \tag{9.117}$$

Noise figure

Figure 9.24 illustrates schematically an amplifier with power gain A, matched at the input to a resistor R and at the output to a resistor R_L.

The amplified thermal noise power arising from R is $AkT\,\Delta f$ at the output. In addition, the amplifier will contribute additional noise power, say P_{nA}, at the output due to noise generation within the amplifier (from electron devices and resistors making up the amplifier). Thus the total output noise power will be

$$AkT\,\Delta f + P_{nA}$$

The noise power $kT\,\Delta f$ in R may be considered to arise from an equivalent noise generator with terminal voltage $e_n = (4kT\,\Delta f\,R)^{1/2}$ in series with an ideal noise-free resistor R.

If a signal voltage e_s is applied to the input, the input signal power under matched conditions is $P_{si} = e_s^2/4R$. The output signal power will be $P_{so} = AP_{si}$ from the definition of power gain A. The noise figure of the amplifier is given by

$$F = \frac{\text{signal-to-noise ratio at input}}{\text{signal-to-noise ratio at output}}$$

$$= \frac{P_{si}/kT\,\Delta f}{AP_{si}/(AkT\,\Delta f + P_{nA})}$$

$$= 1 + \frac{P_{nA}}{AkT\,\Delta f} \tag{9.118}$$

Since the amplifier noise P_{nA} is a definite quantity, the noise figure F will not be unique unless a value for the temperature T of the input noise power in R is specified. By agreement, T is taken as 290°K, or 17°C, which is room temperature. Note also that the noise figure is defined under matched conditions. An ideal noise-free amplifier would have a noise figure of unity. The noise contributed by the amplifier is given by (9.118) as

$$P_{nA} = (AkT_0\,\Delta f)(F - 1) \tag{9.119}$$

where $T_0 = 290°\text{K}$.

At times it is more convenient to describe the noise property of an amplifier in terms of an equivalent noise temperature T_A. This is defined as the temperature an equivalent resistor R must be at in the input to produce the same noise-power output; thus

$$T_A = \frac{P_{nA}}{Ak\,\Delta f} = (F - 1)T_0 \tag{9.120}$$

As an example, consider an amplifier with a noise figure of 1.2, which seems to represent a very good low-noise amplifier, since F is close to the ideal value of unity. The noise power at the output, contributed by the amplifier, is

$$P_{nA} = 0.2AkT_0\,\Delta f$$

If we have a resistor or other source at the input with a noise temperature of 20°K, the resulting noise-power output from this source will be

$$P_n = Ak \, \Delta f \, 20 = AkT_0 \, \Delta f \, \frac{20}{290} = 0.069 AkT_0 \, \Delta f$$

In this case the amplifier contributes more noise than the input source by a factor $2/0.69 = 2.9$, and would be considered a noisy amplifier even though its noise figure were close to unity.

If we had given the equivalent noise temperature of the amplifier, which in the present case is $(F - 1)T_0 = 0.2 \times 290 = 58°K$, we should see immediately that for an input noise source with a temperature of 20°K, the noise from the amplifier would be predominant. Consequently, for low-temperature input noise sources, the equivalent noise temperature T_A of the amplifier is a more meaningful parameter to specify.

Microwave traveling-wave amplifiers have noise figures in the range from 4 to 20 db or more. The value of 4 db is achieved only in traveling-wave tubes with optimum design at frequencies below 5,000 Mc/sec for presently available tubes. A noise figure of 10 db (that is, $F = 10$) is more nearly representative of the majority of available traveling-wave tubes.

Problems

9.1 Consider an electron beam of radius a, velocity v_0, and space charge density ρ_0. The d-c current density is then $J_0 = -\rho_0 v_0$. Show that a magnetic field

$$H_\phi = \begin{cases} -\dfrac{r\rho_0 v_0}{2} & 0 \leq r \leq a \\ -\dfrac{a^2 \rho_0 v_0}{2r} & r \geq a \end{cases}$$

is produced. Verify that the compression force $-e\mathbf{v}_0 \times \mathbf{B}_\phi$ is much smaller than the radial outward force due to the space-charge electric field and may therefore be neglected.

9.2 Show that an electron with velocity $\mathbf{v}$ perpendicular to B_0 executes circular motion at the cyclotron frequency $\omega_c = eB_0/m = \eta B_0$ by equating the centrifugal force to the $-e\mathbf{v} \times \mathbf{B}_0$ force.

9.3 An electron beam has a radius of 0.2 cm. The accelerating voltage is 1,000 volts. The total beam current is 0.03 amp. Calculate the beam perveance, the space charge density ρ_0 and velocity v_0, the number of electrons per cubic meter, and the radial electric field due to space charge. Estimate the radial displacement of an electron located at the beam boundary during the time it takes the beam to move a distance $d = 5$ cm. Use the equation $m \, d^2r/dt^2 = -eE_r$, and assume E_r to be constant and equal to its value at the beam boundary. Is the beam dispersion significant in this case if d is kept less than 5 cm?

9.4 Consider an electron beam with d-c parameters ρ_0, $\mathbf{v}_0 = v_0 \mathbf{a}_z$ immersed in a field $\mathbf{B}_0 = \mathbf{a}_z B_0$. Assume a time dependence $e^{j\omega t}$ and a z dependence $e^{-j\beta z}$ and solve the linearized equation of motion (9.8) for $\mathbf{v}_1 = v_x\mathbf{a}_x + v_y\mathbf{a}_y + v_z\mathbf{a}_z$ to obtain

$$\begin{bmatrix} v_x \\ v_y \\ v_z \end{bmatrix} = -\eta \begin{bmatrix} j(\omega - \beta v_0)/\Delta & -\omega_c/\Delta & 0 \\ \omega_c/\Delta & j(\omega - \beta v_0)/\Delta & 0 \\ 0 & 0 & 1/j(\omega - \beta v_0) \end{bmatrix} \begin{bmatrix} E_x \\ E_y \\ E_z \end{bmatrix}$$

where $\Delta = \omega_c^2 - (\omega - \beta v_0)^2$.

From the continuity equation (9.5e) and (9.11b) show that

$$\mathbf{J} = \frac{v_0 \nabla \cdot \mathbf{J}}{-j\omega} - \rho_0 \mathbf{v}$$

$$J_z = \frac{v_0 \nabla \cdot \mathbf{J}}{-j\omega} - \rho_0 v_z$$

$$\mathbf{J}_t = -\rho_0 \mathbf{v}_t$$

$$\nabla \cdot \mathbf{J} = -j\beta J_z - \rho_0 \nabla_t \cdot \mathbf{v}_t$$

$$J_z = \frac{jJ_0 \nabla_t \cdot \mathbf{v}_t - \omega \rho_0 v_z}{\omega - \beta v_0}$$

9.5 Using the results of Prob. 9.4, obtain solutions for β for waves in an infinite electron beam when all a-c quantities are independent of x and y. Note that, for space-charge waves, $E_x = E_y = 0$ but E_z is finite. For the field waves, $E_z = 0$.

Hint: Note that $\nabla \times \mathbf{H} = -j\beta \mathbf{a}_z \times \mathbf{H}$, $\nabla \times \mathbf{E} = -j\beta \mathbf{a}_z \times \mathbf{E}$, which leads to the equation $j(\beta^2 - k_0^2)\mathbf{E}_t = \omega\mu_0 \mathbf{J}_t = -\omega\mu_0\rho_0\mathbf{v}_t$.

Answer: For field waves, β is a solution of

$$(\omega - \beta v_0)\frac{\omega\omega_p^2}{c^2} - (\beta^2 - k_0^2)\Delta = \pm \frac{\omega_c \omega \omega_p^2}{c^2}$$

where Δ is given in Prob. 9.4. Note that two solutions are given by $\omega - \beta v_0 = \pm\omega_c$. These are the cyclotron waves. For $\beta \approx k_0$, so that $\omega \gg \beta v_0$, four other approximate solutions are

$$\beta = \pm k_0 \left[1 - \frac{\omega_p^2}{\omega(\omega \pm \omega_c)} \right]$$

9.6 Compute the gain of a klystron amplifier of the type considered in the text where the following data apply: Beam radius = 0.3 cm, beam current density = 100 ma/cm². Accelerating voltage = 1,000 volts, frequency = 3,000 Mc. $G_L = G_0$, cavity width $d = 0.2$ cm, cavity conductivity = 5.8×10^7 mhos/m. Compute the gain for $d = 0.3$ cm also, and compare with the earlier calculation.

9.7 Consider a reflex klystron employing a cylindrical cavity of the form shown in Fig. 9.12. The data of Prob. 9.6 apply, with $d = 0.2$ cm. The external loading $G_L = G_0$. The cavity grid-reflector spacing s is equal to 1 cm. Calculate and plot the electronic admittance spiral as a function of reflector voltage V_r at a frequency of 3,000 Mc. Plot also the negative cavity admittance $-Y$ on the same susceptance plane. Determine the reflector-voltage variation to tune across the $n = 2$ and $n = 3$ modes. Evaluate the change in oscillation frequency as the modes are tuned across.

9.8 Consider a cylindrical waveguide of radius a lined by a resistance sheet so that the boundary conditions at $r = a$ are $E_z = -Z_m H_\phi$, where $Z_m = (1+j)/\sigma\delta_s$ is the surface impedance of the wall. Analyze this structure as a traveling-wave tube when an electron beam (axially confined flow) with velocity $v_0\mathbf{a}_z$ completely fills the guide. Determine an appropriate value of Z_m in order to obtain amplification. Find the optimum value of Z_m to give a maximum gain.

Answer: β is a solution of

$$\frac{1}{g}\frac{I_0(ga)}{I_1(ga)} = -\frac{j\omega\epsilon_0 Z_m}{\beta^2 - k_0^2}$$

For ga large, so that $I_0/I_1 \approx 1$, $\beta = (1+\delta)\beta_0$ with δ small, δ^2 is given by

$$\delta^2 = j\frac{2k_0^2(R_m/Z_0)^2(\omega_p/\omega)^2}{\beta_0^2 + j2k_0^2(R_m/Z_0)^2}$$

where $Z_m = (1+j)R_m$.

9.9 Consider a cylindrical waveguide of radius a uniformly filled with a stationary plasma (an ionized gas with an equal number of electrons and ions per unit volume). At high frequencies the motion of the ions may be neglected because their mass is much greater than that of the electrons. Thus the guide will have the same properties as one filled with an electron beam with zero axial d-c velocity. Use (9.33) to show that the guide may be considered as filled with a dielectric medium with permittivity $\epsilon = \epsilon_0(1 - \omega_0^2/\omega^2)$, where ω_0 is the plasma frequency for the plasma. Find a solution for the lowest-order circularly symmetric E mode and show that for $\omega < \omega_0$ such that ϵ is negative the wave impedance is inductive.

9.10 The results of Prob. 9.9 may be used to analyze the beam-plasma amplifier. Consider an electron beam passing through the plasma-filled guide. Use a confined-flow model to describe the beam and show that for a beam completely filling the guide the equations of Sec. 9.3 are valid provided ϵ_0 is replaced by $\epsilon = \epsilon_0(1 - \omega_0^2/\omega^2)$ throughout, where ω_0 is the plasma frequency for the plasma. In particular, Eqs. (9.24) to (9.26) hold. Thus in (9.26), if $\omega_p = (e\rho_0/m\epsilon_0)^{\frac{1}{2}}$ is replaced by $(e\rho_0/m\epsilon)^{\frac{1}{2}} = (e\rho_0/m\epsilon_0)^{\frac{1}{2}}[\omega/(\omega^2 - \omega_0^2)^{\frac{1}{2}}]$, it is seen that β becomes complex for $\omega < \omega_0$ and a growing and decaying pair of waves are obtained. Show that the gain constant is given by

$$\beta_0\left[\frac{m\epsilon_0}{e\rho_0}\left(1 + \frac{p_{01}^2}{a^2\beta_0^2}\right)(\omega^2 - \omega_0^2)\right]^{-\frac{1}{2}}$$

and is very large when ω is close to ω_0. Note also that for a finite-radius beam with unconfined flow passing through an unbounded plasma medium the equations of Sec. 9.4 apply with ϵ_0 again replaced by ϵ. For this model of the beam-plasma amplifier (9.45) may be used in place of (9.26) and will predict a gain constant of the same order of magnitude as does the confined-flow model.

9.11 Consider a two-stage amplifier with power gain A_1 and A_2 for each stage. Let the noise figure for stage 1 be F_1 and that for the second amplifier be F_2. Assuming matched conditions, show that the noise figure of the combination is $F = 1 + (F_1 - 1) + (F_2 - 1)/A_1 = F_1 + (F_2 - 1)/A_1$. Express the equivalent noise temperature of the amplifier in terms of the equivalent noise temperature of each stage.

9.12 The noise figure for a klystron amplifier (due to shot effect) is given by

$$F = \frac{M_i^2 e I_0}{4G_i kT}$$

where M_i is the beam coupling parameter and G_i is the conductance of the input cavity. The temperature T may be taken as 290°K. Evaluate F for the case where $M_i \approx 1$, $I_0 = 50$ ma, and the cavity conductance $G_i = 0.01 Y_0$.

References

1. Slater, J. C.: "Microwave Electronics," D. Van Nostrand Company, Inc., Princeton, N.J., 1950.

2. Pierce, J. R.: "Traveling Wave Tubes," D. Van Nostrand Company, Inc., Princeton, N.J., 1950.
3. Kleen, W. J.: "Electronics of Microwave Tubes," Academic Press Inc., New York, 1958.
4. Beck, A. H. W.: "Space Charge Waves," Pergamon Press, New York, 1958.
5. Hutter, R. G.: "Beam and Wave Electronics in Microwave Tubes," D. Van Nostrand Company, Inc., Princeton, N.J., 1960.
6. Spangenberg, K. R.: "Vacuum Tubes," McGraw-Hill Book Company, New York, 1948.
7. Hamilton, D. R., J. K. Knipp, and J. B. Horner Kuper: "Klystrons and Microwave Triodes," McGraw-Hill Book Company, New York, 1948.
8. Collins, G. B.: "Microwave Magnetrons," McGraw-Hill Book Company, New York, 1948.
9. Chodorow, M., and C. Susskind: "Fundamentals of Microwave Electronics," McGraw-Hill Book Company, New York, 1964.
10. Reich, H. J., P. F. Ordung, H. L. Krauss, and J. K. Skalnik: "Microwave Theory and Techniques," D. Van Nostrand Company, Inc., Princeton, N.J., 1953.
11. Reich, H. J., J. K. Skalnik, P. F. Ordung, and H. L. Krauss: "Microwave Principles," D. Van Nostrand Company, Inc., Princeton, N.J., 1957.

Space-charge wave theory

12. Ramo, S.: The Electronic Wave Theory of Velocity Modulated Tubes, *Proc. IRE*, vol. 27, p. 757, 1939.
13. Ramo, S.: Space-charge and Field Waves in an Electron Beam, *Phys. Rev.*, vol. 56, p. 276, 1939.
14. Hahn, W. C.: Small Signal Theory of Velocity Modulated Electron Beams, *Gen. Elec. Rev.*, vol. 42, p. 258, 1939.
15. Chodorow, M., and L. Zitelli: The Radio Frequency Current Distribution in Brillouin Flow, *IRE Trans.*, vol. ED-6, p. 352, 1959.
16. Rigrod, W., and J. Lewis: Wave Propagation along a Magnetically Focused Cylindrical Electron Beam, *Bell System Tech. J.*, vol. 33, p. 399, 1954.
17. Brewer, G. R.: Some Effects of Magnetic Field Strength on Space-charge Wave Propagation, *Proc. IRE*, vol. 44, p. 896, 1956.

10
Microwave masers

This chapter is concerned with the fundamental principles of operation of a relatively new kind of microwave amplifier called a maser (*M*icrowave *A*mplification by *S*timulated *E*mission of *R*adiation). The maser amplifier is a low-noise device by comparison with microwave electron tubes.

The maser depends on certain quantum-mechanical effects for its operation. The electrons around a nucleus in an atom do not obey classical mechanics, but rather are governed by quantum-wave mechanics. Many similarities exist between quantum-wave mechanics and classical wave problems in electromagnetic theory, of the type already discussed in this book. For this reason it is possible to present a brief but still meaningful discussion of certain aspects of quantum-wave mechanics, of importance for understanding the operation of a maser. Thus the first two sections present a short introduction to wave mechanics and the absorption and emission of radiation in atoms. The primary purpose is to develop some feeling for the significance of concepts such as energy levels, quantum states characterized by wave functions, induced emission and absorption of radiation, spontaneous radiation, etc. In addition, these two sections provide a basis for some understanding of the origin of the Einstein coefficients for induced absorption and emission and the nature of angular momentum. Space does not permit a detailed quantum-mechanical treatment of the maser.† Nevertheless, some acquaintance with basic quantum-mechanical results is deemed necessary for a proper understanding of the operation of masers and the various parameters that affect its operation. With this basic material as a background, we are able to develop an equivalent macroscopic circuit for the maser that lends itself to conventional microwave circuit analysis.‡

† An excellent account of quantum-mechanical aspects of masers is given by A. E. Siegman, "Microwave Solid-state Masers," McGraw-Hill Book Company, New York, 1964.

‡ The reader not wishing to go into the quantum details of the operation of a maser may omit Secs. 10.1 and 10.2. If certain quantum-mechanical results are accepted on faith, no great loss in continuity is incurred.

10.1 Some quantum-mechanical fundamentals

Quantum mechanics was developed because of the failure of classical mechanics to explain correctly and to predict many of the effects observed in the behavior of matter at the atomic level. The development, or piecing together, of a consistent quantum-mechanical theory was a very significant achievement. Analogies with classical mechanics, optics, and electromagnetic theory were judiciously exploited in the early development of the theory. As with any theory, the basic premises are built on hypothesis, the validity of which is determined by agreement with experimental results. The success of the quantum theory in explaining atomic phenomena is such that the theory must be regarded as a fundamental law of nature, just as Maxwell's equations are fundamental laws of nature describing the behavior of electric and magnetic fields.

The fundamental equation obtained for describing the motion of electrons in atoms is a wave equation, first presented by Schrödinger and called Schrödinger's wave equation. In order to present the main concepts associated with this equation we shall consider one of the simplest possible applications, the description it provides of the hydrogen atom.

The hydrogen atom consists of a proton of mass M and charge e and a single electron of mass $m \ll M$ and a charge $-e$. According to classical mechanics, the electron revolves about the proton or nucleus in an orbit with a specific radius r_0. However, according to quantum mechanics, the electron can be located at any radial distance r from the proton with only a finite probability of being at any specific radial distance away. Thus the position of the electron is specified by a function $\psi(r, \theta, \phi)$ such that the probability that it is located at the specific point r_0, θ_0, ϕ_0 in a volume element $dV = r_0^2 \sin \theta_0 \, dr_0 \, d\theta_0 \, d\phi_0$ is

$$P(r_0, \theta_0, \phi_0) \, dV = \psi(r_0, \theta_0, \phi_0)\psi^*(r_0, \theta_0, \phi_0) \, dV \tag{10.1}$$

where ψ^* is the complex conjugate of ψ, and r, θ, ϕ are spherical coordinates, as in Fig. 10.1. The function ψ is a solution of Schrödinger's equation, and is therefore called a wave function. In general, ψ is a function of time, so that the probability of an electron being located at r_0, θ_0, ϕ_0 may be changing with time.

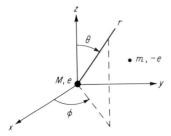

Fig. 10.1 Spherical coordinates used to describe the hydrogen atom.

Schrödinger's wave equation for $\psi(r, \theta, \phi, t)$ is

$$\nabla^2\psi - \frac{8\pi^2 m}{h^2} V(r)\psi + j\frac{4\pi m}{h}\frac{\partial \psi}{\partial t} = 0 \qquad (10.2)$$

where h is Planck's constant, m is the mass of the electron, and $V(r)$ is the potential energy of the electron. In the present case the potential energy $V(r)$ is simply the Coulomb potential energy; i.e.,

$$V(r) = \frac{-e^2}{4\pi\epsilon_0 r}$$

When $V(r)$ is not a function of time, the equation is separable, so that the solution $\psi(r, \theta, \phi, t)$ can be expressed in product form

$$\psi(r, \theta, \phi, t) = \psi(r, \theta, \phi)f(t)$$

When we substitute this form of solution into (10.2) and divide by ψf, we obtain

$$\frac{1}{\psi}\left[-\frac{h^2}{8\pi^2 m}\nabla^2\psi + V(r)\psi\right] = -\frac{h}{2\pi j}\frac{1}{f}\frac{\partial f}{\partial t}$$

The left side is independent of time; so this equation can hold for all values of t only if both sides are equal to a constant. Thus we put

$$\frac{df}{dt} = -\frac{2\pi j}{h}Wf$$

which has the solution

$$f(t) = e^{-j2\pi Wt/h} \qquad (10.3)$$

The constant W has the dimensions of energy. It is determined by a solution of the time-independent Schrödinger equation

$$\nabla^2\psi + \frac{8\pi^2 m}{h^2}[W - V(r)]\psi = 0 \qquad (10.4)$$

Equation (10.4) will have an infinite number of solutions ψ_n with corresponding eigenvalues W_n. Each solution corresponds to a particular stationary state, or configuration, that the electron can have.

The equation satisfied by ψ is similar to the equation a component of the electric field would satisfy in an infinite-radius spherical cavity, where the dielectric constant would be a function of r such that

$$\kappa(r)k_0^2 = \frac{8\pi^2 m}{h^2}[W - V(r)]$$

The boundary condition at $r = \infty$ is vanishing of the wave function. There will be an infinite number of resonant modes, each with a specific resonance frequency, or eigenvalue. In the case of the electron, the eigenvalues W_n correspond to the energy level of the electron when it is in the state described by the eigenfunction ψ_n. Since the electron must be located somewhere in physical space, with unit probability, we must normalize ψ_n so that

$$\int_0^\infty \int_0^{2\pi} \int_0^\pi \psi_n \psi_n^* r^2 \sin\theta \, dr \, d\phi \, d\theta = 1 \quad (10.5)$$

We shall give the solutions to (10.4) without derivation since the detailed derivation is not essential to understanding the meaning of the wave function. For variations with θ and ϕ included, there are a triple infinity of solutions given by (mks units)

$$\psi_{nlm}(r, \theta, \phi) = R_{nl}(r)\Theta_{lm}(\theta)\Phi_m(\phi) \quad (10.6)$$

where n, l, and m are integers $0, 1, 2, \ldots$, and

$$R_{nl}(r) = -\left\{\left(\frac{2}{nr_0}\right)^3 \frac{(n-l-1)!}{2n[(n+1)!]^3}\right\}^{\frac{1}{2}} e^{-\rho/2} \rho^l L_{n+l}^{2l+1}(\rho) \quad (10.7a)$$

with $\rho = 2r/nr_0$, $r_0 = \epsilon_0 h^2/\pi m e^2$,

$$\Theta_{lm}(\theta) = \left[\frac{(2l+1)(l-|m|)!}{2(l+|m|)!}\right]^{\frac{1}{2}} P_l^m(\cos\theta) \quad (10.7b)$$

$$\Phi_m(\phi) = (2\pi)^{-\frac{1}{2}} e^{jm\phi} \quad (10.7c)$$

The functions $L_{n+l}^{2l+1}(\rho)$ are associated Laguerre polynomials, and $P_l^m(\cos\theta)$ are associated Legendre polynomials. The constants in (10.7) are chosen so that the functions are normalized, e.g.,

$$\int_0^{2\pi} \Phi_m \Phi_m^* \, d\phi = 1$$

The first few of the functions involved are written out explicitly below. $\Phi_m(\phi)$ is given by

$$\Phi_0(\phi) = (2\pi)^{-\frac{1}{2}} \qquad \Phi_1(\phi) = (2\pi)^{-\frac{1}{2}} e^{j\phi}$$

$\Theta_{lm}(\theta)$ is given by

$$\Theta_{00} = 2^{-\frac{1}{2}} \qquad \Theta_{10} = \frac{\sqrt{6}}{2}\cos\theta$$

$$\Theta_{11} = \frac{\sqrt{3}}{2}\sin\theta \qquad \Theta_{20} = \frac{\sqrt{10}}{4}(3\cos^2\theta - 1)$$

R_{nl} is given by

$$R_{10} = r_0^{-\frac{3}{2}} 2 e^{-\rho/2}$$

$$R_{20} = \frac{r_0^{-\frac{3}{2}}}{2\sqrt{2}}(2-\rho)e^{-\rho/2}$$

$$R_{21} = \frac{r_0^{-\frac{3}{2}}}{2\sqrt{6}} \rho e^{-\rho/2}$$

Note in particular that the radial functions decay rapidly, so that the probability of an electron being farther than $\rho = 1$ or a distance greater than $r = nr_0/2$ away from the proton is very small. The distance r_0 is the radius of the smallest orbit permitted by the old semiclassical quantum theory.

In order for the solution as presented to satisfy the Schrödinger wave equation, the integers m, l, and n, which are called the quantum numbers, are restricted as follows:

$$m = 0, \pm 1, \pm 2, \pm 3, \ldots \tag{10.8a}$$

$$l = |m|, |m| + 1, |m| + 2, \ldots \tag{10.8b}$$

$$n = l + 1, l + 2, l + 3, \ldots \tag{10.8c}$$

The energy levels, or eigenvalues, W depend on n only, and are given by†

$$W_n = -\frac{2\pi^2 m e^4}{(4\pi\epsilon_0 hn)^2} \tag{10.9}$$

Since W_n depends on the integer n only, there are many (actually $2n^2$) wave functions, or quantum states, that an electron with energy W_n can occupy. These degeneracies are analogous to the degeneracies that occur in waveguides and cavities; e.g., in a rectangular guide the TE_{nm} and TM_{nm} modes have the same propagation constants.

The normal state of the hydrogen atom is the ground state corresponding to the lowest energy level W_1. In the ground state the complete wave function is

$$\psi_{100} e^{-j2\pi W_1 t/h} = R_{10}\Theta_{00}\Phi_0 e^{-j2\pi W_1 t/h}$$

$$= r_0^{-\frac{3}{2}} \pi^{-\frac{1}{2}} e^{-r/r_0} e^{-j2\pi W_1 t/h}$$

The probability that the electron is in the volume element $r^2 \sin\theta\, dr\, d\theta\, d\phi$ at the point r, θ, ϕ is

$$P(r)\, dV = \psi_{100}\psi_{100}^*\, dV$$

$$= \frac{1}{\pi r_0^3} e^{-2r/r_0} r^2 \sin\theta\, dr\, d\theta\, d\phi$$

and is independent of θ and ϕ. The probability that the electron is between r and $r + dr$ is obtained by integrating over θ and ϕ to give

$$P(r)\, dr = \frac{4}{r_0^3} e^{-2r/r_0} r^2\, dr \tag{10.10}$$

A sketch of ψ_{100}, $|\psi_{100}|^2$, and $P(r) = 4\pi|\psi_{100}|^2$ is given in Fig. 10.2, which shows a very high probability that the electron is within one angstrom of the nucleus.

† We have considered the proton to have infinite mass in order to simplify the discussion. The correct result for W_n should have the mass m replaced by the reduced mass $mM/(m + M)$. Note also that m is used to denote a quantum number. In context the meaning should be clear.

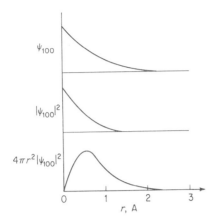

Fig. 10.2 Description of ground state of a hydrogen atom.

When the complete set of wave functions for an atom or a system of atoms is known, it is possible to calculate or evaluate any average value of a classical property of the system. For example, the average value of any function $F(r, \theta, \phi)$ of the coordinates is given by $\bar{F}$, where

$$\bar{F} = \int_0^\infty \int_0^\pi \int_0^{2\pi} \psi_{nlm}^* F(r, \theta, \phi) \psi_{nlm} r^2 \sin\theta \, dr \, d\theta \, d\phi \tag{10.11}$$

for the nlmth state of the atom. Any classical property described by a dynamical function such as

$$G(p_x, x, \ldots)$$

where p_x is the momentum in the x direction, etc., has an average value obtained by interpreting $G(p_x, x, \ldots)$ as a differential operator

$$G(p_x, x, \ldots) \rightarrow G\left(\frac{h}{2\pi j}\frac{\partial}{\partial x}, x, \ldots\right)$$

and finding the average according to the rule

$$\bar{G} = \int_0^\infty \int_0^\pi \int_0^{2\pi} \psi_{nlm}^* G\left(\frac{h}{2\pi j}\frac{\partial}{\partial x}, x, \ldots\right) \psi_{nlm} r^2 \sin\theta \, dr \, d\theta \, d\phi \tag{10.12}$$

The classical expression for angular momentum **S** is $\mathbf{S} = \mathbf{r} \times m\mathbf{v} = \mathbf{r} \times \mathbf{p}$ and has components

$$S_x = yp_z - zp_y$$
$$S_y = zp_x - xp_z$$
$$S_z = xp_y - yp_x$$

These are to be replaced by operators

$$S_x \rightarrow \frac{h}{2\pi j}\left(y\frac{\partial}{\partial z} - z\frac{\partial}{\partial y}\right) \cdots$$

in (10.12) to determine the average angular momentum associated with a given quantum state. By expressing these differential operators in spherical coordinates the integral (10.12) can be evaluated for the hydrogen atom to give

$$\bar{S}^2 = \bar{S}_x{}^2 + \bar{S}_y{}^2 + \bar{S}_z{}^2 = \frac{l(l+1)}{4\pi^2} h^2 \qquad (10.13)$$

for the nlmth state. Note that $\bar{S}^2$ depends only on the angular, or azimuthal, quantum number l, which describes the θ variation of the wave function. The average angular momentum in the z direction is found to be (a meaningful z axis exists only if the atom is placed in an external magnetic field)

$$\bar{S}_z = \frac{mh}{2\pi} \qquad (10.14)$$

and depends only on the "magnetic" quantum number m describing the ϕ variation in the wave function.

The angular momentum may be thought of as a vector having a projection $mh/2\pi$ along the z axis and a projection $\bar{S}_t$ given by

$$[l(l+1) - m^2]^{\frac{1}{2}} \frac{h}{2\pi}$$

on the xy plane, as in Fig. 10.3a. The total length of the vector is thus $(S_t{}^2 + S_z{}^2)^{\frac{1}{2}} = [l(l+1)h^2/4\pi^2]^{\frac{1}{2}}$ as required by (10.13). Since l and m are integers, it is apparent that $\bar{S}$ can take on only certain discrete orientations. This is the quantum basis for discrete orientations of the angular-momentum vector as compared with the possibility of a continuous spread of orientations permitted by the classical theory. However, representing the angular momentum by a vector with a specific orientation is somewhat artificial. The total average momentum given by

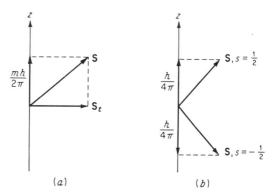

Fig. 10.3 Angular-momentum vector and spin-momentum vector.

(10.13) and its z-axis projection are the only significant parameters, so that the vector picture is merely a device used to bring out the difference between the classical and quantum-mechanical results. For a given quantum number n the allowed values of l are $l = 0, 1, 2, \ldots, n - 1$, since n is given by (10.8c). Thus, if $n = 2$, we must have $l = 0, 1$. If $l = 1$, we can have $m = 0, \pm 1$. Hence the angular-momentum vector can be directed along the z axis, perpendicular to the z axis, or antiparallel to the z axis in this case.

In addition to angular momentum associated with average orbital motion of the electron, the electron has an intrinsic spin momentum of its own. The spin momentum may be directed parallel or antiparallel to the z axis, and its z-axis projection has the value

$$S_{sz} = s \frac{h}{2\pi} \qquad s = \pm \tfrac{1}{2} \tag{10.15}$$

The spin momentum may also be thought of as a vector of total length $[s(s + 1)]^{\frac{1}{2}} h/2\pi$, where $s = \tfrac{1}{2}$. The permitted orientations of the spin-momentum vector are thus two, namely, those giving projections $\pm h/4\pi$ along the z axis, as in Fig. 10.3b.

For atoms consisting of many electrons, say N, and a nuclear charge Ne, the wave equation for the electrons is

$$\left[\sum_{i=1}^{N} \nabla_i^2 + \frac{8\pi^2 m}{h^2} \left(\sum_{i=1}^{N} \frac{Ne^2}{4\pi\epsilon_0 r_i} - \sum_{i=1}^{N} \sum_{\substack{s=1 \\ s \neq i}}^{N} \frac{e^2}{4\pi\epsilon_0 r_{is}} \right) + j \frac{4\pi m}{h} \frac{\partial}{\partial t} \right] \psi = 0 \tag{10.16a}$$

where r_i is the distance of the ith electron from the nucleus, r_{is} is the distance between the ith and sth electrons, ∇_i^2 is the laplacian operator operating on the coordinates of the ith electron, and the wave function ψ is a function of the coordinates of all electrons and time. An exact solution of this equation is not possible. Approximate solutions are obtained by either neglecting electron-electron interaction or replacing the electron-electron potential-energy term by a spherically symmetrical potential function which approximates the effect of $N - 1$ electrons on one particular electron. In this case ψ can be expressed as a product of single-electron wave functions similar to those occurring for the hydrogen atom. For later use we represent the quantity in brackets in (10.16a) by an operator $\mathcal{H}$ so that we have

$$\mathcal{H}\psi = 0 \tag{10.16b}$$

Statistical description of many-particle systems

In systems involving many particles, such as complex atoms and groups of atoms, each particle has a position described by a wave function which is a solution of Schrödinger's equation. However, in these systems the potential energy of any one particle depends on its location relative to that of all the other particles. Consequently, an exact solution of the wave equation is not feasible. It is now more expedient to describe statistically certain average characteristics, or properties, of the system as a whole.

Consider a macroscopic system of atoms. This system can exist in an infinite number of states, the nth state being described by a wave function ψ_n, which is a function of the coordinates of all the individual particles. The corresponding energy of this state is W_n. The probability that the system is in the state n with energy W_n is given by

$$P_n = A e^{-W_n/kT} \tag{10.17}$$

where A is a constant so chosen that

$$\sum_n P_n = A \sum_n e^{-W_n/kT} = 1$$

since the system must be in one of the possible states. In (10.17) T is the absolute temperature in degrees Kelvin and $k = 1.3709 \times 10^{-23}$ joule/deg is Boltzmann's constant. This probability distribution law, known as the Boltzmann distribution, will be useful to us later on in determining the relative number of atoms in any one energy level.

10.2 Absorption and emission of radiation

If an atom in a state with energy W_n makes a transition to a lower energy state, with energy W_m, it will radiate an amount of energy $W_n - W_m$ at a frequency f_{nm}, given by Planck's formula

$$f_{nm} = \frac{W_n - W_m}{h} \tag{10.18}$$

Conversely, if an atom in a state with energy W_m is placed in an electromagnetic field of frequency f_{nm}, it can absorb a quantum of energy and make an upward transition to the nth state with energy W_n. In materials under normal conditions the population, or number of atoms, in the low energy states greatly exceeds those in high energy states, as examination of the Boltzmann distribution (10.17) shows. Consequently, absorption of radiation predominates over the emission of radiation. In a maser it is possible to reverse this situation by creating a population inversion between two energy levels, i.e., by making the number of atoms in the higher energy level greater than that in the lower energy level.

When an atom is placed in an electromagnetic field with frequency f, there is always a finite probability that it will make a transition from some initial state to some other final state with either absorption or emission of radiation. The probability of a transition taking place can be determined by a perturbation analysis and is outlined below.

Consider an atom for which all the stationary quantum eigenstates are known. These are described by an infinite set of wave functions ψ_n with energy eigenvalues W_n. When the atom is placed in an electromagnetic field, its state at any time can be described by a Fourier series involving the complete set of functions ψ_n, with coefficients $a_n(t)$ that are functions of time; thus

$$\psi = \sum_n a_n(t)\psi_n e^{-j2\pi W_n t/h} \tag{10.19}$$

The time-dependent factors $e^{-j2\pi W_n t/h}$ are included only for later simplification in the analysis. We could equivalently consider $a_n(t)e^{-j2\pi W_n t/h}$ as a new coefficient $c_n(t)$. If the applied electric field is in the x direction only and is

$$E_x = E_0 \cos \omega t$$

the energy of an electron in this field is $eE_0 x_i \cos \omega t$, where x_i is the x coordinate of the ith electron in the atom. The wave equation (10.16) is modified by the addition of this extra potential-energy term and thus becomes

$$\left(\mathcal{H} - \sum_{i=1}^{N} eE_0 x_i \cos \omega t\right)\psi = 0 \tag{10.20}$$

If we substitute the assumed expansion for ψ into (10.20) and note that the ψ_n are solutions of $\mathcal{H}\psi_n = 0$, we obtain

$$\sum_{i=1}^{N} eE_0 x_i \cos \omega t \sum_n a_n(t)\psi_n e^{-j2\pi W_n t/h} = -\frac{h}{2\pi j}\sum_n \psi_n e^{-j2\pi W_n t/h} \frac{da_n(t)}{dt} \tag{10.21}$$

The eigenfunctions ψ_n occurring in quantum theory are always orthogonal, so that

$$\int_V \psi_m^* \psi_n \, dV = \begin{cases} 1 & m = n \\ 0 & m \neq n \end{cases}$$

where the integration is over the coordinates of all particles. Thus, if we multiply (10.21) by $\psi_m^* e^{j2\pi W_m t/h}$ and integrate over the coordinates of all electrons, we obtain

$$\frac{da_m(t)}{dt} = -\frac{2\pi j}{h} E_0 \cos \omega t \sum_n a_n e^{-j2\pi(W_n - W_m)t/h} \sum_{i=1}^{N} \int_V \psi_m^* e x_i \psi_n \, dV$$

$$\text{for } m = 0, 1, 2, \ldots \tag{10.22a}$$

The coefficients $a_n(t)$ are slow-varying compared with the exponential terms $e^{-j2\pi(W_n - W_m)t/h}$, and may be replaced by their initial values $a_n(0)$ in order to simplify the equation for $da_m(t)/dt$. Thus we have

$$\frac{da_m(t)}{dt} = -\frac{2\pi j}{h} E_0 \cos \omega t \sum_n a_n(0) p_{x,nm} e^{-j2\pi(W_n - W_m)t/h} \tag{10.22b}$$

where $p_{x,nm}$ is defined by the integral

$$\sum_{i=1}^{N} e \int_V \psi_m^* x_i \psi_n \, dV = p_{x,nm}$$

and is the x component of the electric dipole moment $p_{x,nm}$ of the atom, for a transition from state n to state m.

Let us now assume that at time $t = 0$ the atom was in state n, so that $a_n(0) = 1$ and $a_m(0) = 0$ for all $m \neq n$. Note that

$$\int_V \psi^* \psi \, dV = \sum_n \sum_m a_n^* a_m \int_V \psi_n^* \psi_m \, dV = \sum_n a_n a_n^* = 1$$

since ψ must be normalized; so we must have $a_n(0) = 1$ if all other $a_m(0)$ are zero. It is now a simple matter to integrate (10.22b) to obtain

$$a_m(t) = \tfrac{1}{2} E_0 p_{x,nm} \left(\frac{1 - e^{(2\pi j/h)(W_m - W_n + hf)t}}{W_m - W_n + hf} + \frac{1 - e^{(2\pi j/h)(W_m - W_n - hf)t}}{W_m - W_n - hf} \right)$$

(10.23)

which shows how the coefficient $a_m(t)$ for the mth state will develop in time. It is apparent that only if the frequency f is close to the value $|W_m - W_n|/h$ will the probability of transition to state m from state n be significant. The phenomenon involved here is very similar to that in exciting a cavity by an impressed current at frequency f. Only when f is close to one of the cavity resonances will a mode be excited with appreciable amplitude. If $hf = \pm(W_m - W_n)$, we obtain

$$a_m(t) = -\frac{\pi j}{h} E_0 p_{x,nm} t \tag{10.24}$$

so $a_m(t)$ initially grows linearly with time.

The probability that an atom will absorb a quantum of radiation and make a transition from an energy level W_n to a higher energy level W_m in unit time is expressed as follows,

$$P_{nm,a} = B_{nm} U(f_{nm}) \tag{10.25}$$

where $U(f_{nm}) \, df$ is the energy density per unit volume of the electromagnetic field at the frequency f_{nm} in an interval df, and B_{nm} is called Einstein's coefficient of absorption. An atom in an energy level W_m greater than W_n has a natural tendency to emit radiation and drop down to energy level W_n, with a probability A_{mn} of this occurring in unit time. When an electromagnetic field is present, it will act to stimulate the atom to drop from level W_m to W_n, with a probability given by (10.25), as we show later. Thus the probability of emission of radiation by transition from state m to state n in unit time is

$$P_{nm,e} = A_{mn} + B_{mn} U(f_{nm}) \tag{10.26}$$

where B_{mn} is Einstein's coefficient for induced emission and is equal to B_{nm}. The coefficient A_{mn} is Einstein's coefficient for the probability of spontaneous emission.

We now calculate the value of B_{nm}. From (10.23) we obtain

$$a_m(t)a_m^*(t) = |E_0|^2 p_{z,nm}^2 \frac{\sin^2[(\pi/h)(W_m - W_n - hf)t]}{(W_m - W_n - hf)^2}$$

if we assume $W_m > W_n$ and $hf \approx W_m - W_n$, so that the first term on the right-hand side may be dropped.

Any physical field would not be monochromatic but would consist of a band of frequencies around f_{nm}. Thus we regard E_0 as a function of f such that $E_0(f)\cos 2\pi ft$ is the Fourier component of $E_x(t)$ at the frequency f. We must integrate the right-hand side of the above equation over f. But since the integrand is highly peaked in the vicinity of f_{nm}, we can take $E_0(f)$ equal to $E_0(f_{nm})$ and extend the integration from minus to plus infinity with negligible error. Thus, using the result

$$\int_{-\infty}^{\infty} \frac{\sin^2 x}{x^2} dx = \pi$$

we obtain

$$a_m(t)a_m^*(t) = \frac{\pi^2}{h^2} p_{z,nm}^2 |E_0|^2 t \tag{10.27}$$

If $E_x(t)$ is the electric field, the total energy density in the field is

$$\frac{\epsilon_0}{2} \int_{-\infty}^{\infty} E_x^2(t)\,dt = U_{xt}$$

This can be expressed in the frequency domain also. Thus if we let $E_0/2$ be the Fourier transform of $E_x(t)$, that is,

$$\frac{E_0(f)}{2} = \int_{-\infty}^{\infty} E_x(t)e^{-j\omega t}\,dt$$

we find that

$$U_{xt} = \frac{\epsilon_0}{4} \int_0^{\infty} |E_0(f)|^2\,df$$

so that $\epsilon_0|E_0|^2/4$ is the energy per unit volume in a frequency interval df.

If the incident field had y and z components also with the same amplitude, the energy density in the electromagnetic field would be

$$U(f) = 2\frac{\epsilon_0}{4}(|E_x|^2 + |E_y|^2 + |E_z|^2) = \tfrac{3}{2}\epsilon_0|E_0|^2$$

since the magnetic field contributes an equal amount to the energy density. In place of (10.27) we should have

$$a_m(t)a_m^*(t) = \frac{\pi^2}{h^2} E_0^2(p_{x,nm}^2 + p_{y,nm}^2 + p_{z,nm}^2)t \tag{10.28}$$

in general. If we regard U as the energy density such that $U\,df$ is the energy per unit volume in a small frequency interval df, we can express E_0^2 in terms of U to obtain

$$\frac{a_m(t)a_m^*(t)}{t} = \frac{2\pi^2}{3\epsilon_0 \hbar^2} U(p_{x,nm}^2 + p_{y,nm}^2 + p_{z,nm}^2)$$

This is the probability per unit time that the atom will make a transition from state n to state m. If we compare with (10.25), we see that B_{nm} is given by

$$B_{nm} = \frac{2\pi^2}{3\epsilon_0 \hbar^2}(p_{x,nm}^2 + p_{y,nm}^2 + p_{z,nm}^2) \tag{10.29}$$

For induced emission we again consider $W_m > W_n$ but use the initial condition $a_m(0) = 1$ and all other $a_n(0) = 0$. A parallel development then shows that $B_{mn} = B_{nm}$, so that the coefficients of induced emission and absorption are equal.

The coefficient A_{mn} can be found in terms of B_{nm}, as we show below. If a system of atoms in a field with energy density U is in equilibrium with the radiation field, the number of upward transitions must equal the number of downward transitions. The relative numbers of atoms with energies W_m and W_n are given by

$$\frac{N_m}{N_n} = e^{-(W_m - W_n)/kT}$$

from Boltzmann's distribution. The number of transitions from level W_n to W_m in unit time is given by $B_{nm}U$ times the number of atoms present that can make the required transition—thus,

$$N_n B_{nm} U$$

and the number of transitions from level W_m down to level W_n with emission of radiation occurring is given by

$$N_m(A_{mn} + B_{nm}U)$$

per unit time. For equilibrium we must have

$$N_n B_{nm} U = N_m(A_{mn} + B_{nm}U)$$

and hence

$$\frac{N_m}{N_n} = \frac{B_{nm}U}{A_{mn} + B_{nm}U} = e^{-(W_m - W_n)/kT}$$

so that

$$A_{mn} = B_{nm} U \frac{1 - e^{-(W_m - W_n)/kT}}{e^{-(W_m - W_n)/kT}}$$

But for a system in thermodynamic equilibrium with its own radiation at a temperature T, the density of radiation U is given by Planck's law as

$$U = \frac{8\pi h f_{nm}^3}{c^3} \frac{e^{-hf_{nm}/kT}}{1 - e^{-hf_{nm}/kT}} \tag{10.30}$$

at the frequency $f_{nm} = (W_m - W_n)/h$. Using this result in the expression for A_{mn} shows that

$$A_{mn} = \frac{8\pi h f_{nm}^3}{c^3} B_{nm} \tag{10.31}$$

From the preceding relations it is readily found that, for a system in thermodynamic equilibrium, the spontaneous-emission probability will equal the induced-emission probability at a temperature T given by

$$T = \frac{hf_{nm}}{k \ln 2} = 6.9 \times 10^{-11} f_{nm} \tag{10.32}$$

For a frequency $f_{nm} = 10^{10}$ cycles, this corresponds to a temperature of 0.69°K. It should be noted that the energy density U for a system in thermodynamic equilibrium is the energy density of the radiation field produced by the system of atoms involved. In a more general situation, when there is an externally applied field, the induced emission can greatly exceed the spontaneous emission even if the temperature is greater than that given by (10.32). It is only necessary for the energy density U to exceed the value $8\pi h f_{nm}^3/c^3$ in (10.31) for induced emission to be greater than the spontaneous emission. At a frequency $f_{nm} = 10^{10}$ cycles, this corresponds to an energy density greater than 6.1×10^{-28} joule/m³.

The analysis given above was for the case of electric dipole radiation. In general, an atom may have a net magnetic dipole moment also. In this case there will be an additional potential-energy term, representing the interaction energy between the magnetic dipole moment and the applied magnetic field, introduced into the Schrödinger equation. The coefficients B_{nm} and B_{mn} will then have a contribution from the magnetic dipole moment of the atom similar in form to that arising from the electric dipole moment.

The classical model of electric dipole polarization in materials is that of charge bound elastically to the nucleus plus an artificial damping term that accounts for energy absorption. The quantum-mechanical description of absorption justifies this model, provided the constants in the classical equation of motion are chosen so as to yield a resonance at the transition frequency f_{nm} and to have a damping constant that gives the correct amount of absorption. When several transitions from the normally occupied ground state to higher energy levels are involved, the classical model must be assumed to be a sum of damped oscillators, one for each transition, and with parameters chosen to give agreement with the quantum-mechanical results.

When an atom makes a downward transition from a level W_m to a level

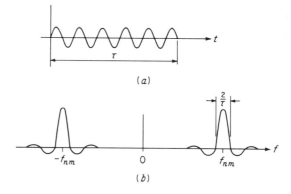

Fig. 10.4 Characteristics of radiation emitted by an atom.

W_n, it emits radiation at the frequency $f_{nm} = (W_m - W_n)/h$. The transition takes a finite length of time; so the wave train emitted is a sinusoidal wave of finite length τ, as illustrated in Fig. 10.4a. The Fourier frequency spectrum of this wave train is given by

$$\frac{\tau}{2}\left[\frac{\sin \pi(f_{nm} - f)\tau}{\pi(f_{nm} - f)\tau} + \frac{\sin \pi(f_{nm} + f)\tau}{\pi(f_{nm} + f)\tau}\right]$$

and is sketched in Fig. 10.4b. The width of the spectrum is essentially

$$\Delta f = \frac{2}{\tau}$$

and is called the natural line width. The natural line width is very small, of the order $10^{-6}f_{nm}$.

In a system consisting of a large number of atoms, the potential energy of the electrons in the different atoms is perturbed by the presence of other atoms whenever "collisions" take place. As a result the emission or absorption of radiation has a very much broader associated line width (collision broadening) because of the spread in values of f_{nm} brought on by the spread in values of the energy levels W_n and W_m.

Spontaneous emission occurs with random phase from one atom to the next. However, for induced emission, all atoms emit radiation that is in phase and coherent with the stimulating radiation field. This property of induced emission is of primary importance in the operation of a maser.

10.3 Description of a maser amplifier

This section presents a qualitative description of a typical maser amplifier. An analytical treatment is given in later sections. We first describe the essential components making up a maser amplifier.

The three-level ruby maser consists of a cavity containing a ruby crystal and immersed in a liquid-helium bath, as shown in Fig. 10.5. The helium dewar is surrounded by a second dewar containing liquid nitrogen, which is used only to prevent the rapid boiling away of the liquid helium. The cavity is designed to be resonant at two frequencies simultaneously, the pump frequency f_p and the signal frequency f_s. The pump power is fed into the cavity through a coaxial line or waveguide. The signal is coupled to the cavity through a three-port circulator. The latter is used to keep the input signal and the amplified output signal separated in order to prevent amplifier instability and the possibility of oscillations.

The active material is a paramagnetic material such as ruby. Three suitable energy levels are obtained by applying a strong static magnetic field across the ruby by means of a permanent magnet. A magnetic dipole of moment **M** has an energy $W = -\mathbf{M} \cdot \mathbf{B}_0 = -MB_0 \cos \theta$ when placed in a magnetic field. Quantum-mechanical considerations show that only certain discrete orientations or values of θ are permitted. Three energy levels corresponding to three orientations of the dipole moment **M** are illustrated schematically in Fig. 10.6. The energy levels are labeled W_1, W_2, and W_3. Under thermodynamic equilibrium conditions the number of magnetic dipoles in levels W_1, W_2, and W_3 is N_1, N_2,

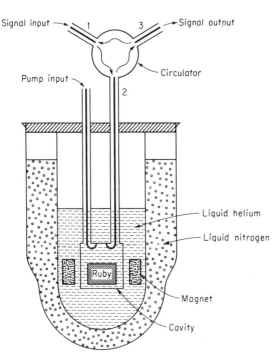

Fig. 10.5 Basic components of a maser amplifier.

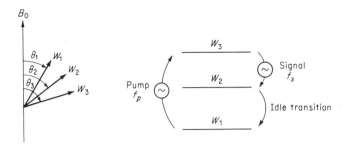

Fig. 10.6 Illustration of energy levels in a ruby crystal.

and N_3, respectively, with $N_1 > N_2 > N_3$, as dictated by the Boltzmann distribution. To achieve maser action it is necessary to invert the population of levels W_3 and W_2. That is, it is necessary to have a greater number of magnetic dipoles in the energy level W_3 than there are dipoles in level W_2. This population inversion is accomplished by applying a strong pump signal at the frequency $f_p = (W_3 - W_1)/h$, which has the effect of inducing transitions from level W_1 to level W_3. The magnetic dipoles pumped up to level W_3 will relax back to level W_2 and then to level W_1 with the emission of radiation at the frequencies $(W_3 - W_2)/h$ and $(W_2 - W_1)/h$. If we also apply a signal at the frequency $(W_3 - W_2)/h$ to the crystal, the effect is to stimulate transitions from level W_3 to level W_2 at a greater rate. When the crystal is kept at liquid-helium temperature, the number of stimulated transitions will be greater than the number of spontaneous transitions. Thus, at low temperatures, the number of transitions from level W_3 to level W_2 per unit time is controlled by the applied signal. One photon of applied signal will trigger, or stimulate, many transitions, so that the emitted radiation at the signal frequency is much larger in amplitude than the applied signal. Since the stimulated radiation is also in phase and coherent with the stimulating signal, the ruby crystal provides amplification.

When the populations of levels W_3 and W_2 are inverted, it has the effect of making the imaginary part of the magnetic permeability of the ruby positive. Thus, instead of having a loss tangent, the material exhibits a gain tangent. If the gain is greater than all other cavity losses, the cavity will act as an LC circuit shunted by a negative resistance. It is apparent that the cavity will then be able to supply power at the signal frequency. A circulator is used to keep the input and amplified signals separated. If this were not done and the output signal were partially reflected in the input line, the reflected part would be amplified again. Successive reflections and amplification could lead to an oscillatory condition, which is undesirable. A more detailed analysis of the maser is given in the following sections.

10.4 Energy levels in ruby

In Sec. 10.1 the possible quantum-mechanical configurations of a simple hydrogen atom were discussed. In this section we present a generalization of these earlier results to more complex atoms. As a first approximation to the solution of Schrödinger's wave equation (10.16a) for an atom with N electrons, the equation is approximated by replacing the electron-electron interaction energy term

$$\sum_{i=1}^{N} \sum_{s=1}^{N} \frac{e^2}{4\pi\epsilon_0 r_{is}}$$

by a spherically symmetric potential energy of the form

$$\sum_{i=1}^{N} eV_i(r)$$

where $V_i(r)$ is the approximate potential produced by all electrons except the ith electron. With this approximation the wave equation separates into a single-wave equation for each electron and has the form

$$\left\{\nabla_i^2 + \frac{8\pi^2 m^2}{h^2}\left[\frac{Ne^2}{4\pi\epsilon_0 r_i} - eV_i(r)\right] + j\frac{4\pi m}{h}\frac{\partial}{\partial t}\right\}\psi_i = 0$$

where ψ_i is the wave function for the ith electron. The total wave function for the atom is the product of all the ψ_i, that is, $\psi = \psi_1\psi_2 \cdots \psi_N$. It is apparent that each electron will have a number of possible quantum configurations similar to those of the hydrogen atom. However, further considerations show that, in an atom, no two electrons can have the same wave function. This condition is called the Pauli exclusion principle. In addition, the number of degeneracies, or wave functions, corresponding to the nth energy level W_n is not the same as for the hydrogen atom. Also, the energy now depends on the angular quantum number l. Corresponding to any given energy W_n, the electron's position is predominantly at a particular radius r_n, with $r_n > r_{n-1}$ for all n. We can

Table 10.1 Shell designation

Energy level	Orbital-momentum quantum number l			
	0	1	2	3
W_1	1s			
W_2	2s	2p		
W_3	3s	3p	3d	
W_4	4s	4p	4d	4f

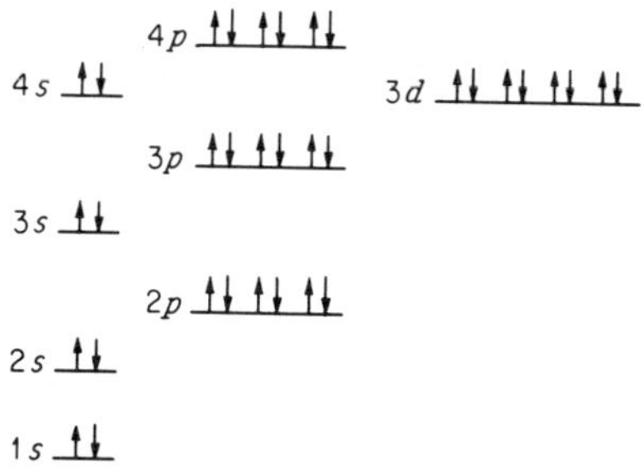

Fig. 10.7 Energy levels and occupation numbers in various electron shells in an atom.

therefore picture the electrons as grouped into shells. The various shells are labeled in the manner shown in Table 10.1. In a hydrogen atom an electron in the 2s or 2p shell or configuration has the same energy W_2. In a more complex atom, the energy of an electron in the 2p shell or configuration has an energy higher than that of an electron in the 2s shell. Thus there are fewer degenerate eigenstates, i.e., fewer wave functions with the same energy, in a complex atom. Figure 10.7 illustrates the relative energy levels for the various shells.

The number of electrons that can exist in any given shell is fixed by the Pauli exclusion principle to one electron for each wave function. Thus, in the 1s shell, two electrons can exist, one with a z-directed spin of $h/2\pi$ and one with a z-directed spin of $-h/2\pi$. The number of electrons permitted in other shells is shown in Fig. 10.7. The arrows indicate the two possible spin-momentum orientations that are permitted to exist. An atom with N electrons will normally exist in the ground state, or lowest energy state, corresponding to the filling up of the various shells, beginning with the 1s shell.

For a given electronic configuration of an atom there is a definite angular momentum, with contributions arising from both the orbital angular momentum and the spin angular momentum of each electron. The magnetic dipole moment is related to the total angular momentum. From a classical viewpoint, an electron with velocity **v** in a circular orbit has an angular momentum $\mathbf{S} = m\mathbf{r} \times \mathbf{v}$. The equivalent circulating current is $-ev$, and hence the magnetic dipole moment is

$$\mathbf{M} = -e\tfrac{1}{2}\mathbf{r} \times \mathbf{v}$$

Consequently, $\mathbf{M} = -(e/2m)\mathbf{S}$, a result which is correct in quantum mechanics also. However, for the electron the magnetic dipole moment $\mathbf{M}_s$ is related to the spin angular momentum by the relation

$$\mathbf{M}_s = -\frac{e}{m}\mathbf{S}_s$$

Since the proportionality factor is different for the two, the total magnetic dipole moment is related to the total angular momentum by some other factor g, called the spectroscopic splitting factor. The numerical value

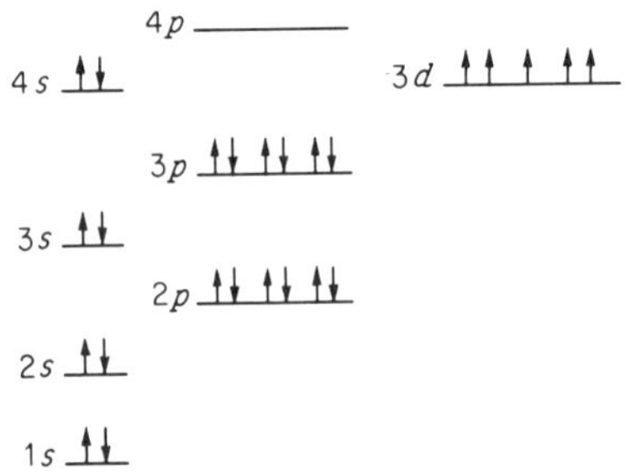

```
4p ─────
4s ↑↓           3d ↑ ↑ ↑ ↑ ↑
       3p ↑↓ ↑↓ ↑↓
3s ↑↓
       2p ↑↓ ↑↓ ↑↓
2s ↑↓
1s ↑↓
```

Fig. 10.8 Electronic configuration of a chromium atom.

of g depends on the relative contributions of the orbital and spin angular momentum to the total angular momentum. If we denote the total angular momentum by $\mathbf{S}_t$, the magnetic dipole moment is given by

$$\mathbf{M} = -\frac{e}{2m} g \mathbf{S}_t \tag{10.33}$$

Since $\mathbf{S}_t$ is given in units of $h/2\pi$, the magnetic moment of an electron is $eh/2\pi m$, a quantity called one Bohr magneton. In many materials the magnetic moment arises mainly from the electron spin, and g will then be close to 2. Such materials are called paramagnetic materials.

If Fig. 10.7 is examined, it will be seen that those shells which are filled have complete cancellation of the magnetic dipole moments arising from electron spin. Only in materials with an incomplete shell will a net spin magnetic moment exist. In the so-called transition group of elements, such as the iron group, which includes Sc, Ti, V, Cr, Mn, Fe, Co, Ni, Cu, and Zn, the two shells $4s$ and $3d$ have equal energies but the $4s$ shell does not fill up completely before the $3d$ shell is occupied. For example, chromium, with 24 electrons, has the electronic configuration shown in Fig. 10.8. In chemical combination with other elements the $4s$ electron and one or two of the $3d$ electrons act as valence electrons to complete the shell of the atom that the chromium is combined with. This still leaves a number of electrons with uncanceled spin magnetic dipole moments in the $3d$ shell, so that the material will exhibit a net magnetic dipole moment. The common maser material is ruby and consists of about 0.1 percent of chromium oxide Cr_2O_3 in a mixture with 99.9 percent of nonmagnetic aluminum oxide Al_2O_3. There are approximately 10^{19} magnetically active chromium ions per cubic centimeter in this mixture.

In the chromium ion Cr^{3+}, there are three unpaired electron spins in the $3d$ shell. These spins can assume the following orientations: all three directed up, two directed up and one directed down, two directed down and one up, and all three directed downward. The corresponding magnetic dipole moment along the z axis is

$$\mathbf{M}_z = -\frac{eh}{2\pi m} g \mathbf{a}_z s \qquad s = \tfrac{3}{2}, \tfrac{1}{2}, -\tfrac{1}{2}, -\tfrac{3}{2} \tag{10.34}$$

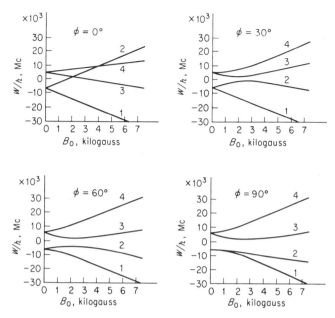

Fig. 10.9 Energy levels in a ruby crystal as a function of field strength B_0 in kilogauss and field orientation angle ϕ relative to the crystal c axis. (*Adapted from "Microwave Solid-state Masers," pp. 115–116, by A. E. Siegman, Copyright, 1964. McGraw-Hill Book Company. Used by permission.*)

When the ruby crystal is placed in a magnetic field, the four different projections of the total magnetic dipole moment on the z axis as given by (10.34) correspond to four different orientation angles θ relative to the z axis. Since the corresponding energy is given by

$$W = -\mathbf{M} \cdot \mathbf{B}_0 = -MB_0 \cos \theta = -M_z B_0$$

the original energy levels are split into new energy levels, depending in absolute value on the field strength B_0. In a free Cr^{3+} ion, the energy for any spin orientation of the electrons in the $3d$ shell is the same. When the ion is incorporated into a crystal structure, there is a zero field splitting of the energy into two levels because of crystalline electric fields. The application of a field B_0 splits these two energy levels into a total of four levels, as shown in Fig. 10.9. In addition, the dependence of the energy on the field B_0 depends on the orientation of the field relative to the crystal axis since ruby is an anisotropic crystalline medium. This energy dependence on orientation angle ϕ relative to the c axis of the crystal is shown in Fig. 10.9 also. The splitting of energy levels by a magnetic field is known as Zeeman splitting. Note that the energy difference between different levels corresponds to frequencies in the microwave range for fields B_0 in the range of 1 to 4 kilogauss (0.1 to 0.4 weber/m^2). In contrast to this, the frequencies associated with the transition of electrons from one shell to a lower- or higher-energy adjacent shell are in the optical range. Thus microwave masers rely on the Zeeman splitting of energy levels in paramagnetic materials. When ruby is used as the active material, the three lower energy levels are employed to obtain maser action. The pump is used to pump spins from level W_1 up to the energy level W_3. The signal controls the transition rate from level W_3 to level W_2. The spins subsequently relax naturally back down to level W_1 to complete the circuit.

10.5 Analysis of maser action

In the preceding section we gave a qualitative description of how the energy levels in a ruby crystal arose when it was placed in a static magnetic field B_0. We now study the dynamics of spin transitions between different energy levels in the presence of electromagnetic fields at the pump and signal frequencies.

Consider three energy levels W_1, W_2, and W_3 as illustrated schematically in Fig. 10.10. In the absence of pump and signal power, let the number of spins per unit volume of crystal in the three energy levels be N_1, N_2, and N_3, respectively. If we let $N = N_1 + N_2 + N_3$ be the total number of spins associated with these three energy levels, under thermodynamic equilibrium the Boltzmann distribution law gives

$$N_i = N \frac{e^{-W_i/kT}}{e^{-W_1/kT} + e^{-W_2/kT} + e^{-W_3/kT}}$$

$$\approx N \frac{1 - W_i/kT}{3} \qquad i = 1, 2, 3; \; W_i \ll kT \tag{10.35}$$

for the number density of spins in each energy level. The denominator in (10.35) is chosen so that $N_1 + N_2 + N_3 = N$. When pump and signal power are also present, the system is not in thermodynamic equi-

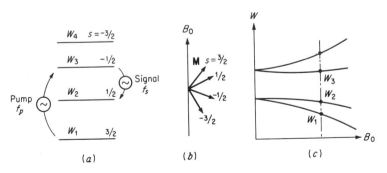

Fig. 10.10 The three energy levels used in a ruby maser.

librium. For this situation we shall let n_1, n_2, and n_3 be the number of spins in energy levels W_1, W_2, and W_3, respectively, per unit volume.

In order to write down the rate equations that will determine the steady-state values of the n_i, we must consider the various factors that cause transitions to take place. First of all, there are the stimulated transition probabilities per unit time that the applied pump or signal power will cause a transition to occur. If the pump frequency f_p is chosen equal to $(W_3 - W_1)/h$, we can have stimulated transitions between levels W_1 and W_3. We shall let C_{13} be the probability per unit time of a stimulated transition between W_1 and W_3 occurring. Thus, since the probabilities for upward and downward stimulated transitions are equal, the net number of upward transitions from level W_1 to level W_3 per unit volume and per unit time is $C_{13}(n_1 - n_3)$. The transition probability C_{13} is essentially Einstein's coefficient B_{13} multiplied by the pump power density in the crystal. However, C_{13} arises from the magnetic dipole moment of the chromium ion, and not from the electric dipole moment, in the present case. Similarly, the net number of upward transitions between levels W_2 and W_3 caused by signal power at a frequency $f_s = (W_3 - W_2)/h$ will be $C_{23}(n_2 - n_3)$, where C_{23} is the Einstein coefficient B_{23} multiplied by the signal power.

The spins will interact with the thermal vibrations of the crystal lattice, with the result that both upward and downward transitions between the different energy levels can occur. These transitions are said to arise from spin-lattice interaction. We shall let D_{ij} be the probability per unit time of a transition between levels W_i and W_j occurring because of spin-lattice interaction. Between any two levels, say W_1 and W_2, the number of upward and downward transitions per unit time will be $D_{12}n_1$ and $D_{21}n_2$, respectively. Under equilibrium conditions we must have $D_{12}N_1 = D_{21}N_2$, and hence

$$\frac{D_{12}}{D_{21}} = \frac{N_2}{N_1} = e^{(W_1 - W_2)/kT} \qquad (10.36)$$

where T is the equilibrium temperature. In spin-lattice interaction the induced transitions are not accompanied by the emission or absorption of a photon of electromagnetic radiation. The energy exchange is instead an increase or a decrease in the vibrational energy of the crystal lattice. This energy exchange is referred to as the emission or absorption of a phonon, which is a quantum of vibrational energy rather than a quantum of electromagnetic energy.

In addition to transitions induced by spin-lattice interactions, there are also transitions induced by the presence of blackbody radiation fields. In particular, there is spontaneous emission of radiation that arises from spontaneous transitions having a probability per unit time given by Einstein's A_{ij} coefficient. If U is the blackbody radiation energy density present, the number of induced upward transitions from level W_i to level

W_j per unit time is $UB_{ij}n_i$ and the number of downward transitions per unit time is $UB_{ji}n_j + A_{ji}n_j$, where $B_{ij} = B_{ji}$. The B_{ij} are determined by the magnetic dipole moments of the chromium ion for transitions between levels i and j. The spontaneous emission as determined by A_{ji} is responsible for the noise properties of the maser since the accompanying radiation is not coherent with either the pump or signal field. In a microwave maser kept at liquid-helium temperatures the blackbody radiation energy density U is so small that the corresponding transition probabilities are negligible compared with the transition probabilities due to spin-lattice interaction. Thus the rate equations for the population-number densities n_i are for all practical purposes independent of transitions induced by blackbody or thermal radiation.

If we take the above considerations into account, we can state that the rate of change of the population density of level W_1 will be equal to the number of transitions from levels 2 and 3 to level 1 minus the number of transitions from level 1 up to levels 2 and 3. Thus

$$\frac{dn_1}{dt} = D_{21}n_2 + D_{31}n_3 - (D_{12} + D_{13})n_1 + C_{13}(n_3 - n_1) \tag{10.37a}$$

Similarly, we have

$$\frac{dn_2}{dt} = D_{12}n_1 + D_{32}n_3 - (D_{21} + D_{23})n_2 + C_{23}(n_3 - n_2) \tag{10.37b}$$

$$\frac{dn_3}{dt} = D_{13}n_1 + D_{23}n_2 - (D_{31} + D_{32})n_3 + C_{23}(n_2 - n_3) \tag{10.37c}$$

Note that terms in C_{12} are absent, since no applied field at the frequency $(W_2 - W_1)/h$ is present. This latter transition is frequently called the idle transition for this reason. These rate equations are the fundamental equations describing the behavior of a three-level maser. A rate equation for level W_4 should also be included for generality. However, the population density n_4 does not vary significantly, so that for simplicity we are justified in considering only three rate equations.

We are primarily interested in the steady-state operating situation. In this case n_1, n_2, and n_3 do not change with time, and (10.37) gives

$$D_{21}n_2 + D_{31}n_3 - (D_{12} + D_{13})n_1 + C_{13}(n_3 - n_1) = 0 \tag{10.38a}$$

$$D_{12}n_1 + D_{32}n_3 - (D_{21} + D_{23})n_2 + C_{23}(n_3 - n_2) = 0 \tag{10.38b}$$

$$D_{13}n_1 + D_{23}n_2 - (D_{31} + D_{32})n_3 + C_{23}(n_2 - n_3) = 0 \tag{10.38c}$$

Before proceeding further, consider a simple two-level spin system with no external energy supplied. The rate equations would be

$$\frac{dn_1}{dt} = D_{21}n_2 - D_{12}n_1$$

$$\frac{dn_2}{dt} = D_{12}n_1 - D_{21}n_2$$

Subtracting these equations gives

$$\frac{d\Delta n_{12}}{dt} = \frac{d(n_1 - n_2)}{dt} = 2D_{21}n_2 - 2D_{12}n_1$$

We may rewrite this equation as

$$\frac{d\Delta n_{12}}{dt} = D_{21}(n_2 - n_1) + D_{21}(n_2 + n_1) + D_{12}(n_2 - n_1) - D_{12}(n_2 + n_1)$$

$$= -\Delta n_{12}(D_{12} + D_{21}) + (D_{21} - D_{12})N$$

since $n_1 + n_2 = N_1 + N_2 = N$. From (10.36) we have $N_1 D_{12} = N_2 D_{21}$; so the term $(D_{21} - D_{12})N = (D_{21} - D_{12})(N_1 + N_2)$ may be written as $(D_{12} + D_{21})\Delta N_{12}$ by adding a term $2(D_{12}N_1 - D_{21}N_2)$. Thus

$$\frac{d\Delta n_{12}}{dt} = -(D_{12} + D_{21})(\Delta n_{12} - \Delta N_{12})$$

and has the solution

$$\Delta n_{12} = \Delta N_{12} + (\Delta n_0 - \Delta N_{12})e^{-(D_{12}+D_{21})t} \tag{10.39}$$

where Δn_0 is the population difference at $t = 0$. The quantity $(D_{12} + D_{21})^{-1}$ is a relaxation time constant τ_{12} and gives the rate at which the population difference approaches the thermodynamic-equilibrium value ΔN_{12}.

If we introduce relaxation time constants $\tau_{ij} = (D_{ij} + D_{ji})^{-1}$ and population differences $\Delta n_{ij} = n_i - n_j$, we can rewrite the steady-state rate equations (10.38) in the form

$$\frac{\Delta n_{12} - \Delta N_{12}}{2\tau_{12}} + \frac{\Delta n_{13} - \Delta N_{13}}{2\tau_{13}} + C_{13}\Delta n_{13} = 0 \tag{10.40a}$$

$$\frac{\Delta n_{21} - \Delta N_{21}}{2\tau_{12}} + \frac{\Delta n_{23} - \Delta N_{23}}{2\tau_{23}} + C_{23}\Delta n_{23} = 0 \tag{10.40b}$$

$$\frac{\Delta n_{31} - \Delta N_{31}}{2\tau_{13}} + \frac{\Delta n_{32} - \Delta N_{23}}{2\tau_{23}} + C_{13}\Delta n_{31} + C_{23}\Delta n_{32} = 0 \tag{10.40c}$$

These algebraic equations can be solved exactly for the Δn_{ij}. However, the solutions are not easy to interpret, and for this reason we shall obtain an approximate solution only, which will still be sufficiently accurate to predict the main conclusions that can be deduced. First of all, we note that all τ_{ij} and Δn_{ij} are finite; so if we increase the pump power indefinitely, the population difference Δn_{13} must vanish. This is because C_{13} is directly proportional to the pump power. Hence the maximum population density of energy level W_3 that can be obtained is just equal to the population density of level W_1; that is, $n_3 = n_1$. This condition is called the spin saturation condition for the pump transition. Most masers are operated with large enough pump power to produce saturation, or at least very nearly so. If we put $\Delta n_{13} = 0$ in (10.40b) and solve for

Δn_{23}, we obtain [note that when $\Delta n_{13} = 0$, $n_1 = n_3$, $\Delta n_{12} = n_1 - n_2 = -\Delta n_{23} = -(n_2 - n_3)$, $\Delta n_{ij} = -\Delta n_{ji}$, etc.]

$$\Delta n_{32} = -\Delta n_{21} = \frac{\Delta N_{12}/\tau_{12} - \Delta N_{23}/\tau_{23}}{\tau_{12}^{-1} + \tau_{23}^{-1} + 2C_{23}} \tag{10.41}$$

If we want level 3 to have a higher population density than level 2, the right-hand side of (10.41) must be positive since $\Delta n_{32} = n_3 - n_2$ will be positive. Thus, for population inversion of levels 3 and 2, we require

$$\Delta N_{23}\tau_{12} < \Delta N_{12}\tau_{23}$$

or

$$\left(1 - \frac{N_3}{N_2}\right)\tau_{12} < \left(\frac{N_1}{N_2} - 1\right)\tau_{23}$$

as obtained by dividing both sides by N_2. From the Boltzmann distribution law we have

$$\frac{N_3}{N_2} = e^{-(W_3 - W_2)/kT} = e^{-f_s h/kT}$$

$$\frac{N_1}{N_2} = e^{f_i h/kT}$$

where $f_i = (W_2 - W_1)/h$ is the frequency of the idle transition. In the microwave range, fh/kT is very small, even when T is the temperature of liquid helium (4°K). For example, for $f = 10^{10}$ cps and $T = 4°K$, we have $fh/kT = 0.12$. Thus the exponential terms can be approximated by $1 - hf_s/kT$ and $1 + hf_i/kT$, and we obtain

$$\frac{f_s}{f_i} < \frac{\tau_{23}}{\tau_{12}} \tag{10.42}$$

If this condition does not hold, the population inversion occurs between levels W_2 and W_1 instead, since the right-hand side of (10.41) will now be negative. In this case the 2-1 transition would be used as the amplifying transition and the 3-2 transition would be the idle transition. Equation (10.42) states that the ratio of signal frequency to idle frequency must be greater than the ratio of the relaxation times for the signal transition and the idle transition.

A parameter of considerable importance is the amount of population inversion that can be achieved since this will be the ultimate limiting factor on the amplification obtainable. The inversion ratio is the ratio of Δn_{32} to the thermal-equilibrium population difference ΔN_{23}. From (10.41), with the signal power assumed absent, so that $C_{23} = 0$, we obtain

$$\frac{\Delta n_{32}}{\Delta N_{23}} = \frac{\tau_{23} \Delta N_{12}/\Delta N_{23} - \tau_{12}}{\tau_{12} + \tau_{23}}$$

But $\Delta N_{12}/\Delta N_{23} = (N_1 - N_2)/(N_2 - N_3)$ can be approximated by f_i/f_s as before; so we obtain

$$\frac{\Delta n_{32}}{\Delta N_{23}} = \frac{(\tau_{23}/\tau_{12})f_i/f_s - 1}{1 + \tau_{23}/\tau_{12}}$$

Now $f_p = f_s + f_i$; so we can also express the inversion ratio in the form

$$\frac{\Delta n_{32}}{\Delta N_{23}} = \frac{f_p/f_s - (1 + \tau_{12}/\tau_{23})}{1 + \tau_{12}/\tau_{23}} \tag{10.43}$$

The largest inversion ratio is obtained if τ_{12}/τ_{23} is zero. In this optimum case

$$\frac{\Delta n_{32}}{\Delta N_{32}} = \frac{f_p}{f_s} - 1 \qquad \frac{\tau_{12}}{\tau_{23}} \to 0 \tag{10.44}$$

To obtain a large inversion ratio, the pump frequency must be large compared with the signal frequency. If $\tau_{12} = \tau_{23}$, the inversion ratio is

$$\frac{\Delta n_{32}}{\Delta N_{23}} = \frac{f_p}{2f_s} - 1 \qquad \tau_{12} = \tau_{23} \tag{10.45}$$

At any rate the inversion ratio is dependent only on the ratio of two relaxation times, so that even if the relaxation times are perturbed, the gain will not change as long as the same ratio is maintained.

The signal power will produce $C_{23} \Delta n_{32}$ stimulated downward transitions from level W_3 to level W_2 per unit time in a unit volume of the crystal. The emitted radiation is coherent and in phase with the stimulating signal. The energy emitted per unit of time or power developed is given by Planck's relation as

$$P_s = hf_s C_{23} \Delta n_{32} \tag{10.46}$$

In order to obtain linear amplification, Δn_{32} must be essentially independent of the signal power since C_{23} is directly proportional to the signal power. From (10.41) we see that the condition for Δn_{32} to be independent of C_{23} is

$$C_{23} \ll \frac{\tau_{12} + \tau_{23}}{2\tau_{12}\tau_{23}} \tag{10.47}$$

If this condition is not met, linear amplification will not be obtained. Furthermore, (10.41) shows that if C_{23} is large, corresponding to a large signal, the population inversion is reduced since Δn_{32} becomes smaller. Physically, this means that the signal induces transitions from level W_3 to level W_2 so fast that a population inversion cannot be maintained. Increasing the pump power will not alleviate this signal-saturation effect since (10.41) is already based on the assumption that the pump has saturated level 3 relative to level 1. That is, the best that can be done by the pump is to make n_3 equal to n_1. If $n_3 > n_1$, the pump will induce downward transitions. Signal saturation usually occurs at power levels less than 10 μw.

Since the signal power developed is proportional to Δn_{32}, it is important to make Δn_{32} as large as possible. From (10.45) this is seen to require a large value of $\Delta N_{23} = N_2 - N_3$ since Δn_{32} is proportional to this quantity. Now $N_2 - N_3$ is proportional to $N_2(1 - N_3/N_2) = N_2(1 - e^{-(W_3-W_2)/kT})$ and will be large only if $e^{-(W_3-W_2)/kT}$ is kept small. This requires that $(W_3 - W_2)/kT$ be large, a condition that can be met only at low temperatures. Consequently, the low operating temperature for a maser is required in order to obtain a significant population inversion. A great advantage that is gained from these low operating temperatures is the very low noise power produced by the maser. Another significant advantage obtained by using a low operating temperature is the much longer relaxation times, which in turn mean that much less pump power is needed to saturate the 1-3 transition, as reference to (10.40a) shows.

10.6 Macroscopic magnetic susceptibility

In order to make quantitative calculations of power amplification and other related parameters, it is necessary to know the value of the stimulated transition probabilities C_{13} and C_{23}. The transition probabilities C_{13} and C_{23} may be evaluated by a perturbation analysis similar to that used to evaluate the transition probability B_{ij} in Sec. 10.2. The essential difference is that, in the present case, the electromagnetic a-c field interacts with the ruby crystal through the magnetic dipole moment of the chromium ion rather than through an electric-dipole term. Since ruby is a paramagnetic material and an applied static field B_0 is present, the macroscopic permeability is similar in form to that of a gyrotropic ferrite medium. Thus the macroscopic susceptibility is, in general, a matrix quantity of the form (Sec. 6.6)

$$\bar{\bar{\chi}}_m = \begin{bmatrix} \chi_{xx} & \chi_{xy} & \chi_{xz} \\ \chi_{yx} & \chi_{yy} & \chi_{yz} \\ \chi_{zx} & \chi_{zy} & \chi_{zz} \end{bmatrix}$$

The absence of zero elements as in the ferrite case is due to the fact that the magnetic field B_0 is arbitrarily oriented relative to the coordinate axis. If there are no losses present, $\chi_{ij} = \chi_{ji}^*$; that is, the matrix is a hermitian matrix. If losses are present, then $\chi_{ij} \neq \chi_{ji}^*$. In this case $\bar{\bar{\chi}}_m$ can be split into a hermitian matrix plus an antihermitian matrix (one for which $\chi_{ij} = -\chi_{ji}^*$). These have elements

$$\chi'_{ij} = (\chi'_{ji})^* = \tfrac{1}{2}(\chi_{ij} + \chi_{ji}^*)$$

$$\chi''_{ij} = -(\chi''_{ji})^* = \frac{j}{2}(\chi_{ij} - \chi_{ji}^*)$$

Thus we have $\chi_{ij} = \chi'_{ij} - j\chi''_{ij}$, or in general

$$\bar{\chi}_m = \bar{\chi}'_m - j\bar{\chi}''_m$$

in which the second matrix $-j\bar{\chi}''_m$ will be the one responsible for power loss in the material, as we now show.

If we consider the complex Poynting vector formulation of Sec. 2.5, we find that the power loss is given by Re $[(j\omega/2)\mathbf{H}^* \cdot \mathbf{B}]$ per unit volume. We can express this as

$$\text{Re}\left(\frac{j\omega}{2}\mathbf{H}^* \cdot \mathbf{B}\right) = \frac{j\omega}{4}\mathbf{H}^* \cdot \mathbf{B} - \frac{j\omega}{4}\mathbf{H} \cdot \mathbf{B}^*$$

i.e., as one-half the sum of $j\omega\mathbf{H}^* \cdot \mathbf{B}/2$ plus its complex conjugate. If we now put $\mathbf{B} = \mu_0(1 + \bar{\chi}_m) \cdot \mathbf{H}$, where $\bar{\chi}_m \cdot \mathbf{H}$ is a condensed notation for the matrix product of the matrix $\bar{\chi}_m$ with a column matrix with elements H_x, H_y, H_z, we obtain

$$\text{Re}\frac{j\omega}{2}\mathbf{H}^* \cdot \mathbf{B} = \frac{j\omega\mu_0}{4}(\mathbf{H}^* \cdot \mathbf{H} + \mathbf{H}^* \cdot \bar{\chi}_m \cdot \mathbf{H} - \mathbf{H}^* \cdot \mathbf{H} - \mathbf{H} \cdot \bar{\chi}_m^* \cdot \mathbf{H}^*)$$

$$= \frac{j\omega\mu_0}{2}(\mathbf{H}^* \cdot \bar{\chi}_m \cdot \mathbf{H} - \mathbf{H} \cdot \bar{\chi}_m^* \cdot \mathbf{H}^*)$$

But the term $\mathbf{H} \cdot \bar{\chi}_m^* \cdot \mathbf{H}^*$ can be written as $\mathbf{H}^* \cdot \bar{\chi}_{mt}^* \cdot \mathbf{H}$, where $\bar{\chi}_{mt}^*$ is the transposed matrix. Hence the power loss per unit volume is given by

$$\frac{\omega\mu_0}{2}\mathbf{H}^* \cdot \left[\frac{j}{2}(\bar{\chi}_m - \bar{\chi}_{mt}^*)\right] \cdot \mathbf{H} = \frac{\omega\mu_0}{2}\mathbf{H}^* \cdot \bar{\chi}_m'' \cdot \mathbf{H} \quad (10.48)$$

Thus the part of the susceptibility matrix that contributes to the loss is the antihermitian part $(j/2)(\bar{\chi}_m - \bar{\chi}_{mt}^*)$, which we earlier called $\bar{\chi}_m''$. If there are no losses, $\bar{\chi}_m = \bar{\chi}_{mt}^*$ and $\bar{\chi}_m$ is a hermitian matrix. Since power absorption (or emission) is also proportional to C_{nm}, we expect a similar relation to be obtained for C_{nm}. Evaluation of the C_{nm} shows that this form is indeed obtained. It is found that†

$$C_{nm} = \left(\frac{eg\mu_0}{4m}\right)^2 L(f)\mathbf{H}^* \cdot \bar{\chi}_{nm} \cdot \mathbf{H} \quad (10.49)$$

where $\mathbf{H}^* \cdot \bar{\chi}_{nm} \cdot \mathbf{H}$ is a condensed notation for the double matrix product of $\mathbf{H}$ with the 3 × 3 matrix $\bar{\chi}_{nm}$. The matrix $\bar{\chi}_{nm}$ has the form

$$\bar{\chi}_{nm} = \frac{1}{4}\begin{bmatrix} \alpha^2 & -j\alpha\beta & \alpha\gamma \\ j\alpha\beta & \beta^2 & j\beta\gamma \\ \alpha\gamma & -j\beta\gamma & \gamma^2 \end{bmatrix} \quad (10.50)$$

The matrix $\bar{\chi}_{nm}$ gives the transition probability from state n to state m. The corresponding magnetic (m) susceptibility is denoted by $\bar{\chi}_m$. We

† For a more detailed discussion see A. E. Siegman, "Microwave Solid-state Masers," McGraw-Hill Book Company, New York, 1964. This reference also contains an extensive tabulation of numerical values for the matrix parameters α, β, and γ.

then have

$$\mathbf{H}^* \cdot \bar{\bar{\chi}}_{nm} \cdot \mathbf{H} = \tfrac{1}{4}[H_x^* \ H_y^* \ H_z^*] \begin{bmatrix} \alpha^2 & -j\alpha\beta & \alpha\gamma \\ j\alpha\beta & \beta^2 & j\beta\gamma \\ \alpha\gamma & -j\beta\gamma & \gamma^2 \end{bmatrix} \begin{bmatrix} H_x \\ H_y \\ H_z \end{bmatrix} \quad (10.51)$$

For (10.50) to be valid, the z axis is chosen as the c axis of the ruby crystal, and the applied field B_0 is taken to lie in the xz plane. The matrix elements α, β, γ depend on the transition being considered, the strength of the static field B_0, and the orientation of the B_0 field. In (10.49) g is the spectroscopic splitting factor and is very nearly equal to 2 for ruby. The quantity $L(f)$ describes the shape of the absorption line (or emission line) for the transition involved, and is normalized so that

$$\int_0^\infty L(f) \, df = 1$$

The line shape is well approximated by the Lorentz line shape, which is given by

$$L(f) = \frac{2\tau}{1 + \tau^2(\omega - \omega_{nm})^2} \quad (10.52)$$

where τ can be considered the decay time for the transition (see discussion at the end of Sec. 10.2). The half width of the line is $\Delta f = (\pi\tau)^{-1}$. Typical values of τ are about 10^{-7} to 10^{-8} sec for ruby.

Siegman gives an extensive tabulation of the values of α, β, and γ in his book. For illustrative purposes we give a few typical values in Table 10.2 (see Fig. 10.11 for definition of the magnetic field orientation angle ϕ).

Table 10.2 Ruby data ($B_0 = 2$ kilogauss)

	W_1/h, Mc	W_2/h, Mc	W_3/h, Mc	W_4/h, Mc
$\phi = 10°$	−14,014	2,363	2,897	8,754
$\phi = 40°$	−12,756	−852	2,868	10,739
$\phi = 90°$	−8,450	−7,005	2,866	12,585

	α	β	γ
$\phi = 10°$			
1-3 transition	1.66	1.66	−0.0947
2-3 transition	0.47	−0.55	0.99
$\phi = 40°$			
1-3 transition	1.49	1.48	−0.341
2-3 transition	0.82	−1.33	1.28
$\phi = 90°$			
1-3 transition	1.75	2.76×10^{-6}	-1.61×10^{-6}
2-3 transition	3.3×10^{-7}	−2.1	1.2

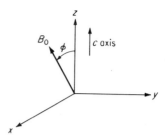

Fig. 10.11 Orientation angle ϕ for magnetic biasing field B_0.

The large variation in the matrix elements α, β, and γ, with orientation angle ϕ of the biasing magnetic field, is due to the anisotropic structure of the ruby crystal. There are large crystalline electric fields that have a significant effect on the spins. The ruby crystal has symmetry about one axis, which is called the c axis. The anisotropic structure of the crystal also shows up in the dielectric constant, which has a value of 11.5 along the c axis and 9.5 in directions perpendicular to the c axis.

A major difference in the magnetic behavior of ruby compared with ferrites is that in ruby the concentration of magnetically active chromium ions is very low (less than 1 percent), and hence the permeability is different from μ_0 by a very small, but important, amount. If the concentration of chromium ions were increased, the spins would begin to interact with each other because of the closer spacing. This has the effect of broadening the energy levels, and hence making the line width much greater. The peak intensity of the line is thereby decreased, and maser performance suffers. Consequently, the optimum concentration of chromium Cr^{3+} ions is less than 1 percent.

In terms of the transition probability C_{23} and the difference in population density of levels 2 and 3, the absorbed power per unit volume of crystal is $C_{23} \Delta n_{23} h f_s$. If there is a population inversion, $n_3 > n_2$ and Δn_{23} is negative, and instead of absorption at the signal frequency there will be a net emission of power. If we compare the quantum-mechanical expression for power absorption with the classical expression (10.48), we see that the macroscopic susceptibility matrix $\bar{\chi}_m''$ is given by

$$\bar{\chi}_m'' = \frac{\Delta n_{23} h}{\pi \mu_0} \left(\frac{e\mu_0}{2m}\right)^2 L(f) \bar{\chi}_{nm} \qquad (10.53)$$

where we have put $g = 2$, and $\bar{\chi}_{nm}$ is given by (10.50), with α, β, γ corresponding to the 2-3 signal transition ($n = 2$, $m = 3$).

In calculating the field distribution in a cavity partially filled with a ruby crystal, it is not necessary to take into account the magnetic susceptibility, since it is very small compared with unity. However, the large dielectric constant of ruby must be taken into consideration. Since the dielectric constant is also anisotropic, the solution of the boundary-value problem for a given cavity mode is quite complex. The antihermitian part of the susceptibility matrix is given by (10.53). In addition, there

524 Foundations for microwave engineering

is also a hermitian part, which, however, vanishes at the resonant frequency. If the maser is operated over a very narrow band (compared with the half width of the line), the hermitian part of $\bar{\chi}_m$ may be assumed to remain zero. For wider bandwidths it is necessary to know $\bar{\chi}'_m$ also, since, when it increases from a zero value at resonance to a finite value, it produces a significant detuning effect on the cavity (equivalent to changing the cavity inductance). When the overall loaded Q is high, this effect can limit the maximum bandwidth of operation.

10.7 Equivalent circuit of a maser amplifier

We have now presented all the basic results needed to derive an equivalent circuit of a cavity-type maser amplifier of the kind discussed in Sec. 10.3. Consider, therefore, a microwave cavity containing a properly biased ruby crystal, as shown in Fig. 10.12. The cavity is coupled to the input and output signal lines through a circulator. In addition, there is a pump signal coupling line which we assume to be located at a point on the cavity surface so as not to couple to the fields of the signal mode. For example, if the pump coupling is a small loop, it can be located at a point where the signal magnetic field produces zero flux linkage through the loop. Thus the pump circuit plays no part in the equivalent circuit for the cavity at the signal frequency.

In the absence of pump power, the ruby is inactive and the equivalent circuit of the cavity is simply that of an ordinary cavity containing a dielectric material, i.e., ruby. Once the fields have been determined, the energy stored in the cavity at the frequency f_s can be computed from the relations

$$W_e = W_m = \frac{\mu_0}{4} \int_V \mathbf{H} \cdot \mathbf{H}^* \, dV \tag{10.54}$$

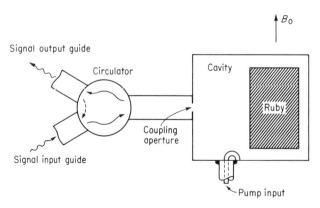

Fig. 10.12 A maser amplifier.

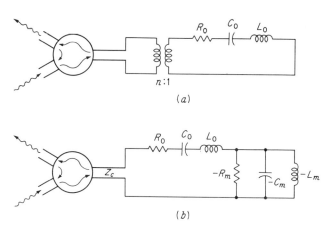

Fig. 10.13 Equivalent circuit of a maser amplifier.

where we can assume that $\mathcal{X}_m$ is negligible in determining the average stored magnetic energy. The power loss in the cavity walls plus that due to dielectric losses in the ruby can also be calculated, and hence the unloaded Q can be determined. We can then represent the cavity by a series (or parallel) combination of an inductor L_0, a capacitor C_0, and a resistance R_0, as shown in Fig. 10.13a. The cavity is coupled to an external load by means of a transmission line with characteristic impedance Z_c in the equivalent circuit. For critical coupling (Sec. 7.4) the cavity resonant circuit is loaded by an additional equivalent series resistance R_0. For noncritical coupling the degree of coupling is described by the coupling parameter $K = Q/Q_e$, where Q is the unloaded cavity Q given by $\omega L_0/R_0$, and Q_e is the external, or radiation, Q. By introducing an ideal transformer of turns ratio $n:1$ in the equivalent circuit, any degree of coupling can be specified if an appropriate turns ratio is chosen. The equivalent cavity loading by the external circuit or load is $R_e = Z_c/n^2$, and hence the external Q is given by

$$Q_e = \frac{\omega L_0}{R_e} = n^2 \frac{\omega L_0}{Z_c} \tag{10.55}$$

Thus the coupling parameter is

$$K = \frac{Q}{Q_e} = \frac{\omega L_0}{R_0} \frac{Z_c}{n^2 \omega L_0} = \frac{1}{n^2} \frac{Z_c}{R_0} \tag{10.56}$$

There would normally also be present some reactance due to the coupling hole or loop. We assume that the coupling reactance has been absorbed into the elements L_0, C_0. In practice we can evaluate the unloaded Q and the resonant frequency (f_s in the present case). When the coupling parameter K is specified and the equivalent impedance Z_c of the input line is given, all other parameters except n^2 in the equivalent circuit are fixed. Thus we have

$$L_0 C_0 = \omega_s^{-2} \qquad R_0 = \frac{\omega L_0}{Q} = \frac{Z_c}{n^2 K}$$

which determines L_0, C_0, and R_0 in terms of Q and K for a given choice of n^2. The turns ratio $n:1$ can be arbitrarily chosen since this corresponds to the arbitrariness in the choice of impedance level for the cavity. For convenience n will be chosen equal to unity, in which case the transformer can be eliminated.

If we now assume that the pump power is turned on so as to produce a population inversion of levels 3 and 2, the maser crystal becomes an active source supplying power at the signal frequency f_s. This power is generated by the emission of an amount of energy hf_s every time a spin is induced to make a downward transition from level 3 to level 2. The total power generated is given by an integral of (10.49), or equivalently of $(\omega_s \mu_0/2) \mathbf{H}^* \cdot \bar{\chi}_m'' \cdot \mathbf{H}$, where $\bar{\chi}_m''$ is given by (10.53), over the cavity volume V_c occupied by the crystal; thus

$$P_g = \frac{\omega_s \mu_0}{2} \int_{V_c} \mathbf{H}^* \cdot \bar{\chi}_m'' \cdot \mathbf{H} \, dV \qquad (10.57)$$

Power is generated rather than absorbed, since $\Delta n_{23} = -\Delta n_{32}$, where Δn_{32} is positive for an inverted energy level. The power generated by the maser action can be represented by an equivalent negative resistance $-R_m$ in the equivalent circuit, as shown in Fig. 10.13b. We may evaluate R_m as follows: When a current I exists in the equivalent circuit, the energy stored in L_0 is

$$W_m = \tfrac{1}{4} L_0 I I^*$$

and the power generated by $-R_m$ is $P_g = -\tfrac{1}{2} R_m I I^*$. Hence

$$\frac{P_g}{W_m} = -\frac{2 R_m}{L_0} = \frac{2\omega_s \int_{V_c} \mathbf{H}^* \cdot \bar{\chi}_m'' \cdot \mathbf{H} \, dV}{\int_V \mathbf{H} \cdot \mathbf{H}^* \, dV} \qquad (10.58)$$

which determines R_m in terms of L_0. The maximum possible value of the numerator is obtained only by having ruby crystal present everywhere in the cavity and having an $\mathbf{H}$ field configuration that maximizes the quantity $\mathbf{H}^* \cdot \bar{\chi}_{nm} \cdot \mathbf{H}$ everywhere. If this ideal maximum value of $\int_{V_c} \mathbf{H}^* \cdot \bar{\chi}_{nm} \cdot \mathbf{H} \, dV$ is denoted by $\sigma^2 \int_V \mathbf{H}^* \cdot \mathbf{H} \, dV$, then in any practical case we have [refer to (10.53)]

$$\frac{\int_{V_c} \mathbf{H}^* \cdot \bar{\chi}_m'' \cdot \mathbf{H} \, dV}{\int_V \mathbf{H}^* \cdot \mathbf{H} \, dV} = \eta \sigma^2 \left[\frac{\Delta n_{23} h}{\pi \mu_0} \left(\frac{e \mu_0}{2m} \right)^2 L(f) \right] \qquad (10.59)$$

where η is called the filling factor. In terms of these parameters we obtain

$$\frac{P_g}{W_m} = -\frac{2R_m}{L_0} = 2\omega_s\eta\sigma^2\left[\frac{\Delta n_{23}\,h}{\pi\mu_0}\left(\frac{e\mu_0}{2m}\right)^2 L(f)\right] \tag{10.60}$$

The value of σ^2 is $\frac{1}{4}(\alpha^2 + \beta^2 + \gamma^2)$. The negative resistance gives rise to an equivalent negative magnetic Q called $-Q_m$, given by

$$\frac{1}{-Q_m} = \frac{-R_m}{\omega_s L_0} = \eta\sigma^2\,\frac{\Delta n_{23}\,h\mu_0}{\pi}\left(\frac{e}{2m}\right)^2 L(f) \tag{10.61}$$

The magnetic Q_m is the most significant single parameter that describes the cavity-type maser amplifier.

The additional negative reactive elements $-L_m$, $-C_m$ come from the additional small amount of magnetic energy ΔW_m arising from the $\bar{\chi}'_m$ part of the magnetic susceptibility, i.e., given by

$$\Delta W_m = \frac{\mu_0}{4}\int_{V_c} \mathbf{H}^* \cdot \bar{\chi}'_m \cdot \mathbf{H}\, dV \tag{10.62}$$

At resonance ΔW_m is zero, and below the resonant frequency it is negative, whereas above the resonant frequency it is positive. Thus the complete macroscopic behavior of the magnetic susceptibility is fully equivalent to a parallel resonant circuit with negative elements $-G_m^{-1}$, $-L_m$, and $-C_m$. That this should be the case is not too surprising since the usual classical model of polarization in materials is a second-order differential equation representing a damped oscillator or an equivalent RLC resonant circuit, as was considered in Sec. 2.5. In the present case the equivalent circuit elements are all negative because of the inverted population density. The energy ΔW_m, which we called magnetic energy, is more correctly interpreted as the difference between the time-average kinetic and potential energy associated with the a-c motions of the spins. Therefore it is not surprising that it can vanish at a resonant frequency and have opposite signs on either side of resonance.

The input admittance of a parallel resonant circuit with negative elements is

$$Y_{in} = -G_m - j\omega C_m\left(1 - \frac{1}{\omega^2 L_m C_m}\right)$$

If we choose $L_m C_m$ so that $L_m C_m = \omega_s^{-2}$, then

$$Y_{in} = -G_m - j\omega C_m\,\frac{\omega^2 - \omega_s^2}{\omega^2}$$

$$\approx -G_m - jC_m 2(\omega - \omega_s)$$

This gives $Y_{in} = -G_m$ at resonance, an effective inductive susceptance $-j2C_m(\omega - \omega_s)$ above resonance, and a capacitive susceptance below resonance. If a current I exists in this circuit, the generated power is

given by

$$\text{Re}\left(\tfrac{1}{2}II^*Z_{\text{in}}\right) = \text{Re}\left[\frac{-G_m + j2C_m(\omega - \omega_s)}{G_m^2 + 4C_m^2(\omega - \omega_s)^2}\frac{II^*}{2}\right]$$
$$= \frac{-\tfrac{1}{2}G_m II^*}{G_m^2 + 4C_m^2(\omega - \omega_s)^2} \qquad (10.63)$$

If we compare this result with the expression $-\tfrac{1}{2}R_m II^*$, we see that the Lorentz-line-shape factor $2\tau/[1 + \tau^2(\omega - \omega_s)^2]$ corresponds to the response of a parallel resonant circuit. If we replace the Lorentz-line-shape factor by the impedance of a parallel resonant circuit, we can correctly account for both parts of the susceptibility matrix. For complete correspondence we must choose $\tau^2 = 4C_m^2/G_m^2$, or

$$C_m = \frac{\tau G_m}{2} \qquad (10.64)$$

We note that (10.63) gives $-II^*/2G_m$ at resonance. Since $L(f) = 2\tau$ at resonance and the generated power is given by

$$P_g = -\tfrac{1}{2}R_m II^* = \frac{\omega_s \mu_0}{2}\int_{V_c}\mathbf{H}^* \cdot \bar{\bar{\chi}}_m'' \cdot \mathbf{H}\, dV$$
$$= \tfrac{1}{4}II^* L_0(2\omega_s \eta \sigma^2)\left[\frac{\Delta n_{23}\, h\mu_0}{\pi}\left(\frac{e}{2m}\right)^2 2\tau\right]$$

from (10.60) we must define the equivalent resistance $R_m = G_m^{-1}$ as

$$-R_m = \omega_s L_0 \eta \sigma^2\left[\frac{\Delta n_{23}\, h\mu_0}{\pi}\left(\frac{e}{2m}\right)^2 2\tau\right] \qquad (10.65)$$

Thus, in terms of the previously chosen value of L_0 and the parameters of the ruby crystal, including the line-width parameter τ, we can specify the equivalent resistance R_m to be used in the equivalent circuit by (10.65). The capacitance C_m is then determined by (10.64), and L_m is given by the relation $L_m C_m = \omega_s^{-2}$. Thus we have completed the construction of an equivalent circuit for a maser amplifier. It might be worthwhile to point out here that the procedure used to account for the hermitian part $\bar{\bar{\chi}}_m'$ of the susceptibility matrix is a correct one since the susceptibility parameters are analytic functions of ω just as impedance functions are. A knowledge of the real or the imaginary part is sufficient to determine the other. Thus it is not mere accident that the Lorentz line shape can be modified to include the contribution of both parts of the susceptibility matrix. The relation between the real and imaginary parts of the susceptibility is given by the Kronig-Kramers relations, which the reader can find discussed in a number of references.†

† See, for example, B. S. Gouray, Dispersion Relations for Tensor Media and Their Application to Ferrites, *J. Appl. Phys.*, vol. 28, p. 283, 1957.

10.8 Gain of a maser amplifier

The gain of a maser amplifier is readily found from the equivalent circuit given in Fig. 10.13b. Let a signal with voltage V_i be incident on the maser cavity. The reflected wave will have a voltage $V_r = \Gamma V_i$, where Γ is the reflection coefficient at the cavity input terminals. The reflection coefficient is given by

$$\Gamma = \frac{Z_{\text{in}} - Z_c}{Z_{\text{in}} + Z_c} \tag{10.66}$$

The cavity input impedance at resonance is simply $R_0 - R_m$, in which case

$$\Gamma = \frac{R_0 - R_m - Z_c}{R_0 + Z_c - R_m} = \frac{R_m - (R_0 - Z_c)}{R_m - (R_0 + Z_c)} \tag{10.67}$$

If R_m is greater than R_0, the reflection coefficient will be greater than unity. Usually, the unloaded cavity Q is several thousand and typical values of magnetic Q_m are 50 to 200. Thus R_m is an order of magnitude greater than R_0 since $R_m = \omega L_0/Q_m$ and $R_0 = \omega L_0/Q$. Consequently, Γ is given very nearly by

$$\frac{V_r}{V_i} = \Gamma = \frac{R_m + Z_c}{R_m - Z_c} = -\frac{Z_c + R_m}{Z_c - R_m} \tag{10.68}$$

which clearly shows that it is greater than unity. Thus a negative resistance terminating a transmission line can give a reflection coefficient greater than unity. Physically, this arises because a negative resistance supplies power to the line instead of absorbing the incident power. The voltage gain of the maser is given by $|\Gamma|$, and hence the resonant-frequency voltage gain is

$$\left|\frac{V_r}{V_i}\right| = \left|\frac{R_m - (R_0 - Z_c)}{R_m - (R_0 + Z_c)}\right| \approx \left|\frac{R_m + Z_c}{R_m - Z_c}\right| \tag{10.69}$$

Voltage gain at the resonant frequency may be expressed in terms of the cavity unloaded Q, magnetic Q_m, and the external Q_e or the coupling parameter K by dividing all terms in (10.69) by $\omega_s L_0$. Thus we have

$$\left|\frac{V_r}{V_i}\right| = \left|\frac{Q_m^{-1} - Q^{-1} + Q_e^{-1}}{Q_m^{-1} - Q^{-1} - Q_e^{-1}}\right|$$

If we introduce the adjustable parameter $K = Q/Q_e$, we obtain

$$\left|\frac{V_r}{V_i}\right| = \left|\frac{Q - Q_m(1 - K)}{Q - Q_m(1 + K)}\right| \tag{10.70}$$

For maximum midband gain the cavity coupling parameter K should be chosen so as to make the denominator very small. If we choose

$$K = \frac{Q - Q_m}{Q_m} \tag{10.71}$$

the denominator will vanish and oscillations will occur because of the infinite gain.

The total loaded Q of the maser amplifier is given by

$$\frac{1}{Q_L} = \frac{1}{Q} + \frac{1}{Q_e} - \frac{1}{Q_m}$$

in terms of which the midband power gain G_0 can be expressed as

$$G_0 = \left|\frac{V_r}{V_i}\right|^2 = \left(\frac{Q^{-1} + Q_e^{-1} - Q_m^{-1} - 2Q_e^{-1}}{Q^{-1} + Q_e^{-1} - Q_m^{-1}}\right)^2 \qquad (10.72)$$

If we have adjusted the coupling so that the amplifier is on the verge of oscillation, then $Q^{-1} + Q_e^{-1} \approx Q_m^{-1}$, in which case G_0 can be expressed as

$$G_0 \approx \left(\frac{2Q_L}{Q_e}\right)^2 \approx \left(\frac{2Q_L}{Q_m}\right)^2 \qquad (10.73)$$

The last result is obtained by using (10.71) to obtain

$$Q_e = \frac{QQ_m}{Q - Q_m} \approx Q_m$$

since $Q \gg Q_m$. For high gain the external Q_e should be approximately equal to the magnetic Q_m. The loaded cavity Q_L is equal to the unloaded cavity Q if we choose $Q_e = Q_m$, so that the power gain can then be expressed as

$$G_0 = \left(\frac{2Q}{Q_m}\right)^2 \qquad (10.74)$$

at midband. Thus, for high gain, a high unloaded cavity Q is desirable. In addition, a low value of magnetic Q_m is needed, and this requires a filling factor η as close to unity as possible plus any other adjustment of parameters that will maximize the transition probability C_{23}. For gain stability the external Q_e must be chosen so that the total loaded Q_L does not become too large. If Q_L is too large, a small variation in some parameter can cause Q_L to become infinite, and oscillations will result.

To find the gain at other than the resonant frequency we need the expression for Z_{in} in the vicinity of resonance. From the equivalent circuit, Z_{in} is readily found to be

$$Z_{\text{in}} = R_0 + j\omega L_0\left(1 - \frac{1}{\omega^2 L_0 C_0}\right) + \left[-G_m - j\omega C_m\left(1 - \frac{1}{\omega^2 C_m L_m}\right)\right]^{-1}$$

Since $L_0 C_0 = L_m C_m = \omega_s^{-2}$ and $\omega^2 - \omega_s^2 \approx 2\omega_s \Delta\omega$, we obtain

$$Z_{\text{in}} \approx R_0 + 2jL_0 \Delta\omega - (G_m + j2C_m \Delta\omega)^{-1}$$

We may now determine the power gain $|\Gamma|^2$, and we find that

$$G = |\Gamma|^2 = \left|\frac{[R_0(1 + 2jQ \Delta\omega/\omega_s) - Z_c](1 + j\tau \Delta\omega) - R_m}{[R_0(1 + 2jQ \Delta\omega/\omega_s) + Z_c](1 + j\tau \Delta\omega) - R_m}\right|^2$$

after replacing $\omega_s L_0$ by $R_0 Q$ and C_m by $\tau G_m/2$ from (10.64). We are primarily interested in frequency deviations $\Delta\omega$ such that the terms $2Q\,\Delta\omega/\omega_s$ and $\tau\,\Delta\omega$ are of the order of unity. For a high-gain maser we have $Z_c \approx R_m$ and $R_m \gg R_0$; that is, $Q_m \ll Q$. We can therefore neglect the term $R_0(1 + 2jQ\,\Delta\omega/\omega_s)$ compared with Z_c, and the expression for G becomes

$$G \approx \frac{(2R_m)^2 + (R_m\tau\,\Delta\omega)^2}{(R_0 + Z_c - R_m)^2 + (Z_c\tau\,\Delta\omega)^2}$$

$$= \left(\frac{2R_m}{R_0 + Z_c - R_m}\right)^2 \frac{1 + (\tau\,\Delta\omega/2)^2}{1 + \left(\dfrac{R_m}{R_0 + Z_c - R_m}\right)^2 (\tau\,\Delta\omega)^2}$$

Now $2R_m/(R_0 + Z_c - R_m) = 2Q_L/Q_m$, and the midband gain is

$$G_0 = \frac{4Q_L^2}{Q_m^2}$$

so we obtain

$$G = G_0 \frac{1 + (\tau\,\Delta\omega/2)^2}{1 + G_0(\tau\,\Delta\omega/2)^2} \qquad (10.75)$$

Since G_0 is much larger than unity, the gain is reduced to $G_0/2$ when the denominator becomes equal to 2. Thus the gain-bandwidth product for the high-gain maser is given by

$$\sqrt{G_0}\,\frac{\Delta\omega}{\omega_s} = \frac{2\sqrt{G_0}}{\omega_s\tau} = \frac{\sqrt{G_0}}{\pi f_s\tau} = \frac{(\Delta f)_L}{f_s}\sqrt{G_0} \qquad (10.76)$$

where $(\Delta f)_L = (\pi\tau)^{-1}$ is the 3-db width of the emission line. Thus we see that the useful bandwidth of the maser is very small if the gain is high. For example, a typical line width would be 100 Mc, and if the gain is 20 db, the useful bandwidth is only 10 Mc. As long as the magnetic Q_m is made small enough to get a high gain, any further reduction of Q_m does not improve the gain-bandwidth product since this is limited by the magnetic-resonance line width.

In principle, a bandwidth approaching the magnetic-resonance line width $(\Delta f)_L$ and still maintaining high gain can be obtained. The simple cavity is a narrowband circuit, and this is what causes the limited gain-bandwidth product given by (10.76). The use of more elaborate cavity networks can give a significant improvement in the gain-bandwidth product. For a discussion of the broadbanding problem, the reader is referred to Siegman's book. An alternative solution to the broadbanding problem is obtained with the traveling-wave maser, in which the cavity is replaced by a slow-wave circuit. A brief description of the traveling-wave maser is given in Sec. 10.10.

10.9 Maser noise

Noise in a maser at the signal frequency arises from the spontaneous emission of radiation as the spins drop from level W_3 to level W_2 spontaneously. The spontaneous emission is not coherent with the signal, and hence represents noise. The number of spontaneous transitions that take place per unit time is given by $A_{32}n_3$, where A_{32} is Einstein's coefficient for the probability of a spontaneous transition occurring, and n_3 is the number of spins per unit volume in level W_3. If the pump power is absent, the maser crystal can still be characterized by an equivalent circuit consisting of a parallel connection of resistance, inductance, and capacitance. However, in this latter situation the equivalent circuit elements will be positive since these elements are defined through (10.61), involving the factor Δn_{23}, now replaced by the thermodynamic-equilibrium population difference ΔN_{23}, which is positive. Since nothing else changes when the pump power is turned off, the new equivalent circuit elements are given by

$$R'_m = \frac{\Delta N_{23}}{\Delta n_{23}}(-R_m) \qquad L'_m = \frac{\Delta N_{23}}{\Delta n_{23}}(-L_m) \qquad C'_m = \frac{\Delta N_{23}}{\Delta n_{23}}(-C_m)$$

Now under thermodynamic-equilibrium conditions the mean-square noise voltage $e_n{}^2$ appearing across a resistance R'_m is given by

$$e_n{}^2 = 4kTR'_m \, \Delta f \tag{10.77}$$

This noise power comes from spontaneous transitions that occur at the rate $A_{32}N_3$, where N_3 is the number of spins per unit volume in level W_3 at the equilibrium temperature T. If we turn the pump power on, the noise voltage produced will be proportional to $A_{32}n_3$. The proportionality constant will be exactly the same as in the thermodynamic-equilibrium case since the mechanism of noise generation has not changed but only the population density has changed. Hence, with the pump on, the mean-square noise voltage across the maser-crystal equivalent circuit will be given by

$$\frac{e_n{}^2}{A_{32}n_3} = \frac{4kTR'_m \, \Delta f}{A_{32}N_3}$$

Substituting for R'_m, we find

$$e_n{}^2 = 4kT \, \Delta f(-R_m) \frac{\Delta N_{23}}{N_3} \frac{n_{23}}{\Delta n_{23}}$$

Now

$$\frac{\Delta N_{23}}{N_3} = \frac{N_2}{N_3} - 1 = e^{-(W_2-W_3)/kT} - 1$$

$$= e^{hf_s/kT} - 1 \approx \frac{hf_s}{kT}$$

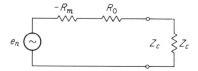

Fig. 10.14 Maser equivalent circuit for noise evaluation.

We can express $\Delta n_{23}/n_3$ in a similar form provided we introduce an equivalent temperature T_e; thus

$$\frac{\Delta n_{23}}{n_3} = \frac{n_2}{n_3} - 1 = \frac{hf_s}{kT_e} = \frac{hf_s}{k(-T_s)}$$

Now, for an inverted population level, $n_2 < n_3$ and Δn_{23} is negative. Thus the spin temperature T_e is a negative temperature, and for this reason we denote it by $-T_s$. When we substitute these relations into the expression for $e_n{}^2$, we obtain

$$e_n{}^2 = 4k(-T_s)(-R_m)\,\Delta f \tag{10.78}$$

Thus the noise properties of a maser can be determined in the same manner as for any other circuit provided that the negative spin temperature $-T_s$ and the equivalent negative resistance $-R_m$ are used to replace the usual resistance and temperature in the noise-voltage formula (10.77).

To determine the noise power delivered to the external load, consider the equivalent maser circuit at resonance, as shown in Fig. 10.14. With the output line properly terminated in a load Z_c, the maser noise power delivered to Z_c is given by

$$P_n = \left|\frac{e_n}{R_0 + Z_c - R_m}\right|^2 Z_c = \frac{4kT_sR_mZ_c\,\Delta f}{(R_0 + Z_c - R_m)^2} \tag{10.79}$$

The maser gain G_0 at resonance is given by

$$G_0 = \left(\frac{R_0 - R_m - Z_c}{R_0 + Z_c - R_m}\right)^2 \approx \frac{4R_m{}^2}{(R_0 + Z_c - R_m)^2}$$

since $Z_c \approx R_m$ and $R_m \gg R_0$ for high gain. Hence the noise temperature of the maser amplifier is

$$T_a = \frac{P_n}{G_0 k\,\Delta f} \approx T_s \tag{10.80}$$

The magnitude of the spin temperature T_s is approximately equal to the temperature of liquid helium, so that the maser is a very low noise device. There is additional noise arising from the equivalent cavity loss resistance R_0, which is at liquid-helium temperature T, but since $R_0 \ll R_m$, this is usually a negligible additional amount of noise.

When a maser is incorporated into a complete receiver system there will be many components involved that are at ambient temperature. Since these components, such as waveguides, the circulator, etc., have some loss, and hence some equivalent resistance, they will radiate additional noise. Although the losses may be small, the temperature is much higher, and a very significant degradation of the overall receiver noise temperature can occur. Thus it is not uncommon to find that the overall receiver equivalent noise temperature is greater by a factor of 5 to 20, or even more, i.e., equivalent noise temperatures of 20 to 80°K or more.

10.10 Traveling-wave maser

In the cavity-type maser the electromagnetic field is essentially a standing wave. However, a standing-wave field is not necessary in order to obtain amplification. A maser crystal can be stimulated to produce power at the signal frequency equally well if the field is a propagating, or traveling, wave. To understand the basic mechanism involved, consider a transmission line, such as a parallel-plate line, filled with a maser material. In the absence of pump power, the line can be characterized by series inductance L and series resistance R plus shunt capacitance C and conductance G per unit length. The attenuation constant is given by

$$\alpha_0 = \tfrac{1}{2}(RY_c + GZ_c) \tag{10.81}$$

where $Z_c = (L/C)^{1/2} = Y_c^{-1}$. When the pump is turned on, the maser material becomes active and produces an equivalent negative series resistance $-R_m$. If R_m is large enough, the new attenuation constant

$$\alpha = \tfrac{1}{2}[(R - R_m)Y_c + GZ_c] = \alpha_0 - \alpha_m \tag{10.82}$$

where $\alpha_m = R_m Y_c/2$, may be negative and then corresponds to a gain constant. Thus the wave at the signal frequency will grow exponentially.

A desirable feature in any traveling-wave maser is a high gain per unit length in order that the size and amount of maser material needed can be kept small. A small size is desirable from the point of view of keeping the magnet used for biasing the crystal small, as well as for keeping the expensive helium bath small. The type of waveguiding structure used for a traveling-wave maser should be one in which the time-average magnetic energy stored per unit length of the wave circuit is large. The reason for this is that the equivalent series resistance and growth constant α_m due to maser action are proportional to the magnetic energy density. The rate at which power is generated by maser action is proportional to the power P which is present; so we can write

$$\frac{dP}{dz} = 2\alpha_m P \tag{10.83}$$

But in a traveling-wave system the power P is given by the average electric and magnetic energy densities per unit length multiplied by the group velocity, or velocity of energy flow, v_g. Since the average electric energy density U_e equals the average magnetic energy density U_m in a traveling wave, we can write $P = 2U_m v_g$. The rate at which energy is generated by the maser crystal in a unit length of the wave circuit is given by a modification of (10.57) and is (integration with respect to z is deleted)

$$\frac{dP}{dz} = -\frac{\omega_s \mu_0}{2} \int_S \mathbf{H}^* \cdot \bar{\bar{\chi}}_m'' \cdot \mathbf{H} \, dS \tag{10.84}$$

where the integration is over the cross section of the waveguiding circuit. The magnetic energy density present is given by

$$U_m = \frac{\mu_0}{4} \int_S \mathbf{H}^* \cdot \mathbf{H} \, dS \tag{10.85}$$

Combining the above results, we obtain

$$\alpha_m = \frac{1}{2P}\frac{dP}{dz} = \frac{1}{4v_g U_m}\frac{dP}{dz} = -\frac{\omega_s}{2v_g}\frac{\int_S \mathbf{H}^* \cdot \bar{\bar{\chi}}_m \cdot \mathbf{H} \, dS}{\int_S \mathbf{H}^* \cdot \mathbf{H} \, dS} \tag{10.86}$$

For high gain per unit length it is necessary to have a small value of group velocity v_g. Thus, for a traveling-wave maser, the waveguiding structure should be one with a small value of group velocity. In contrast to the traveling-wave tube, which requires a structure with a small phase velocity equal to the beam velocity, the traveling-wave maser does not require a critical value of group velocity. It is mainly for practical reasons that the small group velocity is needed. From electrical considerations only, a small growth constant α_m can always be compensated for by a longer length of the traveling-wave system.

If the input and output coupling circuits used in a traveling-wave maser and the group velocity v_g can be kept relatively constant over a frequency band equal to the magnetic-resonance line width, the useful bandwidth for essentially constant gain will equal the line width. These conditions are relatively easy to satisfy; consequently the traveling-wave maser is a broadband amplifier.

Figure 10.15 illustrates a simple traveling-wave maser obtained by

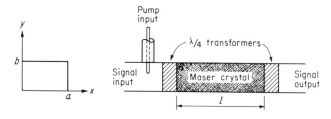

Fig. 10.15 A simple traveling-wave maser.

filling a rectangular guide with maser crystal (the helium bath is not shown). If the c axis of the crystal coincides with one of the coordinate axis, the dominant H_{10} mode is the same as that in a dielectric-filled guide having a permittivity ϵ along the y direction. The maser crystal can be matched to the empty guide at the signal frequency by means of dielectric-slab quarter-wave transformers. The pump power can be coupled in through a coaxial-line probe or by some other suitable means.

The magnetic Q_m of the traveling-wave maser can be defined as

$$Q_m = \frac{2\omega_s U_m}{dP/dz} \tag{10.87}$$

by analogy with the expression $2\omega W_m/P_l$ for a cavity. In terms of Q_m, we find from (10.86) that

$$\alpha_m = \frac{\omega_s}{2v_g Q_m} \tag{10.88}$$

In the maser illustrated in Fig. 10.15, the group velocity v_g is given by

$$v_g = \frac{c^2}{\epsilon v_p} = \frac{c}{\sqrt{\epsilon}} \frac{\beta}{k}$$

where β is the propagation constant and $k = \sqrt{\epsilon}\, k_0$. The propagation constant β is given by $\beta^2 = k^2 - (\pi/a)^2$, and hence

$$\frac{d\beta}{\beta} = \left(\frac{k}{\beta}\right)^2 \frac{dk}{k}$$

If we choose a frequency and guide dimensions such that $\beta = 0.5k$, the change in β for a 1 percent change in frequency, i.e., in k, is still only 4 percent. For ruby the dielectric constant ϵ/ϵ_0 may be taken as 9.5, except along the optical, or c, axis, where it is approximately equal to 11.5. If we choose the former, we get

$$v_g = \frac{0.5c}{\sqrt{9.5}} = 0.162c$$

corresponding to a slowing factor of about 6.2. A typical value of magnetic Q_m is 100. Thus, at a frequency of 10^4 Mc (X band), the growth constant α_m as given by (10.88) is 0.065 neper/cm. If 10 in. of active ruby material is employed, the amplitude gain will be by a factor $e^{\alpha_m l}$, where $\alpha_m l = 0.065 \times 25.4 = 1.65$. The gain in decibels is 14.3 db. If the length is doubled, the gain increases to 28.6 db. This example clearly illustrates the need for a large reduction in group velocity if high gain in a short length is to be achieved. With suitable periodic slow-wave structures, such as a helix, slowing factors of 50 or more can be obtained quite readily. Thus it is possible to obtain gains of up to 10 db or more per free-space wavelength in a practical traveling-wave-maser design.

10.11 Lasers

A laser is the optical version of the microwave maser. The term laser is coined from the phrase "*L*ight *A*mplification by *S*timulated *E*mission of *R*adiation." Two types of lasers that have received considerable study to date are the pulsed ruby laser and the continuously operating neon-helium gas laser. A qualitative description of these two types of lasers is given below. In addition, it has been found possible to obtain laser action in a wide variety of other materials.

Ruby laser

Figure 10.16 is a schematic illustration of a pulsed ruby laser. The ruby rod, typically about 6 in. long and 0.5 in. in diameter, is ground optically flat and silvered on both ends. One end is plated so as to be completely reflecting, and the other end is partially transmitting. The ruby rod with its silvered ends forms a parallel-plate Fabry-Perot (Sec. 7.5) optical resonator. The ruby rod is surrounded by a high-power flash lamp and a reflecting cavity to concentrate the light from the flash lamp onto the rod. When the flash lamp is fired, the chromium atoms in the ruby rod are pumped to a higher energy level. The atoms relax back to an intermediate metastable energy level almost immediately. The frequency difference between the normal low-energy state and the intermediate metastable state corresponds to that of red light. When a few chromium atoms drop from the metastable energy level to the lower level, they emit radiation that stimulates additional atoms to make a downward transition. This regenerative process rapidly triggers all the atoms to make the downward transition. The coherent radiation emitted rapidly builds up, and a high-energy pulse of red light is radiated from the partially transparent end. Since the ruby-rod cross section is very large in terms

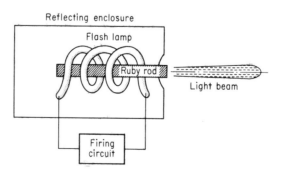

Fig. 10.16 The pulsed-ruby-rod laser.

of optical wavelengths, the radiated light is confined to a narrow beam of the order of 10 to 30 seconds of arc in angular width. Peak power outputs up to 50 Mw and more, with total energy outputs up to 100 joules, have been obtained. The flash lamps needed to obtain these high outputs produced 50,000 joules or more of light output, so that the efficiency is very low.

Gas laser

The gas laser gives continuous output, in contrast to the pulsed output from the ruby laser. It is therefore more suitable for communication uses. A common form of gas laser is the neon-helium gas laser illustrated in Fig. 10.17. A mixture of neon and helium gas is contained in a discharge tube. The electrical discharge can be initiated and maintained by means of a radio-frequency generator connected to electrodes placed around the tube as shown. This arrangement was used by Bell Telephone Laboratories in the first gas laser built. To form a Fabry-Perot optical cavity, the discharge tube is located between two mirrors. The ends of the discharge tube are commonly sealed off by glass windows placed at the Brewster angle to prevent reflections (Sec. 2.8). One mirror can be made partially transparent to permit a beam of radiated light to be formed external to the cavity.

The electrical discharge excites the helium atoms to a very high metastable energy level. When the excited helium atoms collide with the neon-gas atoms, they transfer energy to the neon atoms and cause them to jump to a high metastable energy level. When the neon atoms drop down to their original low energy level, red light is emitted. The emitted light reflecting back and forth between the mirrors stimulates further transitions in the downward direction. The radiation continually builds up, until the rate at which downward transitions occur equals the rate at which upward transitions are produced by the collision with excited helium atoms.

Output powers are about 0.01 watt, with an excitation energy of 20 to 50 watts. It is possible to voice-modulate the output beam by using a Kerr cell or other schemes that can control the light intensity according

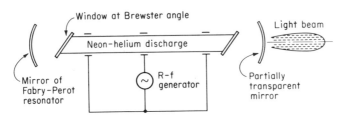

Fig. 10.17 The neon-helium gas laser.

Sec. 10.11   Microwave masers   539

to the modulating signal. Some of these modulation schemes are discussed in the references cited at the end of this chapter.

Problems

10.1 Solve Schrödinger's wave equation for a single free electron (zero potential energy). That is, verify that $\psi = e^{-j\omega t + j(k_x x + k_y y + k_z z)}$ is a solution provided $k^2 = k_x^2 + k_y^2 + k_z^2 = 4\pi m\omega/h$. The frequency ω equals the energy W divided by $h/2\pi$; that is, $\omega = 2\pi W/h$. Thus in terms of W show that the wavelength is given by $h/(2mW)^{\frac{1}{2}} = h/mv$, where $\frac{1}{2}mv^2$ equals the energy W for a particle with velocity v. Show that the group velocity given by $\partial\omega/\partial k$ equals v.

10.2 A maser crystal has three energy levels corresponding to frequencies 10^{10}, 2×10^{10}, and 6×10^{10} cps ($W = hf$). Calculate the relative number of particles in energy levels 2 and 3 (N_2/N_1 and N_3/N_1) at liquid-helium temperature 4.2°K and at room temperature 290°K.

10.3 Evaluate the inversion ratio for a three-level maser for an inverted 3-2 transition when $\tau_{12} = \tau_{23}$ and $T = 4°K$. Assume pump saturation of level 3, so that $\Delta n_{13} = 0$. The energy levels are those given in Prob. 10.2. For $T = 4°K$ or greater (10.43) may be used. Find the ratio of Δn_{32} at 290°K to Δn_{32} at 4°K.

10.4 Evaluate the maximum values of the matrix elements of $\tilde{\chi}''$ given by (10.53). Assume $\tau = 10^{-8}$ sec in the line-shape factor $L(f)$ and $\omega_s = \omega_{32}$. Also, use $B_0 = 2$ kilogauss and $\phi = 40°$ and the tabulated values of α, β, and γ. To evaluate Δn_{23}, assume $T = 4°K$ and an inversion ratio $f_p/2f_s - 1$. The pump and signal frequencies may be calculated from the energies tabulated in Table 10.2. Take the total number of available spins ($N_1 + N_2 + N_3$) to be 10^{20} spins/cm³. N_i may be found from (10.35). The spin density corresponds to approximately 0.5 percent chromium doping of the ruby crystal. The pump frequency is $f_{31} = f_3 - f_1$, where f_3 and f_1 are given in Table 10.2.

10.5 A maser is operated with the parameters given in Prob. 10.4. For a filling factor $\eta = 0.2$, evaluate the magnetic Q_m. For an unloaded cavity Q of 5,000, determine the external Q_e and cavity coupling parameter K to give a gain of 30 db. Determine the percentage change in the coupling parameter K that will result in oscillations.

10.6 For a transmission line terminated in an impedance $R + jX$, let the reflection coefficient be Γ_+. If the load impedance is changed to $-R + jX$, show that the new reflection coefficient Γ_- is given by $\Gamma_- = (\Gamma_+^*)^{-1}$. This relation reduces the maser broadbanding problem to one of making Γ_+ small and constant over the desired frequency band which is a conventional network problem. When Γ_+ is small and constant, the amplifier gain $|\Gamma_-|$ will be large and constant.

10.7 A rectangular waveguide of dimensions $a = 0.6$ cm, $b = 0.3$ cm is filled with a ruby crystal with the optical, or c, axis along the z axis. The field in the filled guide is an H_{10} mode. A magnetic field of 2 kilogauss is applied at 90° to the c axis, i.e., along the y axis. The signal frequency is that of the 3-2 transition, for which the data given in Table 10.2 apply. Assume an optimum inversion ratio $f_p/f_s - 1$ and the total number of available spins $N_1 + N_2 + N_3$ equal to 10^{20} spins/cm³. Take the magnetic-line-width time constant τ equal to 2×10^{-8} sec. Evaluate the magnetic Q_m as given by (10.87) and the growth constant α_m given by (10.88). Calculate the length of guide required to give a gain of 20 db. The pump may be assumed to saturate the 1-3 transition. The operating temperature is 4°K. For the 1-3 transition, α may be taken as zero, and since B_0 is applied along the y axis and not along the x axis, the appropriate form of the susceptibility transition matrix to be used is

$$\bar{\bar{\chi}}_{nm} = \frac{1}{4}\begin{bmatrix} \beta^2 & -j\beta\alpha & j\beta\gamma \\ j\beta\alpha & \alpha^2 & -\alpha\gamma \\ -j\beta\gamma & -\alpha\gamma & \gamma^2 \end{bmatrix}$$

and not that given by (10.50). The H_{10} mode has field components H_x, H_z, and E_y. α, β, γ are given in Table 10.2. Do not confuse β occurring in $\bar{\bar{\chi}}_{nm}$ with the waveguide propagation constant.

References

Two introductory articles on masers are:

1. From, W.: The Maser, *Microwave J.*, vol. 1, pp. 18–25, November–December, 1958.
2. Heffner, H.: Masers and Parametric Amplifiers, *Microwave J.*, vol. 2, pp. 33–40, March, 1959.
3. Siegman, A. E.: "Microwave Solid-state Masers," McGraw-Hill Book Company, New York, 1964. Highly recommended for a comprehensive discussion of solid-state microwave masers.
4. Singer, J. R.: "Masers and Lasers," 2d ed., John Wiley & Sons, Inc., New York, 1963.
5. Troup, G.: "Masers: Microwave Amplification by Stimulated Emission of Radiation," Methuen & Co., Ltd., London, 1959.
6. Vuylsteke, A. A.: "Elements of Maser Theory," D. Van Nostrand Company, Inc., Princeton, N.J., 1960.

Lasers

7. Lenygel, B. A.: "Lasers," John Wiley & Sons, Inc., New York, 1962.
8. Yariv, A., and J. P. Gordon: The Laser, *Proc. IEEE*, vol. 51, pp. 4–29, January, 1963.
9. Special Issue on Quantum Electronics, *Proc. IEEE*, vol. 51, January, 1963. Contains many excellent papers related to masers and lasers.

Quantum mechanics

10. Pauling, L., and E. B. Wilson, Jr.: "Introduction to Quantum Mechanics," McGraw-Hill Book Company, New York, 1935.
11. Schiff, L. I.: "Quantum Mechanics," 2d ed., McGraw-Hill Book Company, New York, 1955.
12. Kompaneyets, A. S.: "Theoretical Physics," Dover Publications, Inc., New York, 1962.

11
Parametric amplifiers

A parametric amplifier is an amplifier utilizing a nonlinear reactance, or a reactance that can be varied as a function of time by applying a suitable pump signal. The time variation of a reactive parameter can be used to produce amplification. This is the origin of the term parametric amplifier. The possibility of parametric amplification of signals was shown theoretically, as long ago as 1831, by Lord Rayleigh. The first analysis of the nonlinear capacitance was given by van der Ziel in 1948.† He pointed out that this device could also be useful as a low-noise amplifier since it was essentially a reactive device in which no thermal noise is generated. The first realization of a microwave parametric amplifier was made by Weiss, following the earlier proposal (1957) by H. Suhl, suggesting the use of the nonlinear effect in ferrites (Sec. 6.6). In the following few years the semiconductor-diode (sometimes called a varactor, for variable reactance) parametric amplifier was developed through the combined efforts of many workers. At the present time the semiconductor junction diode is the most widely used parametric amplifier. For this reason we limit the discussion in this text to this particular type of parametric amplifier. The p-n junction diode has a nonlinear capacitance. If a pumping signal at frequency ω_p and a small-amplitude signal at frequency ω_s are applied simultaneously, the device behaves like a time-varying linear capacitance at the signal frequency ω_s. As we show in later sections, a time-varying capacitance or a nonlinear capacitance can be incorporated into a circuit to give linear amplification of a small-amplitude signal. Before presenting this analysis a brief description of some of the properties of junction diodes is given, followed by a presentation of the Manley-Rowe relations. The latter are a set of power-conservation relations, of considerable value in determining the maximum gain and other performance features of parametric amplifiers.

† A. van der Ziel, On the Mixing Properties of Nonlinear Capacitances, *J. Appl. Phys.*, vol. 19, pp. 999–1006, November, 1948.

11.1 p-n junction diodes

The diodes used for parametric amplifiers consist of a junction of n-type and p-type semiconductor material. An n-type semiconductor has an excess supply of electrons, which is why it is called n-, or negative-, type. An example of an n-type material is a pure semiconductor such as germanium, with a small amount (about 1 part in 10^5) of impurity doping with an element such as arsenic or antimony. Germanium has a valence of 4, whereas arsenic or antimony has a valence of 5. Thus, at each site in the germanium or host crystal where an arsenic or antimony atom replaces a germanium atom, four of the valence electrons are used up to form the bond, and this leaves one excess valence electron. The valence electrons left over are relatively free to move around in the crystal under the influence of applied electric fields and make the material a donor of electrons, or n-type.

In p-, or positive-, type material, the impurity atoms are chosen to have a valence less than that of the host atoms. For example, gallium, with a valence of 3, may be used in a germanium crystal. When a gallium atom replaces a germanium atom, there are only three available valence electrons to form the required bond. A stable bond requires four valence electrons, and consequently, at each site where a gallium atom is located, a hole is created which can be filled by an electron that may be passing by. If an electron from some other bond moves over to fill the hole, the result is the creation of a new hole at some other point. The overall effect is as though the holes were positive carriers of electricity, i.e., equivalent positive electrons, that can move through the crystal. The holes do, in fact, behave as equivalent positive carriers, and thus p-type material can be considered as essentially the same as n-type material except that the signs of the charge carriers are opposite.

Consider now a linearly graded junction of n-type and p-type material, as shown in Fig. 11.1a. In the linearly graded junction, the n-type material changes gradually and in a linear fashion over to p-type material in a distance d. This variation is obtained by gradually reducing the doping, or concentration of donor atoms, down to zero in the region $x = d/2$ down to $x = 0$ and then linearly increasing the concentration of acceptor atoms in the region $x = 0$ to $x = -d/2$. If the number density of acceptor atoms is N_a and the density of donor atoms is N_d, the difference will vary linearly with x across the junction; thus

$$N_a - N_d = kx \tag{11.1}$$

where k is a suitable constant.

When there is a gradient in the impurity-concentration densities, electrons will diffuse from a region of high concentration to one of low concentration. Holes will diffuse in a similar manner. Thus the electrons will diffuse into the p-type side of the junction and holes will diffuse into

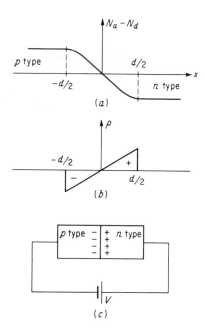

Fig. 11.1 The linearly graded junction.

the *n*-type side. This diffusion continues until a space-charge distribution, together with a resultant electric field, is set up of sufficient strength to produce a force that is equal and opposite to the effective force created by the concentration gradients. When equilibrium has been reached, a small region, called the depletion region, substantially free of charge carriers, is produced at the junction. The space charge built up on either side of the depletion region, together with the electric field existing across the depletion region, constitutes an equivalent capacitor. If a reversed-bias voltage is applied across the junction, the electron distribution and hole distribution will be forced farther apart. This widening of the depletion layer results in a decrease in the junction capacitance. It is now apparent that if an a-c pumping voltage is superimposed on the bias voltage, the equivalent junction capacitance can be varied as a function of time.

In the graded junction the space charge density will vary linearly across the junction so that a depletion layer completely free of carriers is not produced. The effect of having the space charge vary linearly across the junction instead of being concentrated at $x = \pm d$ is, however, much the same. If we consider a linearly varying space charge density (per unit cross-sectional area) $\rho = qx$, where q is a suitable constant, Poisson's equation gives

$$\frac{d^2\Phi}{dx^2} = -\frac{\rho}{\epsilon} = -\frac{q}{\epsilon}x$$

for the potential Φ. Integrating and using the boundary conditions that the electric field, and hence $d\Phi/dx$, is zero for $|x| > d/2$ and that $\Phi = 0$ at $x = 0$ from symmetry considerations, we get

$$\Phi = -\frac{qx}{2\epsilon}\left(\frac{x^2}{3} - \frac{d^2}{4}\right) \tag{11.2}$$

The potential difference across the junction is

$$\Phi\left(\frac{d}{2}\right) - \Phi\left(-\frac{d}{2}\right) = \frac{qd^3}{12\epsilon}$$

Under equilibrium conditions this potential difference must equal the contact potential Φ_c plus the negative applied bias voltage $-V$; thus

$$\Phi_c - V = \frac{qd^3}{12\epsilon} \tag{11.3}$$

The total stored charge per unit area is given by

$$Q = \int_0^{d/2} qx\, dx = \frac{qd^2}{8}$$

Eliminating d by means of (11.3) gives

$$Q = \frac{q}{8}\left[\frac{12\epsilon(\Phi_c - V)}{q}\right]^{\frac{2}{3}} \tag{11.4}$$

Since the capacitance is a function of the voltage, it must be defined as the ratio of an incremental change in Q to incremental change in $\Phi_c - V$. Thus the capacitance per unit cross-sectional area is

$$C = \frac{dQ}{d(\Phi_c - V)} = \epsilon\left[\frac{q}{12\epsilon(\Phi_c - V)}\right]^{\frac{1}{3}} \tag{11.5}$$

As seen from this equation, the junction capacitance C is nonlinear since it depends on the voltage V. If C is a linear element, $Q = CV$. In an abrupt junction diode C is proportional to $(\phi_c - V)^{-\frac{1}{3}}$.

If we denote $\phi_c - V$ by V_0 and apply in addition a pumping voltage $v_p = V_p \cos \omega_p t$, the capacitance becomes a function of time:

$$C(t) = \epsilon\left(\frac{q}{12\epsilon V_0}\right)^{\frac{1}{3}}\left(1 + \frac{V_p}{V_0}\cos \omega_p t\right)^{-\frac{1}{3}} \tag{11.6}$$

We now have a nonlinear capacitance that is also a function of time. The capacitance is a periodic function of time, and can be represented by a Fourier series expansion of the form

$$C(t) = \sum_{n=0}^{\infty} C_n \cos n\omega_p t \tag{11.7}$$

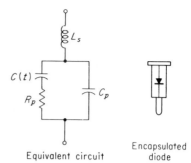

Fig. 11.2 Equivalent circuit of a parametric diode.

The coefficients are given by

$$C_0 = \frac{\epsilon}{2\pi} \left(\frac{q}{12\epsilon V_0}\right)^{\frac{1}{3}} \int_{-\pi}^{\pi} \left(1 + \frac{V_p}{V_0} \cos \theta\right)^{-\frac{1}{3}} d\theta$$

$$C_n = \frac{\epsilon}{\pi} \left(\frac{q}{12\epsilon V_0}\right)^{\frac{1}{3}} \int_{-\pi}^{\pi} \left(1 + \frac{V_p}{V_0} \cos \theta\right)^{-\frac{1}{3}} \cos n\theta \, d\theta$$

where $\theta = \omega_p t$. To evaluate the coefficients would require a numerical procedure. However, we do not need to know the values of the C_n in order to analyze the general properties of a parametric amplifier. The important feature brought out in the foregoing analysis is that $C(t)$ is a function of time that can be represented by a Fourier series involving all harmonics of the pumping frequency f_p. It is important to note that the coefficients are not, in general, linear functions of the a-c voltage V_p or the voltage V_0. Thus, since the junction capacitance $C(t)$ is a nonlinear capacitance, the principle of superposition *does not* hold for arbitrary a-c signal amplitudes. Under small-signal conditions, a Taylor series expansion of $C(t)$ about an operating point may be used and only the linear term in the signal amplitude retained. In this case superposition *does* hold. The situation here is no different from that in any other amplifying device since all are nonlinear for sufficiently large applied signals.

In addition to the capacitance associated with the diode junction, there is a shunt conductance arising from the bulk resistance of the material in the depletion layer. This shunt conductance is proportional to the area, and since C is also proportional to the area, the ratio is independent of the cross-sectional area of the diode. The shunt conductance of the depletion layer is small, and can often be neglected. Of more importance is the series resistance of the n- and p-type semiconductor material outside the depletion layer. When the p-n junction is encapsulated and connecting leads are put on, an additional shunt capacitance C_p due to the cartridge and a series inductance arising from the leads are also present. The overall equivalent circuit is thus of the form shown in Fig. 11.2. Typical values of C_p and L_s are somewhat less than one micromicrofarad and one millimicrohenry, respectively, at microwave frequencies. The junction capacitance C is also about one micromicrofarad, and typical values of R_p are a few ohms.

11.2 Manley-Rowe relations

Manley and Rowe have derived a set of power-conservation relations that are extremely useful in evaluating the performance which can be achieved from a parametric device consisting of a nonlinear reactance.† These relations are also derived below.

The circuit considered by Manley and Rowe is shown in Fig. 11.3. It consists of resistive loads in series with ideal bandpass filters connected in shunt with a lossless nonlinear capacitance. Two sinusoidal signals at frequencies f_1 and f_2 are applied. The nonlinear capacitance causes frequencies at the harmonics of f_1 and f_2 to be generated. Each bandpass filter is considered to pass only one harmonic component $nf_1 + mf_2$. The overall circuit thus isolates all the harmonics and dissipates their power in separate resistive loads. The Manley-Rowe relations establish two constraints governing the conversion of input power at the frequencies f_1 and f_2 into power at other frequencies.

Let the charge Q on C be a single-valued function of the a-c voltage $v = v_1 + v_2 = V_1 \cos \omega_1 t + V_2 \cos \omega_2 t$ applied across it. Thus $Q = Q(v)$. We may expand Q in a Taylor series in v to obtain

$$Q = Q(0) + \frac{\partial Q}{\partial v} v + \frac{1}{2} \frac{\partial^2 Q}{\partial v^2} v^2 + \cdots \tag{11.8}$$

where all derivatives are evaluated at $v = 0$. Since all powers of v occur, it is clear that, because $v = (V_1/2)(e^{j\omega_1 t} + e^{-j\omega_1 t}) + (V_2/2)(e^{j\omega_2 t} + e^{-j\omega_2 t})$, the charge Q will have frequencies at all harmonics of f_1 and f_2. If currents at the various harmonics are permitted to pass through C, the voltage developed across C will also contain all possible harmonics. In this case Q is a function of all voltages present. However, the expansion

† J. M. Manley and H. E. Rowe, Some General Properties of Nonlinear Elements, Part I, General Energy Relations, *Proc. IRE*, vol. 44, pp. 904–913, July, 1956. See also *Proc. IRE*, vol. 47, pp. 2115–2116, December, 1959.

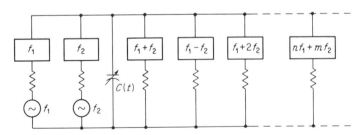

Fig. 11.3 Circuit for illustration of the Manley-Rowe relations.

(11.8) will still be valid, except that now the coefficients will have different values. Consequently, the general expansion of Q has the form

$$Q = \sum_{n=-\infty}^{\infty} \sum_{m=-\infty}^{\infty} Q_{nm} e^{j(n\omega_1 + m\omega_2)t} \tag{11.9}$$

The charge is a real function of time; so we must have $Q_{-n-m} = Q_{nm}^*$ in order that the n, m and $-n$, $-m$ terms will combine to form a real function of time with frequency $n\omega_1 + m\omega_2$.

The total voltage V can be expressed as a function $V(Q)$ of the charge. A similar Taylor series expansion of $V(Q)$ then shows that V can be expressed in a form similar to (11.9); thus

$$V = \sum_{n=-\infty}^{\infty} \sum_{m=-\infty}^{\infty} V_{nm} e^{j(n\omega_1 + m\omega_2)t} \tag{11.10}$$

For V to be real, we must have $V_{-n-m} = V_{nm}^*$.

The current through C is the total rate of change of Q with time, and is given by

$$I = \frac{dQ}{dt} = \sum_{n=-\infty}^{\infty} \sum_{m=-\infty}^{\infty} j(n\omega_1 + m\omega_2) Q_{nm} e^{j(n\omega_1 + m\omega_2)t}$$

$$= \sum_{n=-\infty}^{\infty} \sum_{m=-\infty}^{\infty} I_{nm} e^{j(n\omega_1 + m\omega_2)t} \tag{11.11}$$

where $I_{nm} = j(n\omega_1 + m\omega_2) Q_{nm}$.

Since C is a pure reactive element there can be no net power into or out of C. If we assume that ω_1 and ω_2 are incommensurable, there will be no time-average power due to interacting harmonics. The average power at the frequencies $\pm|n\omega_1 + m\omega_2|$ is

$$P_{nm} = (V_{nm} I_{nm}^* + V_{nm}^* I_{nm})$$
$$= (V_{nm} I_{nm}^* + V_{-n-m} I_{-n-m}^*) = P_{-n-m} \tag{11.12}$$

since the time average of

$$(I_{nm} e^{j(n\omega_1 + m\omega_2)t} + I_{-n-m} e^{-j(n\omega_1 + m\omega_2)t})(V_{nm} e^{j(n\omega_1 + m\omega_2)t} + V_{-n-m} e^{-j(n\omega_1 + m\omega_2)t})$$

is

$$V_{nm} I_{-n-m} + V_{-n-m} I_{nm} = V_{nm} I_{nm}^* + V_{nm}^* I_{nm} = V_{nm} I_{nm}^* + V_{-n-m} I_{-n-m}^*$$

Conservation of power can therefore be expressed as

$$\sum_{n=-\infty}^{\infty} \sum_{m=-\infty}^{\infty} P_{nm} = 0 \tag{11.13}$$

since $P_{nm} = P_{-n-m}$ from (11.12). To obtain the Manley-Rowe relations we multiply each term by $(n\omega_1 + m\omega_2)/(n\omega_1 + m\omega_2)$ and split the sum into two parts; thus

$$\omega_1 \sum_{n=-\infty}^{\infty} \sum_{m=-\infty}^{\infty} \frac{nP_{nm}}{n\omega_1 + m\omega_2} + \omega_2 \sum_{n=-\infty}^{\infty} \sum_{m=-\infty}^{\infty} \frac{mP_{nm}}{n\omega_1 + m\omega_2} = 0 \qquad (11.14)$$

We can now show that each double sum must vanish separately. We can replace each $I_{nm}/(n\omega_1 + m\omega_2)$ by jQ_{nm}, and then $P_{nm}/(n\omega_1 + m\omega_2)$ becomes $-jV_{nm}Q_{nm}^* - jV_{-n-m}Q_{-n-m}^*$ and does not depend explicitly on ω_1 or ω_2. For any choice of ω_1 and ω_2, we can always adjust the network external to C so that the currents which result keep all the voltage amplitudes V_{nm} unchanged. The Q_{nm} are then also unchanged since they depend only on the V_{nm}. When this is done we see from (11.14) that it is possible to change ω_1 and ω_2 arbitrarily but keep the two double sums involving $P_{nm}/(n\omega_1 + m\omega_2) = -jV_{nm}Q_{nm}^* - jV_{-n-m}Q_{-n-m}^*$ unchanged. Consequently, (11.14) can hold for arbitrary ω_1 and ω_2 only if

$$\sum_{n=-\infty}^{\infty} \sum_{m=-\infty}^{\infty} \frac{nP_{nm}}{n\omega_1 + m\omega_2} = 0$$

$$\sum_{n=-\infty}^{\infty} \sum_{m=-\infty}^{\infty} \frac{mP_{nm}}{n\omega_1 + m\omega_2} = 0$$

That is, the coefficients of ω_1 and ω_2 must vanish separately. The above two relations are the Manley-Rowe relations. They are usually written in a somewhat different form, however. We may write the first sum as

$$\sum_{n=0}^{\infty} \sum_{m=-\infty}^{\infty} \frac{nP_{nm}}{n\omega_1 + m\omega_2} + \sum_{n=0}^{\infty} \sum_{m=-\infty}^{\infty} \frac{-nP_{-n-m}}{-n\omega_1 - m\omega_2}$$

where n and m have been replaced by $-n$ and $-m$ in the second term. Since $P_{-n-m} = P_{nm}$, the two parts are equal; so we obtain

$$\sum_{n=0}^{\infty} \sum_{m=-\infty}^{\infty} \frac{nP_{nm}}{n\omega_1 + m\omega_2} = 0 \qquad (11.15a)$$

Similarly, we can obtain

$$\sum_{m=0}^{\infty} \sum_{n=-\infty}^{\infty} \frac{mP_{nm}}{n\omega_1 + m\omega_2} = 0 \qquad (11.15b)$$

These are the standard forms for the Manley-Rowe relations. The Manley-Rowe relations are general power-conservation relations, and do not depend on any specific circuit such as that in Fig. 11.3. This is apparent since no reference to an external circuit was made in the derivation.

For an example of the application of the Manley-Rowe relations, consider a circuit similar to that in Fig. 11.3 with generators at frequencies

f_1 and f_2. Let all harmonics be open-circuited except $f_1 + f_2$. Thus currents at the three frequencies f_1, f_2, and $f_1 + f_2$ are the only ones that can exist. The $n = \pm 1, m = 0$ and $n = 0, m = \pm 1$ and $n = m = \pm 1$ terms in (11.15) are the only ones present. Thus we get

$$\frac{P_{10}}{\omega_1} + \frac{P_{11}}{\omega_1 + \omega_2} = 0 \tag{11.16a}$$

$$\frac{P_{01}}{\omega_2} + \frac{P_{11}}{\omega_1 + \omega_2} = 0 \tag{11.16b}$$

Since power is supplied at the frequencies ω_1 and ω_2, we must have P_{10} and P_{01} positive. Therefore P_{11} is negative, and power is delivered from the nonlinear capacitor C at the frequency $\omega_1 + \omega_2$. If ω_1 is the input-signal frequency and ω_2 is the pump frequency, then $\omega_3 = \omega_1 + \omega_2$ is the output frequency. The maximum signal gain is

$$-\frac{P_{11}}{P_{10}} = \frac{\omega_1 + \omega_2}{\omega_1} = \frac{\omega_3}{\omega_1} = 1 + \frac{\omega_2}{\omega_1} \tag{11.17}$$

as obtained from (11.16a). A parametric amplifier of this type is called an up-converter. Because of losses that are always present in a practical amplifier, the gain will be less than ω_3/ω_1. The Manley-Rowe relations give the maximum gain possible and hence provide a criterion by which a practical up-converter can be judged.

11.3 Linearized equations for parametric amplifiers

Consider a linear capacitance C for which the charge-voltage relationship is $Q = Cv$. The current flowing through C is given by

$$i = \frac{dQ}{dt} = \frac{d}{dt}(Cv) = C\frac{dv}{dt}$$

If C is made a function of time, for example, a parallel-plate capacitor with a plate separation that is varied with time, the current will be given by

$$i = \frac{d}{dt}(Cv) = v\frac{dC}{dt} + C\frac{dv}{dt} \tag{11.18}$$

If instead of a time-varying linear capacitance we have a nonlinear capacitance, where the charge Q is a nonlinear function $Q(v)$ of the voltage, the current is given by

$$i = \frac{dQ}{dt} = \frac{\partial Q}{\partial v}\frac{dv}{dt}$$

If the voltage v is the sum of a pump voltage v_p at frequency ω_p and a signal voltage v_s at frequency ω_s and $|v_s| \ll |v_p|$, we can expand

$$Q(v) = Q(v_p + v_s)$$

in a Taylor series about the point $v_s = 0$. Thus we obtain

$$Q(v_p + v_s) = Q(v_p) + \left.\frac{\partial Q}{\partial v}\right|_{v_s=0} v_s + \frac{1}{2}\left.\frac{\partial^2 Q}{\partial v^2}\right|_{v_s=0} v_s^2 + \cdots$$

For v_s sufficiently small, we can obtain satisfactory accuracy by retaining the first two terms only. The current is then given by

$$i = \frac{dQ(v_p)}{dt} + \frac{d}{dt}\left(\left.\frac{\partial Q}{\partial v}\right|_{v_s=0} v_s\right) \tag{11.19a}$$

Let the quantity $\partial Q/\partial v$ for $v_s = 0$ be denoted by $C(t)$, in which case we can write

$$i = \frac{dQ(v_p)}{dt} + \frac{d}{dt}[C(t)v_s] \tag{11.19b}$$

If we compare this result with (11.18), we see that the nonlinear capacitance behaves like a time-varying linear capacitance for signals with amplitudes that are small compared with the pump signal amplitude. The first term, $dQ(v_p)/dt$, in (11.19) gives a current at the pumping frequency, and is not related to the signal current. If the pumping voltage is also small compared with the d-c bias voltage in a junction diode, we can assume $C(t)$ to have the form [see (11.6)]

$$C(t) = C_0(1 + 2M \cos \omega_p t) \tag{11.20}$$

since $(V_0 + V_p \cos \omega_p t)^{-\frac{1}{2}} \approx V_0^{-\frac{1}{2}}[1 - (V_p/3V_0) \cos \omega_p t]$ for $V_p \ll V_0$. The linearized equations (11.19) and (11.20) are the ones we shall use in the analysis of parametric amplifiers.

The equivalent circuit of a p-n junction diode is illustrated in Fig. 11.2. For the purpose of analysis it is more convenient to use an equivalent series circuit of the form shown in Fig. 11.4. The two circuits are equiv-

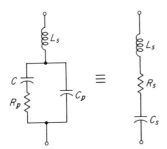

Fig. 11.4 Equivalent series circuit for a junction diode.

alent if we choose

$$R_s = \frac{R_p C^2}{(\omega C C_p R_p)^2 + (C + C_p)^2} \qquad C_s = \frac{(\omega C C_p R_p)^2 + (C + C_p)^2}{(\omega C R_p)^2 C_p + C + C_p}$$

which makes the input impedance the same for both circuits at the frequency ω. In most diodes the resistance R_p is small compared with the cartridge reactance $1/\omega C_p$; so $\omega C_p R_p \ll 1$. In this case we find that

$$R_s \approx \left(\frac{C}{C + C_p}\right)^2 R_p \qquad C_s \approx C + C_p$$

When these approximations are valid, the two circuits are equivalent, independently of frequency. This is a necessary requirement if the series circuit is to be useful for analysis purposes, since in a parametric amplifier currents and voltages at several different frequencies are simultaneously present. When C is a function of the voltage, R_s will also be voltage-dependent. The effect of a voltage-dependent resistance R_s will, for simplicity, be neglected, since it is not too important. In other words, we shall consider R_s to be a constant resistance.

11.4 Parametric up-converter

In the up-converter, a pump voltage at frequency ω_p and a signal at frequency ω_s are applied to the diode, and the output signal is taken at the higher frequency $\omega_s + \omega_p$. Mixing effects take place that give rise to all possible harmonics of ω_p and ω_s. However, in the up-converter, the circuit external to the diode is chosen so as to permit currents only at the signal frequency ω_s, the pump frequency ω_p, and the output frequency ω_0, which is chosen as the sum of the pump and signal frequencies, that is, at $\omega_0 = \omega_s + \omega_p$. There will, consequently, be a voltage across the diode at the three possible frequencies. If we let the diode voltage be represented as

$$v_s = \text{Re}\,(V_s e^{j\omega_s t}) = \tfrac{1}{2}(V_s e^{j\omega_s t} + V_s^* e^{-j\omega_s t})$$
$$v_p = \text{Re}\,(V_p e^{j\omega_p t}) = \tfrac{1}{2}(V_p e^{j\omega_p t} + V_p^* e^{-j\omega_p t})$$
$$v_0 = \text{Re}\,(V_0 e^{j\omega_0 t}) = \tfrac{1}{2}(V_0 e^{j\omega_0 t} + V_0^* e^{-j\omega_0 t})$$

we may generalize (11.19) and (11.20) to give

$$i = \frac{dQ(v_p)}{dt} + \frac{1}{2}\frac{d}{dt} C_0(1 + 2M \cos \omega_p t)$$
$$\times (V_s e^{j\omega_s t} + V_s^* e^{-j\omega_s t} + V_0 e^{j\omega_0 t} + V_0^* e^{-j\omega_0 t}) \quad (11.21)$$

Let the current at the frequencies ω_s and ω_0 be expressed as

$$i_s = \tfrac{1}{2}(I_s e^{j\omega_s t} + I_s^* e^{-j\omega_s t}) \qquad (11.22)$$
$$i_0 = \tfrac{1}{2}(I_0 e^{j\omega_0 t} + I_0^* e^{-j\omega_0 t}) \qquad (11.23)$$

When the time derivative of (11.21) is taken and the terms at frequencies ω_s and $\pm\omega_0$ only are retained, we obtain (a knowledge of the pump current at frequency ω_p will not be required; so we do not need to evaluate it)

$$i_s + i_0 = \frac{C_0}{2}(j\omega_s V_s e^{j\omega_s t} - j\omega_s V_s^* e^{-j\omega_s t} + j\omega_0 V_0 e^{j\omega_0 t} - j\omega_0 V_0^* e^{-j\omega_0 t}$$
$$+ j\omega_0 M V_s e^{j\omega_0 t} - j\omega_s M V_0^* e^{-j\omega_s t} - j\omega_0 M V_s^* e^{-j\omega_0 t} + j\omega_s M V_0 e^{j\omega_s t})$$

Using (11.22) and (11.23 gives

$$I_s = j\omega_s C_0 V_s + j\omega_s C_0 M V_0 \tag{11.24a}$$
$$I_0 = j\omega_0 C_0 V_0 + j\omega_0 C_0 M V_s \tag{11.24b}$$

These two equations show that, for the input signal current i_s and output signal current i_0, the junction capacitance may be represented by an admittance matrix such that

$$\begin{bmatrix} I_s \\ I_0 \end{bmatrix} = \begin{bmatrix} j\omega_s C_0 & j\omega_s C_0 M \\ j\omega_0 C_0 M & j\omega_0 C_0 \end{bmatrix} \begin{bmatrix} V_s \\ V_0 \end{bmatrix} \tag{11.24c}$$

The parameter M is proportional to the pump voltage and gives the coupling between the voltages at the two frequencies ω_s and ω_0.

Figure 11.5 illustrates an equivalent-circuit model of an up-converter. The series tuned circuits are chosen so that the three circuit loops have resonant frequencies of ω_s, ω_0, and ω_p and only currents with these respective frequencies can exist in each loop. Thus, in the input circuit loop, only I_s is present. The three circuit loops are coupled together through the time-varying part C of C_s only. Therefore, for the two frequencies ω_s and ω_0, the equivalent circuit can be reduced to that illustrated in Fig.

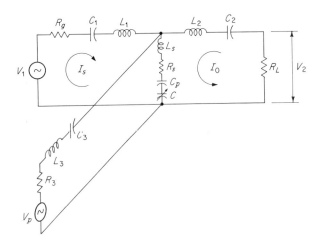

Fig. 11.5 Equivalent-circuit model of an up-converter.

Sec. 11.4   Parametric amplifiers

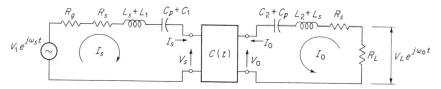

Fig. 11.6   Reduced equivalent circuit for an up-converter.

11.6. The box labeled $C(t)$ in this circuit is an equivalent impedance network that maintains the relationship given by (11.24) between the terminal currents and voltages. The analysis of the parametric amplifier is a conventional network-analysis problem since the diode has been replaced by an equivalent linear two-port network with terminal relations described by (11.24).† Each loop in the circuit is assumed to provide a very high impedance to currents at all frequencies present except the resonant frequency for that loop. The resonant circuits in Fig. 11.5 have been assumed to have zero loss. Circuit losses can be considered included in R_g and R_L. At the end of the analysis, R_g and R_L can then be split into two parts so as to separate the circuit losses from the generator and load impedances. In practice, circuit losses are small compared with the loss arising from the diode resistance R_s and the external loading represented by R_g and R_L.

We may solve (11.24c) for V_s and V_0 in terms of I_s and I_0 to obtain

$$\begin{bmatrix} V_s \\ V_0 \end{bmatrix} = \frac{1}{1-M^2} \begin{bmatrix} 1/j\omega_s C_0 & -M/j\omega_0 C_0 \\ -M/j\omega_s C_0 & 1/j\omega_0 C_0 \end{bmatrix} \begin{bmatrix} I_s \\ I_0 \end{bmatrix} \quad (11.25)$$

For the input circuit we may now write

$$V_1 = I_s \left[R_g + R_s + j\omega_s(L_s + L_1) + \frac{1}{j\omega_s(C_p + C_1)} \right] + V_s$$

$$= I_s \left[R_g + R_s + j\omega_s(L_s + L_1) + \frac{1}{j\omega_s(C_p + C_1)} + \frac{1}{j\omega_s(1-M^2)C_0} \right]$$
$$- \frac{MI_0}{j\omega_0(1-M^2)C_0}$$

whereas for the output circuit

$$0 = I_0 \left[R_L + R_s + j\omega_0(L_s + L_2) + \frac{1}{j\omega_0(C_p + C_2)} \right] + V_0$$

$$= I_0 \left[R_L + R_s + j\omega_0(L_s + L_2) + \frac{1}{j\omega_0(C_p + C_2)} + \frac{1}{j\omega_0(1-M^2)C_0} \right]$$
$$- \frac{MI_s}{j\omega_s(1-M^2)C_0}$$

† Care must be exercised in the analysis since currents and voltages at several different frequencies are simultaneously present.

Let us now assume that the circuits are tuned so that the following conditions hold:

$$\omega_s^2 = \frac{1}{L_s + L_1}\left[\frac{1}{C_p + C_1} + \frac{1}{(1 - M^2)C_0}\right]$$

$$\omega_0^2 = \frac{1}{L_s + L_2}\left[\frac{1}{C_p + C_2} + \frac{1}{(1 - M^2)C_0}\right]$$

We then obtain

$$V_1 = I_s(R_g + R_s) + \frac{jMI_0}{\omega_0(1 - M^2)C_0}$$

$$0 = I_0(R_L + R_s) + \frac{jMI_s}{\omega_s(1 - M^2)C_0}$$

We may solve for I_0 to obtain

$$I_0 = \frac{-jV_1 M\omega_0 C_0(1 - M^2)}{M^2 + (R_g + R_s)(R_L + R_s)\omega_0\omega_s(1 - M^2)^2 C_0^2} \tag{11.26}$$

The maximum available input power from the generator is $\frac{1}{2}(V_1^2/4R_g)$, and the output power developed in R_L is $\frac{1}{2}|I_0|^2 R_L$. The midband power gain is thus given by

$$G_0 = \frac{4|I_0|^2 R_L R_g}{V_1^2}$$

$$= \frac{4R_L R_g M^2}{\omega_s^2(1 - M^2)^2 C_0^2 \left[(R_g + R_s)(R_L + R_s) + \dfrac{M^2}{\omega_0\omega_s(1 - M^2)^2 C_0^2}\right]^2} \tag{11.27}$$

when (11.26) is used to express I_0 in terms of V_1. If desired, circuit losses can be included at this point by replacing R_g and R_L in the denominator in (11.27) by $R_g + R_{1l}$ and $R_L + R_{2l}$, where R_{1l} and R_{2l} represent the loss resistance in the input and output circuits. For simplicity we shall take $R_{1l} = R_{2l} = 0$.

To achieve maximum gain requires adjustment of R_g and R_L. Since R_g and R_L occur symmetrically in the expression for G_0, the optimum values of R_L and R_g are equal. Hence we need to maximize

$$G_0 = \frac{4R_L^2 M^2}{\omega_s^2(1 - M^2)^2 C_0^2 \left[(R_L + R_s)^2 + \dfrac{M^2}{\omega_0\omega_s(1 - M^2)^2 C_0^2}\right]^2}$$

Equating dG_0/dR_L to zero and solving for R_L give

$$R_L = R_s\left[1 + \frac{M^2}{\omega_0\omega_s R_s^2(1 - M^2)^2 C_0^2}\right]^{1/2} \tag{11.28}$$

The effective Q of the diode may be defined as

$$Q = \frac{1}{R_s \omega_s (1 - M^2) C_0} \tag{11.29}$$

We then find that

$$R_L = R_s \left[1 + \frac{\omega_s}{\omega_0} (MQ)^2 \right]^{\frac{1}{2}} \tag{11.30}$$

and the maximum gain is

$$G_0 = \frac{\omega_0}{\omega_s} \frac{\delta}{(1 + \sqrt{1+\delta})^2} \tag{11.31}$$

where $\delta = (\omega_s/\omega_0)(MQ)^2$. According to the Manley-Rowe relations discussed in Sec. 11.2, the maximum gain of an up-converter is ω_0/ω_s. The quantity $\delta/(1 + \sqrt{1+\delta})^2$ may therefore be regarded as a gain-degradation factor. As the diode Q approaches infinity, that is, as R_s goes to zero, δ approaches infinity, and the gain-degradation factor becomes equal to unity. Hence, for a lossless diode, the gain becomes equal to ω_0/ω_s, as predicted by the Manley-Rowe relations. In a typical microwave diode, MQ could be equal to 10. If $\omega_0/\omega_s = 10$ also, the maximum gain as given by (11.31) is 7.3 db.

To achieve high gain with an up-converter requires a large ratio ω_0/ω_s of output-to-input frequency. At the higher microwave frequencies this is not a very practical requirement, and for this reason up-converters are usually restricted to operation at signal frequencies f_s below 1,000 Mc. Higher gain can be obtained from the negative-resistance parametric amplifier, which is discussed in the next section.

11.5 Negative-resistance parametric amplifier

In the negative-resistance parametric amplifier currents are permitted to exist at the signal frequency ω_s, the pump frequency ω_p, and the idler frequency $\omega_i = \omega_p - \omega_s$. The equivalent-circuit model that will be analyzed is shown in Fig. 11.7.

When we replace the voltage v_0 in (11.21) by $v_i = \frac{1}{2}(V_i e^{j\omega_i t} + V_i^* e^{-j\omega_i t})$ and introduce the idler current $i_i = \frac{1}{2}(I_i e^{j\omega_i t} + I_i^* e^{-j\omega_i t})$, we may solve for

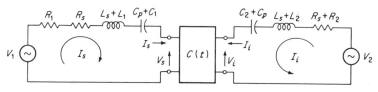

Fig. 11.7 Equivalent circuit for a negative-resistance parametric amplifier.

I_s and I_i in terms of V_s and V_i in the same manner that was used to treat the up-converter. It is readily found that, for $\omega_i = \omega_p - \omega_s$,

$$\begin{bmatrix} I_i \\ I_s^* \end{bmatrix} = \begin{bmatrix} j\omega_i C_0 & j\omega_i C_0 M \\ -j\omega_s C_0 M & -j\omega_s C_0 \end{bmatrix} \begin{bmatrix} V_i \\ V_s^* \end{bmatrix} \quad (11.32a)$$

and

$$\begin{bmatrix} V_i \\ V_s^* \end{bmatrix} = \frac{1}{1-M^2} \begin{bmatrix} \dfrac{1}{j\omega_i C_0} & \dfrac{M}{j\omega_s C_0} \\ \dfrac{-M}{j\omega_i C_0} & \dfrac{-1}{j\omega_s C_0} \end{bmatrix} \begin{bmatrix} I_i \\ I_s^* \end{bmatrix} \quad (11.32b)$$

For the circuit of Fig. 11.7 we can write the following equations:

$$V_1 = I_s \left[R_1 + R_s + j\omega_s(L_1 + L_s) + \frac{1}{j\omega_s(C_p + C_1)} + \frac{1}{j\omega_s(1-M^2)C_0} \right]$$
$$+ \frac{M}{(1-M^2)j\omega_i C_0} I_i^* \quad (11.33a)$$

$$V_2 = I_i \left[R_2 + R_s + j\omega_i(L_2 + L_s) + \frac{1}{j\omega_i(C_p + C_2)} + \frac{1}{j\omega_i(1-M^2)C_0} \right]$$
$$+ \frac{M}{(1-M^2)j\omega_s C_0} I_s^* \quad (11.33b)$$

If we impose the tuning conditions

$$\omega_s^2 = \frac{1}{L_1 + L_s} \left[\frac{1}{C_p + C_1} + \frac{1}{(1-M^2)C_0} \right]$$

$$\omega_i^2 = \frac{1}{L_2 + L_s} \left[\frac{1}{C_p + C_2} + \frac{1}{(1-M^2)C_0} \right]$$

we obtain

$$V_1 = (R_1 + R_s)I_s - \frac{jM}{\omega_i(1-M^2)C_0} I_i^* \quad (11.34a)$$

$$V_2 = (R_2 + R_s)I_i - \frac{jM}{\omega_s(1-M^2)C_0} I_s^* \quad (11.34b)$$

Let us now assume that $V_2 = 0$. We can then determine I_i from (11.34) and evaluate the gain $G_0 = 4R_1 R_2 |I_i|^2 / V_1^2$. We readily find that

$$G_0 = \frac{4R_1 R_2 M^2}{\left[R_1 + R_s - \dfrac{M^2}{(R_2 + R_s)\omega_i \omega_s (1-M^2)^2 C_0^2} \right]^2 (R_2 + R_s)^2 \omega_s^2 C_0^2 (1-M^2)^2} \quad (11.35)$$

The term $-M^2/[(R_2 + R_s)\omega_i \omega_s (1-M^2)^2 C_0^2]$ may be interpreted as an equivalent negative resistance $-R$. Introducing R, we may express

G_0 as

$$G_0 = \frac{4R_1R_2[\omega_i(1 - M^2)C_0R]^2}{M^2(R_1 + R_s - R)^2} \tag{11.36}$$

It is clear that a very large gain can be obtained if R is made almost equal to $R_1 + R_s$. However, care must be taken not to make R too close to $R_1 + R_s$ because a small change in parameters will then cause large changes in the gain and will cause oscillations to occur if R becomes equal to $R_1 + R_s$.

The parametric amplifier discussed above is called a negative-resistance converter. It is possible to take the output at the same frequency ω_s as the input. If we split R_1 into a generator internal resistance R_g plus a load resistance R_L, the power delivered to R_L is $\tfrac{1}{2}R_L|I_s|^2$. The power gain will be

$$G_0 = \frac{4R_g R_L |I_s|^2}{V_1^2}$$

We may evaluate I_s from (11.34) to obtain

$$G_0 = \frac{4R_g R_L}{(R_g + R_L + R_s - R)^2} \tag{11.37}$$

where, as before,

$$R = \frac{M^2}{(R_2 + R_s)\omega_i\omega_s(1 - M^2)^2 C_0^2} \tag{11.38}$$

The effective negative resistance $-R$ arises in the following manner: The application of signal plus pump power to the nonlinear capacitance causes frequency mixing to occur. When current is permitted to exist at the idler frequency $\omega_p - \omega_s$, further frequency mixing of power at the pump and idler frequencies occurs. This latter mixing causes harmonics of ω_p and $\omega_i = \omega_p - \omega_s$ to be generated; in particular, power at the frequency ω_s is generated. When the power generated through frequency mixing exceeds that being supplied at the signal frequency ω_s, the diode appears to have a negative resistance. If idler current is not permitted to exist, the negative resistance vanishes, as reference to (11.38) shows when R_2 is made infinite (open circuit for the idler signal).

The negative-resistance parametric amplifier with input and output at the same frequency is not very stable. The reason is that in a microwave system R_g and R_L are the impedances seen looking into the input and output transmission-line ports. If the loads connected to these transmission lines are not matched, reflected waves occur. Reflected waves in the output line return to the amplifier and are amplified and fed into both the input and output lines. The result is that the gain becomes a sensitive function of the external generator and load impedances. The stability of the amplifier is greatly improved by the use of a circulator, as

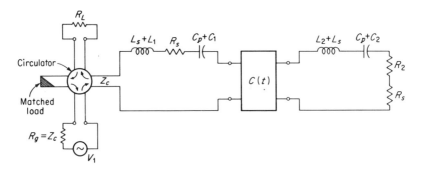

Fig. 11.8 A negative-resistance parametric amplifier using a circulator.

illustrated in Fig. 11.8. The use of a circulator makes the load termination R_L for the amplifier equal to the characteristic impedance of the transmission line independently of the external generator or load impedances Z_g and Z_L. The available power from the generator is still given by $\frac{1}{2}V_1^2/4R_g$. However, the amplifier power is now all delivered to the load $R_L = Z_c$, and none of it is dissipated in the internal generator impedance R_g. Consequently, the power gain is nearly four times greater since I_s is nearly twice as large, because the series resistance in the input circuit is now $R_L + R_s$ instead of $R_L + R_g + R_s \approx 2R_L$, when $R_L = R_g \gg R_s$. The power gain is the square of the voltage reflection coefficient, and is given by (see discussion in Sec. 10.8 on negative-resistance maser amplifiers)

$$G_0 = \left|\frac{Z_{\text{in}} - Z_c}{Z_{\text{in}} + Z_c}\right|^2 = \frac{(R_s - R - R_L)^2}{(R_s + R_L - R)^2}$$

since $Z_{\text{in}} = R_s - R$ at resonance and we are taking $R_L = Z_c$. For high gain, $R \approx R_L + R_s$ and is large compared with R_s. Consequently, the gain can be expressed as

$$G_0 = \frac{4R_L^2}{(R_L + R_s - R)^2} \tag{11.39}$$

The maximum value of negative resistance that can be obtained is determined by the diode that is used. If we make $R_2 = 0$ in (11.38), we see that maximum R is R_m, where

$$R_m = \frac{\omega_s}{\omega_i}(MQ)^2 R_s \tag{11.40}$$

(11.29) having been used to introduce the diode Q. If we regard R as fixed, we see that, for the amplifier without a circulator, we must make $R_g + R_L + R_s \approx R$, as (11.37) shows. But with $R_g = R_L = Z_c$ large compared with R_s, we get $R_g = R_L \approx R/2$, and (11.37) gives

$$G_0 = \frac{R^2}{(2R_L + R_s - R)^2} \tag{11.41a}$$

whereas for the amplifier with a circulator, $R_L \approx R$ and

$$G_0 = \frac{4R^2}{(R_L + R_s - R)^2} \tag{11.41b}$$

Note that when a circulator is used, the optimum value of R_L is twice what it is when a circulator is not used. Thus the denominators in (11.41a) and (11.41b) are the same. These relations then show that the use of a circulator gives 6 db more gain for the same amount of diode loading, and hence will have a gain-bandwidth product twice as great.

If the pump frequency ω_p is chosen equal to twice the signal frequency ω_s, the idler frequency $\omega_i = \omega_p - \omega_s = \omega_s$ is equal to the signal frequency. In this case the amplifier is called a degenerate negative-resistance amplifier. For the degenerate amplifier the signal and idler circuits would be a single resonant circuit. The analysis of the degenerate amplifier is similar to that already carried out, and so will not be presented here (some results on noise properties are given in Sec. 11.6).

There are many different ways of building microwave parametric amplifiers. Transmission lines, waveguides, or a combination of the two may be used to construct suitable cavities to use as resonant circuits. A typical microwave negative-resistance amplifier is illustrated in Fig. 11.9.†
The pump and idler cavities are formed in an X-band rectangular waveguide. The signal cavity is a coaxial-transmission-line cavity. The varactor diode is mounted in the center of an inductive diaphragm located between the pump and idler cavities. Coupling to the signal cavity is achieved by having the diode terminate in the center conductor of the

† W. O. Troetschel and H. J. Heuer, A Parametric Amplifier for 1296 Mc, *QST*, January, 1961.

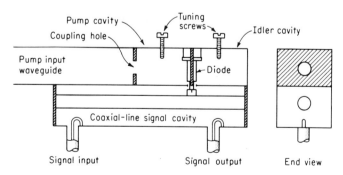

Fig. 11.9 A microwave negative-resistance parametric amplifier.

coaxial-line signal cavity. The pump is coupled to its cavity by an aperture. The pump frequency is chosen around 9,200 Mc, and the idler frequency is 7,900 Mc. The input and output signal frequencies are the same and equal to 1,300 Mc. This amplifier gives a gain of 25 db or more with a bandwidth of about 5 Mc.

The bandwidth over which high gain can be obtained in a negative-resistance amplifier of the type discussed above is small. The negative resistance has the effect of increasing the loaded Q, which results in a high-Q resonant circuit with a narrow bandwidth. To analyze the bandwidth properties we shall assume that the circuit model given in Fig. 11.7 is valid. For high-Q circuits the impedance of the signal and idler circuits may be expressed in the form

$$Z_s = (R_1 + R_s)\left(1 + 2j\frac{\Delta\omega_s}{\omega_s}Q_1\right) \tag{11.42a}$$

$$Z_i = (R_2 + R_s)\left(1 + 2j\frac{\Delta\omega_i}{\omega_i}Q_2\right) \tag{11.42b}$$

where

$$Q_1 = \omega_s \frac{L_1 + L_s}{R_1 + R_s} \qquad Q_2 = \omega_i \frac{L_2 + L_s}{R_2 + R_s}$$

The derivation of these expressions is as follows: From (11.33a) the impedance of the signal circuit at a frequency $\omega_s + \Delta\omega_s$ is

$$Z_s = R_1 + R_s + j(\omega_s + \Delta\omega_s)\left[(L_1 + L_s) - \frac{1}{(\omega_s + \Delta\omega_s)^2}\left(\frac{1}{C_p + C_1} + \frac{1}{C_0 - M^2 C_0}\right)\right]$$

$$= R_1 + R_s + j(\omega_s + \Delta\omega_s)(L_1 + L_s)\left[1 - \frac{\omega_s^2}{(\omega_s + \Delta\omega_s)^2}\right]$$

$$= R_1 + R_s + j\frac{L_1 + L_s}{\omega_s + \Delta\omega_s}(2\omega_s\,\Delta\omega_s + \Delta\omega_s^2)$$

$$\approx (R_1 + R_2)\left(1 + j2Q_1\frac{\Delta\omega_s}{\omega_s}\right)$$

A similar derivation holds for Z_i.

From (11.33) we obtain

$$V_1 = I_s(R_1 + R_s)\left(1 + 2jQ_1\frac{\Delta\omega_s}{\omega_s}\right) + \frac{MI_i^*}{(1 - M^2)j(\omega_i + \Delta\omega_i)C_0} \tag{11.43a}$$

$$V_2 = I_i(R_2 + R_s)\left(1 + 2jQ_2\frac{\Delta\omega_i}{\omega_i}\right) + \frac{MI_s^*}{(1 - M^2)j(\omega_s + \Delta\omega_s)C_0} \tag{11.43b}$$

If $R_1 = R_g + R_L$ and the output is taken at the frequency ω_s, the gain is

given by

$$G = \frac{4R_g R_L |I_s|^2}{V_1^2}$$

When we solve (11.43) for I_s, we find that (note that $V_2 = 0$ in this case)

$$G = \frac{4R_g R_L}{\left| Z_s - \frac{M^2}{(\omega_i + \Delta\omega_i)(\omega_s + \Delta\omega_s)(1 - M^2)^2 C_0^2 Z_i^*} \right|^2} \tag{11.44}$$

Since the pump frequency is fixed, $\Delta\omega_i = -\Delta\omega_s$. The midband gain G_0 is given by (11.37). To determine the bandwidth we equate G given by (11.44) to $G_0/2$ and solve for $\Delta\omega_s$. When the gain G_0 is high, we find that, to a good approximation,

$$4\left(\frac{\Delta\omega_s}{\omega_s}\right)^2 = \frac{(R_1 + R_s - R)^2}{(R_1 + R_s)^2 (Q_1 + \omega_s Q_2/\omega_i)^2} \tag{11.45}$$

Thus the gain-bandwidth product becomes

$$2 \frac{\Delta\omega_s}{\omega_s} \sqrt{G_0} = \frac{2\sqrt{R_g R_L}}{(R_1 + R_s)(Q_1 + Q_2 \omega_s/\omega_i)} \tag{11.46}$$

If we assume as a typical case $R_g + R_L = R_1 + R_s$ and $R_g = R_L$ and note that, for high gain, $R_1 + R_s \approx R$, we get

$$2\sqrt{G_0} \frac{\Delta\omega_s}{\omega_s} = \frac{1}{Q_1 + (\omega_s/\omega_i)Q_2}$$

The smallest possible value of Q_1 occurs if $C_p + C_1$ equals zero. In this case

$$Q_1 = \frac{\omega_s(L_1 + L_s)}{R_1 + R_s} = \frac{1}{\omega_s(1 - M^2)C_0(R_1 + R_s)} \approx \frac{1}{\omega_s(1 - M^2)C_0 R}$$

Similarly, the smallest possible value of Q_2 is obtained if $C_2 + C_p = 0$, and is

$$Q_2 = \frac{1}{\omega_i(1 - M^2)C_0(R_2 + R_s)}$$

If we refer to (11.38) for R, we now find that $M^2 Q_1 Q_2 = 1$. Thus

$$2\sqrt{G_0} \frac{\Delta\omega_s}{\omega_s} = \frac{1}{Q_1 + \omega_s/M^2 \omega_i Q_1}$$

This expression has a maximum value for

$$Q_1 = \frac{1}{M}\sqrt{\frac{\omega_s}{\omega_i}}$$

Hence the maximum gain-bandwidth product is

$$\left(2\sqrt{G_0}\,\frac{\Delta\omega_s}{\omega_s}\right)_{\max} = \frac{M}{2}\sqrt{\frac{\omega_i}{\omega_s}} \qquad (11.47)$$

For 20-db gain we obtain a bandwidth of

$$\frac{2\,\Delta\omega_s}{\omega_s} = \frac{M}{20}\sqrt{\frac{\omega_i}{\omega_s}}$$

Usually M is no greater than about 0.2, and consequently the fractional bandwidth in percent is approximately equal to $\sqrt{\omega_i/\omega_s}$. In a practical amplifier the gain-bandwidth product will be less since the capacitances $C_1 + C_p$ and $C_2 + C_p$ cannot be reduced to zero.

The parametric amplifier may, of course, be broadbanded by using broadband circuits at the signal and idler frequencies. An alternative scheme for obtaining broadband operation is the traveling-wave parametric amplifier, where resonant circuits are avoided entirely. In the traveling-wave amplifier a waveguiding system is loaded periodically with varactor diodes. With the application of pump power and signal power, mixing occurs, with resultant power generation at the signal frequency. For a detailed analysis the references cited at the end of this chapter may be consulted.

11.6 Noise properties of parametric amplifiers

The noise produced by parametric amplifiers is the thermal noise in the resistances in the equivalent circuit. For the up-converter illustrated in Fig. 11.6, the input thermal noise at frequency ω_s is that arising from the generator resistance R_g. The mean-square noise voltage across R_g is $e_1^2 = 4kTR_g\,\Delta f$. If we neglect circuit losses compared with the diode resistance R_s (this is a reasonably good approximation), the only other noise originating in the signal circuit is that generated in R_s across which a mean-square noise voltage $e_2^2 = 4kT_sR_s\,\Delta f$ at frequency ω_s exists. We denote the equivalent noise temperature of the diode at frequency ω_s by T_s. The noise at frequency ω_s in the signal circuit is amplified and converted to noise at the output frequency ω_0. The amplified output noise at midband is given by (replace V_1 by $\sqrt{e_1^2 + e_2^2}$ in Fig. 11.6)

$$P_1 = |I_1|^2 R_L = \frac{R_L(e_1^2 + e_2^2)M^2\omega_0^2C_0^2(1 - M^2)^2}{[M^2 + (R_g + R_s)(R_L + R_s)\omega_0\omega_s(1 - M^2)^2C_0^2]^2}$$

where the noise current I_1 is obtained from (11.26), with V_1 replaced by $(e_1^2 + e_2^2)^{\frac{1}{2}}$. If we introduce the midband gain G_0, we obtain (see 11.27)

$$P_1 = \frac{e_1^2 + e_2^2}{4R_g}G_0 = \frac{k\,\Delta f(R_gT + R_sT_s)}{R_g}G_0 \qquad (11.48)$$

Sec. 11.6   **Parametric amplifiers**

There is also noise generated in the output circuit at frequency ω_0 in the resistance R_s. If the diode noise temperature at the frequency ω_0 is T_0, an equivalent noise voltage $e_3{}^2 = 4kT_0R_s\,\Delta f$ appears across R_s in the output circuit. When we solve for the noise current I_2 that results and compute $P_2 = |I_2|^2 R_L$, we obtain

$$P_2 = \frac{\omega_s{}^2(1 - M^2)^2 C_0{}^2(R_s + R_g)^2 G_0 e_3{}^2}{4R_g M^2}$$

$$= \frac{\omega_s{}^2(1 - M^2) C_0{}^2(R_s + R_g)^2 k T_0 R_s \,\Delta f\, G_0}{R_g M^2} \qquad (11.49)$$

The total output noise power at frequency ω_0 is $P_n = P_1 + P_2$. The available input noise power from R_g is $kT\,\Delta f$. Hence the noise figure is given by ($T = 290°$ in the definition for F)

$$F = \frac{P_1 + P_2}{G_0 kT\,\Delta f} = 1 + \frac{R_s T_s}{R_g T} + \frac{\omega_s{}^2(1-M^2)^2 C_0{}^2(R_s+R_g)^2 R_s T_0}{R_g M^2 T} \qquad (11.50)$$

Note that the gain G_0 was defined as output power divided by the available power from the input signal source, so that F as evaluated above conforms to the accepted definition of noise figure. If we take the diode noise temperature at the two frequencies ω_s and ω_0 equal to T_d, we obtain

$$F = 1 + \frac{R_s}{R_g}\frac{T_d}{T}\left[1 + \frac{(R_s+R_g)^2}{M^2 Q^2 R_s{}^2}\right] \qquad (11.51)$$

after introducing the diode Q from (11.29). For maximum gain,

$$R_g = R_L = R_s\left(1 + \frac{\omega_s}{\omega_0} M^2 Q^2\right)^{\frac{1}{2}}$$

from (11.30). Thus the noise figure under maximum-gain conditions is given by

$$F = 1 + \frac{T_d}{T}\left(1 + \frac{\omega_s}{\omega_0}M^2 Q^2\right)^{-\frac{1}{2}}\left(1 + \frac{\{1 + [1 + (\omega_s/\omega_0)M^2 Q^2]^{\frac{1}{2}}\}^2}{M^2 Q^2}\right)$$

$$\approx 1 + \frac{T_d}{T}\left(1 + \frac{\omega_s}{\omega_0}M^2 Q^2\right)^{-\frac{1}{2}} \qquad (11.52)$$

As a typical example consider $T_d = T$, $\omega_0 = 10\omega_s$, and $MQ = 10$. We then find that $F = 1.36$, or 1.3 db. This example clearly demonstrates the low-noise property of the parametric amplifier.

The diode noise temperature T_d is somewhat greater than the ambient temperature T because of shot noise that arises from random motion of the carriers across the junction. In addition, the resistance R_s is usually not exactly the same at the two frequencies. However, this difference can be taken into account by choosing appropriate values of T_0 and T_s. The thermal noise arising from circuit losses can be included in (11.50) very simply by replacing $R_s T_s$ by $R_{1l}T + R_s T_s$ and $R_s T_0$ by $R_{2l}T + R_s T_0$, where R_{1l} and R_{2l} represent the equivalent circuit resistances. However, since the effective Q of most diodes, that is, $[(1 - M^2)\omega_s C_0 R_s]^{-1}$, would rarely exceed 50 whereas circuit Q's would normally be 1,000 or more, the resistances R_{1l} and R_{2l} are negligible compared with the diode resistance R_s.

The noise properties of negative-resistance parametric amplifiers have been analyzed and measured by Uenohara† and others. The noise theory for a negative-resistance parametric amplifier employing a circulator, and with the output signal taken at the same frequency ω_s as the input, is presented below.

With reference to Fig. 11.10, the following sources of noise are the main ones that need to be considered: (1) input noise from $R_g = Z_c$ at temperature T_1 and frequency ω_s; (2) input noise at the idler frequency $\omega_i = \omega_p - \omega_s$ that arises in R_2; (3) noise arising in the diode resistance R_s at frequency ω_s (equivalent noise temperature T_s); and (4) noise arising in R_s at the idler frequency ω_i (equivalent noise temperature T_i). It is important to consider noise sources at the idler frequency because these noise signals are converted into noise at the frequency ω_s by frequency mixing that takes place in the diode. The equivalent circuit and noise voltage sources are shown in Fig. 11.10.

The noise power input from R_g is $kT_1 \Delta f$. This will appear at the output load R_L as amplified noise of amount

$$P_1 = G_0 k T_1 \Delta f = \frac{4R_L{}^2 k T_1 \Delta f}{(R_L + R_s - R)^2} \tag{11.53}$$

where G_0 was obtained from (11.39) for the high-gain case.

The noise contributed by the amplifier is represented by the voltage sources e_2, e_3, and e_4. Since the three noise sources are uncorrelated, the noise powers add. Hence the effective noise voltage in the idler circuit is $e_i = \sqrt{e_3{}^2 + e_4{}^2}$. The equations describing the noise currents I_1 and I_2

† M. Uenohara, Noise Considerations of the Variable Capacitance Parametric Amplifier, *Proc. IRE*, vol. 48, pp. 169–179, February, 1960.

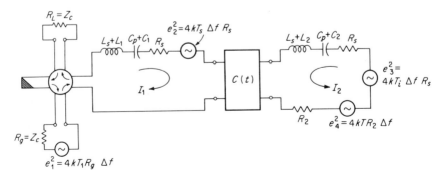

Fig. 11.10 Noise circuit for a negative-resistance parametric amplifier.

at midband are obtained from (11.34). Thus

$$e_2 = (R_L + R_s)I_1 - \frac{jMI_2^*}{\omega_i(1 - M^2)C_0} \tag{11.54a}$$

$$e_i = (R_2 + R_s)I_2 - \frac{jMI_1^*}{\omega_s(1 - M^2)C_0} \tag{11.54b}$$

since the loading R_1 of the signal circuit as seen by the source e_2 is $Z_c = R_L$. When we solve (11.54) for I_1 we can evaluate the noise power

$$P_2 = |I_1|^2 R_L$$

delivered to the external load R_L. Giving the final results, we find that

$$I_1 = \frac{e_2}{R_L + R_s - R} + \frac{jMe_i^*}{\omega_i(1 - M^2)C_0(R_L + R_s - R)(R_2 + R_s)}$$

We must find the noise power contributed by e_2 and e_i separately, since noise voltages do not add. From e_2 we obtain

$$P(e_2) = \frac{e_2{}^2 R_L}{(R_L + R_s - R)^2}$$

and from e_i we obtain

$$P(e_i) = \frac{e_i{}^2 R_L M^2}{[\omega_i(1 - M^2)C_0(R_L + R_s - R)(R_2 + R_s)]^2}$$

$$= \frac{e_i{}^2 R_L R \omega_s}{\omega_i(R_L + R_s - R)^2(R_2 + R_s)}$$

Hence

$$P_2 = P(e_2) + P(e_i) = \frac{G_0}{4R_L}\left(e_2{}^2 + \frac{\omega_s}{\omega_i}\frac{e_i{}^2 R}{R_2 + R_s}\right)$$

$$= \frac{G_0 k\,\Delta f}{R_L}\left[R_s T_s + \frac{\omega_s}{\omega_i}\frac{R}{R_2 + R_s}(R_s T_i + R_2 T)\right] \tag{11.55}$$

If we assume that $T_s = T_i = T = T_d$, we obtain the simplified expression

$$P_2 = G_0 k\,\Delta f\,T_d \frac{R_s}{R_L}\left(1 + \frac{\omega_s}{\omega_i}\frac{R}{R_s}\right) \tag{11.56}$$

The corresponding noise figure F is given by

$$F = \frac{P_1 + P_2}{kT_1 \Delta f G_0} = 1 + \frac{T_d}{T_1}\frac{R_s}{R_L}\left(1 + \frac{\omega_s}{\omega_i}\frac{R}{R_s}\right) \tag{11.57}$$

where T_1 must be taken to be 290°K for the standard definition of F. In (11.57) R may be replaced by $R_L + R_s$ since this is the requirement for high gain. We then obtain

$$F = 1 + \frac{T_d}{T_1}\left(\frac{\omega_p}{\omega_i}\frac{R_L + R_s}{R_L} - 1\right) \tag{11.58}$$

As an example, assume $T_d = T_1$, choose $\omega_p = 1.5\omega_i$, and let R_L be much greater than R_s. We then obtain a noise figure of 1.5, or 1.76 db.

The foregoing analysis is valid only when the two frequencies ω_s and ω_i are spaced by an amount greater than the passband of the signal and idler circuits. When this is not the case, $\omega_s \approx \omega_i$, and the amplifier is a degenerate negative-resistance amplifier. Noise current from the input source resistance R_g at the frequency ω_i will now exist in the signal circuit. Similarly, noise currents at the frequencies ω_i and ω_s arising from R_s will be present in both the signal and idler circuits. In the degenerate amplifier the signal and idler cavities are identical and the equivalent circuit is that shown in Fig. 11.11. If the amplifier passband is symmetrical about $\omega_p/2$, noise at the frequency $\omega_s < \omega_p/2$ is amplified to give noise output at the frequency ω_s, and also converted and amplified to give noise-power output at the frequency $\omega_i = \omega_p - \omega_s$. The two frequencies ω_s and ω_i are symmetrically located about $\omega_p/2$ and fall within the amplifier passband. Therefore we must consider the noise from R_g and R_s at the two frequencies ω_s and ω_i as separate uncorrelated noise in evaluating the total output noise power. The present situation is illustrated schematically in Fig. 11.12. The passband is split into two equal parts Δf_1 and Δf_2, located symmetrically on either side of $\omega_p/2$. Noise in the band Δf_1 is amplified to give noise in the same band Δf_1 at the output. In addition, noise in the band Δf_1 is amplified and converted into noise in the band Δf_2 at the output. The lower sideband Δf_1 will be regarded as the signal sideband.

Since $\omega_s \approx \omega_i$, the noise power in the band Δf_1 arising from R_g and R_s produces the same amount of noise power in the output bands Δf_1 and Δf_2 as does the noise power arising from R_g and R_s in the band Δf_2. Hence we need to consider only the noise in the lower band Δf_1 in detail.

The noise from R_g in the band Δf_1 is represented by an equivalent voltage source e_{1s} in Fig. 11.11. To evaluate the noise power delivered to R_L from e_{1s}, we shall make the approximation of taking the series

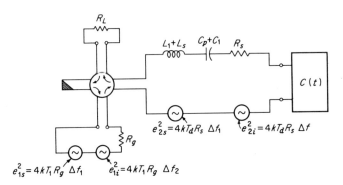

Fig. 11.11 Noise circuit for a degenerate parametric amplifier.

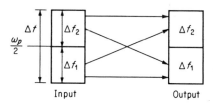

Fig. 11.12 Illustration of noise conversion.

impedance of the circuit as $R_s + R_L$ throughout the passband. We denote the noise currents at frequencies ω_s and ω_i by I_1 and I_2.

At the signal frequency ω_s the available gain is $G_0 = |\Gamma|^2$. Under high-gain conditions $R_L + R_s \approx R$; so (11.39) may be used for G_0. The available input noise power from e_{1s} is $e_{1s}^2/4R_g = e_{1s}^2/4R_L$, since we assume that $R_g = R_L = Z_c$. The noise power P_{1s} delivered to R_L in the band Δf_1 is thus

$$P_{1s} = G_0 \frac{e_{1s}^2}{4R_L} = G_0 k T_1 \Delta f_1 \tag{11.59}$$

When there is no impressed voltage at the frequency ω_i, the currents I_1 and I_2 are related by [see (11.34) with $R_2 = R_L$]

$$(R_L + R_s)I_2 = \frac{jMI_1^*}{\omega_s(1 - M^2)C_0}$$

The power P_{1i} delivered to R_L in the band Δf_2 is given by

$$P_{1i} = |I_2|^2 R_L = \frac{|I_2|^2}{|I_1|^2} P_{1s} = \frac{R}{R_L + R_s} P_{1s} \tag{11.60}$$

since $R = M^2/[\omega_s^2(1 - M^2)^2 C_0^2(R_L + R_s)]$ when $\omega_s \approx \omega_i$.

The source e_{1i} delivers an amount of power equal to P_{1s} to R_L in the band Δf_2 and an amount equal to P_{1i} into the band Δf_1. Hence the total noise power delivered to R_L from R_g is the same in the two bands Δf_1 and Δf_2, and is given by

$$P_1 = P_{1s} + P_{1i} = G_0 k T_1 \Delta f_1 \left(1 + \frac{R}{R_L + R_s}\right) = 2G_0 k T_1 \Delta f_1 = G_0 k T_1 \Delta f \tag{11.61}$$

since $R \approx R_L + R_s$ and $2 \Delta f_1 = \Delta f$.

To evaluate the noise power delivered to R_L from R_s, we consider the noise in the band Δf_1 first. The equivalent voltage source is e_{2s}, and the circuit equations are (we can put $\omega_i = \omega_s$)

$$e_{2s} = (R_L + R_s)I_1 - \frac{jMI_2^*}{\omega_i(1 - M^2)C_0} \tag{11.62a}$$

$$0 = (R_L + R_s)I_2 - \frac{jMI_1^*}{\omega_s(1 - M^2)C_0} \tag{11.62b}$$

The output noise in the two bands Δf_1 and Δf_2 is given by

$$P_{2s} = |I_1|^2 R_L \qquad P_{2i} = |I_2|^2 R_L$$

When we solve for I_1 and I_2, we obtain

$$P_{2s} = \frac{e_{2s}^2 R_L}{(R_L + R_s - R)^2} = \frac{e_{2s}^2 G_0}{4 R_L} = \frac{k T_d \, \Delta f_1 \, R_s G_0}{R_L} \tag{11.63}$$

The output noise in the band Δf_2 from e_{2s} is given by

$$P_{2i} = \frac{R}{R_L + R_s} P_{2s} \approx P_{2s} \tag{11.64}$$

The noise source e_{2i} contributes a noise power P_{2s} in the band Δf_2 and P_{2i} in the band Δf_1. Hence the total noise power delivered to R_L from the internal amplifier noise sources is (in each band Δf_1 and Δf_2)

$$P_2 = P_{2s} + P_{2i} = \frac{2 k T_d \, \Delta f_1 \, R_s G_0}{R_L} = \frac{k T_d R_s G_0 \, \Delta f}{R_L} \tag{11.65}$$

The total output noise from both R_g and R_s is

$$P_n = P_1 + P_2 = k \, \Delta f \, G_0 \left(T_1 + \frac{R_s}{R_L} T_d \right) \tag{11.66}$$

If the degenerate amplifier is used as a single-sideband amplifier (signal input in the lower band Δf_1 only), the single-sideband noise figure F_{ss} is defined by the ratio of the total output noise power in the band $\Delta f = 2 \, \Delta f_1$ divided by the available input noise power in the signal band Δf_1. Thus

$$F_{ss} = \frac{k \, \Delta f \, G_0 T_1}{k \, \Delta f_1 \, G_0 T_1} \left(1 + \frac{T_d}{T_1} \frac{R_s}{R_L} \right) = 2 \left(1 + \frac{T_d}{T_1} \frac{R_s}{R_L} \right) \tag{11.67}$$

It is seen that, for single-sideband operation, the noise figure cannot be less than 2, or 3 db. The signal-to-noise power ratio in the signal band Δf_1 at the input is $P_s / (k \, \Delta f_1 \, T_1)$. The signal-to-noise power ratio in the band Δf_1 at the output, given by $G_0 P_s / P_n$, is worse by a factor equal to the single-sideband noise figure F_{ss}. The noise degradation is due to noise entering in the idler band Δf_2, in which no signal is present.

For double-sideband operation input signal power is present in both bands Δf_1 and Δf_2. In this case the available input noise power is taken to be that in the band Δf. Hence the double-sideband noise figure F_{ds} is

$$F_{ds} = \frac{k \, \Delta f \, G_0 T_1}{k \, \Delta f \, G_0 T_1} \left(1 + \frac{T_d}{T_1} \frac{R_s}{R_L} \right) = 1 + \frac{T_d}{T_1} \frac{R_s}{R_L} \tag{11.68}$$

and is a factor of 2 (3 db) better than for the single-sideband case.

The double-sideband noise figure has been measured by Uenohara[†] for a number of different diodes. It is found that the theory given above is reasonably accurate. Typical noise figures that were measured ranged from 0.9 to 4.5 db. For diodes with $\omega_s C_0 R_s$ less than 0.1, the noise figure was 2 db or better. There was a strong correlation between the meas-

† *Ibid.*

ured noise figure and the diode quality factor Q. This is predicted by the theory as well. Since $R_L + R_s \approx R$ for high gain and R can be expressed by

$$R = \frac{R}{R_L + R_s} M^2 Q^2 R_s \gg R_s$$

from (11.29) and (11.38), we see that the optimum value of $R_L + R_s$ is MQR_s. The factor R_s/R_L in the expression for noise figure may now be replaced by $(MQ - 1)^{-1}$ to give

$$F_{ds} = 1 + \frac{T_d}{T_1} \frac{1}{MQ - 1}$$

showing that the noise figure improves with diode quality factor Q.

Parametric amplifier noise is primarily thermal noise in the diode resistance R_s. The equivalent amplifier noise temperature is given by $T_A = (F - 1)T_1$, where $T_1 = 290°K$ [for the degenerate amplifier with single-sideband operation $T_A = (F - 2)T_1$]. By cooling the amplifier to liquid-nitrogen temperature, noise temperatures below 100°K have been obtained.

Problems

11.1 Consider a square-law mixer for which the output current $i(t) = k[v(t)]^2$, where $v(t)$ is the applied voltage and k is a constant. Let a local-oscillator signal $V_p \cos \omega_p t$ and a signal $V_s \cos \omega_s t$ be applied, with $V_s \ll V_p$. Show that the output current at the sum or difference frequencies $\omega_p \pm \omega_s$ is a linear function of V_s when $V_s \ll V_p$. Thus the square-law mixer is a linear converter for small signal amplitudes.

11.2 Consider a parallel-plate capacitor with capacitance C_0. Let a voltage $V = V_s \cos \omega_s t$ be applied. At time $t = 0$ the plate separation is suddenly increased to change the capacitance from C_0 to $C - C_0 - \Delta C$. Since the charge cannot change instantaneously, the voltage must increase. At time $t = (4f_s)^{-1}$, when $V = 0$ and $Q = 0$, let the plate separation be brought back to its original value. There is no change in V produced since $V = 0$ at this time. When $t = (2f_s)^{-1}$, let the capacitance be suddenly decreased to a value C again. When this process is continued, the resultant voltage across C is amplified and will have the waveform illustrated. This is an example of a linear capacitance varied at a rate twice that of the signal frequency. Evaluate the incremental change in voltage that occurs every half cycle and the power supplied by the pump. To evaluate the latter, determine the change in stored energy that occurs every time C is suddenly decreased.

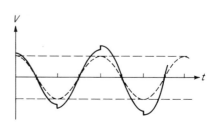

Fig. P 11.2

11.3 A down-converter is a parametric amplifier with an input signal at the frequency $\omega_0 = \omega_s + \omega_p$ and the output signal taken at frequency ω_s. Using the circuit of Fig. 11.6, show that the down-converter gain (actually a loss) is given by (11.27), with ω_0 and ω_s interchanged.

11.4 Derive (11.32).

11.5 Derive the expression (11.35) for the gain of the negative-resistance parametric amplifier.

11.6 Derive (11.49) for the noise power P_2.

11.7 Obtain an expression for the equivalent noise temperature of a parametric up-converter.

11.8 Derive an expression for the gain of the negative-resistance degenerate parametric amplifier illustrated in Fig. 11.11.

11.9 Consider the degenerate parametric amplifier with circulator shown in Fig. 11.11. Assume an input generator voltage V_s at frequency ω_s in place of e_{1s} and e_{1i}. The generator sends a wave with current I_1^+ into the amplifier, where I_1^+ must be equal to $V_s/(R_g + Z_c) = V_s/2Z_c$, when $R_g = Z_c$, since V_s sees a matched load. A reflected wave is set up with a current $I_1^- = -\Gamma I_1^+$. The load current in R_L is I_1^-, apart from a phase angle. The total amplifier current at frequency ω_s is $I_1 = I_1^- + I_1^+$. With this information determine the appropriate circuit equations, analogous to (11.34), for I_1 and I_2.

11.10 A parametric diode has the following parameter values: $C_0 = 2$ μμf, $R_s = 1$ ohm. The modulation index $M = 0.25$. The frequency $f_s = 5,000$ Mc and $f_p = 12,000$ Mc. Evaluate the diode effective Q. Determine the load resistance R_L to give a gain of 20 db for a negative-resistance amplifier of the form shown in Fig. 11.8. Assume $R_g = R_L = R_2$. Calculate R for 20 db gain.

References

1. Blackwell, L. A., and K. L. Kotzebue: "Semiconductor-diode Parametric Amplifiers," Prentice-Hall, Inc., Englewood Cliffs, N.J., 1961.
2. Penfield, P., and R. P. Rafuse: "Varactor Applications," The M.I.T. Press, Cambridge, Mass., 1962.
3. Chang, K. K. N.: "Parametric and Tunnel Diodes," Prentice-Hall, Inc., Englewood Cliffs, N.J., 1964.

Traveling-wave parametric amplifiers

4. Cullen, A. L.: A Traveling Wave Parametric Amplifier, *Nature*, vol. 181, February, 1958.
5. Honey, R. C., and E. M. T. Jones: A Wide-band UHF Traveling Wave Variable Reactance Amplifier, *IRE Trans.*, vol. MTT-8, pp. 351–361, May, 1960.
6. Heilmeier, G. H.: An Analysis of Parametric Amplification in Periodically Loaded Transmission Lines, *RCA Rev.*, vol. 20, pp. 442–454, September, 1959.

Broadbanding techniques

7. Matthaei, G. L.: A Study of the Optimum Design of Wideband Parametric Amplifiers and Up-converters, *IRE Trans.*, vol. MTT-9, pp. 23–28, January, 1961.
8. Gilden, M., and G. L. Matthaei: Practical Design and Performance of Nearly Optimum Wide Band Degenerate Parametric Amplifiers, *IRE Trans.*, vol. MTT-9, pp. 484–490, November, 1961.

appendix I

Useful relations from vector analysis

I.1 Vector algebra

Let vectors **A** and **B** be expressed as components along unit vectors a_1, a_2, a_3 in a right-hand orthogonal coordinate system. Then

$$\mathbf{A} \pm \mathbf{B} = (A_1 \pm B_1)\mathbf{a}_1 + (A_2 \pm B_2)\mathbf{a}_2 + (A_3 \pm B_3)\mathbf{a}_3 \tag{I.1}$$

$$\mathbf{A} \cdot \mathbf{B} = |\mathbf{A}|\,|\mathbf{B}| \cos\theta = A_1 B_1 + A_2 B_2 + A_3 B_3 \tag{I.2}$$

where θ is the angle between **A** and **B**.

$$\mathbf{A} \times \mathbf{B} = \mathbf{a}_1(A_2 B_3 - A_3 B_2) + \mathbf{a}_2(A_3 B_1 - A_1 B_3) + \mathbf{a}_3(A_1 B_2 - A_2 B_1) \tag{I.3a}$$

$$|\mathbf{A} \times \mathbf{B}| = |\mathbf{A}|\,|\mathbf{B}| \sin\theta \tag{I.3b}$$

$$\mathbf{A} \cdot \mathbf{B} \times \mathbf{C} = \mathbf{A} \times \mathbf{B} \cdot \mathbf{C} = \mathbf{C} \times \mathbf{A} \cdot \mathbf{B} \tag{I.4}$$

$$\mathbf{A} \times \mathbf{B} = -\mathbf{B} \times \mathbf{A} \tag{I.5}$$

$$\mathbf{A} \times (\mathbf{B} \times \mathbf{C}) = (\mathbf{A} \cdot \mathbf{C})\mathbf{B} - (\mathbf{A} \cdot \mathbf{B})\mathbf{C} \tag{I.6}$$

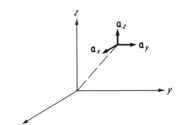

Fig. I.1 Rectangular coordinates.

I.2 Vector operations in common coordinate systems

Rectangular coordinates

$$\nabla\Phi = \mathbf{a}_x \frac{\partial \Phi}{\partial x} + \mathbf{a}_y \frac{\partial \Phi}{\partial y} + \mathbf{a}_z \frac{\partial \Phi}{\partial z} \tag{I.7}$$

$$\operatorname{div} \mathbf{A} = \nabla \cdot \mathbf{A} = \frac{\partial A_x}{\partial x} + \frac{\partial A_y}{\partial y} + \frac{\partial A_z}{\partial z} \tag{I.8}$$

$$\operatorname{curl} \mathbf{A} = \nabla \times \mathbf{A} = \mathbf{a}_x \left(\frac{\partial A_z}{\partial y} - \frac{\partial A_y}{\partial z} \right) + \mathbf{a}_y \left(\frac{\partial A_x}{\partial z} - \frac{\partial A_z}{\partial x} \right) + \mathbf{a}_z \left(\frac{\partial A_y}{\partial x} - \frac{\partial A_x}{\partial y} \right) \quad (I.9)$$

$$\nabla^2 \Phi = \frac{\partial^2 \Phi}{\partial x^2} + \frac{\partial^2 \Phi}{\partial y^2} + \frac{\partial^2 \Phi}{\partial z^2} \quad (I.10)$$

$$\nabla^2 \mathbf{A} = \mathbf{a}_x \nabla^2 A_x + \mathbf{a}_y \nabla^2 A_y + \mathbf{a}_z \nabla^2 A_z \quad (I.11)$$

Cylindrical coordinates

$$\nabla \Phi = \mathbf{a}_r \frac{\partial \Phi}{\partial r} + \mathbf{a}_\phi \frac{1}{r} \frac{\partial \Phi}{\partial \phi} + \mathbf{a}_z \frac{\partial \Phi}{\partial z} \quad (I.12)$$

$$\nabla \cdot \mathbf{A} = \frac{1}{r} \frac{\partial}{\partial r} (rA_r) + \frac{1}{r} \frac{\partial A_\phi}{\partial \phi} + \frac{\partial A_z}{\partial z} \quad (I.13)$$

$$\nabla \times \mathbf{A} = \mathbf{a}_r \left(\frac{1}{r} \frac{\partial A_z}{\partial \phi} - \frac{\partial A_\phi}{\partial z} \right) + \mathbf{a}_\phi \left(\frac{\partial A_r}{\partial z} - \frac{\partial A_z}{\partial r} \right) + \mathbf{a}_z \left[\frac{1}{r} \frac{\partial (rA_\phi)}{\partial r} - \frac{1}{r} \frac{\partial A_r}{\partial \phi} \right] \quad (I.14)$$

$$\nabla^2 \Phi = \frac{1}{r} \frac{\partial}{\partial r} \left(r \frac{\partial \Phi}{\partial r} \right) + \frac{1}{r^2} \frac{\partial^2 \Phi}{\partial \phi^2} + \frac{\partial^2 \Phi}{\partial z^2} \quad (I.15)$$

$$\nabla^2 \mathbf{A} = \nabla \nabla \cdot \mathbf{A} - \nabla \times \nabla \times \mathbf{A} \quad (I.16)$$

Note that $\nabla^2 \mathbf{A} \neq \mathbf{a}_r \nabla^2 A_r + \mathbf{a}_\phi \nabla^2 A_\phi + \mathbf{a}_z \nabla^2 A_z$ since $\nabla^2 \mathbf{a}_r A_r \neq \mathbf{a}_r \nabla^2 A_r$, etc., because the orientation of the unit vectors $\mathbf{a}_r$, $\mathbf{a}_\phi$ varies with the coordinates r, ϕ.

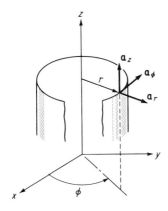

Fig. I.2 Cylindrical coordinates.

Spherical coordinates

$$\nabla \Phi = \mathbf{a}_r \frac{\partial \Phi}{\partial r} + \mathbf{a}_\theta \frac{1}{r} \frac{\partial \Phi}{\partial \theta} + \frac{\mathbf{a}_\phi}{r \sin \theta} \frac{\partial \Phi}{\partial \phi} \quad (I.17)$$

$$\nabla \cdot \mathbf{A} = \frac{1}{r^2} \frac{\partial}{\partial r} (r^2 A_r) + \frac{1}{r \sin \theta} \frac{\partial}{\partial \theta} (\sin \theta \, A_\theta) + \frac{1}{r \sin \theta} \frac{\partial A_\phi}{\partial \phi} \quad (I.18)$$

$$\nabla \times \mathbf{A} = \frac{\mathbf{a}_r}{r \sin \theta} \left[\frac{\partial}{\partial \theta} (A_\phi \sin \theta) - \frac{\partial A_\theta}{\partial \phi} \right] + \frac{\mathbf{a}_\theta}{r} \left[\frac{1}{\sin \theta} \frac{\partial A_r}{\partial \phi} - \frac{\partial}{\partial r} (rA_\phi) \right]$$
$$+ \frac{\mathbf{a}_\phi}{r} \left[\frac{\partial}{\partial r} (rA_\theta) - \frac{\partial A_r}{\partial \theta} \right] \quad (I.19)$$

$$\nabla^2 \Phi = \frac{1}{r^2} \frac{\partial}{\partial r} \left(r^2 \frac{\partial \Phi}{\partial r} \right) + \frac{1}{r^2 \sin \theta} \frac{\partial}{\partial \theta} \left(\sin \theta \frac{\partial \Phi}{\partial \theta} \right) + \frac{1}{r^2 \sin^2 \theta} \frac{\partial^2 \Phi}{\partial \phi^2} \quad (I.20)$$

$$\nabla^2 \mathbf{A} = \nabla \nabla \cdot \mathbf{A} - \nabla \times \nabla \times \mathbf{A} \quad (I.21)$$

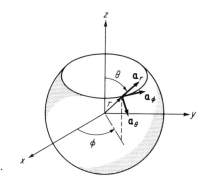

Fig. I.3 Spherical coordinates.

I.3 Vector identities

$$\nabla(\Phi\psi) = \psi\,\nabla\Phi + \Phi\,\nabla\psi \tag{I.22}$$

$$\nabla \cdot (\psi\mathbf{A}) = \mathbf{A} \cdot \nabla\psi + \psi\nabla \cdot \mathbf{A} \tag{I.23}$$

$$\nabla \cdot (\mathbf{A} \times \mathbf{B}) = (\nabla \times \mathbf{A}) \cdot \mathbf{B} - (\nabla \times \mathbf{B}) \cdot \mathbf{A} \tag{I.24}$$

$$\nabla \times (\psi\mathbf{A}) = (\nabla\psi) \times \mathbf{A} + \psi\nabla \times \mathbf{A} \tag{I.25}$$

$$\nabla \times (\mathbf{A} \times \mathbf{B}) = \mathbf{A}\nabla \cdot \mathbf{B} - \mathbf{B}\nabla \cdot \mathbf{A} + (\mathbf{B} \cdot \nabla)\mathbf{A} - (\mathbf{A} \cdot \nabla)\mathbf{B} \tag{I.26}$$

$$\nabla(\mathbf{A} \cdot \mathbf{B}) = (\mathbf{A} \cdot \nabla)\mathbf{B} + (\mathbf{B} \cdot \nabla)\mathbf{A} + \mathbf{A} \times (\nabla \times \mathbf{B}) + \mathbf{B} \times (\nabla \times \mathbf{A}) \tag{I.27}$$

$$\nabla \cdot \nabla\Phi = \nabla^2\Phi \tag{I.28}$$

$$\nabla \cdot \nabla \times \mathbf{A} = 0 \tag{I.29}$$

$$\nabla \times \nabla\Phi = 0 \tag{I.30}$$

$$\nabla \times \nabla \times \mathbf{A} = \nabla\nabla \cdot \mathbf{A} - \nabla^2\mathbf{A} \tag{I.31}$$

If $\mathbf{A}$ and Φ are continuous functions with at least piecewise continuous first derivatives in V and on S (or on S and the contour C bounding S),

$$\int_V \nabla\Phi\,dV = \oint_S \Phi\,d\mathbf{S} \tag{I.32}$$

$$\int_V \nabla \cdot \mathbf{A}\,dV = \oint_S \mathbf{A} \cdot d\mathbf{S} \quad \text{(divergence theorem)} \tag{I.33}$$

$$\int_V \nabla \times \mathbf{A}\,dV = \oint_S \mathbf{n} \times \mathbf{A}\,dS \quad d\mathbf{S} = \mathbf{n}\,dS \tag{I.34}$$

$$\int_S \mathbf{n} \times \nabla\Phi\,dS = \oint_C \Phi\,d\mathbf{l} \tag{I.35}$$

$$\int_S \nabla \times \mathbf{A} \cdot d\mathbf{S} = \oint_C \mathbf{A} \cdot d\mathbf{l} \quad \text{(Stokes' theorem)} \tag{I.36}$$

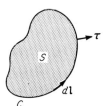

Fig. I.4 Surface S bounded by contour C.

I.4 Green's identities

If **A**, **B**, Φ, and ψ are continuous with piecewise continuous first derivatives,

$$\int_V (\nabla\Phi \cdot \nabla\psi + \psi \nabla^2\Phi)\, dV = \oint_S \psi \nabla\Phi \cdot d\mathbf{S} \tag{I.37}$$

which is Green's first identity. Green's second identity is

$$\int_V (\psi \nabla^2\Phi - \Phi \nabla^2\psi)\, dV = \oint (\psi \nabla\Phi - \Phi \nabla\psi) \cdot d\mathbf{S} \tag{I.38}$$

In two dimensions (I.37) becomes

$$\int_S (\nabla_t\Phi \cdot \nabla_t\psi + \psi \nabla_t^2\Phi)\, dS = \oint_C \psi \nabla_t\Phi \cdot \boldsymbol{\tau}\, dl \tag{I.39}$$

where ∇_t is the del operator in two dimensions and $\boldsymbol{\tau}$ is a unit vector normal to C and in the plane of S. The two-dimensional form of (I.38) is similar.

The vector forms of Green's identities are

$$\int_V \nabla \cdot (\mathbf{A} \times \nabla \times \mathbf{B})\, dV = \int_V [(\nabla \times \mathbf{A}) \cdot (\nabla \times \mathbf{B}) - \mathbf{A} \cdot \nabla \times \nabla \times \mathbf{B}]\, dV$$
$$= \oint_S \mathbf{A} \times (\nabla \times \mathbf{B}) \cdot d\mathbf{S} \tag{I.40}$$

$$\int_V (\mathbf{B} \cdot \nabla \times \nabla \times \mathbf{A} - \mathbf{A} \cdot \nabla \times \nabla \times \mathbf{B})\, dV$$
$$= \oint_S [\mathbf{A} \times (\nabla \times \mathbf{B}) - \mathbf{B} \times (\nabla \times \mathbf{A})] \cdot d\mathbf{S} \tag{I.41}$$

appendix II

Bessel functions

II.1 Ordinary Bessel functions

The wave equation and Helmholtz's and Laplace's equations are separable in cylindrical coordinates. The differential equation describing the radial dependence of the solution is Bessel's differential equation. Bessel's equation is

$$\frac{1}{r}\frac{d}{dr} r \frac{df}{dr} + \left(k^2 - \frac{n^2}{r^2}\right)f = 0 \tag{II.1}$$

When k^2 is real and positive, the two independent solutions of Bessel's equation are called Bessel functions of the first and second kind, denoted by $J_n(kr)$ and $Y_n(kr)$, respectively. These solutions may be expressed as power series as follows:

$$J_n(kr) = \sum_{m=0}^{\infty} \frac{(-1)^m (kr/2)^{n+2m}}{m!(n+m)!} \tag{II.2}$$

$$Y_n(kr) = \frac{2}{\pi}\left(\gamma + \ln\frac{kr}{2}\right) J_n(kr) - \frac{1}{\pi}\sum_{m=0}^{n-1} \frac{(n-m-1)!}{m!}\left(\frac{2}{kr}\right)^{n-2m}$$

$$- \frac{1}{\pi}\sum_{m=0}^{\infty} \frac{(-1)^m (kr/2)^{n+2m}}{m!(n+m)!}\left(1 + \frac{1}{2} + \frac{1}{3} + \cdots + \frac{1}{m} + 1 + \frac{1}{2} + \frac{1}{3} + \cdots + \frac{1}{n+m}\right)$$

$$\tag{II.3}$$

where $\gamma = 0.5772$ is Euler's constant. The subscript n denotes the order of the function and is usually an integer in physical problems. The Y_n functions become infinite at $r = 0$. For large values of kr, the Bessel functions approach damped sinusoids:

$$\lim_{r \to \infty} J_n(kr) = \sqrt{\frac{2}{\pi kr}} \cos\left(kr - \frac{\pi}{4} - \frac{n\pi}{2}\right) \tag{II.4a}$$

$$\lim_{r \to \infty} Y_n(kr) = \sqrt{\frac{2}{\pi kr}} \sin\left(kr - \frac{\pi}{4} - \frac{n\pi}{2}\right) \tag{II.4b}$$

A few of the lowest-order Bessel functions are plotted in Fig. II.1.

To represent radially propagating waves, linear combinations of the J_n and Y_n are formed, called Hankel functions of the first and second kind. Thus the Hankel function of the first kind is

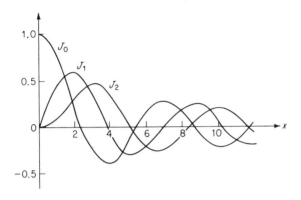

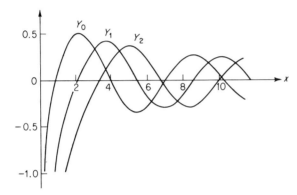

Fig. II.1 Ordinary Bessel functions.

$$H_n{}^1(kr) = J_n(kr) + jY_n(kr) \tag{II.5a}$$

and the Hankel function of the second kind is given by

$$H_n{}^2(kr) = J_n(kr) - jY_n(kr) \tag{II.5b}$$

For large values of kr, the Hankel functions are given by the following expressions:

$$H_n{}^1(kr) \sim \sqrt{\frac{2}{kr\pi}}\, e^{j(kr-\pi/4-n\pi/2)} \tag{II.6a}$$

$$H_n{}^2(kr) \sim \sqrt{\frac{2}{kr\pi}}\, e^{-j(kr-\pi/4-n\pi/2)} \tag{II.6b}$$

Some useful relations that hold for any of the Bessel functions J_n, Y_n, or H_n are given below, where Z_n denotes J_n, Y_n, or H_n.

$$xZ_n'(x) = nZ_n(x) - xZ_{n+1}(x) = -nZ_n(x) + xZ_{n-1}(x) \tag{II.7}$$

where the prime denotes differentiation with respect to x.

$$\int^x Z_n(kx)Z_n(lx)x\,dx = \frac{x}{k^2 - l^2}[kZ_n(lx)Z_{n+1}(kx) - lZ_n(kx)Z_{n+1}(lx)] \quad \text{(II.8)}$$

$$\int^x Z_n{}^2(kx)x\,dx = \frac{x^2}{2}\left[Z_n'^2(kx) + \left(1 - \frac{n^2}{k^2x^2}\right)Z_n{}^2(kx)\right] \quad \text{(II.9)}$$

II.2 Modified Bessel functions

When k^2 is negative, k is pure imaginary. If we let $k = jh$, the solutions are given by $J_n(jhr)$ and $Y_n(jhr)$. However, for convenience, new modified Bessel functions are introduced and denoted by $I_n(hr)$ and $K_n(hr)$. The modified Bessel function of the first kind is $I_n(hr)$, and is given by

$$I_n(hr) = j^{-n}J_n(jhr) = j^n J_n(-jhr) \quad \text{(II.10)}$$

and the modified Bessel function of the second kind is given by

$$K_n(hr) = \frac{\pi}{2}j^{n+1}[J_n(jhr) + jY_n(jhr)] = \frac{\pi}{2}j^{n+1}H_n{}^1(jhr) \quad \text{(II.11)}$$

For large values of hr we have

$$I_n(hr) \sim \frac{e^{hr}}{\sqrt{2\pi hr}} \quad \text{(II.12a)}$$

$$K_n(hr) \sim \sqrt{\frac{\pi}{2hr}}\,e^{-hr} \quad \text{(II.12b)}$$

The first few modified Bessel functions are plotted in Fig. II.2.

A number of useful relations that hold for the modified Bessel functions are given:

$$xI_n'(x) = nI_n(x) + xI_{n+1}(x) = -nI_n(x) + xI_{n-1}(x) \quad \text{(II.13a)}$$

$$I_0'(x) = I_1(x) \quad \text{(II.13b)}$$

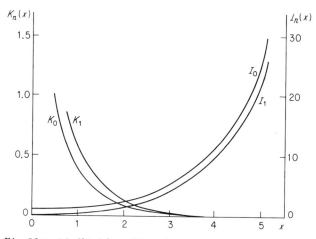

Fig. II.2 Modified Bessel functions.

$$\int^x x^{-n} I_{n+1}(x)\, dx = x^{-n} I_n(x) \tag{II.14a}$$

$$\int^x x^n I_{n-1}(x)\, dx = x^n I_n(x) \tag{II.14b}$$

When $n > -1$, we have

$$\int_0^x I_n(kx) I_n(lx) x\, dx = \frac{x}{k^2 - l^2} [k I_n(lx) I_{n+1}(kx) - l I_n(kx) I_{n+1}(lx)] \tag{II.15}$$

$$\int_0^x I_n^2(kx) x\, dx = -\frac{x^2}{2}\left[I_n'^2(kx) - \left(1 + \frac{n^2}{k^2 x^2}\right) I_n^2(kx) \right] \tag{II.16}$$

$$x K_n'(x) = n K_n(x) - x K_{n+1}(x) = -n K_n(x) - x K_{n-1}(x) \tag{II.17a}$$

$$K_0'(x) = -K_1(x) \tag{II.17b}$$

$$\int^x x^{-n} K_{n+1}(x)\, dx = -x^{-n} K_n(x) \tag{II.18a}$$

$$\int^x x^n K_{n-1}(x)\, dx = -x^n K_n(x) \tag{II.18b}$$

When Re $(k + 1) > 0$, we have

$$\int_x^\infty K_n(kx) K_n(lx) x\, dx = \frac{x}{k^2 - l^2} [k K_n(lx) K_{n+1}(kx) - l K_n(kx) K_{n+1}(lx)] \tag{II.19}$$

For Re $k > 0$

$$\int_x^\infty x K_n^2(kx)\, dx = \frac{x^2}{2}\left[K_n'^2(x) - \left(1 + \frac{n^2}{k^2 x^2}\right) K_n^2(kx) \right] \tag{II.20}$$

References

1. Watson, G. N.: "Theory of Bessel Functions," Cambridge University Press, New York, 1922.
2. McLachlan, N. W.: "Bessel Functions for Engineers," 2d ed., Oxford University Press, Fair Lawn, N.J., 1948.
3. Bowman, F.: "Introduction to Bessel Functions," Dover Publications, Inc., New York, 1958.

appendix III

Physical constants and other data

III.1 Physical constants

Permittivity of vacuum $= \epsilon_0 = 8.854 \times 10^{-12} \approx (1/36\pi) \times 10^{-9}$ farad/m
Permeability of vacuum $= \mu_0 = 4\pi \times 10^{-7}$ henry/m
Impedance of free space $= Z_0 = 376.7 \approx 120\pi$ ohms
Velocity of light $= c = 2.998 \times 10^8$ m/sec
Charge of electron $= e = 1.602 \times 10^{-19}$ coul
Mass of electron $= m = 9.107 \times 10^{-31}$ kg
$\eta = e/m = 1.76 \times 10^{11}$ coul/kg
Mass of proton $= M = 1.67 \times 10^{-27}$ kg
Boltzmann's constant $= k = 1.380 \times 10^{-23}$ joule/°K
Planck's constant $= h = 6.547 \times 10^{-34}$ joule-sec
10^7 ergs $= 1$ joule
1 joule $= 0.6285 \times 10^{19}$ electron volts
1 electron volt $=$ energy gained by an electron in accelerating through a potential of 1 volt
Energy of 1 electron volt $=$ equivalent electron temperature of 1.15×10^4 °K
Electron plasma frequency $f_p = \dfrac{e}{2\pi} \left(\dfrac{N}{m\epsilon_0}\right)^{\frac{1}{2}} = 8.97 N^{\frac{1}{2}}$ cps, where N is the number of electrons per cubic meter
Electron cyclotron frequency $f_c = eB/2\pi m = 28{,}000 B$ Mc/sec for B in webers per square meter; $f_c = 2.8 B$ Mc/sec for B in gauss
10^4 gauss $= 1$ weber/m^2

III.2 Conductivities of materials

Material	Conductivity, mhos/m	Material	Conductivity, mhos/m
Copper (annealed)	5.8×10^7	Steel	0.5–1.0×10^7
Aluminum	3.54×10^7	Water (distilled)	2×10^{-4}
Silver	6.14×10^7	Sea water	3–5
Nickel	1.28×10^7	Quartz (fused)	$<2 \times 10^{-17}$

III.3 Dielectric constants of materials

Material	Frequency, Mc	ϵ'/ϵ_0	Loss tangent ϵ''/ϵ'
Polystyrene	3,000	2.54	0.00025–0.0016
Polystyrene (foam)	3,000	1.05	0.00003
Lucite	10,000	2.56	0.005
Teflon	10,000	2.08	0.00037
Fused quartz	10,000	3.78	0.0001
Ruby mica	3,000	5.4	0.0003
Titanium dioxide	10,000	90	0.002
Mahogany wood	10,000	1.7	0.021

III.4 Skin depth in copper

Frequency, cps	10	60	10^2	10^3	10^4	10^8
Skin depth δ_s, cm	2.08	0.85	0.66	0.208	6.6×10^{-2}	6.6×10^{-4}

$\delta_s = \sqrt{2/\omega\mu\sigma} = 6.6 f^{-\frac{1}{2}}$ cm for copper ($\sigma = 5.8 \times 10^7$ mhos/m).

Index

Adler, R. B., 63
Admittance, electronic, in klystron, 472
  input, for transmission line, 93
Allison, J., 104
Amplification, of klystron, 469
  of maser, 529–531
  of parametric amplifier, 555, 557–559
  of traveling-wave tube, 480, 482
Anderson, T. N., 278
Angular momentum, 498–500
Anisotropic media, 21
Antenna, probe in waveguide, 183–187
Aperture, coupling by, in waveguide, 190–197
  polarizability of circular, 191
  in rectangular cavity, 329–336
Attenuation, for circular waveguide, 110
  for degenerate modes, 129
  for rectangular waveguide, 101–103
  for transmission line, 76–81
  (*See also* Transmission line; Waveguide)
Attenuator, resistive card, 262
  rotary, 263–265
Attwood, S. S., 114
Atwater, H. A., 143
Ayres, W. P., 291

Babinet's principle, 392
Backward-wave oscillator, 485

Bandwidth, of resonant circuit, 314–315
Barlow, H. E. M., 113, 114
Beam, electron (*see* Electron beam)
Beam coupling parameter, 456, 468
Beam, R. E., 143
Beck, A. H. W., 492
Benson, F. A., 104
Bessel functions, 442, 575–578
Bethe, H. A., 273
Bethe directional coupler, 273–275
Bevensee, R. M., 433
Binomial quarter-wave transformer, 227–229
Blackwell, L. A., 570
Bloch wave, 369
  impedance of, 372
  (*See also* Periodic structures)
Bobroff, D., 452
Bolinder, F., 241, 258
Bolljahn, J. T., 433
Boltzman distribution, 501
Boundary conditions, at conducting edge, 38
  at conducting surface, 36–37
  for electromagnetic field, 34–35
  at infinity, 38
Bowman, F., 578
Boyd, G. D., 362
Bradley, E. H., 433
Branch, G. M., 445
Brewer, G. R., 492

Brillouin, L., 433, 439
Brillouin flow, for electron beam, 438–439
Bronwell, A. B., 143
Brown, J., 114, 224
Button, K. J., 312

Carcinotrons, 485
Cavity, coupling parameter for, 334–335
  cylindrical, 326–329
    mode chart for, 328
    Q of, 327
    resonant frequency of, 327
  degenerate modes in, 355–356
  excitation of, 356–359, 464–468
  Fabry-Perot, 337–344
    Q of, 338–339
  field expansion in, 344–351
  filter, 424–430
  loop-coupled, 336–337
  oscillations in, 351–356
  rectangular, 322–325
    aperture-coupled, 329–336
    Q of, 325
    resonant frequency of, 323
Chandler, C. H., 113
Chang, K. K. N., 570
Chebyshev filters, 406–407
Chebyshev polynomials, 230–231
Chebyshev quarter-wave transformer, 229–237
Chebyshev tapered transmission line, 248–251
Chodorow, M., 457, 492
Choke joint, 262
  in variable short circuits, 259–262
Christian, J. R., 65
Chu, L. J., 63, 452
Circular polarized field, 287–288
Circulator, four-port, 304–306
  scattering matrix for, 307–308
  three-port, 307–309
Clarricoats, P., 312
Coaxial transmission line (See Transmission line)
Cohn, M., 113

Cohn, S. B., 65, 224, 258, 409, 428
Collin, R. E., 63, 81, 114, 143, 218, 224, 248, 258, 265, 344, 433
Collins, G. B., 492
Constitutive relations, 17–22
Continuity equation for current, 14
Corrugated plane as periodic structure, 383–389
Coupling, in directional coupler, 271
  of modes in lossy cavity, 355–356
  of modes in lossy waveguide, 124–131
Coupling parameter, for cavity, 334–335
  for electron beam, 456
Cullen, A. L., 570
Culshaw, W., 362
Current, equivalent, in waveguide, 145–147
  linear, excitation of waveguide by, 187–189
  loop, in waveguide, 189–190
  normalized, 147
Cutoff frequency (See Waveguide, rectangular)

Damping of cavity, 316
Delta function, 54
Diaphragm, capacitive, in rectangular guide, 219–220
  inductive, in rectangular guide, 218–219
Dicke, R. H., 202, 273, 312, 362
Dielectric-coated plane, 114–117
Dielectric-coated wire guide, 118–121
Dielectric constant, 20
Directional coupler, Bethe type, 273–275
  Chebyshev, 280–282
  coupling in, 271
  directivity of, 271
  multielement, 278–282
  scattering matrix for, 271–273
  Schwinger reversed phase, 276–278
  two-hole, 275–276, 278
    Moreno crossed-guide, 277, 278
    Riblet T-slot, 277, 278

Dispersion of signal in waveguide, 136
Double-stream amplifier, 484
Double-stub tuner, 212–215
  for waveguide, 220–221
Dow, W. G., 439

E mode, 71–72
  in circular guide, 107–110
  in rectangular guide, 106–107
E-H tuner, 220–221
Einstein's coefficients for radiation, 503–504
Electron beam, a-c power relations for, 452–455
  with axially confined flow, 437–438
  beam coupling parameter for, 456
  Brillouin flow for, 438–439
  d-c conditions for, 435–436
  ion-neutralized, 436–437
  Kinetic-power theorem for, 453–454
  perveance of, 436
  space-charge waves on, 439–452
  velocity modulation of, 455–462
  (See also Space-charge waves)
Electron precession in ferrite, 286–290
Electron spin, 500
Electronic admittance of reflex klystron, 472
Elliott, R. S., 113
Energy, electric, 28–30
  levels of, in ruby, 510–514
  magnetic, 28
  velocity of, in free space, 43
    in periodic structures, 380–381
    in waveguides, 137
Excitation, of cavity, 356–359, 464–468
  of waveguides, 183–197
Exponential taper for transmission line, 239

Fabry-Perot resonator, 337–344
  Q of, 338–339
Fano, R. M., 63
Faraday rotation in ferrites, 296–299

Faraday's law, 12
Ferrite, electron precession in, 286–290
  Faraday rotation in, 296–299
  magnetic permeability of, 22, 291
  in microwave devices, 300–309
  plane-wave propagation in, 292–296
Filters, cavity, direct-coupled, 428–430
  quarter-wave-coupled, 424–428
  frequency transformations in, expansion, 411
    low-pass to bandpass, 413–414
    low-pass to high-pass, 412
    periodic, 415–416
  image-parameter design of, 399–403
  impedance inverters in, 415–422
  insertion-loss design of, 403–410
  low-pass designs for, 407–410
  power loss ratio in, 403
    for Chebyshev, 406–407
    for maximally flat, 405
  for transmission line, 422–424
Floquet's theorem, 381–383
Foster's reactance theorem, 153–155
Fox, A. G., 270, 362
Fox, J., 10
Fresnel reflection coefficient, 46–47
Fresnel transmission coefficient, 46–47
Friis, H. J., 9
From, W., 540

Gain, of klystron, 469
  of maser, 529–531
  of parametric amplifier, 555, 557–559
  of traveling-wave tube, 482
Gain-bandwidth product, for maser, 531
  for parametric amplifier, 561–562
Gauss' law, 13
Ghose, R. N., 143, 202, 433
Gilden, M., 570
Ginzton, E. L., 10
Gordon, J. P., 362, 540
Goubau, G., 65, 118, 362
Gouray, B. S., 528

Group velocity, in periodic structure, 379–380
  in waveguide, 133–136
Gurevich, A. G., 312
Gustincic, J. J., 129
Gyrator, 300–301

H modes, 69–77
  in circular guide, 110
  in rectangular guide, 97–106
Hahn, W. C., 448, 492
Half-wave plate, 267
Hamilton, D. R., 492
Hankel functions, 575–576
Harrington, R. F., 143
Harvey, A. F., 65, 312, 390
Haus, H. A., 452, 485
Heffner, H., 540
Heilmeier, G. H., 570
Helix, general properties of, 396–398
  sheath, 392
    dispersion equation for, 395
    in traveling-wave tube, 476–482
Helmholtz's equation, 26
Helmholtz's theorem, 13, 344
Hensperger, E. S., 224
Heuer, H. J., 559
Honey, R. C., 570
Hopfer, S., 65
Hosono, T., 113
Hutter, R. G. E., 388, 492
Hybrid junction, as balanced mixer, 285–286
  magic T as, 283
  ring circuit as, 285
  scattering matrix for, 283–284

Image parameters of filters, 399–403
Impedance, characteristic, of capacitively load transmission line, 368
  of coaxial line, 76
  of transmission line, 86
  general definition of, 32
  input, even and odd properties of, 155–157

Impedance, input, on transmission line, 93
  matching of, 207–217
    (See also Quarter-wave transformers; Transmission line, tapered)
  matrix, imaginary property of, 160
    symmetry of, 159
  surface, 37
  wave, for circular guide, 110
    for TE waves in rectangular guide, 99
    for TM waves in rectangular guide, 102
  of waveguide elements, 148–151
Impedance inverters in filters, 415–422
Insertion loss in filters, 403–410
Interdigital line, 390–392
Ishii, J., 433
Isolator, 301–304

Johnson, R. C., 258
Johnson noise, 486–487
Jones, E. M. T., 433, 570
Jordan, E. C., 143

Karbowiak, A. E., 113
k_0-β diagram, 377–379
Kerns, D. M., 202
Kerr, D. E., 9
Kinetic-power theorem for electron beam, 453–454
Kinetic voltage, 454
King, D. D., 113
Kleen, W. J., 492
Klopfenstein, R. W., 258
Klüver, J. W., 452
Klystron, reflex, electronic admittance in, 472
  oscillation conditions for, 473
  tuning curves for, 473
  two-cavity, 462–470
    equivalent circuit for, 466–467, 469
    excitation of fields in, 464–468
    gain of, 469

Knipp, J. K., 492
Kohno, S., 113
Kompaneyets, A. S., 540
Kotzebue, K. L., 570
Kraus, J. D., 9, 63
Krauss, H. L., 143, 492
Kronig-Kramers relations, 528
Kuper, J. B. H., 492
Kurokawa, K., 362

Laplace's equation, 23
Larmor frequency, 287, 438
Laser, gas, 538–539
 ruby, 537–538
Lax, B., 312
Lenygel, B. A., 540
Lewin, L., 218, 220
Lewis, J., 492
Li, T., 362
Lorentz condition, 52
Lorentz force, 11
Lorentz reciprocity theorem, 56–59
Loss tangent, 20
Luderer, G. W., 113

McLachlan, N. W., 578
Magic T, 283
Magnetic permeability, 22
 for ferrite, 22, 291
Magnetic susceptibility, 21
Magnetron, 473–476
Manley, J. M., 546
Manley-Rowe relations, 546–549
Marcuvitz, N., 202, 217, 312
Masers, description of, 507–509
 equivalent circuit for, 524–528
 gain of, 529–531
 gain-bandwidth product for, 531
 noise in, 532–534
 ruby material for, 510–514
 susceptibility tensor for, 520–524
 traveling-wave, 534–537
Matsumaru, K., 258
Matthaei, G. L., 424, 433, 570
Maxwell's equations, 15
 phasor form, 17

Meander line, 390–392
Melchor, J. L., 291
Mihran, T. G., 445
Milford, F. J., 63
Miller, S. E., 112
Mixer, balanced, 285–286
Mode chart for cylindrical cavity, 328
Montgomery, C. G., 10, 202, 273, 275, 312, 362
Morgan, S. P., 113
Morrison, J. A., 113
Motz, H., 218
Mumford, W. W., 425

Nakahara, T., 65
Negative-resistance amplifier, 524–528, 555–562
Noise, equivalent temperature of, 487
 figure, 487–489
 Johnson or Nyquist, 486–487
 in masers, 532–534
 in microwave tubes, 485–486
 in parametric amplifiers, 562–569
Noise figure, 487–489
 of degenerate negative-resistance parametric amplifier, 568
 of negative-resistance parametric amplifier, 565
 of up-converter, 563
Normalized current, 147
Normalized load impedance, 90
Normalized voltage, 147
N-port circuits, 157–160

Ordung, P. F., 143, 492
Ozaki, H., 433

Pannenborg, A. E., 202
Parametric amplifier, linearized equations for, 549–551
 Manley-Rowe relations for, 546–549
 negative-resistance, 555–562
 gain of, 557–559
 gain-bandwidth product for, 561–562

Parametric amplifier, negative-resistance, noise in, 562–569
  noise figure, of degenerate negative resistance, 568
    of negative resistance, 565
    of up-converter, 563
  p-n junction diodes for, 542–546
  up-converter, 551–555
    gain of, 555
Pauling, J., 540
Penfield, P., 570
Periodic structures, Bloch waves in, 369
  Bloch-wave impedance for, 372
  energy flow velocity in, 380–381
  and filters, 398–403
  Floquet's theorem for, 381–383
  group velocity in, 379–380
  k_0-β diagram for, 377–379
  matching of, 375–377
  spatial harmonics in, 382–383
  terminated, 373–375
  for traveling-wave tube, corrugated plane, 383–389
    helix, general properties of, 396–398
    interdigital line, 380–392
    meander line, 390–392
    sheath helix, 392–395
    tape ladder line, 390
  unsymmetrical two-ports in, 372–373
Permeability, 22
  for ferrite, 22, 291
Perveance of electron beam, 436
Phase changer, linear, 265–266
  rotary, 266–270
Phase velocity, 42
  in waveguides, 132
Physical constants, 579–580
Pierce, J. R., 435, 439, 492
Planck's constant, 579
Planck's formula for radiation frequency, 501
Plane waves, 38–43
Plasma frequency, 438
  effective, 444–445
Plonsey, R., 63, 81, 344
Poisson's equation, 23

Polarization, circular, 287–288
  of circular aperture, 191
  in dielectric, 18
p-n junction diode, 542–546
Post, capacitive, in waveguide, 220
  inductive, in waveguide, 219
Potential, scalar, dynamic, 51
  static, 23
  vector, dynamic, 51
  static, 24
Power, in circular guide, 110
  in TE waves in rectangular guide, 100
Power loss ratio, in filter, 403
  for quarter-wave transformer, 233
Power orthogonality, for degenerate modes, 124–129
  in waveguides, 121–124
Poynting vector, 31, 33
  complex, 31–32
Probe, radiation resistance of, 187
  in waveguide, 183–187
Purcell, E. M., 202, 273, 312, 362

Quality factor or Q, 314–316
  of cylindrical cavity, 327
  external, 315
  loaded, 315
  of rectangular cavity, 325
  unloaded, 315
Quantum mechanics, 494–507
  absorption and emission of radiation in, 501–507
    Einstein's coefficients for, 503–504
  angular momentum in, 498–500
  electron spin, 500
  for many particle system, 500–501
Quarter-wave plate, 266
Quarter-wave transformers, Chebyshev, exact results, 233–237
  three-section, 236–237
  two-section, 234–235
  N-section, approximate theory for, 226–233
    binomial, 227–229
    Chebyshev, 229–233
    single-section, 221–224

Rafuse, R. R., 570
Ragan, G. L., 262, 312, 362, 404
Ranco, S., 143, 202, 492
Reactive elements in waveguide, 217–221
  shunt capacitive, 219–220
  shunt inductive, 218–219
  stubs as, 220–221
Reciprocity theorem, 59
Reflection, from conducting plane, 47–50
  from dielectric surface, parallel polarization, 44–46
    perpendicular polarization, 46–47
  small, theory of, 233
Reflection coefficient, current, 90
  for tapered transmission line, 238–239
    and Riccati equation, 251–253
  for terminated transmission line, 92
  voltage, 90
Reflex klystron, 470–473
Reich, H. J., 143, 312, 492
Reitz, J. R., 63
Resistance, radiation, of probe in waveguide, 187
  of transmission line, 87
Resistance-wall amplifier, 484
Resonant circuits, bandwidth of, 314–315
  damping of, 316
  Q of, 314–315
  transmission line, antiresonant, 320–321
    open circuited, 319–320
    short-circuited, 317–319
Riblet, H. J., 258, 433
Riccati equation for tapered transmission line, 251–253
Ridge waveguide, 6
Rigrod, W., 492
Ring circuit, 285
Roberts, J., 312
Rowe, H. E., 546
Ruby, energy levels in, 510–514
Ruby laser, 537–538

Scalar potential, dynamic, 51
  static, 23
Scattering matrix, of circulator, 307–308
  of directional coupler, 271–273
  of hybrid junction, 283–284
  for lossless junction, 174
  symmetry of, 172
  and transformation of terminal planes, 172
  for two-port junction, 176–179
  unitary property of, 176
Schelkunoff, S. A., 9, 141, 143
Schiff, L. I., 540
Schlesinger, S. P., 113
Schrödinger's equation, 494–496
Schwinger directional coupler, 276–278
Sensiper, S., 398
Separation constant, 39
Separation of variables method, 39
Sheath helix, 392–395
  in traveling-wave tube, 476–482
Short circuit, choke-type, 259–262
  variable, in waveguide, 259–262
Siegman, A. E., 493, 540
Signal velocity, 136
Silver, S., 9
Singer, J. R., 540
Skalnik, J. K., 143, 492
Skin depth, 36
Slater, J. C., 143, 362, 433, 491
Smith chart, 203–207
Smullin, L. D., 485
Snell's law, 45
Solymar, L., 258
Soohoo, R. F., 294, 312
Southworth, G. C., 143, 312
Space-charge reduction factor, 445
Space-charge waves, a-c power relations for, 452–454
  and kinetic-power theorem, 453–454
  and kinetic voltage, 454
  on axially confined beam, 439–446
  d-c propagation constant for, 441
  effective plasma frequency for, 444–445
  fast and slow, 443

Space-charge waves, reduction factor
    for, 445
  on unfocused beam, 446–451
Spangenberg, K. R., 435, 457, 492
Spatial harmonics in periodic structures, 382–383
Standing waves, on transmission line, 91
Standing-wave ratio, 92
Static fields, 22–25
Stratton, J. A., 63
Stub, matching with, 207–217
  double, 212–215
  single, 208–212
  triple, 215–217
  in waveguide, 220–221
Sugi, M., 65
Surface impedance, 37
Surface-wave guide, 6, 113–121
  dielectric-coated plane as, 114–117
  dielectric-coated wire as, 118–121
Susceptibility, electric, 19
  magnetic, 21

Termination, waveguide, 259
Tischer, F. J., 65
Toraldo di Francia, G., 143
Transmission coefficient, 90
Transmission line, capacitively loaded, Bloch waves in, 369
  characteristic impedance of, 368
  circuit analysis of, 364–369
  eigenvalue equation for, 367
  k_0-β diagram for, 377–379
  wave analysis of, 369–371
  distributed circuit analysis of, 80–84
  field theory of, coaxial line, 74
  lossless line, 73–79
  lossy coaxial line, 79
  with small loss, 76–80
  filter, 422–424
  parameters of, capacitance, 85
    characteristic impedance, 76, 86
    coaxial line, 87
    conductance, 86
    inductance, 85
    resistance, 87

Transmission line, resonant circuit, antiresonant, 320–321
  open-circuited, 319–320
  short-circuited, 317–319
  tapered, Chebyshev, 248–251
    exponential, 239
    reflection coefficient on, approximate equation, 238
    Riccati equation for, 251–253
    synthesis of, 240–251
    triangular, 239–240
  terminated, 89–95
Transmission matrix, for cascade network, voltage-current, 179–181
  wave-amplitude, 181–182
TE waves, 69–71
TEM waves, 67–69
TM waves, 71–72
Transverse resonance method, 115–117
Traveling-wave maser, 534–537
Traveling-wave tube, M-type, 482–484
  O-type, 476–482
    gain of, 480, 482
    periodic structures for, 383–398
Troetschel, W. O., 559
Troup, G., 540
Two-port junctions, 160–167
  equivalent circuits for, 168–170

Uenohara, M., 564
Unger, H. G., 113

Van Bladel, J., 362
Van der Ziel, A., 541
Vartanian, P. H., 291
Vector formulas, 571–574
Vector potential, dynamic, 51
  solution for, 53–56
  static, 24
Velocity, energy flow, in periodic structures, 380–381
  for plane waves, 43
  in waveguides, 137
  group, in periodic structures, 379–380

Velocity, group, in waveguides, 133–136
  phase, for plane waves, 42
    in waveguides, 132
  signal, in waveguides, 136
  wavefront, in waveguides, 136–137
Velocity-jump amplifier, 484
Velocity modulation, of electron beam, 455–462
  beam coupling parameter in, 456
Voltage, equivalent, in waveguides, 145–147
  normalized, 147
Vuylsteke, A. A., 540

Watkins, D. A., 398, 433
Watson, G. N., 578
Wave, classification of, 67–72
  impedance, of TE mode, 70
    of TM mode, 71
  plane, 38–43
    reflection of, from conducting plane, 47–50
      from dielectric surface, 43–47
    transmission matrix, 181–183
  TE, 69–71
  TEM, 67–69
  TM, 71–72
  (*See also* Periodic structures; Space-charge waves; Transmission line; Waveguide)
Wave equation, 25
Wave number, 26
Waveguide, capacitive diaphragm in, 219–220
  capacitive post in, 220
  capacitive rod in, 220
  circular, attenuation in, 110

Waveguide, circular, solutions for, 110
  TE waves in, 110
  TM waves in, 107–110
  equivalent current and voltage for, 145–147
  excitation of, by aperture, 192–197
    by current loop, 189–190
    by linear current element, 187–189
  inductive diaphragm in, 219–220
  inductive post in, 219
  properties of, 96
  rectangular, attenuation in, 101–103
    cutoff frequency of, 98
    dominant TE mode in, 104–106
    power in, 100
    solutions for, 102
    TE waves in, 97–106
    TM waves in, 106–107
    wave impedance for, 99
  ridge, 6
  termination, 259
  velocity in, energy, 137
    group, 133–136
    phase, 132
    signal, 136
    wavefront, 136—137
Weinberg, L., 409
Wenzel, R. J., 433
Whinnery, J. R., 202, 143
Wilson, E. B., Jr., 540

Yariv, A., 540
Young, L., 258, 433

Zitelli, L., 492

$\frac{\lambda}{4}$ transformers

$$A = 2^{-N}\left|\frac{Z_L - Z_o}{Z_L + Z_o}\right|$$

Binomial N section

$$\rho_n = 2^{-n}\left(\frac{Z_L - Z_o}{Z_L + Z_o}\right) C_n^N = A C_n^N$$

$$C_n^N = \frac{N!}{(N-n)!\, n!}$$

$$Z_{n+1} = Z_n \left(\frac{1 + \rho_n}{1 - \rho_n}\right)$$

VSWR

$$\Gamma = \frac{Z_L - Z_o}{Z_L + Z_o}$$

$$r = \frac{1 + |\Gamma|}{1 - |\Gamma|}$$

$$\frac{\Delta f}{f_o} = 2 - \frac{4}{\pi}\theta_m$$

$$\theta_m = \cos^{-1}\left|\frac{2\rho_m}{\ln Z_L/Z_o}\right|^{\frac{1}{N}}$$

N-section Chebyshev

$$A = \rho_m = \frac{Z_L - Z_o}{(Z_L + Z_o)\, T_N\left(\frac{1}{\cos\theta_m}\right)} = \frac{r-1}{r+1} \qquad r = VSWR$$

$$T_N\left(\frac{1}{\cos\theta_m}\right) = \frac{Z_L - Z_o}{(Z_L + Z_o)\,\rho_m}$$

$$\sec\theta_m = \cos\left(\frac{1}{N}\cos^{-1}\frac{Z_L - Z_o}{Z_L + Z_o}\rho_m^{-1}\right)$$

$$Z_n = Z_n\left(\frac{1 + \rho_n}{1 - \rho_n}\right) \qquad \frac{\Delta f}{f_o} = 2 - \frac{4\theta_m}{\pi}$$

For 2-section

$$\rho_o = \rho_2 = \frac{\rho_m}{2\cos^2\theta_m} \qquad \rho_1 = \rho_m\left(\frac{1}{\cos^2\theta_m} - 1\right)$$